THE FOREST FARMS OF KA

To Chai Moy and Kathy and the memory of Achmed Saleh, and to Nim Dorji and to all those village farmers met along the Grand Trunk Road who actually wrote this book while the author tried to take notes of what they said and to Linda Johnson.

The Forest Farms of Kandy

and other gardens of complete design

D.J. McCONNELL
with
K.A.E. DHARMAPALA
G.K. UPAWANSA
S.R. ATTANAYAKE

Routledge
Taylor & Francis Group

LONDON AND NEW YORK

First published 2003 by Ashgate Publishing

2 Park Square, Milton Park, Abingdon, Oxon OX14 4RN
711 Third Avenue, New York, NY 10017, USA

Routledge is an imprint of the Taylor & Francis Group, an informa business

British Library Cataloguing in Publication Data
McConnell, D. J. (Douglas John), 1928-
    The forest farms of Kandy : and other gardens of complete
    design - (Ashgate studies in environmental
    policy and practice)
    1. Agroforestry - Sri Lanka 2. Agroforestry
    I. Title II. Dharmapala, K. A. E. III. Upawansa, G. K.
    IV. Attanayake, S. R.
    634.9'9'095493

Library of Congress Cataloging-in-Publication Data
McConnell, D.J. (Douglas John), 1928-
    The forest farms of Kandy : and other gardens of complete design / Douglas John McConnell.
        p. cm. -- (Ashgate studies in environmental policy and practice)
    Includes bibliographical references and index.
    ISBN 0-7546-0958-8
    1. Agroforestry--Sri Lanka--Kandy Region. I. Title. II. Series.

S494.5.A45 M43 2002
338.1'095493--dc21   2002019582

ISBN 13: 978-0-7546-0958-2 (hbk)
ISBN 13: 978-1-138-26368-0 (pbk)

Contents

Preface

This book is addressed to an economic and ecological examination of forest garden farming systems around the tropical world—small, low-input but productive and sustainable family units of highly diversified trees-palms-bushes-vines and having few if any conventional field crops or livestock. It's inspired by those of Kandy, Sri Lanka. It describes and where possible quantifies their botanical and technical structure, energy flows and socio-economic uses which are of modern and pressing relevance, according to some 34 agrosystem efficiency criteria (Table 1.2). Without claiming these systems as a panacea (no farming system can be) it is suggested that over the millennia forest gardens under one name or other have proved themselves in many respects a superior alternative to modern high pressure agriculture which, in devastating the natural world, is rapidly destroying its own resource base. They seem especially useful in the moist tropics as at least a partial alternative to deforestation for expansion of 'proper' agriculture according to temperate European-North American farm development models.

In official economic planning and agrodevelopment circles there continues much resistance to the concept that traditional agroecosystems in general and forest gardens in particular have useful knowledge to impart, partly because they seem to be so alien to received Western historical and cultural traditions. To establish legitimacy of these systems it has therefore been necessary in three chapters (Shifters, Origins, Genesis) to reconsider the origins of agriculture itself.

> *Ecologists have begun to appreciate the resource management systems of traditional societies...These societies co-evolved with their environment, modifying nature but actively maintaining it in a diverse and productive state. Such indigenous ecosystem approaches, which enshrine in them a pool of human experience spanning many millennia and cultures, have suffered a severe setback under the more recent systems of exploitive management, often called scientific, elaborated largely by colonialists. The irony is that just as scientists are beginning to appreciate folk-ecology and its implications, a monolithic vision of modern resource*

management is engulfing these traditional systems (Chandran and Gadgil 1993).

All prices, costs, monetary amounts in US$ unless otherwise noted.

Chapter 1
Nature

Forest garden farms are probably the world's oldest and most resilient agroecosystem. They are nearly as old as human kind and of all the economic systems so far devised they remain – to steal a phrase from some dead poet – the loveliest and the best.

1.1 Description, Occurrence, Classification

This agroecosystem is one which provides all or most of a household's food, beverage, fuel, construction materials, cash income and often drugs and medicines from a small area of highly diverse tree and vine species. Ground level crops might also be present under the canopy and some form of intensive livestock production might be carried on, but these latter components are always much less important than the trees and vines. Senanayake (1987) has described such farms as analogue rain forests. Jose and Shanmugaratnam (1993) say they are a climax ecosystem where ecological succession is manipulated by human beings. In Java, Otto Sumarwoto calls them gardens of complete design.

The family or clan house, buildings and livestock are usually located within the garden and the household might also possess an outfarm which typically produces bulk foodstuffs not possible under the tree canopy (a padi field, a slash-and-burn swidden) and it might range over some area of public land or littoral from which it supplements its farming activities. Structurally forest garden types extend from the classical intensive units of Kandy, Kerala, Java, to being the sedentary components of extensive hunter-gatherer land use systems in Congo Basin, Amazonia, Kalimantan, Melanesia.

Relevance, Purpose, Modes

Forest gardens are relevant for four broad sets of reasons, or can be approached as farming systems having four operational modes (summarized in parts A, B, C, in Table 1.1). As *production* systems they yield a wide range of foods, fibers, materials, medicinals, industrial inputs and probably do this more efficiently (per ha of land or other units of input) than any other tropical farming system. Simultaneously as *conservation* systems they are also highly efficient in preserving soil, water, biodiversity, human culture and knowledge (notably of medicinal plants and their uses). As *rehabilitation* systems they're probably the most efficient and least cost method of reclaiming eroded or otherwise degraded lands and catchments, after this degradation has been caused by exploitive unsustainable forms of agriculture. As *strategic environmental* systems they offer one least cost land use strategy for countering such emerging universal problems as greenhouse gas emissions, ozone depletion, broad climate change and its consequences as extreme weather events. In these four modes forest gardens (and similar agroforestry systems, notably managed forest extractivism) are certainly not a panacea; but where otherwise feasible and addressed to these social purposes they seem to be the most efficient of all available tropical agroecosystems. Some 34 criteria by which they can be judged and compared with other farming systems are later listed in Table 1.2. For the moment it's necessary only to note that their broad *social* value is much greater than is their *private* value as 'farms'.

Location

Forest gardens occur generally throughout the humid tropics at low-mid elevations in areas which were once rain forest, or close to forests from which economic species/crops could be extracted. They have in recent historical times spread into a few non-forest areas as artifacts. In S-SE Asia they extend from South India-Sri Lanka into South China then east and south through Micronesia and Melanesia into Remote Oceania. In tropical America they occur from Mexico to Peru-Colombia-Brazil and east to the Atlantic seaboard. In Africa most are found on the high rainfall west coast, in Congo Basin with outliers around the central African highlands and in Tanzania. There are quasi-forest gardens in higher rainfall areas of the East African coast. In other places (Madagascar, Laos, Vietnam, Kampuchea, Southern Bhutan) there are few or no data. They are intimately associated with the closed moist forest, that is, the true rain forest; but this has receded in recent times and forest gardens are now found in or close to all kinds of lesser forests, but as a copy or a relic of the three Great Forests of tropical America, S-SE Asia/Oceania, Africa.

Origins

Forest gardens seem to have had two source streams, closely related: the useful trees found in a rain forest and retained there (Genesis); and useful trees added to a forest swidden when it was cleared for shifting agriculture (Shifters, economically enriched fallows).

What's in a Name?

In the absence of interest by mainstream agriculturists and economists 'developed' forest gardens have pretty well fallen within the scope of agroforestry; but ethnobotanists, anthropologists, ecologists, nutritionists have contributed most to knowledge of the wilder ones, still close to their rain forest parentage, as human systems. In Nair's (1993) classification forest gardens are one of 18 agroforestry systems ('practices'). The general type have also been called horti-agriculture, agro-horticulture, aboriculture, compound farms, homegardens...(Budowski 1985, Landauer and Brazil 1985), with *home garden* having almost semi-official status, though locally indigenous names are more expressive and accurate. It obscures the point that these agroecosystems have the wider conservation functions noted. The author and Sri Lanka colleagues used the term *forest gardens* in 1972 because it emphasized these social functions and links them with their parent, the Great Forest of which they are biological remnants or planned analogues. Since the conservation aspects of land use are of increasing importance the name is retained here – but the Spanish 'huertos selvas' or 'huertos familiares' are probably better.

Source Field, Administrative Responsibility

Table 1.1 offers a framework within which forest gardens might be discussed, evaluated, classified, as *farms*. Part A suggests the main source fields when these are compartmentalized as academic disciplines; Part B the several fields of administration under which forest gardens fall. There are several of these, including management authorities of national parks and biospheres within which forest gardens sometimes exist. Part B is often where the forest gardeners' problems start. They produce a multiplicity of products from many species for a range of purposes. These separate products/farm enterprises typically fall within the responsibilities of different ministries, departments, and there are often other special-purpose agencies promoting crops for foreign exchange, fibers and oils for industrialization etc. So there can be 3, 4, 5... agencies or their extension representatives each throwing agro-technical advice at various flywheels of the system from different directions but not understanding how the other bits work – or offering no advice at all, which is the better alternative.

1.2 The Structure of Forest Gardens as Farming Systems

Part C of Table 1.1 lists characteristics by which forest gardens either as a general type or some sub-type might be described, classified, differentiated, according to their structure as operational farm systems.

Forest Environment

Forest gardens extend from those which are practically indistinguishable botanically from their parent forest (Amazonia, Kalimantan, Congo Basin) to those which have lost touch with all but a few remnants of it. Two indicators of continuing proximity are the diversity of plant species present on farms and uses to which they are put. Farm species are greatest where botanical diversity of the forest was greatest (NW Amazonia, SE Asia), and least when the forest was species-poor (Remote Oceania). This is obscured by the extent to which regions have been exposed to plant introductions and subjected to pressures for species-mix adjustments, simplification, by commercialization or increasing population pressures. In many places contact with the parent forest has been nearly lost: many of the forest 'pekarangan' gardens around Megatan in the Madiun Valley of Java now consist of little more than one or other of the weaving bamboos, and in the even poorer villages of Trenggalek the only remnant of the forest is (weaving) pandanus.

Orientation – Purpose

Forest gardens may be arranged on a scale extending from complete subsistence through full commercialization to those whose purpose is primarily human habitat. Most have a mix of these purposes. There are now few gardens left which are completely subsistence-directed but nearly all of them could be self-sufficient if the need arose. Purely commercial forest gardens are equally rare. In the various farm surveys one value attributed to them is 'human habitat'. In this regard the extremes are probably the forest gardens of Puerto Rico analyzed by Clarissa Kimber and those tree gardens of the peri urban 'rich' in West Java (Michon, Reconnaissance). But the primary purpose remains economic subsistence, and they are commercialized only to the extent that it is more convenient to sell or barter produce for items less easily home fabricated.

Farm Enterprise Mix

This refers to the possibility of structuring a farm around trees alone (as in a pure forest garden) or some combination of trees-vines-field crops-animals-fish. (These would accord with Nair's (1993) general classifications of agro, agro-pastoral, agro-silvo-pastoral systems.) Comparisons of farm sub-types could if necessary be made among farms with only trees, trees + padi, trees + livestock, tree mixes with-and-without some dominant species (cloves etc - Kandy).

Livestock In general livestock don't have an important economic role, mainly because of lack of feed under the tree canopy. Nonetheless it is remarkable how many horses, cattle, goats, sheep, buffalo they do support. In Sri Lanka most working elephants are maintained on this type of farm; but since their appetite is substantial this requires that the necessary kitul and banana palms be bought in from other farms in the village. Free ranging poultry are ubiquitous (and pigs where culture permits). In West Java a main 'livestock' type are ponded fish, supplemented in Sarawak by crocodiles. In some systems livestock includes insects, caterpillars, ants, which are to some extent 'managed' (Shifters). Taking Java as example, even many of the smallest farms manage to support a couple of stall-fed goats fed on tree leaves, mainly jak and waste tree produce; slightly larger farms might devote 2-300 sq meters to a forage crop (napier/para grass); and where the home farm also has a padi field component this provides padi straw and seasonally the residues of more nutritious palawija crops (corn, beans). Where the system also has an Out-farm component (below) similar crop residues are carried from this to the stall-fed cattle, goats, buffalo. Universally much feed is collected from the Range (if this exists, below) in the form of weeds, grass, tree leaves, which can be collected – sometimes by small boys for pocket money, sometimes by landless adults who subsist in this way.

   The shortage of livestock feed can be acute, especially towards the end of the dry season, and some of the feedstuffs resorted to would violate most rules in the animal nutrition handbook and test a camel (listed below).

* coconut palm leaves
* pure-stand napier grass from own farm
* top-trimmings of live grass fences around house
* weeds/grass purchased from grass cutters
* pumpkin vines

* cassava tubers, stems, leaves
* sweetpotato tubers, leaves
* whole papaya trees, stems, leaves, fruit
* tree leaves from garden mix

Source: Probowo and McConnell (1993), in village of Sukorame, Solo Valley.

Tree species, enterprises and farm planning Some of these tree-field crop-livestock species and the economic enterprises based upon them can be disaggregated from the rest of the system and analyzed individually (for the purpose of comparing their relative economic benefits/costs and restructuring the farm accordingly). However at the subsistence end of the orientation-purpose scale where the tree mix approximates that of a natural forest, there is increasing interdependence among species in relation to light, temperature, nutrient cycling, wind protection, pest attraction/repulsion, so that abstraction of any species from the others for its comparative analysis would be very difficult; and increase/decrease in its population (enterprise optimization and adjustment in farm economics jargon) might well have serious adverse repercussions throughout the whole system (Energy, architecture).

Table 1.1   Basis for Classification of Forest Garden Types as Farms

A	SOURCE FIELD	AGRI-CULTURE	FORESTRY	LIVESTOCK	NATURAL RESOURCES		
	↓				FISH	ANIMALS	LITTORAL

B	ADMINISTRATION FIELD	AGRICULTURE	FORESTRY	CONSERVATION	OTHER

C	FARM TYPES (STRUCTURE)				
	1. FOREST ENVIRONMENT	REMOVED ...			PRESENT
	2. ORIENTATION - PURPOSE	SUBSISTENCE ... COMMERCIAL ... HABITAT			
	3. FARM ENTERPRISE MIX	TREES/VINES ONLY	WITH GROUND CROPS	WITH ANIMALS	WITH CROPS AND ANIMALS
	4. OUT-FARM TYPE	NONE ...	FIXED-IN-PLACE	...	SHIFTING
	5. OFF-FARM RANGE	NONE ...	INTENSIVE	...	EXTENSIVE

Outfarms

Forest farms can also be classified according to whether the household possesses an outfarm, the number of these and their purpose. These can be fixed in place or revolve in time and place around the home farm. Where they exist their number can range from one (Kandy, Java, the Chagga of Tanzania) to as many as five, serving separate purposes (as in the complex land use systems of the Maya, the Sumatra Minankabau, the Kantu' Iban of Sarawak). The simplest case are those households who in addition to the forest garden have some area of irrigated padi land or dryland (*tegal*) fields. These will be used to grow mainly bulk foodstuffs rice, cassava, maize which cannot be grown efficiently under the canopy. A more complex case (Michon et al 1986) are the Minangkabau farms of West Sumatra with their components at three elevational zones: the home forest garden, below this the attached padi fields, above it a managed forest of tree crops (and above that a public mountain forest which they also no doubt exploit to vigorous degree).

Dove (1993) has described how households of the Kantu' Iban people of West Kalimantan have an average of five fixed-in-place rubber groves scattered throughout their territory (and when visiting each to tap their rubber they also open up a food swidden in the vicinity, Shifters). In Peruvian Amazonia Padoch et al (1985) describe some (nominally) shifting/swidden, outfarms there which, because they are planted to trees and continue to be exploited for up to 50 years, can be regarded as fixed outfarms. On Palawan, Philippines, the previously shifting swiddens described there by Eder (1981) and Raintree and Warner (1986) have similarly been stabilized in place by the planting to permanent tree crops; and in the Solomons Vergara (1987) discusses similar bush swiddens which for the same reason become essentially fixed outfarms very close in structure to forest gardens. Among ethnobotanists perhaps the best known forest outfarms are those resource islands of the Kayapo in Amazonia. They have a desert parallel in Sudan where the outfarms, resource islands, consist of single trees (the tebeldi, Adansonia digitata) developed strategically along the seasonal migration routes (Origins). As structural components of agroecosystems these merge into the next component, Range.

Range

This refers to the presence of some forest, swamp, riverine, maritime area over which household/clan members may range and extract food, fiber, materials, animals, fish, to supplement produce from their forest garden. Such a range usually exists though in developed places one often must look hard to find it. This can vary from a few square meters of roadside grass to hundreds of kilometers of forest. Somewhere along this scale, towards the extensive end, such terms as sedentary, nomadic, shifting, hunter gathering, foraging, even agricultural, lose precise meaning. There are still a large body of forest garden 'farms' which anthropologists would declare to be the (temporary or permanent) bases of the hunter-gatherers.

Extensive range Most examples of forest gardens associated with an extensive range come from Amazonia, Congo Basin, Indonesia. Typically these 'farms' shade into hunter-gatherer systems or commercial forest extractivism. Possibly the extreme is found with the 40 000 people in the 100 000 km^2 Ituri forest in Congo (Ichikawa 1993). These consist of two groups living in symbiotic relationship: the Mbuti are traditionally hunter-gatherers while the

Efe are agriculturists. The economic relationship between the two groups is through exchange of forest produce (game meat, honey) for carbohydrate foods, cassava etc). Ichikawa does not discuss the sedentary components of either group but it seems safe to assume the Mbuti range over their whole forest and manage its resources as if it were their extended forest garden. They use more than 200 animal and 100 plant species as food (Bailey and Peacock 1988). The orthodox will object that such systems are not agriculture; archaeology outside of Europe and Near East increasingly recognizes them as the *first* agriculture (Genesis).

Also in Congo, Takeda and Sato (1993) report multi-component land use systems in which while the sedentary and swidden components provide up to 96% of household energy food requirements they contribute only 66% of protein needs, the balance of which must be obtained from hunting and fishing over the range. Takedo and Sato put the total land resources of one 190-member group at 110 km^2 of which only 5% is occupied by sedentary agriculture and swiddening rotations, leaving 95% of forest as the group's range.

In central Jambi province, Sumatra, the 800 Kubu 'hunter-gatherers' in their extensive 2 500 sq km range have been studied by Sandbukt (1988) and others. No data relating to the sedentary forest garden components of these systems are available (ie tree/plant species tended and garden structure) but it is pretty clear that these components exist in some form. The Kubu also engage, unenthusiastically, in some swidden (shifting, slash-and-burn) agriculture but they see the necessary effort in this as great and the rewards as small, so their preferred activities are hunting-gathering. Again the objection of non-agriculture will be raised. Again one replies from archaeology, in this case Grube and others in Melanesia (Genesis), who have found clear evidence of plant *domestication*, which is conventionally taken as the basic condition of agriculture, extending back some 30 000 years in similar rain forests.

As described by Sandbukt this Kubu system is very complex. 'Kubu' means defensive refuge (in the recent past from slavery, debt bondage, epidemics). Their forest range provided opportunity to create an agro-social system which, though the need has passed, remains one of 'defensive mobility'. Had various attempts to settle them succeeded, first by the Dutch then Indonesian governments, in Sandbukt's view this would have meant the end of the world as the Kubu know it. This Kubu world is enclaved by settled Malay rubber and swidden farmers. The relationships between the two groups are part complementary (some labor by the Kubu on Malay farms; much exchange of forest produce for cloth and salt). They are also partly competitive: 'hunting' includes exploitation of game attracted to the Malays abandoned swiddens; 'gathering' includes pilfering of food from the Malays' swiddens.

On Seram, Indonesia, Ellen (1988) describes another complex system in which 600 Nuaulu people live as farmers in fixed hamlets, with food gardens and 'tree groves' (forest gardens), and who also range over the rain forest which provides hunting-foraging resources to an extent of 1.3 sq km per person. This range provides some 120 animal/fish/insect species and 48 plant species as food (and a 'vast' number of non-food materials). In terms of inputs, 72% of all male productive energy is expended on procuring non-domesticated food. In terms of outputs, 63% of all energy is from a single plant species, the sago palm. Here again distinction between agriculture and gathering begins to blur: half the sago is from wild groves and to this extent their system is not agriculture; but the other half is from palms which have been planted and 'tended', and to this extent it is.

In Amazonia (and elsewhere) most commercial forest extractivist systems also have a sedentary forest garden component, with this supplemented by one, two or even three 'ranges' used for different purposes. Sizer describes one 57-household community of *caboclos* on the Rio Jaú in Amazonas state whose sedentary farm consists of a dense mix of up to 30 tree species with small livestock, supplemented by swidden clearings (one ha of swidden opened up per household annually). In addition they range over a 5-10 km radius for forest foods, but over a larger area for hunting, and over a much larger area still in their primary occupation as extractivists of rubber, sorva latex, brazil nuts, palm fibers. In Middle America similar ranges were always integral components of Mayan and Huantec land use systems (Reconnaissance).

Intensive range In regions of high population pressure only remnants of the original range exist. From Afghanistan through Pakistan and India to Bangladesh where need for productive land is greatest the total range is still extensive, consisting of hundreds of thousands of ha of commons or village 'grazing' lands, or nominal forests now converted to this purpose. But their positive productivity in grass or fuel or other materials is close to zero; their negative productivity in silt and sand to clog the streams and irrigation systems is remarkable; and collectively they lie as one of that region's more notable agro-ecological disasters.

In the more favored parts of South India (and West Africa too) other factors have been at work in removing the range, eg colonial monopurpose foresterism. Chandran and Gadgil (1993) describe the former rain forest range once available to the farmers of Uttara Kannada, of the species contributions it made to forest garden farm development, of its much wider historical social role, how this was subverted by colonial foresters after 1878 who saw in it only a source of commercial timber, and of how this ancient, holistic, forest-and-farm land use system might be reclaimed by once again involving villagers in management of its remnants. Working from old records and folk history they list some of the wild produce of this forest traditionally taken for cash sale, subsistence and farm use: weaving leaves, fuel, palm toddy, bamboo construction poles, bamboo shoots and seed (as grain), palm sago, starch, flour, fruits, oil seeds, honey, mushrooms, cashew nuts, pepper curry leaf and about 15 species of game animals and birds...*(Field) agriculture alone, on the fragile tropical rain-washed soils would not have provided subsistence of a durable nature. Even today after nearly two centuries of state domination over its forests several caste groups of Uttara Kannada show traits of a hunting-gathering way of life.* The range was well-regulated communal property. It was of religious significance and contained the sacred groves. In times of drought it functioned as a forest granary and in one famine 50 000 people came to collect its bamboo seed for food.

In Central and East Java the range is greatly diminished but it still serves two important purposes. In the upper Madiun Valley there are still a few leguminous trees left along the roadsides (with others planted in the *pekarangan* forest gardens for this purpose) which are seasonally hacked about for their leaves as paddy field green manure. On the mid slopes of Mt Lawu the tiny vegetable farms of Blumbang – among the most intensively cultivated relay-planted crop lands in the world – have been traditionally dependent on baskets of ground litter and tree leaves carried down from the 'protective' rain forest. The first use is to maintain soil fertility and structure, the second to partly support stall-fed cattle and through them also to obtain soil fertility (Probowo and McConnell 1993). Elsewhere throughout Java the main use of the range is to supply whatever grass and weeds can be gleaned from roadsides or irrigation ditches for livestock feed. One hazards a guess that fully a third of Java's cattle and goats might be supported in this way.

The littoral The forest gardens of Oceania are of two types: on atolls they exist along the coastal strand; on volcanic islands both there and in the forested (or once-forested) interior. There are, or were, two ranges, the rain forest and the littoral. On some atolls agriculture on the coarse infertile coral sands still depends on marine organic matter (seaweed and sea creatures) harvested from the inter-tidal zone. These traditional uses are decreasing. Nowadays the littoral in many places is only a source of pretty aquarium fish (Sri Lanka), live coral and seashells (Moluku, Bali, Philippines), the reef itself (for lime kilns on the Hikkodua coast). These things now contribute much more to the economies of dwellers of the littoral, and to its devastation. In Philippines only 10% of the reefs and their biosystems now remain intact;

globally with Greenhouse warming and agricultural run-off the coral reefs, like the rain forests, will soon be gone.

1.3 Perceptions

A first condition for the continuation and possible expansion of these ancient agroforestry systems is that governments, aid agencies, banks and mainstream agriculturists understand them, their strengths and weaknesses and the specific roles which they might play in agricultural development and resource conservation. Agroforesters count and measure trees. Botanists find new species. Ethnobotanists list product uses and traditions. Nutritionists know the hours it takes to gut a sago palm. Climatologists measure humidity in the canopy and soils surveyors dig holes beneath it. Knowledge accumulates in all these fields. But agricultural economists show little interest in putting these pieces together to formulate holistic man-plant-animal agroecosystems of contemporary relevance.

Administrators and planners remain confused. These forest farms are too primitive, simplistic. Too close to the forest, the natural world, so they can't be sound. As development vehicle they will not bear big-ticket items – no fertilizer factories, grain silos, milk plants or tractor pools so they lack visual appeal...or worse, political appeal, but the two are generally equated. They purport to solve present day problems with the folk technology of the past and so they must be a step backward down the agro-development ladder...But on the other hand they are too complex, sophisticated. No one understands them. They compress too many socio-agro-economic-ecological relationships into too small a space: *The major constraint of the system is that it is the least understood scientifically* (Tajwani 1987)...And then again, they're too elusive, no bureaucrat can claim them. They fall somewhere among the ministries, departments, agencies sketched above, and there's increasing threat that bits of them will be grabbed by hydrologists (for catchment protection), engineers (for soil conservation), by fisheries (for reef protection), plant breeders (for genetic conservation), by ecologists...for God knows what.

So let's be frank about it. These forest gardens are not preferred by governments or development agencies. Banks are not keen on them either. Fertilizer salesmen approach them with distaste. Agricide distributors hate them. To the Agricultural Orthodox they carry the stink of heresy about them.

1.4 Criteria...The Private and Public Efficiency of Farms

There are at least 34 criteria of contemporary relevance by which individual farms and wider farming systems can be evaluated, judged, ranked, planned

(Table 1.2); and the reader will think of others to meet specific circumstances. These apply to all kinds of farms, modern and traditional, and other land use. These criteria also serve as a basis for measuring the full consequences of converting a natural ecosystem of forest, swamp, savanna, into an agroecosystem and to farms of different types – cattle v rice v agroforest etc.

The criteria fall into three groups. Group A1-12 are relevant mainly at farmer or household, village, community level, dominated in modern agriculture by profitability with ever-increasing productivity and/or cost-minimizing as its basis. Depending on its operational mode (McConnell and Dillon 1997) most Western farm economics (and this applied also to Eastern commercial agriculture) seeks with use of such tools as budgeting, response analysis, linear programming to maximize profit subject to available resources. As the planning *milieu* moves further away from this simple objective towards the subsistence end of the farm-type scale the other criteria 3-12 as planning conditions or constrains become increasingly important – things like income stability over time, diversity, flexibility in product uses, sustainability. Towards the subsistence end of the scale, ie for the bulk of the world's farmers, 'profitability' recedes into the background, perhaps disappears altogether, and is replaced by such concerns as sustainability or how efficient this farm is in providing human habitat, disposing of household waste, avoiding the need for debt. In the traditional or 'developing' world most of these criteria 1-12 will operate more-or-less simultaneously and be ranked according to local circumstances and culture (but always based on some minimum level of productivity). Their specific acknowledgement here is considered to be of great importance. When traditional agriculture is approached by Western farm planners – or local trained in the Western tradition – such farming systems because they might score poorly according to the familiar profitability criterion are usually judged to be backward, primitive, undeveloped and sorely in need of modernization. This is in fact the premise on which most development aid is based. While bad farmers and inefficient farming systems can be found anywhere it is also true that within the traditional world there are agroecosystems which are far more efficient if all the criteria are allowed than the modern Western systems which seek to replace them. One of these is forest gardens.

The criteria of B are primarily of concern beyond the farm gate, from community to regional to national level and view a farm or system in terms of how efficient it is according to the attributes listed. Does it contribute to socio-economic equity? To beyond-the-farm employment? To urban food supply? These considerations are reviewed below as they apply to forest gardens. The criteria of C address problems which arise partly or mainly in agriculture or agricultural expansion and are now universal – conservation of plant/animal species, of human culture, of fossil fuel energy and above all the impacts of specific agroecosystems through atmospheric pollution and climate change.

To this partial list of the criteria one might want to add such social efficiency measurements as: Does this crop or system mitigate or compound the risks of war over the scarce resources it needs, now particularly irrigation water? Does it require slavery or indentured labor, or their modern equivalent of underpaid and forced female and child labor? This latter can't be dismissed as irrelevant to the modern world; much of the 'success' of such export crops as tea, cinnamon, coffee, cotton…is based on it.

Summary – externalities The efficiency criteria of Table 1.2 from about 12 on are also a list of potential externalities or indirect unintended social costs which might arise from selection and operation of any farming system. The best of systems would minimize or avoid or actively counter them. Here again these most efficient systems are often found in the tropical traditional world. Many of the criteria are mutually incompatible: eg maximum profitability is almost antithetical to sustainability; bulk food output for urban populations is difficult to reconcile with biological conservation. It used to be primarily the task of economics and especially agricultural economics to reconcile these conflicts and somehow optimize the use of 'land,' mainly with its market pricing mechanism. That was when land use was seen as having only one or a few economic dimensions – 1, 2 and perhaps 3, 15, 16 – and these were quantifiable in a few simple measures like net farm income or yield per hectare or rate of return on capital. Just who or what agency or discipline does it now in this vastly more complex world where all 34 of the criteria are relevant is not too clear. Probably nobody. What one can say with some confidence is that in agricultural planning in the 1st World – and especially when its aid is directed at developing the 3rd – the economic paradigm which guides it is still stuck on 1 and 2; more short-term production, greater profit with little regard for the others. The criteria are briefly reviewed as they pertain to forest gardens.

1.5 Private Household, Farm, Village Efficiency Criteria

Productivity – 1

Stamford Raffles observed long ago that the productivity of Java's forest gardens appeared to be at least as great as that of her padi fields. Probably on a per unit of land and certainly on a per unit of labor basis forest gardens are more productive than alternative farm types. This is sometimes difficult to substantiate. The multiplicity of species and the greater number of their separate products – foods, beverages, drugs, building materials – defy aggregation into a single economic productivity measure (Diversity, below). But if all of these are represented by the ecological measure of net primary production (NPP) there is little doubt that forest gardens are the most 'productive' agroecosystem because they are analogue of a rain forest, itself biologically the most productive of all natural terrestrial ecosystems.

Table 1.2   Private and Public Criteria for Planning Farming Systems or Evaluating their Efficiencies

A.   Private – Household, Farm, Village

1.	Productivity/yield levels	... of food, fibers, industrial outputs
2.	Profitability	... of these directly or through their processing
3.	Stability	... of production and/or income over time
4.	Diversity	... of species, crops, types, as income sources
5.	Flexibility	... in the uses of products (a dimension of economic diversity)
6.	Time dispersion, concentration	... of products, production, farm income
7.	Sustainability	... over time, indefinitely
8.	Compatibility	... of system activities internally; system with outside world of the system with others and the larger world
9.	Need for credit, debt	... for production, consumption, emergency
10.	Human habitat	... contribution to quality of life, aesthetics
11.	Health hazards	... arising from or contained by system (internal, external)
12.	Effluent, waste disposal	... from system, village, community

B.   Mainly Public, Economic and Social

13.	Equity	... in distribution of system outputs, benefits
14.	Employment	... generated by system, direct and indirect
15.	Raw materials output	... for farm, local, national industries
16.	Bulk food output	... for urban populations
17.	Foreign exchange earnings, costs	... required or generated by system
18.	Water (irrigation) required	... by system
19.	Fertilizer required	... by system
20.	Agricide/chemicals required	... by system
21.	Research support required	... by system
22.	Foreign aid required	... developmental and emergency

23.	Soil erosion/conservation	...	caused by/contained by system
24.	Water conservation, flood mitigation	...	contributed by system
25.	Soil/water salinization, acidification	...	caused/reversed by system
26.	Forests, agroforests	...	requiring deforestation, allowing reforestation
27.	Rehabilitation	...	of land and water sources

C. Of Global Concern

28.	Biological conservation	...	of flora, fauna, species, landraces
29.	Social conservation	...	of human culture, folk knowledge
30.	Energy conservation	...	mainly of fossil fuels
31.	Greenhouse impact	...	directly/indirectly by system
32.	Ozone impact	...	directly/indirectly by system
33.	Hydrological cycle	...	impact by system, direct and indirect
34.	Extreme weather/climate events	...	contribution to floods/droughts/fires (El Niño)

Those few measurements which have been attempted (Kandy) invariably measure only the main economic commodities (number of jakfruit, kg of coffee etc) which are harvested; but much produce in excess of household requirements is not harvested. Also, productivity or yield comparisons between this and other systems usually refer to single crops extracted from the context of the all-species tree mix. Here it is often found that the *per tree* yields of some crops such as cacao, coffee, citrus etc are lower on gardens than on some standard farm growing these same species as monocrops. An example of such comparisons is shown in Table 1.3. On the farm growing these eight species, with their populations as shown, yields are lower for four crops and higher for four others than they would be on farms growing these crops as pure stands. Averaged over all eight crops yield levels are about the same; but the mixed garden's total plant population is 35% greater, which is also about the total relative *production* advantage of the forest garden. But *aficionados* of the separate crops (eg coffee with a relative disadvantage of 0.33 on the forest garden) don't make this adjustment. They insist that if their crop of coconuts or coffee or citrus etc is to be optimized then forest gardens are not an efficient vehicle for this. They often go further and recommend that the whole mess be chopped down and converted into a mini monocrop estate. If their planning criterion is eg maximum production of some crop for foreign exchange earning then they're right (for only some of the crops). But if it is any of the criteria 3-through-10, or item 24 the preservation of species and genetic diversity, they are wrong. Nonetheless the biggest threat facing forest gardens is their simplification and 're-development' based on limited and technocentric evaluation criteria to make them more 'productive'.

Table 1.3   Species Yields on a Mixed Farm in Comparison with Pure Stand Yields

Crop	Standard yields as pure stand crops			Comparative yields on a forest garden		
	Pop/ha (no)	Units	Per tree yield (units)	Pop (no)	Avg. tree yield (units)	Relative per tree yield
Coconuts	150	No	70	60	50	50/70 = .71
Coffee	800	Kg	3	260	2	0.67
Jak	100	No	200	20	220	1.10
Durian	100	No	200	1	250	1.25
Cloves	200	Kg	10	25	9	0.90
Areca	1600	No	400	200	400	1.00
Kitul	400	Bot	15	25	20	1.33
Citrus	300	No	400	25	300	0.75
Total population	456	(mean)	...	616		

Limits Productivity is a general condition for profitability, but it can also take on an independent momentum of its own, in the Third World as well as the First – making two ears of grain grow where one grew before, though sometimes for reasons verging on the moronic. In the hills of East Bhutan the main crop is corn and the subsistence farmers there can always grow enough, primitively, to feed themselves and through lack of markets they then convert the surplus into the local firewater, often with the most appalling social consequences. Yet through the 1980s there were agricultural 'development' projects aimed at teaching these poor people how to grow more corn. Production becomes an end, a good thing in itself. Similar technocracies masquerade as development through much of the tropical world.

This difficulty of deciding on just what constitutes 'productivity' follows modern agriculture into the swamps. In Southern Sudan since colonial days there have been attempts to convert the bio-rich ecosystems of the River Lol's floodplains into intensive Taiwan-type paddy field agrosystems to grow improved HYV rice. Though millions of dollars have been thrown at them intermittently, over this 30 years or so by governments and aid agencies, these attempts have always failed. On the other hand the flood plain-riverine ecosystems which this technology has sought to liquidate are of great richness in everything from *wild* rice, the natural panicum grains (Origins, Africa) and in every animal resource from elephants through crocodiles to water fowl; and these natural systems of the Lol, linking up downstream with that wetland world the Sud, have fed the local Dinka people for millennia – at least since their emergence as H. sapiens and no doubt since their days as H. erectus. Yet agronomy is not content with this and would sweep it all away in favor of the greater 'productivity' promised by proper farming. (For decades there have been engineering plans to drain the Sud itself for farms, and carry its 'surplus' water off to an Egyptian captivity.)

Profitability – 2

Those few economic studies which have been made indicate this agroecosystem is usually at least as profitable as alternative systems in the same area. On subsistence-oriented farms, and on those still close to their forest parentage the concept of profit or net returns is largely irrelevant; but most farms now have some degree of commercialization. From their survey of 30 Kandy gardens McConnell and Dharmapala report them invariably more profitable than the other main farm type in this area, wet rice production. But since only the comparatively wealthy own land of any sort, a better yardstick is money income several times that of rural wage labor in the same area (Kandy). Also in the Kandy hills Perera and Rajapakse (1991) reported from their survey of 50 smaller and even minute farms of 0.1-0.25 ha that 80% of them had incomes above the national per capita income of Rs 12 000:

Cash Incomes on a Sample of Kandyan Farms

20% of farms ...	above Rs – 12 000	18% of farms ...	below Rs 48 – 72 000
36	12 – 24 000	6	above 72 – 72 000
20	24 – 48 000		

Source: Perera and Rajapakse (1988). $US = Rs25.

Bavappa and Jacob, Alles (1981, 1987) developed models of the same traditional farm type under the assumption that these were developed and commercialized (Table 1.4). By year 7 of re-development gross value of output would be Rupees 73 700 (net returns about Rs 60 000 or five times the average national income in Sri Lanka's monetized sector). These 'developed' and monetized forest gardens would then be similar to the mainly commercialized forest gardens of South India; but they would differ from those forest gardens in most of the tropical world which have primarily a subsistence orientation.

As summarized in Table 1.5, Michon et al (1983, 1986) report from West Sumatra that money incomes on forest gardens in two Minangkabau villages there, with rice fields attached, are comparable to or higher than on pure rice farms ($400 - 1 200 v $500 - 800 on pure rice farms), with much lower year-to-year income variation.

Table 1.4   Annual Return from a One-Hectare Mixed Cropping Model at Delpitiya, 1987

Crops/Yield years after	Yield (kg/plant) or no. of fruits/plant							No. of plants	Total yield	Unit* price	Total* value
	1	2	3	4	5	6	7				
Pepper (kg)	-	0.2	0.3	0.5	0.8	1.0	1.0	1 296	1 296	25	32 400
Coffee-R (kg)	-	-	0.2	0.4	0.6	0.75	0.75	220	165	30	4 950
Coffee-S (kg)	-	0.1	0.2	0.3	0.4	0.5	0.5	1 914	957	30	28 710
Clove (kg)	-	-	-	0.1	0.2	0.5	1.5	12	18	100	1 800
Nutmeg (no.)	-	-	-	-	25	100	500	12	6 000	0.2	1 200
Coconut (no.)	-	-	-	5	15	25	50	36	1 810	1.5	2 700
Mango (no.)	-	-	10	30	40	50	50	3	150	0.5	75
Jack fruit (no.)	-	-	-	5	10	15	15	3	45	2	90
Breadfruit (no.)	-	-	-	5	15	30	50	3	150	1	150
Avocado (no.)	-	-	-	10	30	50	50	3	3	0.2	30
Arecanut (no.)	-	-	-	10	50	100	100	24	2 400	0.2	480
Banana (bunches)	-	1	1	1	1	1	1	24	24	10	240
Lime (no.)	-	-	-	10	20	50	100	44	4 400	20	880
12 crops								3 594			73 705

*Value in Sri Lanka rupees.   US $ = SL Rs 25.
Source: Jacob and Alles 1987.

Table 1.5   Typical Farm Incomes in Two Minangkabau Villages having Forest Gardens-plus-Rice Field

Product	Home use - sale %			Village total, Rp million	
				Village 1	Village 2
Cinnamon	0	-	100	41.1	133.8
Nutmeg	10	-	90)		
- mace	0	-	100)	3.6	15.0
Coffee	10	-	90	16.0	30.0
Durian	10	-	90	17.5	165.5
Firewood	70	-	30	4.0	13.5
Timber	60	-	40	20.0	25.0
Rice		var.		285.0	192.0
Others				(not specified)	
Totals per family	Rp			931 000	1 222 000
	$			931	1 222

These farms are discussed in Chapter 2. Values of many minor products and livestock are omitted. US$ = Rupiah (Rp) 1 000 (1985).
Source: Michon, Mary, Bompard 1986.

Elsewhere on Palawan, Philippines, Raintree and Warner (1986) found that on small highly mixed tree crop farms converted from previous swidden plots incomes are five times higher from the tree crops than from the swidden agriculture per unit of land and 3.8 times higher per unit of labor. In Peruvian Amazonia Padoch et al (1985) note that value of produce annually taken from the forested ex-swidden plots (essentially forest gardens) is well above the incomes of families growing rice. In SW Nigeria the value of output from the forested component of 'compound' farms there is reported 5-10 times that from the cultivated portions of these farms (Reconnaissance).

Income Stability – 3

On both developed and 'wild' forest gardens, and in traditional agriculture generally, this is a more important planning criterion or dimension of farm performance than productivity or profit. It measures year-to-year stability of yields and the prices attached to them. It is important, among other things, as an indicator of the household's need to incur debt (Time-dispersion, below). Because of geographical-climatic conditions throughout the forest garden zone farm total productivity is usually stable. But yields of individual species (and their prices) are often not. The data for annual clove yields on 9 Kandy farms in Table 1.6 will illustrate the hazard: yields can go from 100% of potential to zero in two consecutive years. (Clove prices also are unstable.)

Table 1.6 Kandy Farms: Variation in Annual Clove Yields

Farm No. 1	2	3	4	5	6	7	8	9
1972	-	(100)	(100)	(100)	(100)	(100)	(100)	(100)	(100)
1971	(100)	2	18	400	0	40	167	40	600
1970	0	457	80	250	5	35	150	13	0
1969	70	71			30			100	0
1968	93								75
1967	70								
1966	0								
1965	52								

Source: McConnell and Upawansa, Peradeniya 1972.

Forest gardens achieve income stability through the diversity of species grown, and the diversity of uses of their separate products. In this they are much superior to monocrop (cotton, sugar cane, coffee...) farms producing only one product for which there is only one use, and often having to sell it to only one buyer...the sugar *central*, the gin, the tea factory.

Farm Diversity – 4

On any farm there are two aspects of diversity, biological/botanical and economic. The first is a necessary but not a sufficient condition for economic diversity. This latter has five dimensions:

1. species   -   the number of plant/animal species present
2. products -   the numbers of their respective direct products
3. uses   -   the numbers of ways in which these products can be used including the possibility of re-cycling some of them back to the system as inputs (Flexibility below)
4. time   -   the time/seasonal pattern over which the products of 2 naturally occur (or can be induced to occur)
5. transfer -   the time pattern which these products can be given by processing and/or storage.

In all these dimensions except sometimes 5 forest gardens are the most diverse of agrosystems. The bases of this diversity are the large number of plant/tree species present (Reconnaissance, Kandy), each constituting a separate activity or enterprise in farm economics jargon.   Some of these diversity data are consolidated in Table 1.2 below. To readers used to Western corn or wheat farms, or even the finely integrated 10-species crop and livestock systems of Sind or the Pathan farms of NW Frontier, the forest farms of Santa Rosa and the Maya farms of Central America with 50, 60, 70... significant species might seem like an agronomic impossibility – or a management nightmare. The causes of forest garden diversity in general might be listed:

Initial forest species diversity The parent forests provided a matrix into which these agroecosystems were integrated. Many forest species were retained. It is no accident that presently the most species-diverse farms of SE Asia and tropical America occur where the parent rain forests achieved greatest diversity, and the least diverse farms of Oceania are where the parent forests were least species-diverse (Genesis).

Exposure to historical (or prehistorical) plant imports This has been important in South India, Sri Lanka, Java, but less so in Outer Indonesia, Africa, Oceania, and also until historical times in tropical America (but here there apparently was considerable species interchange *within* the Central/South American region).

Agroecological requirements As in the parent rain forest, many tree/vine crops when extracted from it do best as part of a plant association, providing mutual shade-humidity-light-moisture-nutrient cycling conditions. Forest gardens attempt to maintain this interdependent ecology, to varying degree; the optimum performance of some subject species requires the presence of others. (Other tropical agroecosystems also have this need, eg cacao-under-coconuts, but find it impractical or unprofitable to go beyond 2- or 3-species combinations.)

Geographical (and economic) isolation This leads to the need to develop farms as comprehensive life-support systems, as distinct from monodimensional commercial agrosystems. This favors both a large number of species and, of these, preference for multipurpose species.

Farm size Other things equal smaller subsistence-oriented farms are usually more diversified than larger ones.

Culture-religion This relates to the degree to which certain plants (and animals) become and remain associated with the spiritual dimension of existence. It pertains mainly to indigenous plants but can be extended to exotics (Reconnaissance, the temple groves). The superficial way of acknowledging this is to say that some cultures 'like' plants while to others they have only material significance (Reconnaissance; Posey on the Kayapo, Rocas on the Maya).

Comparative crop economies Concern with the financial economics of species is nearly always a negative contributing factor; the greater the exposure to commercial opportunities the greater the tendency towards farm simplification, specialization.

Measurement of diversity It is often necessary to make diversity comparisons within a group of farms, or between different farm types. Eg Table 3.1 of Chapter 3 lists the species grown on 30 Kandy farms. Which farm is the most diversified? If all these farms are approximately similar in enterprise/species structure it would be enough simply to count the number of species present. Farm 1 with 12 species (S) appears more diversified than farm 3 with only 8. But this does not reflect the total plant populations on these farms or relative sizes of the separate species populations. Several methods are available – Simpson's Index, Shannon-Wiener's Diversity, Margalef's H, or coming from the other direction, Sorenson's Similarity Index. The most user-friendly is probably Simpson's Index (Simpson 1949).

$$D = [1 - \overset{S}{\Sigma} (n_i / N)^2]$$

where S is the number of species/crops present, n is population of each species ($i = 1$ to S) and N is the total plant population on the farm, and diversity D takes on values from lowest (zero) to highest (one). In Kerala, Kumar et al (Reconnaissance) applied this to compare the diversities of groups of mixed farms in 17 *thaluk* (sub-districts), obtaining diversity measures ranging from 0.25 to 0.74. Applying it to the farm (1) of Table 3.1 (Kandy), $S = 10$, $N = 216$ and n is successively 4, 5, 0, 15, 150... and the obtained diversity here could be compared against those on the other 29 farms etc, or as a group with other types of farms. There are still two problems. First, many of these species can be quantified only on an area basis (vegetables, tea bushes, grass). This might be overcome by applying $values to all the species and giving diversity an economic rather than a botanical basis. But many of the species on forest gardens don't *have* a $value because they don't have a market price reference. Second, there are too many possible valuation bases. Consider a *caboclo* housewife with four plants representing all the species on her farm. The first is a food or spice and can be valued readily enough because it's sold in the market. The second has no material use but looks pretty, contributes to habitat and is subjectively valued on that account. The third is a cure for snakebite and a dozen other maladies and represents the pharmacopoeia. The fourth has no use whatsoever; but old people say its possession once promised access to the Sun God, and this is the most prized of all. In short, it is sometimes necessary to pretend that measurement of diversity-in-use, as distinct from objective botanical diversity, is possible but the results can only be suggestive.

Flexibility (in uses of primary and secondary products) – 5

Flexibility in product disposal is a dimension of farm diversity, the number of ways in which some first product can be disposed of by sale, use/consumption, barter. The possibility of processing this first product might also be present, and some of these secondary products might also offer these same alternatives. Further, the possibility of storage of the first or the derived products adds a time dimension to flexibility. This adds the possibility of haggling for price advantage, or fitting second stage processing into the household work schedule. Flexibility-in-disposal is not unique to forest gardens; it exists in most traditional agriculture. It just happens to be more pronounced here because there are so many species yielding so many first products to work with. This situation might be compared with that on a small tea farm: this produces only one primary product; there is no possibility of farm processing for secondary products; the green leaf must be sold at only one time, within a day of picking; it can be sold to only one buyer, the local tea factory and at only one price, that set by the factory. But consider the jakfruit (Figure 1.1). Apart from the tree's other products of leaf as goat feed, litter as fuel and eventual use of the tree as timber, the fruit can be sold, used, bartered, processed. If kitchen processed its pulp must be soon used but its seeds can be used or bartered at this time, or water-stored in a jug for future use or barter. The possibility of this type of flexibility in disposal of much forest garden produce has important implications for cottage scale village industrialization (below).

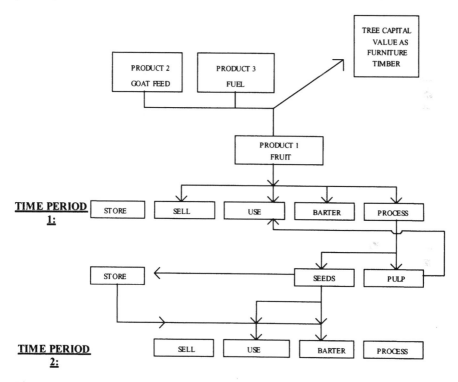

Figure 1.1   Disposal of a Jakfruit

Time-Dispersion (of production and income) – 6

This dimension of diversity refers to the time pattern of income receipt – whether this occurs as a single lump sum once in the year (a perfectly time-concentrated income), or as three or four amounts during the year, or as a uniform stream throughout the year. A monocrop sugar farm with one annual harvest is an example of the first; a relay cropped rice farm an example of the second; small dairy farms selling a few bottles of milk around the village every day and forest gardens are examples of the third.

Table 1.7   Monthly Distributions of Total Annual Production of Some Tropical Crops

	Jan	Feb	Mar	Apr	May	Jun	Jul	Aug	Sep	Oct	Nov	Dec
				Percentage of total annual production by month								
Cacao; Perak, Malaysia	17	10	15	7	4	2	3	4	7	8	11	12
Cacao; Matale, SL	3	2	2	5	9	5	3	2	5	20	29	15
Arecanut; Kandy, SL	5	35	5	15	0	0	0	0	0	5	5	10
Breadfruit; Kandy, SL	3	3	0	0	4	15	36	30	7	0	1	1
Cardamom; SL	6	5	3	4	4	10	10	10	10	12	15	11
Cloves; Kandy, SL	25	28	32	10	2	0	0	0	0	0	0	3
Coconut; Negombo, SL	6	6	6	7	10	11	11	12	10	9	7	6
Nutmeg; Kandy, SL	2	1	45	10	1	1	35	1	1	1	1	1
Pepper; Matale, SL	20	10	2	0	2	7	12	3	2	3	14	25
Kapok; Matale, SL	0	0	0	0	80	20	0	0	0	0	0	0
Rubber; southern SL	13	9	9	8	5	4	7	7	11	7	11	9
Tea; SL high country	8	8	11	12	13	8	7	5	4	6	7	11

Source: McConnell 1986.

Other things equal dispersion is preferable to concentration. In subsistence situations a large seasonal glut of produce will be of little use unless it can be processed. Much produce cannot be stored or if it can, eg grains in mud bins on African farms, losses to fungus, weevils, rats, can account for easily one third of the crop. The major problem with time-concentrated production patterns on seasonal rice crop farms in South/SE Asia is the frequent necessity to incur bridging subsistence debt between harvests. These food advances from the village storekeeper or the money lender carry usurious interest rates; 60, 70...even 100% on an annual basis are common. Then if the planting monsoon arrives late, or El Niño intervenes and it does not arrive at all, this debt can become perpetual bondage from which the family might never escape.

Forest gardens are immune to this kind of debt (but not to some others, Kandy); their total output occurs as a time-dispersed stream. The monthly distributions of production of common SE Asian tree crops are shown in Table 1.7. Some are uniformly dispersed (coconut); some are bimodal (breadfruit); only one is time-concentrated (kapok). Aggregating these and many other species to compose farms, the result is usually a uniform income stream.

Sustainability – 7

This refers to the ability of an agroecosystem to remain viable over some very long time period – indefinitely. There are five main groups of reasons why systems become *unsustainable*; often these reasons exist in combination, one prime cause leads to others.

Physical and chemical – soils The soil base wears out, washes away or become too infertile to permit continued farming (Energy, erosion).

Physical – water The water supply gives out. The aquifer gets pumped dry; or the streams from denuded catchments become a trickle alternating seasonally with a torrent.

Biological Monocrops invite pests and disease. For a few years these can be contained by agrochemicals. Resistant strains appear. Stronger chemicals are formulated. They cost too much. The main equity holder in the system becomes the insecticide salesman. The hunt is on for genetic resistance or for the pest's natural predators. Too late. The forest where the crop once came from, and where its genes and the natural predators of the pest might still have been found, has been cut down and sold for chopsticks. If the subject crop was a profitable one and can't be easily replaced (as with tea replacing blighted coffee in Sri Lanka) the farm's value declines. The system might not be abandoned entirely and to this extent it remains nominally sustainable. It's downsized, and the other bits of the economy which once depended upon it are downsized too.

Environmental – social Examples of this cause of unsustainability are provided by the banana and cotton industries. Subject to the eventual futility of chemical pest control, aerial spraying is usually the most cost-effective method of application. It ceases to be feasible when urban sprawl brings schools and shopping centers into the spray drift zone, or when idle troublemakers start reworking statistics on birth defects and cancer incidence. If these industries (and others such as orchards, feed lots, battery pigs) have nowhere else to go, or can relocate only at high cost, they cease to become locally sustainable.

Economics This cause consists of all those factors underlying shifting demand and price. Farms producing only one or a few commodities are most exposed; diversified farms are least exposed. If adjustment into alternative crops is not possible such farms might simply fade away, revert to bush. The once thriving sisal estates of Kenya and the gum arabic gardens of Kordofan are cases in point.

Compatibility – 8

In farming systems this refers to the absence of internal or external conflict. There are three aspects:

1. compatibility among the separate production steps, processes;
2. compatibility of this crop with other crop/livestock enterprises on the farm;
3. compatibility of this crop or the farm as a whole with the wider society.

Both 1 and 2 are best approached by examples.

Compatibility of production processes In the 1980s a Middle East country with a large river, which shall otherwise be nameless, received a two billion dollar grant for its agricultural development, specifically to increase its main food crop. First item on the shopping list were tractors, lots of them. So many that these replaced the previous work oxen. With these gone and the fertilizer of their manure for the Main Food Crop no longer available this was soon in trouble, avoided only by spending a good part of the two billion balance on importing fertilizer, then fertilizer plants, then replacement parts for the plants as these wore out. All this in an ancient land in which the river, the crop, the cattle, the farmer, the sun, had been compatible components of a sustainable system for 9 000 years.

Compatibility of enterprises When sugar cane prices crashed in Australia, in the late 1980s many cane farmers looked for alternative enterprises, including beef cattle. These were turned loose to glean the cane fields, sometimes on the standing cane itself. For young cattle there was no problem, but market cattle could not be sold because of meat contamination with cane pesticide residues. (The same occurred in the drought of 1995 when pesticides in cane and cotton stover fed as emergency rations to dairy cows turned up in milk.) It is not clear that forest gardens could claim to be superior in these aspects of compatibility to other traditional systems, but they would not rank lower.

Debt and Credit – 9

This criterion quantifies the degree to which a farm household must or might incur debt (Kandy). The most onerous kinds of debt are so-called production credit associated with seasonal field crop farming under risky climatic conditions (eg drought, flood, pest-disease outbreaks in S. Asia), emergency debt when the crop fails, and consumption debt if the farm is insufficiently productive relative to family needs. Because of their regular flow of produce from a multiplicity of species/crops forest gardens are relatively immune to these hazards, especially in the true wet equatorial belt and outside the hurricane zone.

Credit This refers to the possible need for governments or aid agencies to provide a banking system or equivalent credit mechanisms. These can be very expensive to set up and operate, especially in poor countries. Good examples are those countries which have most enthusiastically adopted the Green Revolution package (of high yielding variety HYV seed, fertilizer, agricides) and applied it to a monocrop (eg commercial corn in Malawi). For decades much of the proper agriculture in this and similar countries has led a precarious hand to mouth existence, with trucks and trains rolling in at planting time with the imported inputs Package, then after the harvest rolling out again with the crop to pay for it, with everyone keeping their fingers crossed that:

1. the inputs can be found somewhere at reasonable cost;
2. the ship doesn't sink or get diverted;
3. once on the train the inputs don't get blown up by rebels;
4. once the crop is planted it doesn't wither with drought;
5. once harvested it can be sold;
6. once sold enough trucks could be found to take it out;
7. the residual profits which eventually trickles down to the *campesinos* are sufficient to induce them to embark on the same trip next season...with failure of any of these always threatening to bring the whole elaborate structure tumbling down.

There is much to be said for an agroecosystem which by its nature does not generate these headaches. Another hazard is that these credit-agrodevelopment agencies can become monolithic bureaucracies in their own right, responsive not to small-farmer needs but to their own momentum.

Human Habitat – 10

High subjective values attached to this criterion are a recurring theme through forest garden surveys (Reconnaissance). Some aspects of habitat can be readily measured – efficiency in waste disposal (below), absence of dangerous chemicals, climatic factors like wind and heat and their modification by trees, plants kept for decorative purposes. Others can't. To the extent that habitat is valued for subjective non-economic reasons the more likely is the farm to contribute to flora and its associated fauna conservation. In the tropical world a large percentage of plants kept for 'habitat' do in fact have useful/economic purposes, especially medicinal, or had them in the recent past. But habitat can mean many things. In countries colonized by Europeans it often means duplication of a pretty English rose garden, or manicured grounds of some French chateau torn from a calendar, of no other than nostalgic value.

Health and Waste Disposal – 11, 12

These two considerations are closely related, with health also including hazards arising from necessity of the system to use dangerous chemicals and agricides. Conscious avoidance of these risks is usually taken for granted in the 1st World but they can become acute in the 3rd when products with indecipherable labels are urged on illiterate peasants with perhaps only pictured representations of their marvelous efficiencies offered. The hazard is to themselves and, more important, to consumers of their fresh produce, especially vegetables and fruit. The risks are increased when the products themselves are dumped on 3rd World farmers because they're banned in the countries of manufacture. Other dimensions of health in farming systems are their need for or avoidance of ponded water or soil/vegetation disturbance with these providing habitat for mosquitoes and their diseases. Waste disposal refers to both farming operations and households. These problems become more pronounced as human populations increase, especially in peri urban areas, and an attribute of the system becomes how efficiently it cycles and disposes of such wastes internally (Reconnaissance; Java, Kandy, urban farms West Africa).

1.6 Public Economic and Social Efficiency of Systems

Equity – 13

Equity, defined properly as 'the common fairness,' describes the way in which the total benefits of any agroecosystem are socially distributed, whether concentrated or widely dispersed. Among the criteria of Table 1.2 she must fight hardest for a place. This is generally resisted by planners and financial economists for whom equity is a distraction from the main game, criteria 1 and 2; but when denied and pushed to anger she draws a line which even the Tyrant may not cross. There are four main paths to her door. Once there, classical economists proposed and rationalist economists insist that the best way of wooing equity is through the market price mechanism:

1.   Suppliers of the factors of production to the farm obtain their equity

(share) in the benefits of its outputs by the income they receive in this capacity.

2. Consumers obtain *their* equity as access to the farm's produce at reasonable (ie minimum necessary) prices. Except in theocracies this economics apparatus is generally considered the most efficient way of winning equity. The problem is that in the developing world the model often does not work, or it yields grotesque results. Equity is not satisfied. Unpleasantness ensues.

3. A third way of pursuing equity is through the downstream use of farm raw materials as inputs for either local/village industry or more formal industrialization at national/regional level, with such industry offering indirect equity in the farm's output as wages.

4. When all these fail a fourth way is to appropriate some portion of gross farm benefits by taxing inputs or outputs and redistributing these as social welfare payments or jobs...in building hospitals or pyramids, according to the inclination of the Prince.

All agroecosystems can be ranked according to how well or poorly they offer themselves for equity's capture. Forest gardens turn in a mixed score. First, they use few or no external inputs, including hired labor, so cannot distribute equity to the rest of the village by this mechanism. (But in fact in SE Asia large quantities of non-family labor *are* employed, though for cultural and psychological reasons – Kandy.) Second, many of their products are perishable and don't get beyond the village; so in terms of providing bulk foodstuffs to urban consumers path 2 to equity is partly blocked, in comparison with this contribution made eg by grain farms. (On the other hand where transport exists forest gardens *are* often the main suppliers of perishable produce to urban centers.) Third, providers of equity through raw material supply, industrialization and jobs, forest gardens are sometimes the most efficient agroecosystems at national level. (Eg 80% of the timber in Kerala and most in Bangladesh, and most of the cabinet timber in Sri Lanka, and generally most fibers come from forest gardens.) In this region also they are outstandingly the most important basis for equity through *village* industrialization (below). Fourth, they are inefficient in providing a tax base. They have few inputs to be taxed; and their outputs are so diverse and of such small quantities at individual farm level that most of these also easily slip through the tax net – if one is cast at all.

Employment and Raw Materials – 14, 15

In most highly populated parts of South and SE Asia the limits of agriculture have been reached. Further social and economic progress at village level through agricultural intensification is a delusion. The alternative strategies are formal industrialization in those countries rich in natural or social resources (Malaysia, Indonesia, Philippines, Sri Lanka, Viet Nam) or village scale industrialization in those countries which aren't. Where both are possible planning emphasis is invariably given to the first.

The Java example Rural Java illustrates the size of the unemployment and poverty problem – richer than most, more populous than most, more agriculturally developed than most and, with all-Indonesia, having one of the highest rates of formal industrialization and economic growth (GDP, GNP) in the world (until 1998). Java has also been long subjected to the full impact of the Green Revolution, enthusiastically embracing each successively superior strain of rice and its associated technological package as these come on stream.

Five Java provinces are compared in Table 1.8. Taking Central Java as representative there are about 31 million people, 90% are in villages, 62% live below the official poverty line (defined as an annual income of 320 kg of rice equivalent).

Table 1.8 Socio-Economic Parameters of Five Indonesian Provinces

Province	Population	% below poverty line	Illiteracy	Infant mortality
Yogyakarta SI (Java)	3 million	69%	35%	65/1 000
Central Java	31	62	37	81
East Java	35	59	41	84
West Nusa Tenggara	4	53	49	129
South Sulawesi	7	46	43	93

Source: World Bank/International Fund for Agricultural Development – IFAD (1984-7). Data from Indonesia Government Census (1984).

Land, people and the limits of agriculture Going *within* Central Java the following sample of villages across the Solo Valley illustrates the futility of agricultural development as a vehicle for socio-economic progress (Figure 1.2). (The data are from traverse surveys of villages along an 85km track between the two volcanoes, Merapi and Lawu, with Solo City in the middle of the valley.) Village populations range from 15 to 42 persons *per hectare* of cropland. However only four villages have (good) rice land, the others have only (poor) dry cropland.

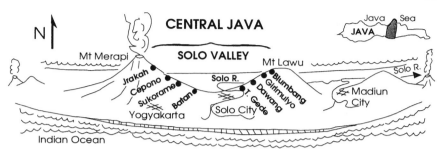

Figure 1.2   Villages and Farming Systems across the Solo Valley, Central Java

Table 1.9   Farmland and Population Pressures across the Solo Valley

	Village	Total crop land (dry and wet)	Pekarangan land (houselots)	Persons per ha cropland
(W: Merapi)	Jrakak	195 ha	117 ha	15
	Cepono	270	60	23**
	Sukorame	170	86	17**
	Batan	60*	39	42**
(Solo City)	Tegal Gede	248*	115	25**
	Dawang	182*	61	22
	Girimulyo	200*	111	18
(E: Lawu)	Blumbang	120	39	28

*villages with some rice land; **villages in which some non-farm work opportunities exist (for some people).

Source: From survey by McConnell and others 1974; the same villages surveyed by Probowo and McConnell 1987, reported 1993.

Degrees of poverty   Going *within* the typical village of Miri Sriharjo in neighboring Yogyakarta Special District, David Penny (of ANU 1986) ranked its population by degrees of poverty in comparison with all rural Java. This village had been exposed to all the benefits of the Green Revolution since its inception, yet 69% of its people remained 'poor' and of these 18% were destitute.   (Those 31% who were not poor were primarily landowners, government officials, and as such chief beneficiaries of that agronomic Great Leap Forward.)

Sriharjo is a prosperous village with good land and able to grow rice and sugar cane. Penny found that conditions there had not changed much since a previous study by Singarimbun (1973) who did a noon-time house-to-house survey and found that 10% of the population of this rich village had no food in the house at all, had not eaten that day, and had no idea if they could find a few hours work to buy food that night.

Comparative Degrees of Poverty in a Javanese Village

		Population in each 'poverty' class, cumulative	
	Rice income equivalent	M-Sriharjo	All rural Java
poor	320 kg/year	69%	60%
very poor	240	44	35
destitute	180	18	20

Percentages cumulative from bottom. Penny 1986. The *cucupan* level, 'barely enoughness' is 240 kg per person per year. The poverty level is 320 kg.

Rational economics visits Miri Sriharjo As noted, rational economics has it that best way of capturing equity is by paying the 'factors of production', including labor, that competitive price necessary to attract them to some economic activity rather than another, and that this can only happen if free and flexible markets exist for each of these factors. Penny (1984) investigated how well these markets actually worked in the rich village of Sriharjo. The price of labor was around 0.5 kg of rice per day. But most of these 'factors' could find a market for themselves less than 200 days annually. Their income levels then were below the *cucupan* or enoughness level of 'very poor' (previous box) or slightly above the destitution level. The second factor of production is land. For any of Sriharjo's landless who might have attempted to avoid starvation by renting a few square meters to grow their own food the market price of this was 70-to-90% *of the crop*. The third factor market is that for capital. For those factors who were able to borrow a little money to get their crop under way the interest rate (price) of this was 5-7%...*per day*. Penny sums up the factor markets in this neo-classical economics paradise of Sriharjo: *...They are indeed so effective that many people are now well below the cucupan (enoughness) level...For people who must rely on sale of their labor this market is a terrifying place.* Equity would be displeased and look for alternatives.

Alternative strategies The broad possibilities for rural development and increasing socio-economic equity, in most of rural South and Southeast Asia, are represented by these eight villages across Solo (Figure 1.2).

1. *Formal agriculture* Most official efforts are still directed at conventional field crop agriculture – by governments, aid agencies, banks. But in the villages of Java, and only slightly less so in most of S and SE Asia, with available labor already at 20, 30, 40...persons per hectare even some miraculous doubling of per hectare crop yields would have no impact on unemployment, and through this on equity. All it would do is increase income disparity.

2. *Formal industrialization* Batan and Tegal Gede represent those villages helped by industrialization. These are close to Solo City. But such opportunities for non-farm non-village work are not available in the distant villages. Further, they are not available to the elderly, uneducated, women with dependent children or otherwise socially disadvantaged, ie to the bulk of the rural populations. And most rural areas do not have such dynamic growth centers as Solo.

3. *Intervention* Possibilities for interventionist job creation are represented by Sukorame where there is a dairy milk collection center and processing plant. This small government sponsored industry generates some employment for the poorest of the village who have no land. There are also good forest gardens.

4. *Cottage industry* The distant village of Cepono represents cottage scale industry as another strategy. About 350 of the 1 300 households make their living as copper and tinsmiths. (No one remembers why this has happened in Cepono and not in other villages.) Most of Cepono's steep eroded volcanic sands are too poor for anything except cassava; but the houselots on the same soil (pekarangan lands) support rich forest gardens. So Cepono with the least resources of all the villages lives tolerably well by beating pots and picking tree crops.

5. *Forest gardens, raw materials and jobs* By mid 1980s Indonesian Government was becoming aware of these alternative employment generation strategies and their limitations, especially that of Green Revolution rice. With FAO and UNDP assistance it began a pilot program of village income generation based on Prof. Mohd. Yunis' model in the Bangladesh Grameen (Village) Bank. This extended small loans without material security to informal men's and women's groups to enable these to apply the only resource they have, their labor, in processing whatever raw materials could be found in the village. As in Bangladesh this program was aimed at the village poorest-of-the-poor, including female heads of destitute families. Group rather than individual lending, and group responsibility for repayment, resulted in nearly as good loan recovery rate in Java as the 98% achieved in Bangladesh. (This unsecured repayment rate is in fact higher than repayment rates on bank secured loans in rural areas.) Loan amounts in the Indonesian program have been very small – $50, $60, $100 per group are common – but not as small as in Bangladesh where loans as small as one dollar have been extended to 5-member groups. This illustrates the size of the poverty problem (and suggests why Mohammed Yunis, an economist who prefers to deal with small numbers, has been proposed – though only yet by public sentiment – for a Nobel Prize).

Before it provided finance for an Indonesia-wide extension of this program the International Fund for Agricultural Development (IFAD) surveyed the kinds of activities in which these villagers were involved (McConnell et al 1990). Among the men's activities: quarrying limestone; rounding up the village cobras for sale to Yogyakarta restaurants; collecting edible birds' nests. Squeezing down through limestone cracks then out into caverns beneath the ocean off the south Java coast. Building bamboo scaffolds to get at the nests, in bamboo torch light, in darkness for several days when the earth moved with the volcanoes. Sometimes forever. Women's groups do women's things, sewing and renting out traditional bridal outfits, fattening two or three goats, petty trading. But most of the 68 activities enumerated in this 1990 survey through Java, Bali, Sulawesi, South Kalimantan, had to do with processing village foodstuffs and fibers (eg the hundred and one uses of bamboo and pandanus). None of the on-going or proposed activities concerned *agricultural* development as such: its potentials have already been exhausted in most villages, and those that remain can be left to 'rich' landowners and line departments.

This introduces a new dimension to village agriculture, underlying the importance of raw materials production for village processing and job generation as a criterion in selecting from among alternative agroecosystems. Scope for these non-agricultural survival activities is determined largely by the *diversity* of crops present and *flexibility* in use of their products. At one extreme is the planners' favorite, rice, but apart from grain this produces nothing – even the straw is burned unless there is a paper mill in the area. Corn, oilseeds, milk are better but not to be compared, in terms of their potential for village job and equity generation, with the range of raw materials from forest gardens.

Bulk Food Output – 16

Fattening goats, exploiting the 156 uses of the coconut and catching cobras for a living must be good fun and no doubt it contributes to village equity. But is it *serious*? Central policy economists think not; they have more important things to worry about. One of these is arranging agriculture, selecting farming systems to maximize the supply of cheap bulk foodstuffs to hungry and threatening urban populations. In this regard forest gardens in comparison with grain, oilseed and sugar cane farms seem hopelessly inefficient. The best they can claim, where transport is available, is ability to supply most fruits, much coconut and such carbohydrates as breadfruit and jak to those urban consumers who have not yet advanced on to imported wheat flour, canned mutton and similar superior produce. There are exceptions. When Sri Lanka, a notorious sweet tooth, ran into foreign exchange problems in the 1970s and could no longer import cane sugar the village *kitul* palms (Caryota urens) were rediscovered overnight. Cassava, breadfruit and *borassus* and the other old foods were rediscovered too. Hardly any of the previously imported foodstuffs did not have a traditional local substitute. (One of the most effective FAO projects the author has encountered was a one-woman operation transferring knowledge of traditional food crops and their uses from Philippines where it still exists, to Sri Lanka where at official level it had largely been lost.)

Foreign Exchange Earnings and Costs – 17

This is another property of agroecosystems which often keeps the planning economists from sleeping soundly. They occupy the goblin hours by drawing up lists of farming systems according to how efficient these are in attracting honest dollars or good German marks for their exports. The problem of foreign exchange became important for that plethora of tropical countries which achieved independence in the mid twentieth century; they have looked for twentieth century systems to solve it. These do not include forest gardens. However in such regions as Sri Lanka-South India the bulk of produce from these small mixed farms has always been an important generator of foreign exchange through colonial times and this continues – most or all of the pepper, cloves, nutmeg, coffee, palm fibers and sugar, kapok, essential oils, some rubber and tea and a hundred other miscellaneous commodities (Reconnaissance, Kandy). Even so, this farm type can suffer three deficiencies. The individual lots of produce are small. This requires a network of village collectors, larger assemblers, traders. These exist in South Asia but not in much of Melanesia and Africa. Buyers are not amused by having to lope round the countryside picking up a shoebox of nutmegs here, a cattie of kapok seed there, half a peck of cardamom somewhere else.

Quality is often variable, standards lacking. These are hard to introduce and enforce. A good part of the produce is not grown for export anyway; only the surplus might enter external commerce. Since its quality is good enough for domestic use there is little incentive for improvement. Produce might be assembled from a miscellany of species types, cultivars, varieties, and it will vary in size, shape, recovery ratio, color, chemical composition. Development agronomists would fix this problem by distributing some superior standardized cultivar to the *campesinos*. The planning economist agrees. But before they can act some irate plant breeder starts shouting that it is precisely this variability which makes these mixed up farms so valuable, then hurls abuse at the economist for messing up his gene pool.

To chambers of commerce product adulteration is a four-letter word. Forest gardeners are not immune to its temptations. Adding a few rocks in a case of export bananas to bring up their weight is on page one of the rural adulterer's handbook. Padding out the papain crystals with a little quartz gravel is also suggested. Polishing nutmegs with kerosene to improve their gloss is harmless. For those who would improve the economics of their red chili pepper enterprise a few handsfull of red laterite dirt are usually available. All these are reasons why economists on the Exports Enhancement Desk might not sleep so well. Their's is a great relief when some agribusiness entrepreneur turns up, arguing that as foreign exchange earners these simple forest gardeners are an anachronism – but that he or she, the agribusinessperson, can soon fix it up if given a few hundred hectares of fresh rain forest to do the thing properly.

Foreign exchange costs   It often happens in their tête-à-tête that the agribusinesspersons do not confide the full truth about their proposed venture. Farm projects set up specifically with an eye to their foreign exchange earnings also incur foreign exchange costs, initial or recurring or both. Modern agrosystems require modern inputs – machinery, fuel, spare parts, agricides, fertilizer, research; and if in an estate environment then also a docile labor force, communications and professional management (and if this then probably pensions and marketing trips to Venice). Whether this game is worth the candle – whether it is likely to result in net foreign exchange earnings, or a running sore on the economy if the terms of trade turn sour will be determined more by chance than the economists' budgets. They can be reasonably sure of two things. First, if the terms of trade turn at all it will be in favor of the overseas suppliers of these inputs – the cost of imported tractors and fuel will increase faster than the export price of citronella oil or copra. Second, forest gardens producing the same export commodities incur few or none of these foreign exchange costs. Their folk technology does not need them.

The economists will strive mightily in their objective analysis of these alternatives but in the end their results are often set aside. The imported technology model is also a symbol, psychologically valuable in its own right, and greatly comforting to the Prince as proof of progress in his Nation Building.

Water Use, Demand – 18

By this criterion an agroecosystem which requires no or little external water – mainly irrigation – is superior to one which needs much. This is based on several considerations. First, the world is running out of water faster than any other agricultural input, though in the regions at different rates. Second, apart from fossil water from some aquifers and desalinization plants (with these constrained by energy costs) the hydrological cycle on which rainfall and streamflow depend is itself changing as one manifestation of climate change. There will probably be no reduction in total water transferred in the cycle – more likely an increase – but changes are probably occurring to its geographical distribution, together with increase in intensities of rainfall events, storms (global warming, climate change). Third, and closely associated with both these, water is emerging as a major cause of global conflict over catchments, stream flow, diversion. Fourth, irrigation has historically been and still is a major cause of land degradation and productivity loss to salinization; and historically it's always been a health hazard by providing habitat for insect vectors and their diseases (mosquitoes, snails, malaria).

More recently irrigation is associated with large scale chemical pollution of aquifers and streams, with the ecological and economic costs (externalities) this carries. The more water used the greater this contamination is likely to be. Finally, irrigation as basis for agricultural intensification is closely associated with greenhouse gas emissions and global warming. The most direct of these relationships is through expansion and relay-cropping intensification of irrigated rice, with this crop the major anthropogenic global source of methane, third most important of the greenhouse gases (Energy). Similar adverse impacts are through nitrogen gases N_2O and NO_x emitted by fertilizers and disturbed soils for the irrigated crops, in addition to carbon dioxide (second most important greenhouse gas) emitted in manufacture of fertilizers and other inputs needed to justify the high cost of irrigation scheme development.

Fertilizer Use, Demand – 19

Again this criterion says that a farming system which uses no or little artificial fertilizer is superior to those which need much. It's based on several considerations, again externalities or indirect social costs. First is environmental contamination particularly of surface waters and aquifers; eg taking USA in 1970s as representative of high fertilizer-using countries, nitrogen fertilizers and animal manures were then adding nitrogen at the rate of 6.8 million tons/year in some places to US waterways and phosphorus 0.54 million tons, with agriculture by far the major source of this pollution (Tivy 1990). Globally these contamination rates would have greatly increased since then as artificial fertilizer use expands into the developing world and farmers in the 1st World come under increasing pressure to maintain yield and income levels. In the late 1980s Europe was using about five times more nitrogen fertilizer (117 kg/ha) than was Africa (27 kg/ha). A closely related contamination is through nitrogen fertilizer as one cause of soil acidification. (Taking Australia as example, such acidification, mainly by nitrogen fertilizer, accounts for around 17% of all lost production through all forms of land degradation.)

A third externality is fertilizer as one cause of global warming. This occurs in two components. First is direct emission of the global warming gas N_2O from applied fertilizer itself, about one million tonnes annually world wide. Second is warming caused by the carbon dioxide emitted from fossil fuels in fertilizer manufacture, around 150-160 million tonnes CO_2 from manufacture of all fertilizer types (N,P,K...). Another cost, impossible to measure, is the effect of quick-fix nitrogen fertilizers especially to entice farmers to abandon previously sustainable farming systems or to continue to flog worn out lands beyond their point of possible rehabilitation.

In summary, while world food supply and modern agriculture generally now depend heavily on artificial fertilizers (and agricides) these benefits have come at considerable indirect costs. On these grounds also there is much to be said for those traditional agroecosystems which don't need this input. These are mainly agroforestry systems. The internal nutrient arrangements of one of them, forest gardens, are examined below (Nutrients; Jensen's farm).

Agricides – 20

This input too carries severe externalities, now becoming known in the 1st World but generally less appreciated in the 3rd where usage is expanding rapidly. The main problems are farmer and consumer health (for reasons noted previously) and contamination of soils, waterways, aquifers, with these costs also borne by downstream eco and economic systems. In some developing countries (Indonesia, Philippines, Pakistan, Sri Lanka) usage has been increasing at phenomenal rates – more than 10% annually. Most usage is associated with export cash crops (notably cotton), and with those crops which require high inputs of fertilizer, seed, water, management intensity, are expensive to grow and in which agricides are seen as a necessary complementary input to protect this investment. While they do usually succeed in this on a field-by-field basis, globally over time they've had little success in pest control. Insects, pests, weeds, diseases still claim 30-40% of all global crops, about the same as 50 years ago before these miracle fixes became available. About the only thing certain – apart from chemical industry profits – is ability of the pests to mutate, evolve or otherwise develop immunity as fast as the miracle fix comes on stream. Pests as problems are largely self-inflicted; they're generated by the same mechanistic technology which sets out to quash them. Traditional tropical agriculture is awash with alternatives, ranging from the fanciful to the fully efficacious (Shifters, Diversity). Unlike agribusinesspersons the forest gardeners are not terrified by pest attack, finding some of these as predators control worse ones, or as pollinators, or in their preference for plants already sick or debilitated thereby protecting their own future food supply – anticipating Darwin by millennia though in modesty they never would have written these reasons down.

Genetic engineering of plants to withstand pests is a similar more recent category of defensive inputs (though proponents prefer the less confrontationist term 'modification', GM). The global market for these engineered organisms is now worth billions, again led by cotton this time engineered to carry a Bt gene (of Bacillus thuringiensis bacterium) against the cotton bollworm, both to increase net crop yields by reducing crop losses by 60% and reduce conventional agricide spraying theoretically by up to 80%. The innovator, Monsanto Corporation, has staked most of its marketing campaign for GM generally on these claims for cotton. China with about 1.6 million ha of GM cotton has adopted this technology most enthusiastically. However China's Nanjing Institute of Environmental Science now has compelling evidence that the Bt gene is actually harming the natural parasitic enemies of the bollworm, and pests other than bollworm are now emerging as the main threat to yields; in any case the Bt cotton will probably lose most resistance to bollworm in a few more generations and the need for agricides will be greater than ever.

Research Support – 21

In suggesting that an agroecosystem which requires little supporting research is preferable to those needing much, this criterion is based on the proposition that the bulk of agricultural research is reactive, mechanistic, muddled and without holistic vision of what it's trying to achieve. The problems it addresses are largely those it causes. It's self-feeding. The more problems it claims to solve the more problems it generates and the more research is needed. Probably 80% of it would come under the heading of what economists call defensive expenditures, needed to correct past blunders and excesses. Billed as development and progress it's really about rehabilitation from the effects of past research and development. These problems and their past research-and-development origins would take a page to list and a library to substantiate but include: soil erosion; acidification; soil and water contamination; many animal diseases (including those able to cross to humans); species and genetic erosion; control of feral plants, weeds, animals; breeding salt and pest tolerant crops; farm adjustment planning to find feasible and affordable solutions – most caused by past research or agricultural development policies based on the recommendations of past research. Taking Australia as example most research until the present has been directed in one way or other at how to run more sheep or cows or grow more wheat per acre. The environmental and economic fallout from this is only now starting to be realized. Rehabilitation from these past 'successes', including remedial research, is estimated to cost $40-50 billion if it's possible at all, little short of rehabilitating a continent. Yet from here and elsewhere this is the same technology and agro-ideology which is being exported by the 1st World as model for developing the 3rd.

Foreign Aid Required – 22

Insofar as foreign development aid in general is aimed at extending the Western economic model to the undeveloped and its rural sector in particular, and that Western model is deeply flawed according to most of the criteria of Table 1.2, this criterion claims that any reasonable indigenous agroecosystem which needs no aid props is superior to one which does. There are pragmatic reasons. One of these is aid as a threat to economic followed by political independence then to loss of cultural identity. Another is observation that much bilateral aid by circumlocution is intended primarily to aid the donor, notably that which provides them with convenient least-cost means of disposing of outmoded technology, unsaleable hazardous agricides, surplus fertilizer and vehicles, inappropriate machinery, and probably most important, a base from which to later commercialize continuing supply of these inputs once they've become accepted as part of the modernized production process. Considered from the other direction many aid projects accepted or requested by recipient countries have little basis other than they aid the urban governing elite or the career prospects of some ambitious local bureaucrat. There's also adverse aid feedback. Students sent abroad as part of the aid package will often go to institutions where the model they work to, though it might be technically excellent, can be wildly irrelevant to the country's real needs. Returning home there might be much discontent until their newly acquired model is adopted, the indigenous model left unexplored, its folk knowledge devalued.

Where aid takes the form of credit extended at normal commercial or concessional terms this often simply increases the recipient country's debt burden without increasing its real production and debt repayment capacity. If such capacity is generated it then must be strained through the urban elite.

Debt repayment pressures mount. Structural adjustment is demanded. School fees will rise, some will close. Hospitals will be boarded up as unnecessary drain on the treasury. For several years now 3rd World debt repayments to the 1st have annually been greater than new development aid received and these consequences in loss of social infrastructure are common, especially in Africa.

The case for promoting those agoecosystems which minimize need for emergency aid is also strong. Put bluntly, and with exception of most NGOs or where bilateral (country-to-country) aid also serves the donor's economic or geopolitical interest, the 1st World is aid-fatigued. Due certainly to population growth and their exposure and probably to climate change the number and severity of such emergencies as flood, drought, wildfires, disease, which they're called on to confront are increasing globally. (At the moment some 4.5 million people are rendered homeless and their food crops lost throughout NE Indian states, Bangladesh, Bhutan, Nepal.) But attention shifts as rapidly as tabloid headlines do. Hurricane Mitch killed 7 000 across Central America and those countries won't recover within a generation. The European Union announced $350 million in emergency aid. The citizens down there are still waiting for any of it. That was two years ago.

So criterion 22 says the hell with it, let's do our own thing…or more precisely in the context of farming systems, select-activate-retain those systems which by their nature minimize such risks. The most efficient of these are mixed tree-palm-vine systems. The worst are monocrops of commercialized agriculture. (Crop resiliency to hurricanes in Oceania is discussed elsewhere – McConnell and Dillon 1997; and those species which best withstood the NE India floods of August 2000 are the palms.)

Soil and Water Demand, Conservation – 23, 24, 25

These are important attributes of any land use system, the extent to which it will intensify demands on soil and water resources or conserve them. Soil erosion and other types of land degradation are discussed below (Energy) where it's concluded that the most efficient soil conservation systems are those which are closest analogues of natural forests. High water use (and inefficiency in *in situ* water conservation) is a feature of modern agriculture, most notably an indispensable ingredient of The Package—along with HYV seed, agricides, fertilizers, mechanization, small farm amalgamation, redundant farm populations—when this modernization is extended to the 3rd World via Green Revolution techno-ideology. Undoubtedly the world is exhausting its water resources faster than any other inputs, including soils. Meyer (1996) has estimated global increase in irrigated area since 1800, expansion by a factor of 12 in water use or 27 in terms of irrigated area.

A good part of the area increase after 1800 would have come from colonial era schemes (Pakistan, India, Sudan, Egypt) for export crops, mainly cotton, then movement of US agriculture to western aquifers, the Texas High Plains, then more recently into Mexico. Still more recently 1950-85 the volume expansion especially in Asia is associated with Green Revolution year-round relay cropping. Myer doubts if there's now any net increase in global irrigated area as the table might imply because expansion into new areas is about offset by salinization, decline in productivity or abandonment of old irrigated areas. Global water use for *all* purposes has about doubled since 1960 and annual use increase rate is around 6% (Cambridge Univ. Conservation Monitoring Center). About half of accessible global fresh water (outside polar ice) is now used by humans. This appropriation of water-as-habitat, especially the wetlands, also represents pressure on non human species of like magnitude.

Global Increases in Irrigation Water Use and Area

	1800	1900	1950	1985
*Global irrigation water use, *cubic km:*	226	550	1 080	2 700
Global irrigated area, *1 000 sq km:*	80	480	1 400	2 200

*cubic km used annually; about 85% of this use consumed, 15% returned to streams and aquifers.
Source: Meyer 1996.

Irrigation water (storage-use-tail water) has always had a classical set of externalities associated with it. The first is soil/stream salinization. *Having destroyed Carthage in the Punic wars the Roman victors sowed it with salt to ensure it would never be settled again. Irrigation inadvertently can have much the same effect* (Meyer). The extent of salinization globally, in Near East, Pakistan, Australia, is discussed below (Energy). A second externality is irrigation as transport agent for carrying agro pollutants into aquifers and far downstream to coastal estuaries and offshore coral reefs. (This, with El Niño coral bleaching, is now a major threat to Australia's Great Barrier Reef and the tourist-fishing industries based on it and the situation is similar throughout SE Asia, particularly Philippines and Outer Indonesia.)

Disease also is an ancient cost of irrigation, probably going back to Babylon and in Sri Lanka certainly to King Parakrama's Great Seas, dams, and his abandoned cities of the wide rice plain (Kandy) – leaking drains, stagnant tailwater ponds, mosquitoes and malaria, exodus. A contemporary reminder of these agro-engineering delusions is found in the shallow tube wells for both domestic and irrigation supply now poisoning a good part of

Bangladesh population:

In Bangladesh some one million tube wells have been sunk by government, international aid agencies and privately in the past 20 years. Their water is now absorbing natural ground arsenic. Within 10 years this will be causing 10% of all deaths in Southern Bangladesh. A $32 million World Bank loan (February 1998) will seek a solution but treat only 10% of the wells. British Geological Survey is investigating the technical reasons for this arsenic contamination, including the possibility that accelerated irrigation pumping by lowering water tables is permitting oxidization of arsenic-bearing pyrites... Most people have been drinking contaminated surface water (through the record floods of August 1998), and as the floods recede they will again be drawing water from their wells, swapping one poison for another...(Summarized from (UK) Guardian, Sept 1998).

Another externality is exposure to risk when irrigation is based on large dams. China's Three Gorges dam on the middle Yangtze represents those projects consisting of complicated mixtures of positive and negative elements and much uncertainty. The main positive is that it will generate about 8% of national electricity and in this reduce greenhouse gas emissions from 50 million tonnes/year coal replaced by hydropower. The main negatives (apart of construction costs of $24.5 billion, forced resettlement of a million people and loss of priceless heritage sites) are rate of siltation behind the dam wall, its final rate hotly disputed but calculated by dam opponents to actually increase flooding incidence within 15 years after dam completion (2009), and increased risk to 500 000 people downstream should the unthinkable happen—earthquake, war. Main uncertainty arises from climate change to the regional hydrological cycle, whether increased future rain amounts/intensities will occur upstream from the dam (a hydrological regime on which the dam is planned), or downstream from it, in which case from a flood mitigation viewpoint the dam-as-protection would—like its engineering predecessor the Great Wall as protection from barbarians—be equally as pointless.

At best and given rates of population increase and pressures for ever more production from ever more irrigated lands, these mega dams can be only a stopgap measure. Even with its Aswan High Dam 90% of Egypt's Nile water is now used and the other 10% is too polluted to be used for anything.

A final externality now emerging as the most important is increased risk of wars over dwindling supplies. India-Pakistan's sharing of the Indus system supplies, and Turkey's Gap scheme (which could with ill will be used to shut off vital supplies to Iraq and Syria) are particular flash points. Water has become a strategic weapon, in the language of real politik the Water Bomb. All these externalities seem to offer reasons why those farming systems not demanding artificial irrigation should be superior to those that need much, and polluting much with it.

Forests, Agroforests and Land Rehabilitation – 26, 27

This system attribute refers to its efficiency in bringing land back into a productive condition after it's been degraded. Forests, including some agroforests, are closely related to rehabilitation in that they're often the only feasible or'least-cost rehabilitation strategy. Globally the need and scope for land rehabilitation is enormous. Land rehabilitation also usually involves that of associated water supplies, ie the hydrological efficiency of farms as parts of catchments. Rehabilitation also includes dependent communities. Physical rehabilitation often coincides with the need anyway to diversify an agricultural economy (in mini states this is equivalent to diversifying the national economy) and to extend equity through land reform and increasingly to positively discriminate in favor of giving women and marginalized groups a more meaningful role in the economy. In short, land rehabilitation goes far beyond just improving the productivity of land.

As with conservation/prevention, the rehabilitation challenge is particularly great in impoverished and highly populated 3rd World countries where rehabilitation needs are greatest and highly capitalized TVA-type schemes are not feasible. Again the only practical rehabilitation strategy is selection and activation of appropriate farming systems which by their nature incidentally achieve these rehabilitation and reconstruction objectives. In the wet tropics by far the most efficient systems for this purpose are small highly mixed tree-palm-vine-bush agroforests, forest gardens; and because they're basically *gathering* systems women and children are not at a production-labor disadvantage (Kandy). By rehabilitating or re-creating a quasi forest habitat they also incidentally serve biodiversity conservation (Reconnaissance).

Probably the first national attempt at using forest gardens in a rehabilitation role was in Sri Lanka, 1977 (McConnell 1977). The specific project objectives were to rehabilitate eroded and uneconomic tea estate lands by settling a maximum number of landless villagers on them at minimum public cost. It incidentally met most of the other criteria of Table 1.2 which have since emerged as important issues – biodiversity enhancement, habitat restoration, minimization of farm (fossil fuel) energy inputs, Greenhouse amelioration through carbon sequestering in forest garden biomass etc. (In those uncertain times its executing agency first called itself MIDAS, Mid-country Development and Settlement Authority, in a blatant attempt to convince its sponsors that the forest garden settlers would – if turned loose on the wreckage of 19th century agribusiness, the eroded and non-viable tea estates – quickly turn this garbage into gold – Kandy.)

Senanayake and Jack (1987, 1998) have described continuation of similar rehabilitation work at research level elsewhere in Sri Lanka by establishing forest gardens in the *mana* grasslands (Imperata, below), with these grass wastelands also resulting from abandonment of uneconomic tea which was planted about a hundred years ago. The first step is to plant quick growing tree species (Erythrina, Gliricidia, Cassia) to shade out and kill this difficult grass and follow this with planting a suite of bush/tree/palm/vine species to establish an economic analogue rain forest. These model farms are in fact attempts to recreate an entire ecosystem (the term 'rehabilitation ecology' is now in vogue), with the plants/crops also selected for their capacity as niche creators for faunal/bird/insect habitat as well as provide micro climate for orchids, anthurium (flowers), ferns, small palms etc for the export trade. These models are directly relevant for rehabilitating millions of ha of degraded lands elsewhere in the tropical world. (They're also a warning of the time and effort required to re-create a forest ecosystem, even an analogue one, once it's been destroyed by or for proper agriculture.)

The cogon lands   It seems an axiom that rehabilitation needs (and opportunities) are greatest in poorest lands with highest populations and least prospects for rehabilitation by large scale capital intensive projects. This is so with the Imperata cylindrica and similar grass wastelands of India – China – SE Asia – Melanesia – Australia, called variously spear grass, blady grass, cogon/kogon (Philippines), alang-alang (Indonesia), lalang (Malaysia), mana or illuk (Sri Lanka). Historically most have been caused by forest slash-and-burn farming (Shifters) or more recently by unsustainable cash cropping without soil protection (Conservation). From Garrity et al (1997) they occupy about 35 million ha in the region (excluding Australia-Melanesia) or 4% of the land area (Table 1.10), but up to 17% and 23% in Philippines and Sri Lanka where populations and land demand are increasing fastest. Being practically impossible to eradicate by hand cultivation (though responsive to weedicide spraying which peasants can't afford) the only permanent solution is eradication by shade under forests or, better still, agroforests, which takes about six years but can be accelerated (Shifters).

As well as being wastage of a resource which these countries can ill afford, expansion of the 'cogon' lands (and of populations which can't rehabilitate them) are a main reason for continuing deforestation. De Foresta and Michon (1997) have offered planning principles which favor rehabilitation by family-scale agroforestry. These are similar to the strategy which guided the 1977 Sri Lanka rehabilitation project (except that was aimed at soil erosion not grass infestation). These appropriate agroforest systems include residential forest gardens; family-scale mixed rubber agroforests

(Sumatra and Kalimantan); mixed damar forests (Sumatra); similar mixed agroforests producing rattan or *illipe* nuts or durian fruit (also Kalimantan). Taking the mixed damar forest as representative (Shorea javanica, for commercial resin), this would also contain 10-12 commercial fruit species per ha, about 350 cubic meters/ha of standing timber, require 130 labor days/year and return $1 200/ha annually—as well as permanently eradicate the Imperata.

Table 1.10   The Imperata Grasslands of S and SE Asia

*Indonesia	...	8.5mil ha	... 4% of country	Thailand	...2.0mil ha	...	4%
Philippines		5.0	17	Myanmar	2.0		3
Malaysia		0.2	1	B'desh	0.3		3
Vietnam		3.0	9	India	8.0		3
S. China		3.0	2	S. Lanka	1.5		23
Laos		1.0	4				
Cambodia		0.3	1	Region	34.7m ha		4%

*All provinces but mainly Nusa Tenggara, Sumatra, K'tan, Sulawesi.
Source: Garrity et al 1997.

1.7 Factors of Global Concern

Group C of the criteria in Table 1.2 extend beyond the farm, village, nation, and are of global concern.

Biodiversity Conservation – 28

This attribute or lack of it continues from criterion 4 of Table 1.2 – the number of species, cultivars, landraces, wild types which might be on a farm as foods, beverages, drugs, decoratives, fibers, condiments, medicinals, magicals etc, or no reason at all (Kandy). Here forest gardens (and extractivist systems) have social conservation as distinct from private economic farm value. In this respect all agroecosystems range between the extremes of biodiversity run riot (the wilder kinds of forest gardens and the forest extractivist systems of Amazonia, Kalimantan, Congo basin) through commercialized monocrop systems where biodiversity is barely tolerated.

A first approximation in measuring such diversity on farms is to count the number of distinct species they support. Some of these counts for forest gardens around the world are summarized in Table 1.11 (taken from the farms encountered in Reconnaissance). From Table 1.11 forest gardens are probably the most diversified (biologically and economically) agroecosystems in the world, with farms in Amazonia, Java, Kandy, Kerala of little more than an acre having 50, 60, 70... distinctly different large economic plants, comparable to the biodiversity of rain forests (see also Shifters). These are incomplete counts; they omit the minor plants such as epiphytes, ferns, mosses, parasites and the aerial gardens (Energy, architecture); and they also omit the faunal species associated with the plants for which on forest gardens there are no data. They probably have a distribution similar to that which Klinge (1975) found in rain forests, dominated by microfauna and insects.

soil microfauna	-	40%		other insects	-	18%
ants & termites	-	24		other arthropods	-	10
mammals	-	5		amphibians	-	1
birds	-	1		reptiles	-	1

Similarly forest gardens could hardly exist without their 'non economic' fauna, although their mammals, birds, reptiles, amphibians along with the larger insects and their larvae can also be important human food resources. The larger species are also the system's pollinators. In this regard one consequence of the lack of wind in equatorial forests, particularly as this might affect plant species of the lower strata (Shifters) is that these must depend on the faunal agents birds, bats, larger insects for pollination. Myers (1979) offers the example of a 40 hectare forest in Brunei in which only one tree species out of 760 is wind pollinated. Dependence on these faunal pollinators is probably similar on forest gardens. (On Kandy farms the only species sometimes hand pollinated is the passionfruit.) Another essential economic input by birds, reptiles, amphibians and larger insects is their service in controlling insect pests. In brief, forest gardens differ from other agroecosystems in that the zoomass present are usually the unseen suppliers of *inputs* rather than – as in conventional livestock systems – suppliers of some final economic output.

Beyond this, whether forest gardens play a *refugia* role depends on human culture and need as modifying factors. In (Buddhist) Sri Lanka there is a general reluctance to take life so such megafauna as monkeys, bats, birds, reptiles, find tolerance in the gardens although no great warmth of welcome. On the other hand in Central Java such refugees are regarded primarily as source of protein. (The need for this is sometimes acute: while wandering around the bare hills in Central Java the only visible biology the author encountered over a period of several months, apart from paddyfield rats, was one large serpent. When related to the village chief the tale was dismissed with derision, because – and the *lurah* and his mates were adamant on this – it was well known that all such resources had been eaten long previously.) In West Java Michon and Mary (Reconnaissance) list birds, bats, insects,

Table 1.11   Plant Species Diversity on Forest Gardens around the World*

Kandy:	5-18 main *economic* species on individual farms in a sample of 30 farms (McConnell and Dharmapala 1992).
Kandy:	135 tree/shrub species of all kinds in a 60-farm sample (Perera and Rajapakse 1991).
Kerala:	127 species (97 indigenous) on a sample of 252 extremely small farms (Kumar et al 1993).
Java:	60 species on one 920-sq m surveyed farm (Jensen 1993).
Java:	250 species on a sample of very small (300-500 sq m) farm pekarangan gardens; around 50 species on individual gardens (Michon and Mary 1994).
Sumatra:	48 tree/bush species on most Minangkabau farms; 12 of these in the houselot, 36 in the attached farm agroforest (Michon 1986).
Tanzania:	111 species (53 trees) in a sample of Chagga farms (Fernandes et al 1984).
Nigeria:	60 indigenous species on a sample of 'compound' farms, with another 110 species sometimes used in nearby forests (Okafor and Fernandes 1987).
Sierra Leone(Freetown) and Nigeria (Ibadan):	over 30 tree/vine/herb species on minute 'farms' established on urban wastelands (Tricaud 1987).
Costa Rica:	83 species on one inventoried farm (Gliessman 1990).
Peru (Tamshiyuca):	15 main economic trees/palms/food field crops plus 'many other' multipurpose species (Padoch et al 1985).
Peru (Santa Rosa):	168 species on 21 sampled household farms, up to 74 species in individual gardens (Padoch and de Jong 1991).
Amazonas (upper):	typically 50-100 multipurpose species around houses of *ribereño* settlers (Lathrap 1977).
Amazonas:	60 useful species in a small village; typically 30 main species plus herbs per household (Lescure and Pinton 1993).
Amazonas (Goritori Indian Post):	140 species in Kayapo Indian 'resource islands' (Posey 1985).
Amazonas (Rio Negro):	27-30 tree species around most houses plus spices/condiments in containers above flood reach (Emperaire and Pinton 1993).
Mexico (Yucatan):	135 woody and herbaceous species in a sample of 44 farms (Rico-Grey 1990).
Mexico (Yucatan):	301 useful species in a sample of Maya farms; of these, 123 species found *only* around the houses (Rico-Grey 1985, 1991).

Mexico (San Louis Potasi):	over 300 species inventoried on the forested portions of Huatec Indian farms, including more than half of all traditional medicinal plants (Alcorn and Hernandez 1984).
Mexico (Vera Cruz):	338 species in a sample of 71 house gardens (Rocas et al 1989).

*References relate to Chapter 2 (Reconnaissance).

squirrels and civets as still sheltering in the forest gardens and although these...*are only a pale shadow of the original (species) once found in the area they play an essential role in biological processes such as pollination, natural hybridation and fruit dispersal.* In areas of vast and contiguous monocropping such as the rice plains of Java and India the only significant reservoir of biodiversity remaining, plants or animals, are in the village forest gardens and the sacred groves.

Relic forests, sacred groves   Of these as biodiversity repository there are many kinds. Those of Cherripunji provided Khiewtam and Ramakrishnan (1993) with opportunity to study litterfall and fine root dynamics in an area where the rest of the forest has disappeared (Nutrients). Pei's account of plant diversity in Dai temple courtyards and Holy Hills in South China is discussed below (Reconnaissance). Of the courtyards: ...*With the exception of fruit trees and some ornamental plants these introduced species are grown almost exclusively in the yards of Buddhist temples.* Of the Holy Hills: ...*These are a kind of natural conservation area founded with the help of the gods and all animals, plants, land and sources of water within it are inviolable.* Globally Unruh (1994) offers a reminder of the many trees and lesser plants which are kept, encouraged, dispersed, for their religio-spiritual connotations, and the contribution of these and their associated plants/zoosystems to biodiversity: *karite* (B. paradoxum) in West Africa; *pippul* (Ficus religiosa) in India; *palmyrah* (Borassus) in India; *ceiba* in Africa.

Throughout the great rice plains of Java where there are otherwise no trees, except those which cluster as forest gardens in the villages, the sacred groves continue. There are thousands of them, remnants of Hindu shrines or reminders of forest gods more senior. Some are a quarter hectare or so, some only a single large old tree. Sometimes they occur within the village itself. Children don't play there. One instinctively avoids their darkness for the same reason Beccari (1904) gives for the Dyaks of his time avoiding their mountain tops...*they are the residence of supernatural beings termed kamang which it is neither right nor prudent to disturb.* If biodiversity exists in the rice lands it must be in the forest gardens and in these tangled and melancholy places.

Social Conservation – 29

This suggests that agroecosystems can also be considered according to their efficiency in conserving human culture. It's based partly on subjective values which can't be approached by economics but also on pragmatic social self-interest. One of these factors is preservation of folk knowledge regarding the uses of plants, their edaphic/nutrient requirements and best methods of producing them, especially the so-called New Crops, and this extends to their ecological and micro climatic requirements. There are literally thousands of these potential new economic crops under investigation, though they're actually ancient (Genesis), most from the rain forests and have been included in their economic suite by forest gardeners for millennia. Traditional people have been accumulating knowledge about them for thousands of years. This can serve as pointer to the new crop's agronomic requirements when it's rediscovered for modern uses – but only if the people as well as their crops are preserved.

This desirability of folk knowledge conservation extends to the medicinals as sources of new drugs. There's evidence that while some 40% of these folk cures are fanciful the other 60% are efficacious to varying degree. The medicinals can range to more than 250 species in some tropical American villages for 10-20 uses (usually led by malaria and snakebite cures). These continuing traditional uses are a valuable guide to those species which warrant scientific investigation and further development. Forest gardens are an important repository of this knowledge (Diversity, cures and curanderos).

Energy, Global Warming and Climate Change – 30-34

This group of the criteria, led by energy conservation, claim that an agroecosystem which uses less fossil fuel energy and contributes least to atmospheric pollution and consequent global warming, climate change—or actively counters this phenomenon—is preferable to those systems which cause it (Energy, greenhouse, atmospheric pollution).

1.8 Conclusions

It could not be seriously suggested that forest gardens are a panacea for the problems discussed under criteria 1-34 of Table 1.2, or a practical substitute for existing modern agriculture; it's too late for that. On the other hand they're probably the most efficient agroecosystems in terms of most of the criteria; and they *are* a viable alternative to tropical deforestation which is usually for the expansion of modern agriculture to which many of these problems can be attributed.

Some General Guiding Principles

The natural world on which agriculture depends is visibly eroding, due largely to the impact of agriculture itself. There's urgent need to re-define agriculture in terms of its sustainability rather than extracting more—more productivity for more profit; but at the same time because of rising population which will double within 70-80 years there will be more need than ever to develop—or rediscover—farming systems which don't destroy their own productivity base. Kiley-Worthington (1981) has offered seven guiding principles. With some elaboration these are listed. They serve as a summary of the efficiency criteria discussed above and an introduction to following chapters. They're a reasonable description of the attributes of forest gardens, though these are certainly not the only systems which would win Kiley-Worthington's approval.

1. The system must be *self-sustaining*, especially in energy. For this it must maximize the potential of incoming solar radiation, insolation, generate its own plant-animal nutrients, re-cycle these and make maximum use of farm/household wastes. Where this nutrient cycle must be broken (for export of products from the system) nutrients they represent should be replaced from organic sources. Reliance on fossil fuels and artificial fertilizers should be minimized. (Here Kiley-Worthington points out that in terms of net energy balance intensive British agriculture consumes six times more energy than it produces and in the US this can go as high as 10 times.) In short, a farm system should be designed to produce its own energy inputs, in addition to producing that energy in farm yield which is exported from the system.

2. The system must be *diversified*: The economic rationale of this was discussed above (equity and employment). The ecological rationale is that farm diversification is equivalent to maximizing the number of separate niches in an ecosystem in which the different plant/animal species can exist in mutually supportive relationships which maximize *total* system production.

3. Net yields *per unit area* must be high: The Great Frontier of abundant soil and water is gone forever. This requires intensification on existing land, but *without* external energy subsidies to the farm. (This ecological objective is difficult to achieve except in farming systems which mimic such natural ecosystems as forests, swamps.)

4. Farms should be *small*: The technical reason (in the West) is that this

would force more careful husbanding of resources, and the social reason (in the East) is that this would maximize access to land for those large populations without other means of subsistence or realistic prospects of industrial employment (Java, above). This principle runs counter to agro-giantism of commercial agriculture in the West and formerly of state agriculture in the East.   It is antithetical to the principles of agribusiness. It also runs counter to the dogma of economics of scale. But this psalm is only properly sung on the assumption that labor and technical inputs are scarce and therefore expensive while the natural resources incorporated in land are plentiful and therefore cheap, disposable.   The exact opposite is true;   but this is obscured by distortions caused by the pricing mechanisms on which conventional economics is based.

5.   Farms must be *economically viable*: Here farm economists will object that small farms almost by definition cannot be viable. But Kiley-Worthington points out that conventional agriculture (eg Europe, Australia, USA, Israel, India, Middle East, formerly USSR) has always in recent times been the *least* viable of the economic sectors and could not exist without massive public subsidies.   Farm subsidies have traditionally been by far the largest single item in the European Economic Community's budget and in 1998 amounted to US$900/ha of farm land in Europe and $110 in USA.   The current 2002 US Congressional Farm Bill will subsidize US farmers to the extent of $360 billion over the next ten years;   and the bulk of these are in fact agribusiness corporations held up as models of agro economic 'efficiency' to the developing world. (You jest, Sir? Surely you jest...)

Globally subsidies since the Industrial Revolution are even greater in terms of agriculture's consumption of socially underpriced fossil fuel and through this of all modern technological farm inputs and their external socio environmental impacts which only now are starting to be realized. One must also consider the huge subsidies historically enjoyed by proper agriculture in its devastation of such social resources as soils, streams, aquifers, plant and animal species...for which it is not charged, resources which grow more precious with their depletion.

6.   A maximum amount of produce should be *farm- or community-processed*: One reason has been noted (employment, equity). A second reason is to minimize the social and financial costs arising from drift (stampede?) of unprepared rural masses to the mega cities – Calcutta, Surabaya, Mexico City – and the social costs which this implies and which also are not charged against the kinds of agricultural development which generate these refugees.

7. The farm system must be *aesthetically* and *ethically* acceptable: The first was noted above (habitat). The second refers to that increasing number of consumers who find moral objection in the way battery animals are maintained, as well as culinary objection to their produce, and more recently, health hazard in diseases able to jump the animal-human species boundary when they're approached only as profit-maximizing objects.

Pastoralism? Ghandian escapism? Or desperation. Certainly the Prince believed her wrong when he had her banished from his realm, finding her subversive under a section he found in an economics textbook (brought to his attention by Janzen 1973)...*What is wrong with subsistence agriculture is that everything produced is used up by the people.* It's probably too late, but the following chapters wonder if a defense might still be possible.

References

Bailey, R.C. and N.R. Peacock (1988) Subsistence strategies in the Ituri forest. In Garine I. and G.A. Harrison (eds) *Coping with Uncertainty in Food Supply*. Oxford University Press.

Bavappa, K.V.A. and V.J. Jacob (1981) *A model for mixed cropping*. Ceres May-June, FAO.

Beccari, O. (1904) *Wandering in the Great Forests of Borneo*. Constable, London (1904). Oxford (1986).

Budowski, G. (1985) Home gardens in tropical America. *Tropical Home Gardens Landauer and Brazil*. UN University Press, Tokyo/Paris.

Chandran, S. and M. Gadgil (1993) State forestry and the decline of food resources in the tropical forests of Uttara Kannada, South India. *Tropical Forests, People and Food, Man and the Biosphere* (13). UNESCO.

De Foresta, H. and G. Michon (1997) The agroforest alternative to Imperata grasslands: when smallholder agriculture and forestry reach sustainability. *Economic Botany* (36):105-120.

Dove, M.R. (1993) Smallholder rubber and swidden agriculture in Borneo: a sustainable adaption to the ecology and economy of the tropical forest. *Economic Botany* 47(2):136-147.

Eder, J.F. (1981) From grain crops to tree crops in the Cuyunon swiddening system. In H. Olofson (ed) *Adaptive Strategies and Change in Philippine Swidden-based Societies*. Forest Research Institute College, Laguna, Philippines.

Ellen, R. (1988) Foraging, starch extraction and the sedentary lifestyle in the lowland rain forest of central Ceram. In T. Ingold, D. Riches, J. Woodburn (eds) *Hunters and Gatherers 1: History, Evolution and Social Change*. Berg, NY/Oxford.

Garrity, D.P., M. Soekardi, M. Van Noordwijk, R. De La Cruz, P.S. Pathak, H.P. Gunasena, N. Van So, G. Huijun, N.M. Majid (1997) The Imperata grasslands of tropical Asia: area, distribution, typology. *Economic Botany* 36:3-29.

Ichikawa, M. (1993) Diversity and selectivity in the food of the Mbuti hunter-gatherers in Zaire. *Tropical Forests, PEOPLE and Food, Man and the Biosphere* (13). UNESCO.

Jacob, V.J. and W.S. Alles (1987) The Kandyan gardens of Sri Lanka. *Agroforestry Systems* 5:123-127.

Janzen, D.H. (1973) Tropical agroecosystems: habitats misunderstood by the temperate zones, mismanaged by the tropics. *Science* 182:1212-1218.

Khiewtam, R.S. and P.S. Ramakrishnan (1993) Litter and fine root dynamics of a sacred grove forest at Cherrapunji in NE India. *Forest Ecology and Management* 60:327-344.

Kiley-Worthington, T. (1981) Ecological agriculture: what it is and how it works. *Agriculture and Environment* (6):349-381.

Klinge, H., W.A. Rodrigues, E. Brunig, E.J. Fittkau (1975) Biomass and structure in a Central Amazonia rain forest. In F.B. Golley and F. Medina (eds) *Tropical Ecological Systems: Trends in Terrestrial and Aquatic Research.* Biographical and Ecological Studies. Springer-Verlag, Berlin.

Landauer, K. and M. Brazil (eds, 1985) *Tropical Home Gardens.* UN University Press, Tokyo.

McConnell, D.J. (1977) *Land Rehabilitation in the Mid Country of Sri Lanka.* Ministry of Agriculture and Lands. Peradeniya Sri Lanka.

McConnell, D.J. (1986) *Economics of Tropical Crops: Some Trees and Vines.* Midcoast, Coffs Harbour, Australia.

McConnell, D.J. and G.K. Upawansa (1972) Cloves in Sri Lanka. In D.J. McConnell (1986) *Economics of Tropical Crops: Some Trees and Vines.* Midcoast, Coffs Harbour, Australia.

McConnell, D.J. and V. Rasmussen, D. Probowo, A. de Mautort, J. Wall (1990) *Income Generating Project for Small Farmers and Landless People, Indonesia.* IFAD, Rome.

McConnell, D.J. and W.J. Surjanto, Abu Yamin Nasir (1974) *Farming systems and land use along a traverse of the Solo Valley.* UNDP/FAO Land Capability Appraisal, Bogor. FAO Rome.

Michon, G. (1983) Village-forest gardens in West Java. In P.A. Huxley (ed) *Plant Research and Agroforestry.* ICRAF, Nairobi.

Michon, G., F. Mary, J. Bompard (1986) Multistoreyed agroforestry systems in West Sumatra. *Agroforestry Systems* 4:315-338.

Myers, N. (1979) *The Sinking Ark: a New Look at the Problems of Disappearing Species.* Permagon.

Nair, P.K.K. (1993) *Introduction to Agroforestry.* Kluwer Academic Publishers, Dordrecht, Netherlands.

Padoch, C., J. Chota Inuma, W. De Jong, J. Unruh (1985) Amazonian agroforestry: a market-oriented system in Peru. *Agroforestry Systems* 3:47-58.

Penny, D.H. (1986) *Starvation and the Role of the Market System.* Australian National University, Canberra.

Penny, D.H. and Singarimbun (1973) *Population and poverty in rural Java: some economic arithmetic from Sriharjo.* International Agricultural Development, Mimeo 41, Cornell University, NY.

Perera, A.H. and R.M.N. Rajapakse (1991) Baseline study of Kandyan forest gardens of Sri Lanka: structure, composition, utilization. *Forest Ecological Management* 45:269-280.

Pimentel, D. and W. Dazhong (1990) Technological changes and energy use in US agricultural production. In C. Carrol, J.H. Vandermeer, P. Rosset (eds) *Agroecology.* McGraw-Hill, New York.

Probowo, D. and D.J. McConnell (1993) *Changes and Development in Solo Valley Farming Systems, Indonesia.* FAO Rome.

Raintree, J.B. and K. Warner (1986) Agroforestry pathways for the intensification of shifting agriculture. *Agroforestry Systems* 4:39-54.

Sandbukt, O. (1988) Tributary tradition and relations of affinity and gender among the Sumatran Kubu. In T. Ingold, D. Riches, J. Woodburn (eds) *Hunters and Gatherers 1: History, Evolution and Social Change.* Berg NY/Oxford.

Senanayake, F.R. (1987) Analog forestry as a conservation tool. *Tiger Paper* 15:25-28.

Senanayake, F.R. and J. Jack (1998) *Analogue Forestry: an Introduction.* Monash University

Melbourne.

Simpson, E.H. (1949) Measurement of diversity. *Nature* 163:688.

Takeda, J. and H. Sato (1993) Multiple subsistence strategies and protein resources of horticulturists in the Zaire Basin. *Man and the Biosphere* 13. UNESCO.

Tejwani, K.G. (1987) Agroforestry practices and research in India. In Gholz, H.L. (ed) *Agroforestry: Realities, Possibilities, Potentials.* Nijhoff, Dordrecht, Netherlands.

Tivy, J. (1990) *Agricultural Ecology.* Longman, UK.

Unruh, J. (1994) Role of land use pattern and process in the diffusion of valuable tree species. *Journal of Biogeography* 21:283-295.

Vergara, N.T. (1987) Agroforestry: a sustainable land use for fragile ecosystems in the humid tropics. In H. Gholz (ed) op. cit.

Williams, D.R. (1990) *Salinity: an Old Environmental Problem.* Division of Water Resources, CSIRO, Australia.

Chapter 2
Reconnaissance

This chapter goes in search of forest gardens around the world.
It learns much from travelers met along the way but is
embarrassed by the many others missed.

The general locations of various types of forest gardens throughout the tropical world are becoming reasonably well known, due partly to the 1982-1987 Agroforestry Systems Inventory of all forms of agroforestry including forest gardens (there designated 'home gardens') compiled by the International Council for Research in Agroforestry (ICRAF, Nairobi). But few of these systems have been described, fewer analyzed in depth, and in the absence of interest by agricultural economists apparently none have been quantified as holistic integrated man-plant-animal agroecosystems. Consequently one must glean crumbs of knowledge about them from the tables of other disciplines — forestry, horticulture, anthropology, botany, ethnobotany. This agroecosystem has received most attention in South India, Sri Lanka, West Java, Nigeria, Central America, parts of Amazonia; but little or nothing has been reported from Outer Indonesia, Vietnam, Laos, Madagascar, Philippines, Melanesia, Malaysia, Bhutan, South China, Myanmar, except that forest gardens there exist in various forms and are a traditional agroecosystem.

2.1 Sri Lanka

In the villages around Kandy, Dharmapala and McConnell (1973, 1992) surveyed 30 traditional forest gardens as part of a UNDP-FAO tea and rubber land crop diversification project. The separate tree crops were also studied, extracted from this forest milieu. Their orientation was farm economics, their purpose to find a farming system which would serve as a vehicle for settlement of a large number of poor rural people and which would at the same time be conducive to the rehabilitation of some 100 000 acres of eroded and submarginal tea land and rubber land. They concluded that the best possible farming systems for these purposes were forest gardens; the best development strategy was simply to duplicate what the Kandyans had been doing for centuries on those lands which had not been alienated for the colonial tea and rubber plantation crops. This farm development-rehabilitation-settlement project was later financed by World Bank (Kandy).

In the same general area Perera and Rajapakse (1988, 1991) later conducted an expanded survey of 50 farms and reported their botanical composition. Where McConnell and Dharmapala had based their analysis on only the 20-30 most economically important tree/plant species Perera and Rajapakse enumerated 125 and listed their uses. Perhaps the most important result of this survey is the large number of species (over 50%) which had no reported economic or household 'use'. That is, in addition to their *farm* role, these gardens function as a repository of species and genetic material which has been lost or is under threat elsewhere... *These farms are banks of unaltered germ plasm and could be regional centers of in situ germ plasm conservation.* The Taoist-Buddhist 'utility of the useless', no economic price itself but of great value through the interdependence of all things.

A previous worker Senanayake (1987) had called attention to their soil conservation role as *analogue rain forests.* Krishnarajah (1982) estimated soil loss under these forest gardens as 1/360—1/1400th the erosion rate under clean-tilled agriculture in the same area (Conservation). In Sri Lanka this is particularly important because many of these farms lie within the catchment of the country's largest river, the Mahaveli Ganga, on which depends Sri Lanka's largest irrigation project in the northern dry zone. Agronomic models for improving Sri Lankan forest gardens have been prepared by Bavappa, Jacob, Alles (1981, 1987). Pressures for such further development are increasing. The danger is that when such models are prepared according to agronomic or farm economics criteria the wider and more valuable social roles of these farms, especially as biodiversity reservoir, must inevitably be discounted.

In the wet zone of SW Sri Lanka Charon (1995) described the adjustments in food production and procurement of the villagers of Pitikele (19 households) made necessary by the creation of the Sinharaja Biosphere, a UN-sponsored rain forest reserve of world significance. Traditionally this village had free access to these forests for *chena* (slash-and-burn) food crop cultivation and extraction of food and commercial produce. (Chena cultivation is now banned.) Although extractivism continues to some extent this also is now illegal, except for tapping the kitul palm (Caryota urens) for jaggery-sugar under licence. There are about 50 similar villages adjacent to the Sinharaja reserve.

The main response of villagers to these restrictions has been to intensify the use of their houselot forest gardens. Average land ownership is one hectare, including forest garden, a rotating share in the village padi fields and usually a small area of monocrop tea. A second response is greater reliance on purchased food, paid for mainly by tapping the kitul palm (both in the gardens and with licence in the forest), and selling green tea leaf, rubber, cinnamon bark and a range of minor garden produce. All this produce except tea is from the trees and understorey crops of the forest gardens. Charon listed 55 species found on the gardens and their parts used as food. There is heavy reliance on breadfruit, jakfruit, coconut, plantain and on winged beans, yams and other rootcrops, primarily as bulk food substitutes for rice in which the village is deficient. (One small jakfruit is equivalent to 1-1½ kg of rice.)

Also in Sri Lanka, Wagachchi and Wiersum (1997) have described the integration of buffalo ponds into these systems which add to them an aquatic dimension and still more ecological and economic diversity, a combination of trees-water-animals-crops having certain parallels with the forest gardens-plus-fishponds of West Java and the trees-crops-lakeshore economic landscape of *chinampa* agroecosystems of SE Mexico. One or a series of these small (10-100 sq m) farm ponds are excavated along a gully, through the forest garden zone, and their banks planted with up to 20 economic tree/palm species and the pond itself planted with 8-10 aquatic species for fibers, medicinals, food, religious use (eg lotus, pandanus). The immediate purpose is to provide an essential wallow for the farm's paddy field buffalo and the trees for their midday resting shade, but ramifications extend through and beyond the agroecosystem. A healthy buffalo provides field work energy (and eventually meat); excavated pond mud provides fertilizer; water provides a farm micro irrigation system; attracted dragon flies are predators of harmful paddy (rice) field insects, as are the birds, frogs, lizards which the pond attracts; micro climate change through pond evaporation and humidity from it are beneficial to the surrounding plants. (A potential negative effect of increased mosquito populations and malaria is largely controlled by fish and the other aquatic fauna.)

Downstream public benefits from the buffalo ponds are probably more important through reduction in storm runoff and siltation rates which threaten long term viability of the multi million $ investments in hydro-irrigation works on which Sri Lanka now heavily depends (Kandy). As elsewhere these ancient integrated and ecologically based farm pond micro systems are threatened by two kinds of technocentric simplification, modern agronomy and hydro engineering giantism. First...*agricultural extensionists often advise reduction in the number of trees and regulation of shade in order to increase production of export-oriented cash crops. This not only reduces the diversity of the system but also its multipurpose utility.* Second...*single-minded development of large scale hydrological structures* (to feed rice monocropping in Sri Lanka's dry zones) has directed attention away from alternative strategies for using water – possibly more economically and probably more sustainably – on farms in these hill-country catchments.

2.2 India

In India forest gardens are concentrated in Kerala and Tamilnadu. (They must also exist in Orissa, A. Pradesh, Bengal in various forms but no comprehensive surveys seem to have been done.) In Kerala they are particularly important for three reasons: this state has the highest population density in India (average 655 persons per sq km but in parts 1 500); farms are small (around 1.00 ha) or minute (200-300 sq m); there is great social need for forms of intensive highly mixed land use; and in its temperature, high humidity, high and well distributed (bimodal) monsoon rainfall (2 960 mm) and wide botanical base Kerala has the physical resources to make intensive forest gardens possible. More than 50% of agricultural land in Kerala is under this system: about 3 million of a total of 3.5 million farms are less than one ha and their average size is 2 200 sq m. Nair and Sreedharan (1986) have described these systems in general terms.

Table 2.1   Cropping Intensity and Tree Populations in relation to Farm Size, Kerala

Farm size				Mean tree cropping intensity*	Number of trees/ha
Very small	200	-	2 000 sq m	51%	621
Small	2 000	-	8 000	38	277
Medium	8 000	-	20 000	27	212
Large	20 000+			23	121

*% of total land occupied/overcrowned by tree canopies.
Source: Nair and Sreedharan 1986.

Structural Variability

Wide variation in intensity of tree cropping and composition of the tree crop mix occurs from farm-to-farm, attributed to socioeconomic factors, eg tenant vs owner status. On small farms there is a greater tendency for the less profitable single-product species (eg coffee) to be culled out and replaced with other trees, ie more evidence of management response to fluctuating commodity prices. As in Sri Lanka there is a direct inverse relationship between farm size and total tree population or mix density:

Farm Economics

Budgeted expected income from one particular 1 200 sq m farm was Rs 9 200 (US$750, 1986) which may be compared with the Sri Lanka farms (Kandy). Elsewhere Tejwani (1987) has called attention to the need for more information for objectively evaluating these Indian farms. Chemical fertilizers are increasingly used but not agricides. Because of the large number of commodities produced much marketing has to take place in the local village (as in Kandy there are no formal or organized markets for many of the products). The Kerala farmers quickly react and sometimes overreact to temporarily high prices for particular commodities, eg cacao, and substitute this species for others, then again overreact by cutting the cacao out when relative prices again change. [This extreme sensitivity to market prices is not common on forest gardens elsewhere. Such behavior on Kerala farms might limit their usefulness as stable reservoirs of genetic and species diversity which elsewhere is a main attribute of this system.]

Labor Requirements and Technical Efficiency

These authors put typical labor requirements for one ha of tree-vine mix (plus ground crops and livestock) at 1 000 adult-days annually and see virtue in this labor absorption capacity (especially if the multiplicity of products are used for farm or village level industries). In spite of the relative intensity with which this particular sample of Kerala farms are operated there is still much scope for productivity improvement, especially on 'larger' farms, through architectural modification for closer management of light, mutual complementarity of species etc (Energy).

Timber Production and Species Diversity

Also in Kerala Kumar, George, Chinnamani (1994) surveyed 252 forest gardens in 17 subdistricts and evaluated their role in producing commercial timber and fuelwood, a role which in areas of high population and land shortage which is often overlooked. Kerala's small forest garden farms were by far the most important sources of timber (gardens 80%, agricultural estates 10%, imports 2%, with only the small balance coming from official government forests). On the surveyed farms the volume of standing timber ranged from 7 to 50 cubic meters, usually 20-30 cubic meters in most districts. About 63% of this was in the form of coconut palms, the common timber of that area. The capacity of any farming or other land use system to produce timber and especially domestic fuel as well as food is of critical importance in much of South Asia and Africa. While forest gardens are not the only farming system which can generate these 'byproducts' they are the most efficient in this regard and, where they exist, in making this most direct contribution they also relieve pressure on those remnants of the public forests (in West Africa, India, Bangladesh, Java). These Kerala forest gardens also carry a standing stock of 40-80 cu m fuelwood.

This study also lists 127 tree/shrub species on sampled farms and their uses, all but 30 indigenous to South India. Like other commentators Kumar et al draw attention to the role of these farms in preserving species and genetic diversity, noting that different selected groups of the totally available species are relatively favored (or not favored) by households in different locations within a given region, on grounds of human cultural differences, ecological and economic factors. This means that since the different species are valued for different reasons the overall likelihood of their preservation is enhanced. However again the main threat comes from commercialization and tendency to farm specialization. In Kerala these authors warn that the 'teak boom' could reduce the traditional farm species diversity (as has the clove boom in Sri Lanka and Java, and following writers say the same thing about rubber).

Again in Kerala in their study of a sample of 80 forest gardens in Trivandrum, Jose and Shanmugaratnum (1993) procede from an architectural description of botanical composition to socioeconomic considerations of *why* the households have structured their gardens in different ways. Coconut forms the architectural basis 'around which the other components are orchestrated'; cocoa is planted at the center of the squares formed by four coconut palms; arecanut is similarly positioned; large canopied tall trees are near the borders of the farm; the climbers are planted close to their live (tree) trellises; the smaller fruit trees (jambu, bilimbi) planted away from under the main canopy and close to the house; ornamentals in the courtyard; shade tolerant banana cultivars wherever there is space in the mix, and the ground-level crops (yams, cassava, vegetables) planted anywhere there might be light gaps through the canopy. This description might imply a higher degree of order than actually exists; after initial planting which might have occurred decades or even centuries ago subsequent replenishment of the populations is described as an *ecological succession,* though in Kerala it is a highly man-modified one.

Land use intensity increases with decreasing farm size. Tree populations increase from 245 per hectare on the largest farms to 890 on the smallest. Another dimension of intensity is tree age distribution. There is a remarkable correlation between tree age and farm size which represents a continuing effort on the smaller farms to keep trees only through their years of maximum productivity (table). There are practically no external inputs; the planning strategy is to make maximum use of solar radiation, even if a few agronomic principles regarding 'overcrowding' are bruised to this end.

Table 2.2 Distribution of Tree Ages in relation to Farm Size

	Percent of farms by tree age (years)				
Farm size	0-5 years	5-10 years	10-15 years	15-20 years	20+ years
Below 0.2 ha	46%	29%	9%	5%	11%
0.2 - 0.4	23	41	12	6	16
0.4 - 1.0	32	35	7	18	8
1.0 - 2.0	7	21	48	7	16
All	30	32	16	9	13

Species selection criteria Families refer to a wide range of criteria in selecting their crop mix, listed below in order of their importance to the aggregated 80 surveyed families. Profitability is of only moderate importance. In pre-commercial times these forest gardens of Kerala

Species selection criteria	Criteria used by % of households
physical suitability	27%
spread risk	24
spread income seasonally, over time	22
higher profit	10
greater food security	9
consistent with family labor, management	6
institutional intervention (extension advice etc)	2

perfectly served the needs of small farmers, and they continue in the modern world mainly because they are a proven strategy to provide income and food security at minimum risk, in comparison with such possibly more profitable alternatives as rubber monocropping; and considerations of quality of human habitat are also important to these households in structuring their garden mixes. These benefits would be jeopardized if the traditional systems were to be modernized. Planning officials take a different view, their criterion of success is money income; so pressures for farm development and restructuring are mounting.

Tamil Nadu

Jambulingam and Fernandes in Tamil Nadu (1986) have described some of the many local rain forest species which over the centuries have found a domestic role on the small farms there: Palmyrah (Borassus flabellifer) for *nera* (palm wine), sugar, fiber, tamarind (Tamarindus indica) for food pods, condiment, industrial and household gum; the kapok (Ceiba pentendra) for lint, seeds, oil; Delonix elata for green leaf manure for the padi fields...

Uttara Kanada

Chandran and Gadgil (1993) traced the historic relationship between many contemporary farm tree crops and their rain forest parentage, discussed previously (Nature, range).

Islands of the Bay of Bengal (Andaman and Nicobar)

Dagar and Kumar (1992) discuss severe land degradation and soil erosion following from the abuse of forests by settlers who were allocated 4 ha each on these islands after national independence. They outline a rehabilitation strategy consisting of various forms of agroforestry, using some of the remaining forest as a botanical matrix in which agrosystems are integrated:

1.   multistorey palms-trees-ground level systems;
2.   alley cropping using Leucaena and Gliricidia (green manure), Morus spp (sericulture) Cajanus cajun (pigeon pea) as the hedgerow crops, with vegetables planted in the alleys;
3.   fodder trees (Trema, Erythrina, Bauhinia, Sesbania) introduced into the remaining woodlands;
4.   similar fodder trees as live fences; and
5.   forest gardens around the houses.

2.3 Bangladesh

Leuschner and Khaleque (1986) offer a general profile of the *bari* systems which dot the low-lying rice plains of Bangladesh. Here the farm house and compound are located on raised earth mounds above the flood prone fields and these *bari* areas support a more or less dense mix of trees and shrubs (palms, timber spp, fruit, fuel). But there is much potential for further intensification of these minute *bari* farm components, and much need because land scarcity is such that there are few other areas of the country capable of generating tree products (domestic fuel, food, construction materials).

Also in Bangladesh the structure of the *bari* forest gardens are explored by Hocking and Islam (1994). There has been a government program since 1986 to intensify tree plantings on farms, mainly in the houselots and primarily for timber and domestic fuel. (In this latter the situation is desperate with the poorer people dependent on anything combustible which can be gleaned: crop stubble, dung, jute sticks, bagasse etc.) The greater part of the official forests disappeared decades ago. Although the *bari* farmstead gardens account for only 5% of the total agricultural area, and are only 10% of the area of the official forests, they now account for 70-90% of Bangladesh's total forest production of timber and fuel as well as tree food crops.

Again in Bangladesh, Hocking and Islam (1996) explored some of the gender and cultural factors which influence the survival or mortality of young trees when newly planted on the household *bari* lands (forest gardens). These household areas and their plants are the domain of women, while the *khet* or field lands are the domain of men. Women are constrained to stay generally within their bari areas by *purdah*, seclusion. One result is that the tree species they select and tend here have a much better chance of survival than do trees chosen and planted by men outside the *bari*. When a species is chosen by men for *bari* planting it may not be one in which women are economically interested (alternative foods, and food v timber species) and therefore neglect. Similarly, tree mortality is higher on those portions of the farm to which women may not extend their attention. And again, their mortality is higher on large 'rich' farms than poor farms because the trees there are often neglected by disinterested hired labor. Compared with these sociological factors, agrotechnical ones in tree survival are secondary. Nonetheless, and although the NGO organizations which support these forest garden enrichment programs are supposed to be directed at the household *women*, the men often have a dominant role in tree species and location selection, with the indicated consequences.

These authors also list the 38 main (multipurpose) trees and palms which comprise *bari* forest gardens, and estimate their total standing timber: 53 800 000 cubic meters for trees, 1 176 000 000 tonnes of bamboo and another 748 000 000 standing palms of unknown volume. Given these timber resources as well as their food and industrial (eg fiber) products on the one hand, and this country's huge population relative to land resources on the other, it is possible that forest gardens – although historically neglected and still comparatively species poor – are more important in Bangladesh than anywhere else in the world. But these and other writers, supporting forestry commentators in Malaysia, India and West Africa, note that there is still much entrenched resistance to agroforestry, as distinct from inherited notions of nineteenth century foresterism. In countries with little land and exploding populations this latter is decreasingly an alternative.

2.4 Thailand

A wide variety of forest gardens exist in all the agrozones of Thailand but there is little information. In one project reported by Boonkird, Fernandes and Nair (1984) shifting (slash-and-burn) forest cultivators are given a 1 600 m² houselot on which to establish a mix of largely subsistence tree crops, in addition to the use of another 1.6 ha of *taungya* land for planting groundlevel food crops (cassava, maize etc) for two years in exchange for reforesting this area to teak on behalf of the forestry department. Apart from commercial forestation the environmental objective is to stabilize the swiddeners in place as an alternative to their continuing destruction of the few remaining forests through their traditional slash-and-burn practices (Shifters). Boonkird's survey illustrates the range of species and their multiplicity of uses:

Areca catechu .... masticatory nut
Artocarpus spp (jak) .... fruit, vegetable
Bambusa spp (bamboo) .... mats, furniture, building
Calamus spp (rattan) .... baskets, furniture
Citrus .... fruit
Cocos nucifera (coconut) .... food, oil, thatch, fuel
Dendrocalamus spp .... construction, mats
Desmodium pulchellum .... insect repellent
Mangifera indica mango) .... fruit, shade
Moghania strobilifera .... insect repellent
Morus spp .... sericulture, fuel
Psidium quajava .... fruit
Saccharum spp .... food
Thyrsostachys .... construction, mats, furniture

These and similar *taungya* schemes in West Africa are a colonial legacy from Burma. One wonders how realistic it is, in face of rural population pressures and demand for small living areas, to persist with often monopurpose official reforestation of timber species (usually teak) when the same forestry production and rehabilitation objectives could be more easily and cheaply achieved, with greater social equity, through settlements based on forest gardens.

In Thailand also there have been many projects for improving rural nutrition and household income by distributing improved planting material. This is usually improved high yielding variety (HYV) field vegetable seed — beans, melons, corn etc. Though well intended this approach to village agrodevelopment often has adverse consequences. The new high yielding varieties are so attractive they can drive out the old landraces, pest and disease resistant local varieties whose disappearance goes unnoticed. This results in genetic erosion. The HYVs also require fertilizer inputs and agricides, the safe management of which is hardly comprehended by village folk. When the hazards of these new inputs become realized it is not unknown for these people to refuse to consume their own 'improved' produce and instead dispose of it on some unsuspecting urban market. As Okafor says of Nigeria (below) the possibility of trees as a main food source — including bulk carbohydrates — would strike most agricultural planners with a Euro-North American background as a novel concept.

2.5 West Malaysia

The dominant type of forest garden (village *kampung*) is similar to the forested *pekarangan* village gardens of Java and grows much the same tree species. An important subtype are coastal farms structured around such salt-tolerant pioneering species as the native mangrove and nipah palm. Through their silt-entrapment capacity on tidal mud-flats these raise land elevation until other species such as coconut, mango, sago, areca palm, cashew, citrus can be added to the mix. Collecting and weaving roofing panels, *atap*, from the native nipah is an important cash income source for these farm families in Trengganu and Johor (Diversity).

As in Indonesia the forest garden farms in Malaysia emerged historically, spontaneously, without much if any government input or planning. As a system they have not yet been officially recognized as a potential vehicle for rural or national economic development. Since independence farm settlement projects have taken two forms: small family units on the new irrigation schemes primarily intended for padi/rice production, and larger units on the Federal Agricultural Land Development Authority (FALDA) schemes, most aimed at producing a commercial tree monocrop (oil palm, rubber). The latter settlement farms take as their model the monocrop estates. All of these schemes have been based on extensive forest clear-felling at immense ecological loss. The possibility of using these rain forests as a botanical matrix into which some other more benign form of farm settlement could be integrated has apparently not been considered. As in Indonesia the model for this is present in the traditional Malay *kampung*.

2.6   Indonesia, the Pekarangan Lands ... *those gardens of complete design*

Forest gardens occur in many forms throughout Indonesia. In Java they are a more important farming system in terms of land area occupied and number of people supported than in any other country except possibly South India. Those of West Java have been intensively studied and reported; a few types in Sumatra have been described; but those of Kalimantan have only been noted in passing or — as in West Irian and the other outlying provinces — there is no information at all.

Java

In Java the forest gardens occur on the *pekarangan* lands. These rural village houselot areas take three forms, two of which support forest gardens. In the first the houselot area might be the family's only land resource and it can be devoted entirely to a forest garden; these include gardens in peri-urban areas. In the second the main farm consists of some area of irrigated (padi) or dryland fields and here the usually forested houselot supplements this main outfarm. (The third use of the pekarangan area is simply as an extension of an (outfield) farm: this occurs in those areas such as mid-elevation mountain areas of Central Java where population pressure is so great, land of any type so scarce or is so suited to high value commercial vegetable crops that if the farm grows beans, cabbage, onions etc all land in the houselot also will be put under these crops.) These days the term pekarangan is used somewhat loosely to refer to rural houseyard areas. But Soemarwoto (1991) says it really means 'complete design', a thoughtful and artistic use of limited house land and the arrangement of plants on it to enhance human habitat as well as to produce household food, fuel, fibers, building materials, pharmaceuticals and cash income.

Significance   The pekarangan lands of Indonesia in general, and their smallest forest gardens in Java in particular, have great economic significance. The number of farm/village households in Java are about 16 million. Some 38% of these have land resources of less than ½ hectare; 15% or 2.4 million village households (12 million people) exist entirely or mainly on their small forested houselots of typically 600—1 000 sq meters. In addition those 10 million households with some area of crop land proper, usually padi-rice, also depend to varying extent on the tree gardens around their houses. At a national level there are about 5.1 million hectares of pekarangan lands; these account for 13-14% of Indonesia's total agricultural land (but this increases to 40%, 50%...in some places). Viewed as a food source they amount to one quarter of the area devoted to rice; 30% of them are in Java where they occupy 18-19% of all farm lands. Size is inversely correlated with population pressure: the largest are in Maluku (8 700 sq m) and Riau Islands (8 000 sq m) and the smallest in West Nusa Tenggara (400 sq m) and West Java (660 sq m). Until recently very little was known, officially, about the pekarangan forest gardens but by 1990 policy makers were becoming aware of four things:

1. the total area under these farms was too significant to justify continued neglect;
2. only marginal increases in productivity could be expected from continued concentration on the rice lands;
3. the miscellaneous minor food produce from the forest gardens did in fact account for the *bulk* of supply of such items as poultry, goat meat, fruits; and
4. as a self-sufficient system forest gardens were independent of expensive inputs from the outside world. The first attempt to obtain quantitative national data regarding forest gardens was by a Government of Indonesia — FAO/TCP survey project in 1991 (McConnell 1989, Daw 1991).

Origins Possibly the first English reference to the pekarangan forests of Java was a 1817 note by Stamford Raffles, once governor of Batavia (Jakarta) who mentioned that in one province they appeared to occupy 10% of the land area. Abdoellah (1985) finds Javanese references to them in the 12th century. Soemarwoto and Conway (1991) cite earlier workers who found references in a Javanese royal charter, 860 AD. In recent times these farms were in fact studied in several Dutch land use surveys from 1900 and the opinion then was that the forest gardens of Java could be more productive than the rice lands.

One theory, based partly on Terra (1954) is that this system originated in Central Java and spread from there into West Java; several workers find a link between presence or absence of a matriarchal social system and local strength or weakness of this land use system. If this is correct it would run counter to geographical factors, chiefly the higher rainfall of West Java which is more conducive to quasi-rain forest agriculture and where the gardens are observably more species-rich than in Central Java. In any case the debate ignores the emergence of forest gardens elsewhere in Outer Indonesia, outside the old Central Javanese sphere of cultural influence. Hutterer (1984) puts their real origins at 10 000 years ago as a result of hunter-gatherers discarding collected jungle produce around their dwellings and eventually protecting ('farming') the useful volunteer plants. (This is very close to Edgar Anderson's earlier theory of the origins of agriculture in the tropics generally, Genesis.)

Uses, functions   The main private and social uses of pekarangan forest farms in Java are listed:

1.   *Direct subsistence food life support*   This applies particularly to those several million households (many with only female household heads or consisting of aged people) whose only economic resource is their garden.

2.   *Supplementary income*   This applies to those families who have an insufficient area for field crop farming, or who can find employment for only 100, 150... days annually and who also are substantially dependent on their garden for subsistence food.

3.   *Peri-urban agriculture*   Intensive tree crop gardening is one of the few forms of agriculture which are socially acceptable in heavily populated near-urban areas.   Such urban agriculture is becoming increasingly important as a survival mechanism in and near cities which otherwise offer insufficient employment opportunities (West Africa, below).

4.   *Domestic sewage and waste disposal*   It is a common practice in Java and elsewhere to direct all such waste into pits or furrows among the trees (as nutrient and irrigation supply). The engineering alternative of public sewage systems is often prohibitively expensive.   The exact ways in which forest gardens fulfil this waste removal-sewage disposal role have not received much attention.   In West Java the household bathroom and toilet are often located over the fishpond.   The animal husbandry parallel of this (in SE Asia generally) is to maintain caged chickens over the fishpond.

5.   *Raw materials for village industrialization*   The great range of primary and secondary products from the mix of tree crops can supply the necessary materials (fruits, sugars, oils, fibers, wood) for village- or kitchen-scale processing and income generating activities (Nature, criteria).

6. *Soil and water conservation* This role is particularly important in protecting catchments of municipal or irrigation reservoirs (Conservation).
7. *Irrigation water demand* Practically all of Java's irrigation water is used for rice (and sugar). With increasing demand for this water on the one hand and limited dam storage sites on the other (for topographical reasons), Java will exhaust its water before it runs out of soil. Forest gardens appear to be the only agrosystem which requires no irrigation resources for high productivity.
8. *Plant species and genetic conservation* This role is the same as that noted for Sri Lanka (Perera and Rajapakse). With increasing commercialization and genetic erosion this conservation role is important also in Java. In a survey of 350 gardens in West Java, Karyono (1981) found 602 plant species including 70 varieties of bananas and 30 of mango. But Soemarwoto and Conway (1991) warn that with commercialization, simplification, it might be necessary in future to establish *ex situ* repositories (regional botanical gardens) to assume this traditional but incidental role of these small farms.
9. *Wildlife refuge zones* With dwindling forested areas these gardens are emerging as one of the few wildlife refuge zones, at least for birds. In the intensively relay-cropped rice villages on the plains they are the only one.
10. *Human habitat* This role is always mentioned as a desirable and valuable attribute in studies around the world, from Kerala to Mexico. At the frivolous end of the scale eg the gardens of the peri-urban rich around Jakarta-Bogor, they've even been described as 'status symbols'. These social aspects of the system have been described by Karyono, Christanty et al 1985, 1986; Michon (below).

Evolution, dynamics From descriptions of the forest gardens of West Java at their several evolutionary stages (Abdoella 1986, Weirsum 1982, Linda Christanty et al 1986) it is possible to catch a glimpse of agriculture's emergence from the Great Forest – thousands of years compressed into just a few.

1. Forest swidden (Shifters) no longer existing in Java.
2. *Kebun* the swidden becomes stabilized as a permanent upland field growing a mix of dryland crops (corn, rice, beans...; in Central Java this is described, vaguely, as a *tupang sari* mix).
3. *Kebun campuran* perennials and tree crops are introduced.

4. *'Talun*   the trees become dominant;   ground inter-cropping ceases. The talun develops into a 'wilder' kind of non resident agroforest.

5. *Pekarangan* houselot garden, if this talun becomes residential. The forest mix is disciplined, becomes less wild, a thoughtful creation though an economic one – a garden of complete design.

But now also 6: the garden might be simplified to concentrate on a few commercial tree species (or in fact cleared altogether to grow that most simple of crops cassava, if the owner has found an outside job and can't bother with it except at planting/harvest times).

Farm nutrient balances   Jensen (1993, 1993a) in West Java constructed annual chemical/nutrient balances for one minute 920 sq meter forest garden containing 169 individual trees of 60 species (1 830 per ha), more than a rain forest. His main purpose was to obtain scientific data to support use of these farms in place of conventional agriculture which is causing soil erosion and other ecological damage to such downstream infrastructure as hydro systems, reservoirs, offshore coral reefs and fisheries. In terms of human nutrition (food, energy, protein, vitamins, minerals) the miscellaneous tree foods produced by this forest garden exceeded that from monocrop rice, cassava, corn farms, except in energy (box). (His second balance, the recycling of plant nutrients (N P K etc) among different components of the farm, is discussed in a later chapter – Nutrients, Jensen's farm.)

Table 2.3   Comparative Food Value Productivity on a Javanese Forest Garden

System/crop	K Cal	Protein	Vit A mg	Vit B mg	Vit C mg	Iron mg
forest garden	1 387	29	3 836	12	114	24
*rice (white)	909	16	0	6	0	5
*rice (brown)	1 412	24	41	22	0	6
*cassava	3 983	17	0	11	0	94
*maize/corn	727	19	562	5	0	6

*Comparisons based on yields of 4.1 t/ha for rice, 11.6t cassava, 2.4t maize. These are for single crops; on good soils with irrigation. Two and sometimes three rice/maize crops annually are possible (but only one cassava crop).

Architecture and socio-economic dynamics   Also in West Java Michon et al (1983) surveyed and mapped the vertical profiles of these farms, finding them resembling a rain forest in their four 'ensembles' or botanical strata, though this is also subject to random impacts, storms. Soemarwoto and Conway (1991) surveyed farms up a catena, 100 to 1 000 m, and measured decreasing farm size with elevation, decreasing number of species but increasing plant populations. Michon and Mary (1994) studied reasons why the species mixes and architecture have been changing. Their sampled farms contained 250 species, trees and herbs, about 50 on any farm. The general impact of commercialization and/or smaller farm size due to population pressure is to convert these quasi forests to...*artificial tree structures for a forced and selective biomass production.* Some of the larger true forest species are eliminated, their sun and space are needed for the commercial sun-lovers, eg papaya, cloves. Other species yielding fuelwood, building poles, fibers become obsolete. On the other hand these commercial changes often result in enrichment of some of the strata. Comparatively few species are tolerant of the deep shade at ground level in a natural forest. More of these will be added: edible tubers and those for edible leaves, other leaves as packing materials, shade-tolerant varieties of pineapple.

This now selected biomass, typically for commercial fruit, requires regular management control; it is exported from the site; nutrient cycling is impeded; the system ceases to be self-sustaining; fertilizer is needed; more commercial species are introduced to pay for it. These bring pests. Insecticides are needed... Mary and Michon note that in the Bogor area these structural changes, transition from subsistence to commerce, have resulted in 40-80% of the initial tree species being eliminated. Social consequences also follow. Traditionally these semi wild gardens were accessible to the poorest villagers who were free to take minor produce as they needed it. They were seldom fenced off; this would have displayed 'conceit' but this is changing too. Also in West Java, Otto Soemarwoto (1985) listed the directions of change in the techno-social functions of a garden in response to urbanization:

Techno-social attribute		Effect of urbanization
layering of botanical architecture	. . .	decreases
genetic diversity	. . .	decreases
microclimatic impact of biomass	. . .	decreases
social function, produce sharing	. . .	decreases
subsistence production mix	. . .	decreases
household waste disposal	. . .	increases
private aesthetic function, value	. . .	increases
commercial orientation	. . .	increases

Forest gardens, trees and Koranic credit   In West Java near Bogor, Dury, Vilcosqui and Mary (1996) studied the role of one major forest garden species, the durian (Durio zibethinus), as security for farm loans. In this credit system, *gadai*, a family's durian or other trees may be pawned by verbal agreement and their crops taken by the lender as loan interest. There's no time period attached to these loans, they can bubble on for years. The system also extends to Sumatra and Sulawesi. Since money interest is not charged, only the harvested crop, this village finance mechanism safely avoids the Koranic injunction against usury; and a second Koranic safeguard is that proceeds of such loans are used (or deemed to be used) for productive purposes. Oddly the borrowers here are quite rich (though not as rich as lenders), because of proximity to the Bogor-Jakarta metropolitan axis and its markets, but their wealth is largely in fixed assets, especially land. Here the main purpose of *gadai* is in consumption smoothing. [In keeping with the second Koranic injunction a good part of the loans are directed to such productive purposes as buying TV sets and refrigerators.] Computed actual interest rates are modest by Indonesian standards, zero to 210% annually with a mean of 65% for durian trees, and 80% for cacao trees in Sulawesi. (*Gadai* is quite different from *tabasan* credit, or forward sale of standing crops; both operate in West Java.)

Bamboo and pandanus forest gardens   Java is essentially a west-to-east chain of volcanoes, active, dormant, defunct. There is a general declining west-to-east rainfall gradient and species richness of the pekarangan forest gardens declines with this. Between any two of these adjacent mountains, proceding across the open 'U' of a valley, the land use, farming systems, farm size, tree species in the pekarangan gardens are more-or-less symmetrical. The upper zone (1) is a volcano flank or protective forest; the first inhabited zone (2), cool with locally high rainfall, will be devoted to intensive relay planted temperate vegetables, even in the houselots; zone (3) below that to field crops and the houselots to mixed forest gardens; and the hot and relatively dry valley bottom zone (4) to irrigated rice or sugar cane and the pekarangan gardens to tree species which tolerate seasonal drought. The pattern repeats itself, in reverse, going up the next mountain (McConnell and Nasir 1974; Probowo and McConnell 1993).

Bamboo   Although mixed tree gardens can be found anywhere along such a traverse the pekarangan lands of dry middle Java are often dominated by the bamboos and with these too there is a species gradient. Across the Madiun Valley (Nature, map) the large green mountain types (*pring petung)* are found in zone (2) and used for house construction and irrigation water pipes. Here land is in such short supply that *petung* is grown where no other crop is possible. Some villages of zone (3) grow smaller weaving types and their houselots are almost entirely devoted to these *(pring arpus, p. java, p. wulung)*. The large coarse general purpose type, *pring orie*, occupies the houselots of the hot bottom lands, again often to exclusion of other plants. Pring orie is so widely used as construction material it plays the same role that timber does elsewhere, and its dense thorny habit also makes it a useful live fence.

In the villages around Megatan, Madiun Valley (McConnell 1986) pekarangan houselots are devoted almost entirely to the weaving bamboos, with a scattering of jakfruit trees. Here up to 70% of the households subsist as weavers. Both the houselots (pekarangan lands) and the dry upland fields (*tegal* lands) are planted to these types, rather than to food crops. The bamboos are either used directly by the land owners or sold in two meter sections to wholesalers who sell these on to landless weavers. The common products are place mats, hats, vermin proof food baskets. One section is enough for two hats. A woman can weave one hat per day. Regardless of product she makes about Rp 100/day (US 25 cents, 1974, or the equivalent today).

Pandanus In the next valley east from Madiun, the upper Brantis, pàndanus takes the place of bamboo, and again the 'forest' gardens have been, through poverty, simplified to pandanus forests, with these extending to pandanus planted on any wastelands, roadsides, encouraged by village councils (McConnell 1986). Around Trenggalek and Tulungagung 40-50% of households subsist as weavers, most landless (with others further down the socio-economic scale collecting leaf from wild groves and selling it to the weavers). When planted as a sole crop (with suckers from the base of a mature plant) the spacing is 1.5 x 1.5 meters, or 3 x 3 meters if interplanted with coconut, plantains, cassava. Regardless of these initial agronomics, as the pandan grows it becomes top heavy, falls over, sprawls along the ground for three meters or so then recovers to send up its leaf bearing head to a height of one or two meters; and with all the other pandans doing this the field becomes a twisted mess of trunks and heads within 10 years or so. (Leaves are harvested individually each three months from the 4th year; but leaf collectors from wild plants often chop off the whole head. Others emerge from the trunk, but slowly and the method is a wasteful one.) Main products are sitting and soft sleeping mats, these made from leaf strips which have been boiled for fifteen minutes before weaving then sewn together into double thickness, and again the household's income is about

25 cents (updated) per day, per weaver. Not much of course, but both the bamboos and the pandans represent those forest garden species which offer the only life raft to many millions of rural people. For these, progress in official agriculture on the rice lands, and prospects of jobs in distant urban areas as these industrialize, are irrelevant (Nature, equity).

West Sumatra

Michon, Mary and Bompard (1986) have described Minangkabu systems, within an extinct volcano in West Sumatra, which consist of three zones: a forest garden around the house planted to 12 or so of the most valuable trees;   a larger removed agroforest with 30-40 other species (coffee, nutmeg, cassia) intermixed with the original forest, its most useful trees retained;   and some small area of wet rice fields.

These Minangkabu systems are notable for clan ownership and decision-making and matrilineal descent, and these contribute much to stable continuation of the system. Yields of individual species are low in comparison with commercial monocropping.   Families generally prefer sustainability to higher profits. There is pressure for commercialization but clan decision-making has acted as a break on individual agroadventurism. High species diversity represents a valuable gene bank of both the domesticated and forest timber species. The domesticated species represent many generations of improvement through selection.   Inputs are farm-generated, no agricides, artificial fertilizers used. These writers make the point that government administrative structure is not well suited to this highly mixed farm type: it falls under the jurisdiction of three ministries (Agriculture for the padi-rice;   Plantation Crops for the economic tree species and Forestry for the timber species), with extension effort consequently fragmented on a commodity-by-commodity basis and no one too interested — except the Minankabu ladies — with how the system works as a whole (Nature).

Riau Province Sumatra

Holden and H Voslef (1990) analyzed reasons for the failure of transmigration settlements in Seberida district and weaknesses of this program generally in Indonesia. Since 1981 fifteen farm settlements have been established in Seberida based on two farm models: (a) an 0.25 hectare houselot with one hectare food crop area and a ¾ hectare cash crop area (outfarm) removed from the house; and (b) a 'tree crop' model of 0.4 hectare house garden, a 0.6 hectare food crop area and two hectares of rubber trees. All these settlements were based on clearing evergreen rain forest. Both models largely failed because they were ecologically inappropriate, and resulted in...*a high incidence of poverty, 64% of the migrants falling below the poverty line, and this caused 30% of them to return to Java.*

The 'tree crop' farms performed better than did the field crop farms, but the bulk of these tree crops were in fact monocrop rubber and was adversely affected by disease and over-tapping. (In both cases the home garden consisted of a mix of tree and ground level food crops.) These authors recommend conversion of both models to mixed tree crops...*the unsustainability of (field crop) production, particularly upland rice, has in ten years changed Seberida from a rice surplus to a rice deficit area.* Apart from alleviation of settler poverty the main reason for making these farms more viable is to reduce the threat, caused by land degradation under the models followed so far, to Seberida's remaining rain forests. (These transmigration projects are a major cause of deforestation in Indonesia.)

West Kalimantan

De Jong (1990) described complex Dyak farming systems in Sanggau district, West Kalimantan, which operate six different types of agroforestry (in addition to upland swiddening and wetland rice cultivation):

1. *Hutan tutupan* communally maintained forests.
2. *Sompuat* forest areas left and maintained specifically for honey bees.
3. *Palau rimba* maintained forest 'islands', left when their owners cleared agricultural fields around them. In their natural state these yield construction timber and rattan. Apart from this disturbance they are equivalent to natural rain forests. However sometimes a few coffee or rubber trees are interplanted either for commercial reasons or to indicate ownership.
4. *Tembawang* 'fruit tree' forest gardens containing high densities of domesticated and undomesticated trees. Though primarily tree food gardens these also contain many other species (bamboo, ironwood, *meranti* trees for construction). These *tembawang* might be established by interplanting directly into a patch of forest and also tending or tolerating the naturally occurring species, or artificially by planting some species mix on swidden crop land.
5. *Kebun karet* predominantly rubber groves (the main tree cash crop) but this is often intermixed with other trees.

6. *Bawas tua* Old growth secondary forests in process of regeneration from previous swidden crop cultivation (Shifters).

These village agroforests occur along an economic gradient from forest extractivism (as in Amazonia) to agriculture (the rubber groves). Five *tembawang* surveyed had an average of 426 individual trees per hectare (diameter greater than 10cm) in comparison with 560-570 trees of the near natural *palau rimba* forests. The *tembawang* gardens contained 376 tree species in comparison with 379 species on the *palau rimba* forests. *Tembawang* forest gardens are...*complex forests with structure and plant species diversity which can be almost equal to mature natural forests.* These Dyak systems have important implications for plant (and animal) species conservation.   Because   of   increasing   population pressure...*conservation of species rich tropical forests will increasingly have to look beyond natural reserves and include 'woodlots' in which some exploitation is allowed.* De Jong suggests these Dyak *tembawang* forest gardens as one such alternative, but with increasing population pressures this general Dyak land use system is breaking down. Traditionally when forest was cleared for swidden rice only one crop was taken before plot abandonment for forest regrowth and fertility restoration, now several successive crops are planted, soil fertility and the natural seed bank in soil storage are reduced; the regenerated vegetation is often only ferns and coarse *alang alang* grass (Imperata cylindrica). Natural forest regeneration ceases (Shifters).

Also in West Kalimantan Salafsky (1994) surveyed the agroforestry systems of 88 households in Sukadana district. The zonal arrangements of these are similar to the Minangkabau systems of Sumatra with a houselot 'forest garden' proper close to the padi fields then a second agroforest zone outside this, then a zone of forest used for extractivist activities, with this giving way to the natural forest. In this scheme the outer agroforests are more specialized in cash crops than are the immediate houselot forest gardens. They are larger and not as well tended. Although some of the (removed, zonal) agroforest occur as monocrop stands of coconut, rubber, citrus...most of them are structured around three tree crop combinations: durian plus 10-15 other fruits; durian plus coffee; rubber plus sugar palm (Arenga pinnata). Durian (D. zibethinus) is a major cash crop in this area, most of the fruit being husked in the field, the arils then sold by the bucket to local traders who manufacture durian candy *(lempok, dodol)* for export.

Borneo is (after NW Amazonia) the world's fruit tree epicenter, many species not yet commercially developed. Salafsky does not offer a complete inventory on these Sukadana farms but lists the following, all now well known:

durian .... (Durio zibethinus)
pekawai.............. (D. spp, wild durian with red flesh)
langsat .... (Lansium domesticum)
dukuh .... (L. domesticum var)
bedara................ (Baccaurea motleyana, similar lychee)
keranji................ (Dailium sp)
jakfruit .... (Artocarpus intiger)
mangosteen .... (Garcinia mangostana)
rambutan............. (Nephelium spp, several wild types
mango .... (Mangifera spp, several)
belimbi............... (Averroa bilimbi)

Among the non-fruit species are ironwood (Fusi deroxylon zwageri) which produces a silica impregnated termite resistant timber; *gaharu* (Aquillaria genus) which yields a valuable aromatic resin; and bamboos (for shoots used as a vegetable as well as construction poles).

Until the 1970s these (removed) agroforests of the second type were not much valued by their Malay owners because (except for durable produce such as rubber) there was no ready access to the urban markets of Pontianak and the volume of perishable produce from the trees far exceeded subsistence requirements. Now with the new market roads the garden values have soared and even individual durian trees are a tradable commodity. Interestingly although these are predominantly Malayanized-Dyak villages many Javanese and Balinese people have drifted in as unofficial settlers from failed transmigration schemes elsewhere in Kalimantan which were based on unsustainable 'official' cropping systems (as in Riau Sumatra above).

Again in Kalimantan Godoy and Tan (1991) in search of the rattan cultivators of Dadahup in Central Kalimantan found that like Indonesian smallholders elsewhere the people there cultivate...*many food and medicinal crops in backyard gardens* (coconuts, fruit, cloves, cinnamon underplanted with cassava, maize, chili and have fishponds) but the structure of these gardens is not explored. These are later discussed as forest swidden systems (Shifters). But this is now a requiem. A good part of Kalimantan's forests have been destroyed by Suharto loggers and plantation developers and their wildfires.

The Eastern Islands

Raintree and Warner (1986) called attention to the prosperous tree-based village economies of the Roti and Suva islands, centered around the multipurpose lontar palm (Borassus sundaicus) the sugar of which is used among other things to fatten pigs. Raintree also recalls the tales of a previous traveler, Fox in 1977, who put the population carrying capacity of these tree-based systems at several times that of the degraded swidden systems of the nearby islands of Sumba and Timor.

On Seram the subsistence systems of the Nuaulu which include a forest garden component have been described by Ellen (1988), primarily as hunter-gatherer systems centered around the sago palm. The tree crops component has not been described. Similar extensive or intensive mixed tree crop systems also exist on Ambon, Timor, Bouru, but data are lacking; and parts of Jambi province in Sumatra are said to be particularly rich in forest gardens.

Forest Gardens and Transmigration Villages

One incidental effect of Indonesia's transmigration program, designed to settle poor rural people from Java-Madura to the less densely populated outer islands, is to spread this type of pekarangan land use, although these settlement farms are based on clearing rain forest for conventional field crop agriculture, especially rice. Bogor Agricultural Institute (IPB) has prepared models of tree crop mixes for the houselot portions of settlement villages. These consist of recommended mixes of 25-30 tree species (Guhardja 1985). This Javanese extension should not obscure the fact that these 'virgin lands' in Sulawesi, Sumatra, Kalimantan, Irian Jaya have been occupied by their indigenous peoples for millennia and for an equal length of time they have had their own tree based agrosubsistence systems. But just what these were, or are, we are little wiser than when Wallace left us that day in 1854 at the village of Semabang in Sarawak, saying only that...*all the people were away after edible birds nests and beeswax...and the mountain was close by covered with a complete forest of fruit trees (1869)*. In fact no wiser than after Odoarto Beccari (1904) took us to the village of Senna and asked the people there to bring samples of their tree crops so that he might inventory their mode of subsistence. But when the Dyaks had done this and brought him *bua paya* and *sintol* and things like the *lagnier* fruit which they also used as soap, and many others, Beccari was still so engrossed with five new species of the rambatan which he had just found in another village that he forgot to record the results of his inventory.

2.7 Philippines

Philippines embarked earlier, and with more enthusiasm than other SE Asian countries, on the trip to modernize its agriculture, notably under US occupation which built upon and compounded the socio-environmental mistakes of a plantation system left by the previous colonialists. The ecological results have been disastrous. To support and expand this proper agriculture Philippines has long had one of the highest deforestation rates in the world, 90% of its littoral ecosystems – notably the mangroves – were destroyed by 1950, and by 1999 only 10% of its coral reefs remain intact. With this past and continuing devastation of the resource base, its population growth rate of around 2.5% and approaching 100 million will ensure there's little left within fifty years except a *cogon* wasteland. More benevolent indigenous agroforestry systems have always been present (as well as not so benevolent slash-and-burn), but these are not part of the national development model. For this reason discussions of Philippines' forest gardens occur either as a cry of protest or description of something stumbled upon when other exploitive land use systems fail (Shifters).

Luzon

Fujisaka and Wollenberg (1991) offer case studies of the accidental evolution of quasi forest gardens on Luzon through a succession of 10

Table 2.4 Changing Efficiency Parameters through the Evolution of Forest to Agroforest

Successive phase	Productivity	Stability	Equitability	Sustainability
primary forest	3	3	-	3
logged forest, legal	1	2	-	2
agricultural - forest mosaic	1	1	-	1
secondary forest	3	3	-	3
logging by settlers, illegal	3	-	1	1
*forest extractivism	3	-	2	2
farm trees planted	2	3	3	3
cereal crops on cleared land	1	1	3	1
cash crops (tomato)	3	1	2	1
**agroforestry, mixed, permanent	2	3	3	3

Subjective phase scoring from 1 (worst) to 3 (best). Here equitability is narrowly defined to refer to direct users of the land (Criteria). *rattan, ferns, charcoal, firewood taken by settlers. **this final result falls short of being a true forest garden; of the species established only 15 are trees, 27 are ground crops.

different land uses — logging, forest produce extractivism, proper farming, failure, and eventually establishment of a sustainable tree-based system. They rank each stage in this saga of economic misadventures according to four criteria — productivity, stability, equitability, sustainability – before the system eventually stabilized as a quasi forest garden. These and the evolution of other Philippines tree systems are discussed in a following chapter (Shifters).

Also in Luzon Delbert Rice compared the sustainability of two land use systems in the indigenous Ikalahan's hill lands with another indigenous system found in Southern Mindanao.

System (A) 'single crop intrusive' is applied by lowland settlers when they follow the logging trails into the forests to establish their small subsistence/cash farms according to their lowland (intrusive) practices. After the soil has worn out the land is converted to pasture/cattle, and when it becomes unproductive for this, because of regular grass burning to induce fresh growth, it is abandoned for 30-40 years;

Table 2.5   Efficiency Comparisons of Three Farming Systems in Philippines

System comparative performance (Subjectively ranked on a scale of 0 (worst) to 10 (best))	A	B	C
erosion protection	6	8	9
fertility maintenance	5	7	9
fertility recovery after cropping	0	4	5
rapidity of storm water runoff	1	5	6
Average system effectiveness	30%	60%	73%
Quantitative comparison			
period under ground crop cultivation, years:	5	4	5
after-crop pasture use, years:	40	0	0
required fallow period for recovery, years:	30	14	11
total use cycle, years:	75	18	16
necessary cultivated land per family, ha:	1.4	1.0	0.7
total agricultural + catchment land/family, ha:	420	50	25
families supported per 1 000 ha of total land:	2.4	20	40

System (B) is traditional local Iklahan practice differing from (A) in that as fertility declines under the monocrop (sweet potato), fruit and some restorative trees are planted before a bush fallow phase begins;

System (C) is that of some of the T'boli indigenous hill people of Southern Mindanao. This is closest to forest gardening...*The T'boli ignore a few trees from an area when they clear the land for cultivation* and they also extract forest produce for handicrafts materials.

There are probably many other indigenous tree based systems similar to C in the remaining forests of Philippines but appear not to have been objectively evaluated, at least by agricultural economists. Official preoccupation is with proper agriculture of the rice lands and monocrop plantations – Or how fast the *indigenes* can be cleared out of the way as obstruction to Nation Building.

2.8 China

In their account of Chinese agriculture, Xiong, Yang and Tao (1995) see it as having arisen in the forests and for millennia as having been inseparable from what has been called, only from the 1970s, agroforestry. They place the beginnings of this broad subsistence system in the middle or late periods of the Old Stone Age. Initially such 'agroforestry' meant little more than seeking shelter in the forests and subsisting on its plants and animals. The first technological revolution was invention of slash-and-burn cultivation at some unknown time in the past; these writers also refer to this as a form of agroforestry. (It continues today with minority nationalities such as the Du Long, Nu, Xi, Wa and Yao in Southwest China.) The second revolution, the emergence of sedentary field farming from slash-and-burn, occurred in the forests c 10 000 BP (before present). Perhaps not coincidentally this resulted in a strong central government (the Xia Dynasty 4 000 BP), population increase, collapse of clan society and introduction of slavery.

This latter relaxing to become rural serfdom made possible the 'nine squares' system of land use a large central square occupied by the lord and the other eight squares each worked by a serf family. This in turn ushered in the golden principle of household self-sufficiency which has guided Chinese agriculture ever since and classically was called *Man farming Woman weaving*, ie the concept of an integrated, ordered, holistic system with many components: timber, fuel, incense (sandalwood) and fruit trees planted around the house, field food crops, livestock, ponds for lotus and fish and turtles, willow trees for silk production and all these producing primary or byproducts for the 'woman weaving.' In this total economy these elements of earth and heaven support each other, circling endlessly in perfect harmony, ManWoman lubricating the Wheel with quotations from the Book of Songs from the central square. Many will agree this is the best that humankind has yet achieved. A few might sit with Paul Theroux and look angrily out the window of that train through Sichuan to see the natural world of China — its biosystems, forests, swamps, hills — reduced to a buffalo and a cabbage patch, its wildlife – like the sun bears of Cambodia – to a culinary freak show in the market. Man farming Woman weaving.

Chaos is antithetical to order. Contact with the forest was lost long ago. Mulberry and silk worm rearing, surely among the most civilized of activities, is included as agroforestry. The authors group this as five types. Only one of these, 'agroforestry with forest dominant', partly resembles in its architecture the forest gardens of SE Asia. But these also are sophisticated combinations of long domesticated species: coffee and pepper under rubber, mushrooms under bamboo, tea interplanted with pine trees. Nonetheless forest gardens do exist in the South. A few are noted.

The Autonomous Prefecture of Xishuagbanna

In the Autonomous Prefecture of Xishuagbanna in tropical Yunnan, Saint-Pierre (1991) has described lowland farming system of the Dai people of which forested house gardens are one component. The other components are wetland padi fields, planted fuelwood lots, public forests and certain village sacred forested hills. (These latter play no direct role in the economy of the Dai but their religious significance and species conservation importance are great, Pei below.) The forest gardens are minute, averaging 200 sq meters. They are multistorey, rising to 15 m height or 30 m if a sacred Buddhist tree (Ficus altissima) is present as the emergent. In the wet season they are undercropped to vegetables but in the dry the vegetables are harvested in the form of buds and leaves from the trees. No cash crop trees are included in these forest gardens; they produce only household necessities. (No species inventory is offered but according to Saint-Pierre Chinese botanists have listed 300 trees and herbs.) Saint-Pierre believes these small forest gardens to be eroding by pressure of prosperity: increasing incomes from the cash cropped fields are often spent on house extension, there is no room for the trees.

Another system described by Saint-Pierre in the same region is a Jinuo highland village where 'all the houses are surrounded by trees' (unlisted). In addition the Jinuo have wet padi fields and a range of quasi forest tree crop mixes as removed outfarms. One of these is tea-under-forest in which the forest trees are thinned to around 100 stems (consisting of some of 100 total species) per hectare and tea is then planted at the very low density of 250 bushes per hectare but allowed to grow to three meters height. (This contrasts with plantation tea in Sri Lanka at around 12 000 plants per hectare kept pruned to a little more than one meter height.) These retained forest trees are multipurpose: the women collect leaf and bud vegetables, fruit and medicinal requirements while plucking the tea. Other uses are timber and fuelwood, and as in the swidden-forest gardens of Amazonia the forest tea gardens also attract game.

The temple and sacred hill components of Dai agro-social systems have been described by Pei (1985). Their agricultural components (house forest gardens, padi fields, forest-crop mixes) are much as Saint-Pierre reported them: Pei was concerned with the broader more important matter of the religio-cultural role of trees in Dai society and their contemporary ecological relevance. There are some 750 000 of these non-Han people in Yunnan, about a third of them in Xishuangbanna prefecture. As well as being farmers, some semi-nomadically, they have always depended on the forests for its plant products and game.

Polytheists until their conversion to Hinayana Buddhism, the Dai have inherited two streams of beliefs and practices concerning plants. The second of these in time, the Buddhist stream, exists as forested groves in Dai temple courtyards. By religious injunction certain species must be kept in these groves and according to the classic text, 'Records of the Buddhas Coming into the World for 28 Generations', the Buddhas of each of these generations designated one plant species which is to be adored. In addition there are utilitarian reasons why these temples are repositories of trees. Pei surveyed 20 temples and identified 58 species: 21 ritual plants, 17 fruit-food trees, 20 ornamentals. Pei also lists them all according to their practical uses by the monks and their geographical origins. Oddly many are from India and tropical America, the former probably arriving as seed with pilgrims through the Yunnan corridor, the travels of the latter from Mexico and South America a mystery. Pei's list reads like the inventory of a tropical horticulture station.

The first stream in time are the trees of the Holy Hills. These are untouched forested hills throughout the Dai range where the pre-Buddhist gods still reside. They are of 10-100 hectares, Pei counted 400 of them and they occupy about 40 000 hectares of the prefecture. *All the plants and animals that inhabit the Holy Hills are either companions of the gods or living things in the gods' gardens.* Protected by these beliefs the Holy Hills are, in more practical vein, repositories of biodiversity...*for example there is a Holy Hill called Mangyangguang of 53 hectares...which contains 311 species belonging to 108 families and 236 genera and (of these) 283 are vascular plants.* With consent of these gods and their companions the Yunnan Institute of Tropical Botany has measured the main parameters of their microenvironment.

2.9 Oceania

Of forest gardens in the four Great Forests or once-forested regions, those of Oceania are poorest in crop diversity and have least contemporary economic importance. On the other hand this system is here the most appropriate one for these fragile ecologies (and economies).

Background

...the earliest Austronesians left the richness of the small mammalian fauna (of SE Asia) behind them and then the hunter-gathering mode of subsistence gave way to agriculture (Yen 1983). Such agriculture consisted of two contending streams, both vegeculture, the first based on manipulating and exploiting trees, the second removing them. Both were carried from SE Asia into Melanesia, the first as aboriculture and the second as slash-and-burn cultivation for root crops. The agroforestry subsistence systems were first in time; those of western Melanesia are very old, at least 40 000 years (Genesis). According to archaeology (Sommers 1985, Leach 1984) those which were later carried by the Lapita voyagers beyond Santa Cruz into Remote Oceania, to the dispersal points of Samoa-Tonga and Society Islands, are comparatively recent, about 4 000 years. Such tree based systems were environmentally benign (Genesis).

But as the voyagers continued further eastward to smaller and less fertile islands, and these with successively fewer indigenous economic plant species, their increasing populations required ever greater reliance on the root crops which they had also carried. This stream proved to be malign...most malign indeed. If Clarke and Thaman's (1992) interpretation of the evidence is correct these oldest immigrants voyaging east to their new paradise — with a slash-and-burn technology appropriate to the unlimited land resources of SE Asia and Melanesia but hopelessly inadequate for small islands — quickly rendered a good part of it a wasteland: the fire climax *niaouli* savannas of New Caledonia, the degraded *talasiga* lands of Fiji, extreme soil erosion in the Cooks, Wallis, Futura, Kahoolawe... Species extinction rates under human impact have been greatest on these islands. As Edward Wilson summarizes it, the 'voyagers ate their way through the Polynesian fauna'. Between the

Polynesian arrival on Hawaii in 400 AD and European arrival in 1778 more than half of its 98 bird species had been exterminated by hunting, slash-and-burn clearing and livestock — the introduced dog, rat and pig. Since 1778, largely through European plantation agriculture and import of more domestic animals (especially cats) a further 17 birds have gone extinct, bringing to total avian extinctions so far to 70% of endemic birds. Of endemic plants about half have gone extinct or are under threat (Primack 1993). On smaller atolls these impacts have been much higher.

Some find in Oceania a condensed account of ManWoman's voyage through time and the natural world. From archeological evidence their ETAs at all the main island groups have been given by Nunn (1990). That for Easter Island is 1260 AD ...*It was then covered in rich forest. By 1500 the population was 7 000, divided into little kingdoms with virtually no forest left, but over 1 000 statues. Then came civil war, fighting over resources, with not a single tree left to be made into a canoe to escape. The Dutch explorer Roggerveen arrived in 1722. By then the population had fallen to 500. All were in miserable condition. They could not even remember their own history, or· what the statues around them were for* (Crispin Tickell 1993).

Natural species diversity   While the forest gardens of Oceania are in general species-poorest, Yen (1974) has also pointed out a declining species diversity gradient eastward from Melanesia into Remote Oceania. While some 420 agroforestry species are found throughout the region (65% of them indigenous) 300 of these occur in the western islands towards SE Asia, only 80 or so on the eastern atolls. Of these 420, some 300 have been or are cultivated but most of them now 'fallen into disrepute'.

Contemporary systems and development policy   Quasi forest garden farms occur in two main forms. On the littoral they exist as agrohouselots structured around the coconut and breadfruit interplanted with comparatively few and often introduced species — citrus, mango, papaya, yaquona, banana. The limiting factors are usually salt environment and sandy soil. Many other indigenous species are also present (pandanus, hibiscus) but these typically have little modern relevance and continue as ornamentals.   On better soils on high islands subsistence agriculture continues, mainly as bush swiddening for rootcrops (yams, sweet potato); and here also trees might be left, or deliberately planted, as an element of declining economic importance (Santa Cruz, Tonga, Micronesia below). Where markets exist the thrust of agricultural development is to convert such mixed fields to pure stands of cash crops — pineapple, banana, ginger, market vegetables — and dispense with the trees altogether.

Since colonial times neither of these tree based systems of the upland forests or the littoral have attracted official interest. This has instead been directed at three kinds of development. The first extracting from the indigenous forests those (very few) species which appeared to have commercial and especially export prospects, and intensification of these as plantation monocrops (notably coconut and more recently and to a smaller extent yaquona/kava). Second the introduction of other tree species for the same purpose (rubber, coffee, citrus, cacao) with indifferent results. Third has been the introduction of 'proper' agriculture, field crops as well as the above tree crops, requiring the clearing of rain forests for a plethora of development 'schemes' which environmentally are very much in the old Kahoolawe tradition.

Schemes   ...The schemes of cynical and irresponsible speculators added to the general confusion...(Osmar White, 1965, in describing the general economic development of Oceania 1850-1914.) These schemes emerged as a major rural development strategy throughout much of Oceania. They litter the Pacific from the highlands of Papua New Guinea to the beaches of Niue...The citrus development schemes of Cooks, Fiji, Niue. The older banana, cacao and pineapple schemes of Tonga, Samoa, Fiji. Schemes for import substitution (rice in Solomons) and export schemes for entrepreneurial profit and Nation Building (tea, coffee, pyrethrum in Papua New Guinea). Currently and posing greater environmental hazard than any of the others, *zebu* cattle hamburger schemes wherever there's a patch of rain forest to be cleared and a barge can nose its way through the reef. Post-WWII schemes which aimed at reducing the forests of Vanua Levu to colonial oil palm estates. That post-WWI rubber scheme in Samoa which once threatened to reduce the proud villagers of Upolu to the socioeconomic status of rubber tappers. A similar scheme in pre-independence Fiji established an Australian coconut aristocracy on the Savu Savu cost, these expatriates rivaled in their agronomic incompetence only by the German-Australian plantation oligarchy which earlier (1914-1942) had established itself on the Gazelle Peninsula. By whatever name (corporate venture, commodity program, land development project, modernization campaign) and under whatever sponsorship (New Guinea Kompagnie, colonial office, planning department of now independent states, bilateral aid partner, international development bank) these schemes, the contemporary following the historical, have many common threads. None found or cared to look for elements within indigenous agroecosystems which might be of relevance for solving contemporary problems. All have been intrusive, imposing instead an alien agro-technology in a physical (and social) environment they have not bothered to understand. Their agronomy is simplistic and immediate, their dependence on artificial imported inputs for it remarkable. Their economics when independently scrutinized are usually found to be partial, selective and deceitful. With national independence these schemes continue to be proffered to small states as a quick fix to all problems ranging from sopping up the unemployed to righting the balance of payments. Their policy thrust is to attempt to generate export earnings, via the schemes, and use these to import manufactured and refined food (with the nutritional consequences noted below).

The antecedents of this national version of the old cargo cults are worth a note. They can be traced back beyond the Westec estates of Samoa, and beyond the Kompagnie's disastrous first agroschemes in 1885 around Finschhafen (...*stifled by a murderous climate, fevered by a particularly virulent strain of malaria, consoled only by alcohol, the most frequented place the cemetery*...Souter 1963). Their true origins probably stem from the feverish mind of that earliest Pacific agricultural planner, the Marquis de Rays. His scheme was to colonize the malarial forests of New Ireland in the now Bismarks (which for his marketing purposes he designated Nouvelle France), a place he'd chanced to read about, by cutting these rain forests up into 50 hectare estates, each of which was to have a manor house and to be supported in its implementation phase by six months free food supply. He floated these delusions off as $150 shares to the gullible of France, Italy and Belgium; and these eager for a more exalted social status than they could aspire to in Europe were further enticed by the issuance of 'patents of nobility' to go with their estates. It may be taken as measure of those times — and encouragement for those projects which have since followed in unending stream, though now with more sophisticated sponsors — that the launch of this first agrodevelopment scheme was wildly successful. The Marquis collected the equivalent of several million dollars from his incipient aristocrats, loaded them aboard two old ships and in 1880 dumped them ashore among their surprised Melanesian tenants. (The survivors were eventually rescued by a ship from Sydney Town; the Marquis continued to formulate his development projects to the last, but from the safety of a French lunatic asylum.)

Economics One of these modern schemes, export bananas in Western Samoa was examined by McConnell (1972). He found that although it contributed to the equity of those farmers involved and to agency employees and extended at least some opportunities to earn cash income into remote villages, the main beneficiaries were European-Australian-New Zealand suppliers of the input factors of production — of everything from fertilizers and shovels through packing boxes to shipping services. Results of this analysis were not welcomed. However Felemi (1991) later analyzed a similar scheme in Tonga and came to even less palatable conclusions. He subjected the Tonga scheme to three kinds of analysis — financial, economic and social. Of this latter: *...the social net present value (NPV) of the scheme is even more negative than the economic analysis and the scheme not only results in inefficiency losses but might also exacerbate social problems.* Participants in such schemes are often the more progressive (or aggressive) land owners: *...hence these schemes can worsen income distribution even though island nations pay lip service to rural development as a process of alleviating the plight of the poor...*

Schemes, sociology and war   In the Solomons, Frazer (1987) discussed the detrimental effects of monocrop export schemes there, variously labeled programs, projects, commitments, campaigns... On Malaita there was a great increase in cash monocropping over 1971-1985. Cocoa area increased by a factor of three as a result of a 1960s-1970s campaign and coconut by a factor of four (following government commitment to export crops). Combined with population increases these schemes have sometimes severely distorted local land supply, management practices, food production patterns and household task equity. Land shortage for food production has resulted in reduced bush fallow periods and land degradation. Of task equity, Frazer points out that food land is increasingly scarce and often far away from the village: *...imposing additional work burden on the women who do most of the gardening.* Bourke (1990) has described similar consequences in Papua New Guinea. In the Eastern Highlands establishment of four cattle projects in 1975 pushed food gardens further from the hamlets...*resulting in an increased burden in transporting food, poorer garden maintenance and the location of more gardens in the zone of primary danger from sorcery...* In Solomons a civil war has dragged on through 1998-2002, mainly between Malaitans and Guardalcanalese. Causes of this conflict – basically over population then migration of Malaitans from their own island to Guadalcanal, following its historical denudation by slash-and-burn (Shifters) – are now reinforced by 'forest development' by mainly Asian companies on the other western islands of that country. In some of these concessions such as on Vella Vanella, obtained from indigenous owners by promises of jobs, roads, bridges and selective extraction only – promises which are never kept – the forests are being entirely razed to the extent that not a single large tree is left for making the traditional fishing canoes. Subsequent or parallel agricultural development then procedes, taking the form of Malaysian style oil palm estates, including one of 50,000 acres on Vangunu island on what was to have been – based on its rain forests, lagoons and hornbills – a World Heritage scientific reserve. Local objections, if not smothered by bribery, are overcome by the greater power of the urban elite in Honiara, only too willing to fall in with IMF's and development banks' concepts of progress.

Nutrition, consequences As a result of rapidly increasing populations, urbanization, preference for agricultural commodity export schemes, abandonment of the older systems, Oceania has gone in 60 years from self-sufficiency with a largely intact natural environment to one of the highest food-import dependency rates of any region in the world (Fairbaine 1971, Thaman 1985). This has been paid for by a remarkable depletion of the natural world. It has also been paid for in the currency of declining health and nutrition due to the imported refined foodstuffs. From Thaman's work (1983, 1985) some the highest and most rapidly increasing incidence of nutrition-related disorders and diseases (diabetes etc) are now to be found among the food importing peoples of Pacifica.

Ontong Java

On Ontong Java in this regard, Tim Bayliss-Smith (1982, 1986) has charted changes in food production patterns on these atolls over a 16 year period (below). This Polynesian outlier of Solomons Republic is home to some 1 400 people on its 39 small islets, some minute and supporting as few as four persons.

The previous economic system consisted of some deliberative agriculture, fishing, but much collecting from seven (agro) ecological zones – reef, lagoon, mangroves, pandanus nut groves etc; then with population growth and commercialization it became dominated by export beche-de-mer fishing. Three kinds of balances changed: growing dependence on imported refined foods (box); the atolls' total energy balance in which imported fossil fuels (for fishing boat motors) largely replaced human work energy in farming/collecting; and the traditional social system of mutual help and sharing in times of crisis (hurricanes) corroded as food became an economic commodity and kinship ties loosened. With these increasing population and economic pressures (and now rising sea levels from global warming) Bayliss-Smith sees little future for atolls like Ontong Java, without more commercial exploitation of remaining marine resources and out-migration. His account of these dwellers of the atolls seems hauntingly familiar: paradise found, depleted, abandoned, for one last voyage out beyond the Gates of Santa Cruz. The old saga of Pacifica as flight from Sunda, four thousand years ago, not yet completed (Genesis). The lesson of Easter Island not yet digested.

Table 2.6   Relative Food Production, Procurement, Ontong Java

	1970	1985
Population	850	1 400
Production—		
Taro	29%	18%
Coconuts	21	15
Other agriculture, (sw. potato, banana, papaya, sugarcane, fruit)	4	1
Fish, shellfish, turtles, crabs, seabirds/eggs	21	15
Imported (flour, sugar, canned fish, rice, biscuits, soft drinks)	25	51

Tonga

...Seen from the air Tonga looks like a forest of coconut palms. Traditional tree-based systems of which these palms are a pillar have been described by Kunzel (1989, below). There are some five million palms on   40 000 hectares,   360   on   any   representative   *Api   'uta*   (farm)   of 3.3 hectares. Traditionally new fields were opened up in the forest by clearing only those open spaces required by the sun-loving ground crops, especially yams; all other useful trees and (to minimize effort) many others having no apparent use were left standing in a random pattern. In this Tonga economic system (*Api 'uta*, farm), trees have always been taken for granted...*tolerated rather than actively promoted.* On the one hand to farmers with sufficient land trees mean...*less weeds, increased soil moisture, better soil structure, and lower insect populations.* But on smaller farms trees mean...*lower yield of annual crops, drier soils, higher work loads*...Also, with agricultural modernization, which means clean tillage and tractorization, the old method of leaving trees spaced randomly is increasingly not feasible...*Only the deliberate planting of trees in a pattern which is compatible with tractor cultivation can ensure continuance of the old agroforestry system* — but such tree planting is not part of Tongan culture. In consequence...*tree densities in Tonga are decreasing on*

a broad scale. One specific result is that B. javanica, needed for dyeing tapa cloth for the valuable tourist handicrafts trade, is disappearing rapidly.

Disinterest in aboriculture extends to the main pillar of this system, the coconut. From Kunzel's survey 16% of all palms in Tonga are senile and average farm yields are only 25 nuts per palm (in comparison eg with 50-70 in Sri Lanka). The animosity of smaller farmers towards trees has been captured by Kunzel's fictitious letter from a Tongan farmer: *...I have farmed all my life and it has been a constant battle against the trees.* This 'letter' has now passed into folklore. In addition, this Tongan kept a fictitious diary:

Tue: Adviser from banana scheme tells me to convert my farm to a banana plantation.

Wed: Adviser from coconut scheme offers to clear land free if I will plant that crop.

Thu: A visiting foreign agroecologist says No; it's better to stick with the old mixed systems.

Fri: Visit local research station. They advise me to plough up the farm and grow vegetables.

Sat: Village meeting; all agree to a cooperative scheme to grow commercial pumpkins.

Sun: Go to church and pray for guidance from God.

Mon: Eat the vegetables, cuts down the bananas, burn the coconuts, go fishing...

Table 2.7   Trees and Shrubs on a Tonga Farm (Kunzel)

Carica papaya	22	food
Artocarpis altilis (breadfruit)	13	a major carbohydrate, bulk food
Bishofic javanica	13	wild, for vegetable dye, numbers declining
Mangifera indica (mango)	7	declining, not planted, fruit, timber
Pandanus spp	7	for mats, baskets, perfume but nuts not eaten
Morinda citrifolia (beach mulberry)	3	bush fallow volunteers; medicinal
Pometia pinnata (Pacific lychee)	2	fruit, medicinal
Inocarpus fagifer (Tahiti chestnut)	2	fruit, bark as medicinal bandages
Citrus (6 species)	2	
Aleurites moluccan (candlenut)	2	soap substitute
Macaranga harveyana	2	medicinal leaves, timber, fuel
Rhus taitensus	2	medicinal leaves, timber, fuel
Other trees	8	

Understorey
*Broussonetia papyriferia

(paper mulberry)	for tapa cloth
Piper methysticum (kava)	
Bananas	most important cash crop
Vanilla	second important cash crop
Yams (on live forest tree supports)	in sunlight; most valued food

*dyed with B. javanica extract. Together with pandanus mats, tapa cloth is a main handicraft product for the tourist trade.

Tokelau

In Tokelau the gardens of these atolls have been described by Whistler (1987, Genesis). They are primarily stands of coconut, the main economic species, but when neglected for any length of time these groves revert to the indigenous species-poor forest. There are only about 35 indigenous plants, plus introduced, but they all have several traditional economic uses. (Others have estimated that the 400 or so tree/bush species of the region had an average of nine recorded uses.) On Tokelau the main indigenous foods are coconut, screw pine nuts (Pandanus) and tubers of the giant aroid *pulaka* cultivated in pits, but apart from the coconut the introduced species such as bananas are now more important foods. Village knowledge of plant uses is disappearing, which Whistler regrets...*it might be needed again some day.*

Santa Cruz

In Santa Cruz islands (east Melanesia), Yen (1974) has described both the inland and littoral farms (Genesis). Trees are much more important in Santa Cruz and Melanesia generally than in Polynesia. There are three main types of inland farms: some are temporary slash-and-burn forest swiddens for the root crops only; some are initially swiddens but in which the regrowth of useful tree species is encouraged before the swidden is abandoned; the third is a swidden in which breadfruit are planted among the root crops in which case the swidden is converted to a permanent pure-stand breadfruit plantation. Forest gardens proper are found around the littoral houses as...*virtually tree gardens* (Genesis).

2.10 Micronesia

The agricultural systems of the high islands and atolls of Micronesia, Table 2.8, have been described by Ayres and Haun (1983) as basically...*a root crop-tree crop complex* [ie vegeculture] consisting of three subsystems:

aboriculture, upland cropping based traditionally on slash-and-burn, and permanent wetland taro fields. These are believed to have had two source origins (in addition to the contributions of indigenous tree/bush species): the first from SE Asia spreading eastward and the second arriving from Polynesia and spreading back westward, on linguistic evidence to as far as Tobi island south of Palau. The three subsystems vary in importance, between atolls and the high islands, with the high island of Ponape being exceptional: ...*The diversity of cultivated plants and soil fertility have caused some to call Ponape the garden isle.* Ayres and Haun refer to 42 indigenous cultivated plants, including 200 varietals of breadfruit and 150 of yams. The traditional system is ...*of farmsteads about the house taken up by two tiered cultivation of tree crops (breadfruit, coconut, banana etc) underplanted by root crops...and the situation today still follows the farmstead pattern.*

Table 2.8   Relative Importance of Agricultural Land Use Systems in Micronesia

| | | Relative importance | | |
		aboriculture	dryland	wetland
Palau	(high islands)	10%	30%	60%
Yap		25	30	45
Truk		70	15	15
Ponape		50	45	5
Kosrae		60	25	15
Central Carolines	(atolls)	50	20	30
East Carolines		45	10	45
Marshalls	(atoll chains)	55	30	45
Kirabat/Gilberts		70	10	20

Source: Estimated from Ayres and Haun's 1983 graphs.

Ponape

On Ponape Ayres and Haun (1990) also constructed a picture of prehistoric food production systems from archaeological evidence, c 1 000 BC to 1 800 AD. The earliest dominant system was apparently slash-and-burn of the primary forest for root crops. This gradually gave way to aboriculture (breadfruit, coconut, banana, yaquona) which became the main system by 800 AD and continued as the dominant one until 1800. Slash-and-burn continued but became dependent on much shorter restorative phases under secondary forest. There were also other complementary agrosystems over this 2 800 year period — animal husbandry based on dogs and wetland cultivation, but these are relatively important.

Pohnpei

On Pohnpei in Micronesia, Raynor and Fownes (1991, 1991a) report agrobotanical studies of 54 multistorey tree gardens. These occur as forest gardens around the houses and also as non-residential units which dominate the landscape...*Despite increasing emphasis on the cash economy the traditional agroforestry system still supports most of the 28 000 people on this island...It is the result of more than 2 500 years of development and refinement.* A total of 161 plant species were identified occupying three storeys and averaging 26 per farm, with livestock as chickens, pigs and dogs scavenging in the understorey. These agroforests now occupy about 33% of the island and though increasing population are extending into the forests proper which still occupy 56% of the island. (But because of the antiquity of this system it is possible that in the distant past it also occupied what are now apparently the 'natural' forests.) The core food/beverage species of the vegiculture complex are coconut, breadfruit, banana/plantain, and yams and taro: ...*These supply the basic subsistence and social needs of the people.* All are multipurpose but in this regard Hibiscus tiliaceus has a remarkable range of uses: construction poles, live trellises for yams, leaves as green manure in yam pits, bark for rope, inner bark as bast fiber for straining the ceremonial kava or *sakau* (P. methysticum).

While it is...*not possible to estimate the true genetic diversity in the agroforest,* that this is great is implied in the names for specific cultivars: there are 177 'varieties' of Diascoria, 55 of Musa (banana), 155 of breadfruit, 24 of Cyrtosperma taro, 16 of Colocasia taro, 10 of Alocasia taro, 9 coconut, 16 sugarcane, 3 *sakau*, which these authors believe to be genetically distinct. This is reflected in the different uses and properties of these cultivars, differentiated according to yield, produce size, quality, seasonality, disease resistance etc. Raynor and Fownes conclude: *The documentation, collection and preservation of this genetic resource should become top priority for research and development in the region...*which echoes Perera and Rajapakse in Sri Lanka, Michon in Java, Okofor in Nigeria.

Yap

On Yap in Carolines Falanruw (1985) reports about half the island's land is under agriculture and one third of this is under forest gardens, typically consisting of fruit and other food trees but also timber and other wild residual species from the original forest. They have an understorey of shrubs and herbs supplying fibers, condiments, pharmaceuticals. Gardens are concentrated around the houses and along village pathways but many of the cultivated species have been widely dispersed by bats. Falandruw also calls attention to the wider value of these farms as genetic repositories, listing 21 varietals of coconut, 28 of breadfruit and 37 banana/plantain.

A second type of traditional farm on Yap, slash-and-burn gardens for field cropping, away from the houses, also has an element of aboriculture insofar as tree-vines-bushes emerge as volunteers among the yams and sweet potato. At one time the US administration attempted to modernize this slash-and-burn system by clear felling the bush for Western style vegetable growing, with artificial fertilizers in place of a bush fallow phase. The new system caused considerable soil erosion and was soon largely abandoned for return to bush swiddening. But although this is more environmentally benign than systems based on complete forest clearing, this ancient swiddening system is itself prone to problems — in comparison with the tree gardens — caused by uncontrolled burning, soil exhaustion and build-up of pest populations (Shifters).

2.11 Mexico

Yucatan

In Yucatan forest gardens, *huertos familiares* (family orchards), were always part of Mayan land use systems and remain so. Some 80% of these domesticated species are selections from Central American forests, a 'New Crops development program' which has been occurring for thousands of years, preceding then in parallel with that other Mexican stream of 'proper agriculture' (corn-beans-squash, one source of Mainstream B, Origins). The Maya had both: trees on the *huerto familiare* and corn in the removed (swidden) *milpa* fields. Vera Niñez (1985) has described these dual systems once supporting Maya civilization over about 100 000 sq miles (and there is archeological evidence this civilization collapsed when population pressure exceeded capacity of the swidden forests to regenerate and continue their fertility supply to the milpa fields). In their floristic surveys of 44 Maya *huertos familiares*, Rico-Gray et al (1985, 1990, 1991) describe them as an orchard annex close to the house for production of fruit, medicinal and multipurpose products. From 44 sampled farms, areas are 1 800 – 2 200 sq meters, one typical *huerto* having 135 woody and herbaceous species. Of the 301 botanical species around this surveyed area

186 were found on these forest gardens. Mendieta and del Armo (1981), from other surveys, have listed the medicinals.

Rico-Gray et al also studied the forests proper, concluding that the best way to conserve these was to follow traditional holistic Maya land use practices. As in Oceania the main thrust of agricultural development in Yucatan has been through monocrop 'schemes' for cattle, sisal, market orchards, requiring forest clearing. Many have failed. Even when this clearing is for timber extraction only a few of the higher value logs are recovered. The Maya, by contrast, treated their 'public' forests as extensions of their *huertos familiares* and extracted from them a multitude of products on a sustainable basis. Yucatan is Mexico's largest honey producer; one main commercial value of the remaining forests is as bee range. Removal of the forest diminishes this industry and when it is followed by development 'schemes' addicted to agricides the impact can be devastating.

San Luis Potasi

In the state of San Luis Potasi the ethnobotanical dimensions of Huastec Indian systems, on the eastern slope of the Sierra Madre, have been investigated by Alcorn and Hernandez (1984). They recommend them as a model for agricultural settlement, and like Rico-gray with the Maya, for sustainable forest management. They would be particularly relevant as an alternative to settlement systems requiring deforestation, or to distributing chain saws to hungry *campesinos* and turning them loose in a rain forest. The main thrust of this argument is that strict conservation is often just not possible, for political, social equity and economic reasons. While agroforestry settlement farms on the Huatec model are not a full substitute for exclusive conservation reserves they are a reasonable compromise. (The argument has much relevance for land settlement 'schemes' in Malaysia, Indonesia, Amazonia, but authorities in these places could draw the same conclusions from their own traditional agroforestry.)

The Huastec live in an area at 50-500m elevation in what was once moist forest. Their farms are of about 6ha and usually consist of four components: swidden-maize rotated fields for subsistence carbohydrates (½ the area); a few small sugarcane fields to support a sugar-products cottage industry; a house garden and an adjacent managed forested plot or *te' lom* which occupies about ¼ of the farm area. The *te' lom* component is equivalent to a forest garden though not always adjacent to the house. Here aboriculture and natural forest management are integrated: management interference is casual and consists of introducing desired species into the *te' lom* if they do not generate there spontaneously, removing unproductive trees and occasional weeding. The *te' lom* may be managed as several distinct compartments: one for mainly fruit trees, another for timber, another for a cash crop such as coffee, but if this specialization occurs it still allows a range of species in each compartment. Over 300 species have been recorded on these *te' lom* including 221 or more than half of all Huatec traditional medicinals.

Vera Cruz

In Vera Cruz, Alvarez et al (1989) analyzed the forest garden components of 71 farms of Belzipote ejido (agricultural cooperative settlement). These settlement farms are relatively large, from 40 ha down to one ha, the larger running cattle and growing corn as their main enterprise. All have home forest gardens which are a combination of trees-understorey crop-livestock. The smallest of the ejido's farms consist of only a forested garden around the house. Although of recent origin (1950-1960) these forest gardens are 'traditional' in that the settlers brought with them and apply a rich agro-technology from different places in Mexico. Most farms also contain some remnant of the primary forest or permit its secondary regrowth. In this, as in Yucatan, they serve as repository for at least some of the original forest genetic resources. Only 12% of the forests in the area had escaped land clearing. The garden components of the otherwise often monocrop settlement farms...*are a means for maintaining diverse subsistence production against the impulse to specialize completely in monocultures.* (So at least this component of their system, if not the cornfields and cows, would be approved by the Minangkabu ladies of Sumatra.)

On this *ejido* garden size ranges 230-3 400 sq meters and these contain a total of 338 plant/tree species and are arranged in three zones around each house. That immediately adjacent is largely bare, for grain drying, but supports a few ornamentals; beyond that a 'formal' jardin supports medicinals and ornamentals; then the *huerto* (orchard) occupies the rest of the houseyard supporting a more or less dense mix of fruit and multipurpose species. As elsewhere (Sri Lanka, Amazonia) it is given its basic structure at the time of forest clearing: useful individuals are left standing as are a scattering of larger (15m) trees to provide microenvironment for the future garden or as a standing stock of construction timber. Alvarez et al list 62 local rain forest tree and shrub species which are commonly left in this way and their several uses; but a more general 'use' is to enhance the farm as human habitat. (Non-local or exotics are later added.) There is no clear architectural stratification.

Subsequent management is traditional, often involving ritual (hanging red ribbons to intimidate evil spirits, or outsmarting them by operating according to phases of the moon). Thirty four pests have been recorded but if eradication measures are taken these also are traditional (painting tree trunks with lime, burning small fires). Fertility maintenance is from decomposition of litter fall, household wastes, animal manure. As in Kandy this is essentially a *collecting* system: maintenance labor is only 20 days annually. Livestock consists of dogs, chickens, hogs free-ranging under the forest and subsisting largely on fallen fruit.

Tabasco and Tlaxcala

In Tabasco and Tlaxcala, surveys of lowland-v-upland *huertos familiares* by Allison (reported by Gliessman 1990) showed that in their technical respects (leaf area index etc) both systems, particularly the lowland one, closely resemble natural forest ecosystems (box).

Also in the Tabasco lowlands Gliessman, Garcia, Amadora (1981) report a farm development program using agro-ecological models borrowed from ancient Indian practices. These contain a forest garden component (for food production, shelter belt, microenvironment for field crops and fishponds). Desirability for this return to pre-Conquest technology was based on the fact that it was historically highly successful throughout Middle America (see Yucatan, San Luis Potasi), whereas modern agricultural technology in the Tabasco-Campeche lowlands — cattle grazing and monocrop schemes — has often failed after forest clearing when the soil fertility store became exhausted. (Gliessman also notes, not surprisingly, that these attempts to apply what can be learned from the past have enjoyed little official support.)

Table 2.9   Structural Parameters of Some Huertos Familiares

		Lowland (Tabasco)	Upland (Tlaxcala)
Size	ha	0.70	0.34
useful species per garden	No.	55	33
perennial species	%	52	25
tree species	%	31	12
ornamentals	%	7	9
medicinals	%	2	3
cover	%	97	85
light transmission	%	22	31
leaf area index		4.5	3.2

2.12 Central America ... *the people with many trees*

Guatemala

Jacobs (1982) says that the Maya of Guatemala refer to themselves and their language as Quiché, *qui* meaning many and *ché* tree, the people with many trees. Modern versions of some of these ancient Maya gardens around Santa Lucia were studied by Edgar Anderson (1950). Again these are not the multilevel forest gardens of the true wet tropics but more mixed household orchards with ground level food crops grown throughout. Anderson described one such garden as...*covered with a riotous growth so luxuriant and so apparently planless that any ordinary visitor, accustomed to the...puritanical primness of north European gardens would have supposed that it must be a deserted one.* He prepared diagrammatic sketches, possibly the first time this had been done for this system. The garden was at once an orchard, vegetable-medicinal-flower garden, bee yard, compost heap and garbage dump. There was no soil erosion and few weeds. Production was high and continuous...*I suspect that if one were to make a careful study of such a garden one would find it more productive than ours in terms of pounds of vegetables and fruit per man hour or per foot of ground. Far from saying that time means nothing to an Indian I would suggest that it means so much more to him that he does not wish to waste it in profitless effort as we do.* (Anderson was so impressed by the household garbage dump component of these systems as vehicle for accidental plant domestication that his garbage-dump theory of the origins of agriculture is largely based on it — Genesis.)

Other tree gardens in Guatemala have been studied by Wilken (1977) and Maya gardens in the north by Stavrakis (1978). In the Department of Petan, Gillespie, Knudson, Gielfus (1993) analyzed a sample of four small mixed tree farms in terms of photosynthetically active radiation (PAR) penetrating to the garden floor. This can be taken as a measure of the amount of light not being used by the canopy and a guide to potential improvement for greater physical efficiency through garden restructuring. (On these same grounds Karyono (1990) has suggested there is much scope for such improvement to Javanese gardens.) The PAR values obtained differed markedly among the gardens studied (Energy, light). More interesting from a non-technical viewpoint are these authors' comments on the role these tree gardens played in classical Maya times. In this part of Guatemala, together with slash-and-burn milpa cropping and forest gathering, they once supported 400-500 people per sq km. This declined to only 5 per sq km; but settlers are again moving into the remaining forests and population has increased tenfold within the past 25 years. The significance of the old Mayan tree based system is that it is possibly the only type of agroecosystem which could support such population pressure in a sustainable way (Kandy, farm populations, farm size). Turner and Miksicek have described the agro-technology of these classical Maya systems (Origins).

Costa Rica

In Costa Rica Gliessman (1990) has listed the species found on small houselot tree gardens there, with their understorey species and their uses, finding them subject to change in response to economic factors. Also in Costa Rica, near Puerto Viejo, Fleitner (1985) has described how Don Ignacio's traditional mixed garden was rapidly simplified into a commercial coconut-yuca-pineapple plantation in response to expanded market opportunities opened up by a new road...vindicating Rico-Gray's previous thoughts on this in Mexico.

2.13 West Indies

The few studies from this region have been reviewed by Budowski (1985). They include case studies from Martinique in the field of cultural anthropology (Horowitz 1967);   Pilot (1980) in Haiti has attributed deterioration of gardens there to increasing population pressures and Brierley (1976) has described the gardens of Grenada. Innis (1961) and Hall (1969) explored the link between modern West Indian gardens of all types and the 18th century introduction of slaves and their African agro-technology into this region.

Puerto Rica

Clarissa Kimber (1973) studied 80 of the 'dooryard' gardens of Puerto Rica, supplementing her data with aerial reconnaissance of 2 000 more. She found six types, ranging from unorganized *jibaro* gardens of the poor to manor gardens of the rich. In general they are similar to suburban garden spaces and now have little economic significance. Kimber, a cultural geographer, finds an interest in them as *units of social space*. They did function as serious multistorey, multipurpose systems in former times.

Jamaica

In Jamaica the West African origins and structure of small hill farms have been outlined by Innes (1961). These consist of small areas (three hectares) of highly diversified crop and livestock subsistence enterprises: field food crops (taro, cassava, corn, peas), cattle and small stock (pigs, goats, poultry, bees), napier grass fodder, all as a 'haphazard' mix under or among such tree crops as allspice, coconut, citrus, avocado, coffee, banana, rose apple, 'dandelion trees' and timber and fuelwood trees. As with many African farms (eg of Burundi, Tanzania, Rwanda, Uganda) the ground level bulk food crops, with bananas rather than tree crops, are dominant, so these mixed Jamaican farms fall short of being true forest gardens (except in rocky areas which are devoted exclusively to trees). Innes notes that while this 'higgledy-piggledy' system is deplored as primitive and unscientific by official agricultural advisers it is probably the result of several thousand years of previous experimentation in the African homelands.

2.14 South America

In tropical America generally tree house gardens are so important that they comprise the core of what Lathrap calls the Tropical Forest Tradition (Genesis). Vera Niñez (1985) refers to them as a ubiquitous feature of this region. Perhaps for this reason they have been taken for granted (except by botanists, ethnographers and geographers) and not considered as an appropriate agro-technology for national development. In Nair's (1989) world map of the various agroforestry systems this type occurs in the NE corner of Brazil. This considerably underestimates their distribution also throughout the Amazon basin of Brazil and Peru and in Guiana, Venezuela, Colombia, Ecuador, lowland Bolivia and extending up through Central America to Mexico. But there are few concrete data relating to their exact locations, numbers or agroeconomic structures.

Upper Amazonia

The general structure of the house gardens in this region was early described by Lathrap (1977). Typically some 50-100 tree/bush/ground level-species are present. He discusses the uses of a few: Crescentia and Ceiba for shade and materials; the peach palm Guilielma gasipaes and many other palms for food, thatch, timber; avocado; Anadenanthera peregrina for hallucinogenic snuff; Theobromo cacao (but for fresh fruit rather than for cocoa beans); Anacardium occidentale (cashew, also for fruit rather than for nuts); the dye plants eg Genipa americana; kidney seeded cotton tree (Gossypium barbadense); papaya; grasses and herbs as perfumes; twenty or so medicinal and drug plants; and a wide range of condiment and ground level food crops eg capsicum, tomato, pineapple. The house garden is often combined with a slash-and-burn outfarm, the *chacra*, which is used to produce a narrow range of bulk food crops, typically manioc/yuca/cassava.

In Peruvian Amazonia Padoch and de Jong's (1987, 1991) account of the forested house gardens of Santa Rosa is for several reasons a classic. This village was founded only about 30 years ago by families of detribalized *ribereños* who previously worked on a nearby fundo (semi-feudal commercial estate) until that was abandoned. At Santa Rosa they developed a complex agricultural system with four components:

1. rice and beans on the mud of the annual flood plains;
2. other short term food crops and some trees on the natural levees;
3. forest swiddening; and
4. highly diversified forest gardens around the houses.

It is possible that these forest gardens of Santa Rosa (component iv) are the most species-diversified agroecosystems in the world. Parameters for a few of them are extracted.

A total of 168 species were found on the 21 sampled gardens (which indicates that they are at least as diversified as those of Java on which Soemarwoto (1975, 1979, 1984) recorded 179 species but in a sample covering two districts. Variation among gardens is great, with 65 of the species occurring respectively in only one garden. The village had been founded on forest land that had for long been subject to periodic swidden agriculture into which trees had been planted, so many of the present garden trees had pre-dated settlement. Others were recent natural volunteers (Shifters, managed fallows).

Table 2.10   Areas and Number of Species by Use Category of Santa Rosa House Gardens

House-hold	Area (m^2)	Species	Food	Medicine	Orna-mentals	Miscell-aneous
1.	5 622	46	38	6	1	6
4.	638	39	23	12	0	5
6.	572	34	21	10	0	6
11.	468	47	27	15	1	6
21.	274	47	31	11	0	8

These writers see significance in such diversity for two reasons. First, it has been proposed by other observers that forest gardens develop to their maximum only where land is in short supply (Java, Kandy). Indeed some make this a condition for forest garden development, but there is plentiful land at Santa Rosa. Second, it has also been proposed that diversity would reach such a high level in remote areas where there is greater *need* for self-sufficiency; but Santa Rosa settlers are commercially oriented and have access to the markets of Iquitos.

High species diversity is apparently attributable to four factors. This general region of Peruvian Amazonia is thought to contain the most species-rich rain forests in the world (except parts of Colombia); these therefore provide a rich botanical matrix into which yet other collected plants may be inserted to establish farms. Second is the factor of culture and tradition. Padoch and de Jong describe these *ribereños* of Santa Rosa as being...*very active plant seekers...and inheritors and developers of rich Amazonian traditions of plant and ecosystem management.* Third, geographical conditions differ among and within the communities; some households maintain plants which are impervious to the annual floods while other households out of flood reach might have a quite different tree/plant mix. Fourth, the species mixes also reflect the economic purposes of different households: the poor will possess mainly food species, the mix of *curer* families will contain many medicinal plants etc. (The *curanderos* and their plants are discussed elsewhere, Diversity.)

In Peruvian Amazonia also, at Tamshiyuca near Iquitos Padoch and others (1985) have described a forest swiddening system which, at one of its phases, is close to a sedentary forest garden. The cycle starts with clearing a patch of forest, leaving the useful species (each household operates several plots simultaneously), converting some species to charcoal, planting annual crops. Then in the second year a second annual food crop is planted, along with trees, or useful trees are simply allowed to regenerate. Weeding is done for the first few years then discontinued as the swidden develops into a tree crop garden. This continues for 25-50 years. It is then cut down (some for charcoal) and then finally abandoned to bush fallow phase. The whole land use system (field crops as well as hunting-gathering) generates many products but those from the tree crops predominate (table). The 'abandoned' garden still generates produce in the form of birds, animals which are attracted to it for garden hunting. Because of the nature of the forest produce (table) and proximity to Iquitos markets these gardens were found to be highly profitable, returning up to $5 000 per household annually. This compared with the average of $1 200 per household in the area, and is well above incomes of other families who, with government loans, went into rice production (Shifters, managed fallows).

Table 2.11   Major Sources of Income in Tamshiyacu Swidden – Forest Garden Systems

% of annual income

1. *Cultivated fruits* 63%
    Umari (Poraqueiba sericea)
    Peach palm, pijuayo (Bactris gasipaes)
    Star apple, caimito (Chrysophyllum cainito)
    Uvilla (Pourouma cecropiaefolia)
    Inga (Inga edulis)
    Cashew (Anacardium occidentale)
    Brazil nut (Bertholletia excelsa)
2. *Intensively managed crops* 21%
    Manioc (Manihot esculenta)
    Plantain (Musa paradisiaca)
    Rice (Oryza sativa)
    Papaya (Carica papaya)
    Pineapple (Ananas comosus)
    Cocona (Solanum sessiliflorum)
    Tumbo (Passiflora mollissima)
3. *Animal products* 9%
    Meat (deer, peccary), skins
4. *Charcoal* 3%
    Many tree species
5. *Forest fibers, handicrafts* 2%
    Chambira fiber hammocks (Astrocaryum chambira)
    Tamshi fiber, baskets (Heteropsis jenmani)
6. *Forest fruits, palm heart* 1%
    Aguaje (Mauritia flexuosa)
    Ungurahui (Jessenia bataua)
    Huasai, chonta (Euterpe precatoria)
7. *Medicinal plants* 0.5%
    Chuchuashi (Maytenus krukovii)
    Clavohuasca (Mandevilla scabra)

Amazonas, Brazil

In Amazonas state in Brazil, Lescure and Pinton (1993) compared the benefits of forest garden components of household economic systems with that other common form of 'agricultural' use of tropical forest, commercial extractivism of forest produce, and recommend the former, on the Javanese model, partly on grounds of its greater efficiency in terms of produce recovered per unit of human energy expended. The home base of most extractivists is in fact a small forest garden. On the middle Rio Negro...*Aboriculture takes place in polyspecific and multistratified orchards close to habitations. Each generally contains about 30 species, up to 60 if taken on the scale of the community...(in addition) horticulture produces herbs and spices. Small-scale chicken, duck, pig and sometimes sheep or cow raising is practised in the orchards.* Also on the Rio Negro Emperaire and Pinton (1993) describe a similar system where...*Each family needs at least two 2-ha swiddens (rotated slash-and-burn clearings) for manioc. In addition some 27 species of fruit trees are planted around the house...and the spices are planted in containers above flood level.*

On the Rio Solimoes (Amazon) Pereira dos Santos and Lescure (1993) found the household trees component of these systems an alternative to slash-and-burn. In some areas of Amazonia with high population pressure the restorative phase of swidden agriculture is now down to only five years. It is impossible to reduce this further without irreparably damaging the soil resource base. More benign forms of land use are urgently needed (Shifters).

On the Araquaia and Tapajos rivers of Brazilian Amazonia, Posey (1985) recorded the complex traditional land use systems of the Kayapo Indians. Some of these involve the deliberative structuring of two types of forest gardens. The most interesting involves the creation of forest 'resource islands' (*apete*) both in open country and in the natural forest which the Kayapo then have access to both locally and in their frequent and extended journeys around their range. Posey identifies eight types of *apete* differing in species contained, purpose, shape etc. (The type of interest here served much the same purpose as would a series of supply dumps established by explorers.) As well as food trees and bushes, the resource islands contain species for medicines, weapons, cordage, insect repellents, poisons, containers and others intended to attract game, in fact all the uses for which sedentary forest gardens are intended — and many more, such as refuges in times of war or epidemics (the last as recently as 1983). Posey thinks this strategy of establishing resource islands in the Great Forest was once widespread throughout Amazonia. Kayapo gardens of a second type are along the forest trails themselves; these were planted to all kinds of useful trees as well as tubers in a sort of Kayapo agro-ribbon shopping mall. Posey estimates that near one of the 11 modern Kayapo villages studied there were 500 km of such trails, bordered by their forest resource malls.

By other estimates the Kayapo use 250 tree/foods and 650 medicinal plants, a good number of these planted in the resource islands and trail malls. For planting the more difficult savanna resource islands they carefully prepare each planting point with a mound of compost designed to induce microbial activity then introduce termites from enemy camps to continue adding gleaned nutrients to the mounds, with the termites too busy fighting each other to attack young roots of the planted trees. These are transplanted from the forest and the Kayapo continue tending the site until its canopy has closed over. (Hunter-gatherers or farmers? Agronomy or ecology?)

Caboclos of the floodlands – farmers or extractivists? Forest gardens are essentially managed extractivist systems in that they receive few inputs and the main operation is harvesting/collecting their produce (Kandy). On the flood island of Combu, Amazon mouth near Balem, Anderson and Ioris (1992) have described a community of *caboclos* who operate a similar system on larger scale. With a population of 650 persons on 15 sq kilometers, Combu is typical of much of the lower Amazon river system, covered by rain forest and completely inundated by tide-driven floods during the wet season. This forest consists of relatively few tree/plant species but an unusually large proportion of these are of economic value. Human habitations are 'hamlets perched out of flood-reach along the river banks'. On these flood islands there are four main forms of forest/agroforest land use: (a) small houselot forest garden proper of about one ha and planted mainly to trees for food, beverages, fibers, habitat, condiments, medicines; (b) some households also have 15 ha of forest as their private extractivist reserve for products listed in the box; (c) access to the public forests for hunting and more casual collecting/extractivism; and (d) some households also operate forest swiddens for bulk foods (but only 5% of the island's forests have been cleared for these).

On Combu the 15 ha extractivist reserves are operated like large forest gardens, differing from pure extractivism elsewhere in Amazonia in that they are enriched by planting rubber, cacao, acai palm etc, and these are sometimes weeded, fenced. Altogether forest extractivism, wild and managed, provides 92% of Combu family incomes without significantly altering the integrity of the forests, or necessity for seasonal migration to look for jobs in Balem.

Extractivism for forest conservation In Amazonia (and SE Asia) forest extractivism – especially if managed as on Combu – has great potential importance for rain forest conservation. Globally from FAO data some 80-

90% of rain forest destruction is attributable to agricultural clearing (Diversity). Throughout Amazonia the area required in both public and private settlement schemes is 50-100 hectares of forest cleared per family. This might be justified if such farms were permanently sustainable; but most are productive as cultivated fields for only 5-8 years after which they degenerate into unproductive cattle pasture. Anderson and Ioris on the *caboclo* managed extractivist forests of Combu, and many other writers throughout tropical America, have suggested extractivism as a superior alternative to deforestation both on ecological and economic grounds: as noted only 15 hectares of extractive forest are required per family on Combu and although this is modified to some extent it is not destroyed. The main commercial extractivist produce taken from forests in Brazilian Amazonia, the bulk of it 'unmanaged', is summarized in Table 2.12, supporting about 1.5 million people. Counting produce of all kinds taken for local/non commercial uses, and that from non Brazilian Amazonia (Peru, Colombia, Ecuador etc), it would be many times the levels shown in the table.

Table 2.12 Guide to Commercial Produce Extracted from Brazil Amazonia, 1986

Product		Vernacular	Tonnes	$US 1 000
Rubber:	Hevea brasiliensis	borracha	28 400	28 400
	Castilla ulei	caucho	200	136
Gums:	Manilkara bidentada	balata	20	10
	Manilkara elata	macaranduba	380	170
	Couma utilis	sorva	3 000	1 160
Fibers:	Mauritia flexuosa	buriti	890	80
	Leopoldinia piassabe	piacava	300	130
Oils:	Orbignya phalerata	barbacu	40	10
	Copaifera langsdorffii	copaiba	40	40
	Dipteryx odorata	cumaru	460	760
	Scheelea huebneri	urucuri	4 640	30
Foods:	Euterpe oleracea	acai fruits, palm hearts	133 850	41 600
		acai palm hearts	124 300	6 400
	Bertholletia excelsa	brazil nuts	35 560	6 990

Source: Anderson and Ioris 1992.

Forest extractivism in Amazonia has been largely ignored at official planning level as an alternative to forest clearing for proper agriculture. Although this policy has been changing since the mid 1980s and some 2.5 million ha of official extractivist reserves have been set aside in Brazil

(1988), this area is minute in comparison with land cleared for orthodox farming. One reason for lack of interest is that for nearly four centuries it has been invisible, carried out...*in extensive tracts of forest by dispersed and economically (and socially) marginal populations* (Anderson and Ioris). This invisibility has also permitted the fiction that these forests are...*empty, unowned, useless and abandoned, which facilitates their appropriation for other uses* (Diversity; Indonesia, transmigration).

Forest gardens as commercial estates In the state of Bahia (NE Brazil) where all but about 5% of the Atlantic rain forests disappeared long ago for sugar and cattle pasture, Schultz, Becker, Gotsch (1994) have described how one large (500 ha) commercial cacao and banana farm – but also with about 50 other tree crops – has been developed according to principles borrowed from the Kayopo Indians (Posey, above) and the Huastec of Mexico (Alcorn). The challenge was to rehabilitate this once-forested farm land and make it commercially viable. The object was to recreate the *structured chaos* of Indian farms, *selvas huertos*, by incorporating existing trees in the commercial cacao and banana planting, whether or not these had any 'economic' value, and introducing other species which were *eco-physiologically similar* to the rain forest species which once occupied the site. Subsequent management was to...*maintain an open self-regulating system of coexistence and related organisms and their abiotic environment.* For this the only external input is lime, necessary to correct the highly acidic soil condition which developed after the original forest was cleared. No artificial fertilizers are used; but because the system cannot become self-sustaining in its nutrient requirements until it can provide these through re-cycling its litter (Energy, Nutrients) this is temporarily achieved by growing short term crops and slashing/spreading this to the trees at the rate of 8-16 tonnes/ha (dry matter). This is about the amount of nutrients–organic matter–litter which a rain forest (or an Indian *selva huerto*) feeds to itself annually. With the rediscovered Indian technology, commercial cacao bean yields have been 220-370 kg/ha, in comparison with 225 kg from other farms using the modern standardized agronomic package (of fertilizers, agricides, weeding etc) in the same area. Nitrogen use alone would have run to 130 kg/ha/year if the modern model had been used, and this modernized *selva huerto* requires no agricides. (Another unexpected product was reemergence of 17 water springs after the land was reforested.)

The Japanese Farmers of Tome Accu

Around the town of Tome Accu near Balem, Japanese immigrants have established what are possibly the most carefully managed forest gardens in the world, combining the 'structured chaos' of indigenous American systems with Oriental order and frugality (Nutrients).

Colombia-Ecuador

On the Pacific Andean slopes of Colombia-Ecuador Orejuela (1992) has offered a comprehensive inventory of the plants in household forest gardens of the Awu Indians, describing this system as a: *...polyculture of crops, palms and fruit trees around the farm houses in a heterogeneous and very efficient pattern in excellent accordance with the local environment.* Some 134 species of several types/cultivars are listed together with their primary uses. Most are indigenous to the local rain forests. With up to 8 000 mm (335 inches) of rain annually these are among the wettest in the world. In addition to these tree gardens the Awu economic system includes shifting cultivation for the traditional American corn-beans-squash complex, but this based on slash-and-mulch rather than slash-and-burn; forest gathering and hunting, with some plant species deliberately planted in the outfields to attract game; and fishing for which other fish-stunning plants such as *barbasco* (Banisteriopsis sp) are grown in the gardens. This is also used as an agricide.

From Orejuela's inventory and observations made in passing it is possible that these Awu garden systems supplemented by forest produce could be even more diverse than Santa Rosa. Medicinals-drugs-poisons are particularly rich. Among these are a group of very important plants called *contra* used for treatment of bites from different species of venomous snakes which abound in this area. [These include the *surucucu* which Fawcett's diaries so often shudder from as...*that double fanged abomination the bushmaster.*] Another important economic plant is Perebea sp (Moraceous) the boiled latex of which is used as arrow poison for hunting, equivalent to Amazonian *curare*. The other main arrow poison is obtained by scraping from the skin of the equally abominable *kokoe* or poison-arrow frog, Dendrobates histrionicus. Among the main Awa food plants are five types of plantain which are also cooked and fed to chickens, ducks, pigs; and peach palm here called *chontaduro* (elsewhere in tropical America *pejibaye*, Bactris gasipaes) which yields large quantities of pulp for bulk food rich in starch, protein and carotene (Diversity, palms).

2.15 Africa

There have been two streams of agriculture in Africa (Origins) – the savanna grasses which became field crop farming as granoculture and the rain forest roots and trees which became vegeculture/ aboriculture. The first stream early came to dominate, consolidated in recent times because this stream – grain and cows – accorded most closely with the agro-stereotypes which colonialists took with them from Europe and which are now almost synonymous with proper farming. If the data below are any guide it has not served Africa well. Some 260 million of 600 million people in the sub Saharan countries now fall below the World Bank poverty

Table 2.13 Economic Condition of Africa in the 1980s

Number of countries	53
Population	435 million (1985)
	% changes 1980-85
population growth	+3.2%
food production per head	-2.0
volume of food imports	+4.0
external food aid per head (3.4 to 6.4 kg)	+90.0 (1975-83)
food consumption per head	-1.2
all exports volume	-2.7
all imports volume	-2.5

Source: Grove 1986.

line of $1/day equivalent and their governments spend four times more on 1st World debt repayment than on health. It would be naïve to ascribe this condition solely to inappropriate farming systems and land management technologies. Over-population, war, disease, drought, over-stocking, soil erosion, desertification, urban elitism and corruption are more visible causes. Nonetheless many stem from or are stimulated by the kind of rural 'development aid' and farming systems offered to (or imposed on) Africa, more often according to the donor's short term *real politik* agenda than with long term socio-ecological consequences of this aid in mind. There are curious parallels with Oceania: official disinterest in more appropriate indigenous agroecosystems, disdain for trees (except as fuel), faith in monocrop schemes, addiction to The Package (of imported HYVs, fertilizers, tractors, agricides), reliance for economic equity on shifting global markets, fate determined by the aberrations of some commodities exchange.

The data listed paint only a broad fuzzy picture now known to be inadequate as detail emerges:- 1st World debt repayments which can never be met and keep many countries in perpetual economic bondage; lack of education (African school enrollments are now less than in 1982 and projected to account for 75% of global illiterate children by 2015); loss and cost of the AIDS-afflicted generation and their AIDS orphans (more than one million in Uganda alone) and emerging negative impact of these on future productivity through cost of their support – if any – and lost education and crime; political corruption; accelerating food productivity loss through deforestation, soil erosion, desertification, water shortage and contamination; recurring famine and war... These also provide essentially the list of UN-proposed measures to somehow at least stabilize Africa by 2015, falling under the main headings of debt relief, expanded education, (agro) economic development. But what *sort* of education? (In the Christian south it includes exclusive boys schools where a future governing elite are drilled in the finer points of Latin grammar: in the Moslem north it's a tree under which 'students' regurgitate easy passages from the Koran – girls excluded.) Continuing this first tradition, one serious UN proposal is to replace Latin with computer literacy, put them on the net. And what *sort* of agricultural development?, since Africa's recent woes largely start and end with imposition of flawed EuroAmerican land use systems, technologies, on that continent, usually compounding the accumulated effects of unsustainable indigenous agrosystems. Africa's own Mainstream B (Origins).

By contrast Okafor (1987) says of the forest stream...*these complex agrosystems of tropical Africa have hardly been studied.* Sholto Douglas (1973) was one of the first to raise alarm at these developments. Like Okofor he argued that various forms of aboriculture were in fact the traditional agriculture of much of Africa until quite recent times. But this was abandoned under pressure from agriculture departments, agroentrepreneurs, incoming European and Asian settlers who brought with them a North American agro-ideology and technology. Sholto Douglas foresaw that this would prove futile in solving Africa's emerging food problem and predicted...*a food crisis of calamitous proportions before year 2000.* In much of sub-Sahara Africa that crisis arrived several decades before its due date: 40% of the population of sub Sahara are chronically malnourished or starving, but these generalities depend on the vagaries of rainfall (Energy, climate change, El Niño).

Matthias Igbozurike's (1971) similar prediction was based on the proposition that since some form of agriculture is obviously inevitable, and since...*all agriculture is antithetical to the natural order, then the most rational system would be that which most closely approximates nature.* He found varying elements of this ideal in three traditional mixed cropping systems of West Africa. The best of these is mixed tree crop farming (as on the compound farms of Nigeria, below) which...*bears a strong affinity to the forest from which it was carved.* Indeed he describes a rain forest as an ecosystem in which 'mixed cropping' appears in its most elaborate form. Igbozurike's standard by which the ecological desirability of farming systems can be measured is a model of the energy flows and trophic structure of a rain forest. Official modern agriculture departs from this ecological ideal and the traditional mixed systems which mimic it. The modern reality Igbozurike sketches is official indifference to underlying food production problems and overriding concern with cash monocropping for export markets. This land use is demonstrably unsound and...*it is not difficult to see that with a rapidly multiplying population and an agricultural system which is not ecologically stable...what is now a relatively mild nutritional problem will soon turn into a major disaster.* That was Igbozurike 30 years ago; how right he was; and 50 years ago Aubreville was witnessing the drying up of Africa due to these same land mismanagement causes (Shifters).

Tanzania

In Tanzania the Chagga forest gardens of Mt Kilimanjaro are the best known example of this general system in East Africa, although some critics see them as having degenerated in recent times into basically coffee-and-banana farms. They have been described by Fernandes and others (1984). They are at 900—1 900 m asl elevation, cover a total area of 100—120 000 hectares on the south and east slopes of Kilimanjaro and range in size from 0.2 to 1.2 hectares. They comprise the *vihamba* portions of these farms, the other portion consists of an outfarm of dryland fields (for bulk food crops) down on the plains. Chickens, goats, pigs and usually 2-3 cows are kept in association with the trees. The trees/crops consist of four or five layers or storeys:

Highest:	Level 5	(15+ meters)	. . .	Highgrowing valuable timber
	Level 4	(5 - 20m)	. . .	Timber, charcoal, lopped fodder
	Level 3	(2.5 - 5m)	. . .	Bananas, fruit, fodder trees
	Level 2	(1 - 2.5m)	. . .	Coffee, medicinal shrubs
Ground:	Level 1	(0 - 1m)	. . .	Taro, beans, medicinal/fodder shrubs, grass

There is much variation among farms, fusion between storeys, and most of the species have multiple uses. A total of 111 species were found on one sample of farms, including 53 trees. Primary tree uses are domestic fuel (90% of species), human and livestock medicinal use (30%), construction poles (25%), crop shade (24%), commercial timber (23%), fodder (10%). Uses of tree produce range from beer making to ablution sponges to cures for snakebite.

Other writers have praised further attributes of these Chagga systems: deliberate manipulation/pruning of the tree canopy to manage light for some understorey species (especially for coffee); optimization of plant spacing, especially of bananas; the intermixture of various species for their pest control properties; preservation of a large number of indigenous forest species and thus of local biodiversity. Also contributing to biodiversity are the wide range of cultivars/types kept for different specific purposes, eg 15 types of banana/plantains are kept respectively for fruit, vegetables, beer brewing, cattle fodder and crop mulch. Fernandes et al recommend these Chagga systems as models for establishing a more diversified and non-exploitive agriculture elsewhere in East Africa, but warn that the necessary sociocultural base might not be so easy to translocate. As with the *campesinos* of Santa Rosa (and General MacDowell) some people are keen seekers of plants, others not.

In NW Tanzania a sample of 72 quasi-forest gardens of Bukoba district were surveyed by Rugalema, Okting'ati, Johnsen (1994). These small farms of about 0.6 hectares consist of...*numerous woody species in intimate multistoreyed association, with herbs and annual and perennial food crops and livestock (cattle) all on the same piece of land.* They contain 32 ground level and 25 tree species. In addition to the tree houselots most households also have some *omusiri* outfields or cultivated plots. Again economically these systems are basically banana and coffee farms. However the 'tree' component is more diversified than these species numbers imply: there are over 100 Musa (banana) species/varieties in these villages and usually 20 different types on most individual farms. Nonetheless these systems are said to be breaking down, partly because of land fragmentization with increasing population, out-migration of young people, and also because concentration on coffee and 'bananas' has resulted in increasing incidence of their pests and diseases. These authors do not discuss diversification into other tree species as a strategy for pest/disease risk reduction. Lack of interest possibly illustrates the essentially conservative nature of African agriculture: except in rain forest zones (below) it invariably structured around two pillars: the mindless accumulation of cattle and what can be produced (by a woman) with a hoe.

Elsewhere in East and Southern Africa

Elsewhere in East and Southern Africa there are few forest gardens and little information. One must depend on observation. In colonial times there was in fact much effort directed at importing tropical tree crops but for the purpose of developing these as cash export crops, usually in monospecies stands. The forest gardens which emerged, based on some mix of these, did so largely by accident. In Kenya they exist in the Malindi-Mombasa coastal strip, structured around coconut, containing citrus, mango, banana, guava and undoubtedly much more; but no survey data are available. Elsewhere in Kenya there are other units of this general type around the shores of Lake Victoria. They extend north into Uganda although the predominant form of agriculture here is more organized small-scale plantation cash cropping (tea, coffee, plantains).

In NE Tanzania there are many around Tanga, containing the same species listed for Mombasa, again built around coconut and probably also cashew. They await exploration. South into Mozambique the situation is unknown. In Malawi there are none, but a few vestiges of rain forest remain on the otherwise cleared lands of the lakeshore north from Nkata Bay. This small inland area is unique in its high rainfall and humidity and good soils and there are rubber estates in the vicinity (since 1900), an indicator that tropical tree crops would probably thrive in spite of the altitude (1 600 m asl). But regardless of what agroforestry systems might have been in the distant past there is no knowledge of them now and no official interest in other than clean tilled field crops. (In 1982 a FAO mission recommended a tropical tree crop development project here, partly to counter the accelerated soil erosion which was occurring under corn, McConnell 1982.) This project would have established tree mixes similar to those of SE Asia; it was not proceeded with for the same reasons outlined by Okafor in Nigeria. Oddly at this elevation there are also coconuts around the lake though no one now knows what to do with them. There are also many humid tropical species on research stations, though these seldom succeed in escaping over the fence. (The Prince of those times was into export corn and cotton, marching to a different drum, shipping all farm inputs in and product out to pay for them, according to World Bank nations of agro development, and with Hesiod's 'latest scientific ideas from Greece' to guide this corn/cotton mono-dimensional agronomy – Origins.)

Nigeria

In Nigeria forest gardens have traditionally been a main farm system type in the low elevation, high rainfall, high humidity, rain forest or previously forested areas of East Nigeria in Anambra, Imo, Cross River and River states. Some also extend inland towards the drier savanna. They are one of five main farm system types in this part of West Africa: (a) 'shifting' of both households and crop swiddens; (b) shifting of crop lands only; (c) sedentary with short one or two year fallows; (d) intensive sedentary farming on terraced croplands; and (e) intensive sedentary with a significant forest garden tree component, ie forest gardens.

Okafor and Fernandes (1987) refer to the latter as *compound* farms of two main subtypes, with and without an outfarm of tilled fields. Most also have some livestock. These have also been discussed by Lagermann (1977). Sedentary field outfarms also contain some trees but these are fewer and of different species than on the home forest garden proper, of lesser economic importance or used less frequently. Continuity of food production is achieved partly by having a mix of tree species which mature at different times throughout the year, partly by seasonal scheduling the food field crops (yams, maize, cassava, millet). One main reason for this production scheduling is to avoid the necessity for post harvest storage. Throughout Africa post harvest food losses to rats, insects, fungus etc in mud/adobe village storage 'silos,' storage bins, usually range 30-40% of the harvested crop, frequently higher. (Such losses are often overlooked by agricultural planners who focus on high yielding varieties, more fertilizer, crop mechanization, irrigation etc without thought to how much of this extra output will be consumed by vermin.)

Considerable import substitution potential exists on these Nigerian farms but Okafor and Fernandes note this is not generally exploited by governments. The local species Monodoro myristica is a substitute for imported cloves (M. fragrans), Piper guinense for imported pepper (P. nigram), drying oil from Tetracarpidium canophorum for linseed oil... There is much potential for further species/crop enrichment, especially with tree/vine species from South and SE Asia; but even without these imports there are some 60 (mainly) tree indigenous species commonly present on these compound farms and a further 110 still in the forests thought to have economic potential if domesticated. These would both broaden the farm income base and support local agroprocessing industries (Nature, employment).

Perceptions These Nigerian forest farms have been in process of evolution as holistic agroecosystems for thousands of years (Origins). As with similar indigenous systems (Kandy, Java, Mexico, Tanzania, Kerala) they have been proposed on scientific grounds as models for highly productive sustainable land use elsewhere in the wet tropics (de Forester and Michon, 1993). But Okafor and Fernandes found them viewed officially as a primitive form of subsistence and this perception is said to be the major barrier to their improvement and geographical expansion. Indeed their continued existence is under threat as some owners flatten them with bulldozers in order to institute more advanced Western cash cropping. This seems regrettable; the rate of deforestation in Africa generally and of the rain forests in particular is such that these farms are becoming — as Michon says of Java and Rajapakse of Sri Lanka — a final species reservoir, a last genetic bank.

Ghana

In Ghana also forest gardens are widespread in the high rainfall zone. Asare (1985) and co-workers studied 40 of them in and around Kumasi municipality finding three main types: (a) extensive multistoreyed units with livestock; (b) intensive multistoreyed units; and (c) extensive home gardens with mixed ground level cropping. Type (a) closely resembles Asian forest gardens and consists of emergent coconut (to 18m) with avocado and breadfruit; a middle storey (to 12m) of oil palm and fruit trees; and a ground level mix of cocoyam, taro, pepper, okra, and sheep, goats, pigs, poultry are usually present under the canopy. As in parts of the Pacific (Thaman 1985) both the intensity/diversity and relative economic importance of these small units are increasingly associated with urban and semi-urban residential plots where, because of lack of employment opportunities and high food costs, households are obliged to try to maximize food self-sufficiency on whatever small plot they might possess. In Ghana this movement and home gardens of all types generally has been greatly stimulated by such programs as Operation Feed Yourself. Again, Asare notes, progress is hampered by lack of appropriate research and an extension service overly formal and commodity specific, not geared to consider farms as complex bioeconomic *systems*.

Cameroon

In NW Cameroon lands in Bamenda Valley are under rapidly increasing population pressure (3% increase annually). Prinze and Rauch (1987) have described one farm development program here aimed at finding a more sustainable land use system. This Bamenda model is not a forest garden *per se* but has a strong agroforestry component; 0.6 ha of these model six ha farms is devoted to a mix of tree crops, structured around the emergent oil palm in place of coconut, with raffia palms Raphia hookerii, albizia and eucalyptus, many fruit species, coffee, papaya, plantain as well as understorey food crops taro and pineapple. There are two other tree components: trees (unspecified) are scattered throughout the associated pasture sections and Musa spp and coffee are planted along the soil conservation bunds to achieve 'induced terracing' (Conservation). One of the planning objectives is to achieve a 'closed circuit' in farm energy flow.

Urban Farms, Poverty and Waste

In Sierra Leone Tricaud (1987) called attention to the potential and need for expansion of *urban* agriculture, specifically in Freetown (and Ibadan in Nigeria) but also in African cities generally. He surveyed the extent and nature of existing urban 'farms', their social and economic benefits, and constraints to this kind of agriculture on urban wastelands. Such farming is carried on as house gardens, as private food gardens utilizing temporarily or permanently unoccupied lands (vacant allotments, road and railroad rights-of-way) or as larger 'fields' of rice, maize, cassava in municipal swamps or on hillsides for which there is no other present use. Many of the farmers are the poorest of urban squatters, others street gangs of unemployed youths. While the general emphasis is on vegetable production for consumption/sale and smallstock (pigs/poultry) the garden types include units with up to 30 tree/palm/vine species on 1 000 sq m, but these found only on owned plots or where the gardeners can arrange security of tenure.

The private reasons for urban farming (food, income, self-employment) require no discussion but Tricaud's findings concerning official attitudes are worth considering. Local politicians being close to the people and dependent on votes of even the poorest slum dwellers are sympathetic. But attitudes of the city bureaucracy...*vary from outright rejection to tacit toleration with only a few who encourage it*. While the concerned ministries of agriculture, food supply, health are also supportive, Tricaud found the planning bureaucracy usually incapable of reconciling these agro-social needs with received concepts of how a modern city should appear. (In this they are not alone. Pigs, ducks, urine-inspired cabbages and other symbols of reality have long since been banished from the Yamaguchi city of Singapore.) In the wooded parklands of such other model cities as Lilongwe small agriculture can sometimes gain a presence by bribery, but it is always a furtive and nervous kind.

Among the wider social benefits which Tricaud identifies for these urban farms is their role as economic waste sinks and organic trash recycling units. In this, as with the forest garden farms of peri-urban Sri Lanka and Java, they are a least-cost alternative to municipal waste removal or — since these services are often not affordable in developing countries — to dumping it in the streams.

2.16, Summary

This brief review of forest gardens leaves many gaps; but sufficient examples have been offered to suggest that this general system, far from being an agricultural oddity, has a universality throughout the wet tropics probably unmatched by any other. From this basis it is possible to consider one of these systems in more detail, the forest garden farms of Kandy. But one must concede these systems, like Santa Rosa and Posey's forest resource islands, are thought of as belonging in the past. There is temptation to follow Tricaud back to the slums because this is an agriculture of the future, though usually an insecure and unconventional one. By 2000 about half the world's population were living in urbanized areas, to increase to two thirds within 25 years or an increase of 1.5 billion by 2025 (World Bank/UN data) with most of this increase to be in Asia and Africa, propelled from the countryside by farm amalgamation, mechanization, erosion of ecosystems (as Latin America's *favela* dwellers were) and attracted to the city's fringe as refuge, having no skills except as peasant farmers. That they will seize whatever scraps of urban wasteland they can find, as modern resource islands, is inevitable.

References

Abdoellah, O.S. (1985) Home gardens in Java and their future development. In K. Landauer and M. Brazil (eds) *Tropical Home Gardens.* UN University Press, Tokyo.

Aiken, S.R. and C.H. Leigh (1992) *Vanishing Rain forests: The Ecological Transition in Malaysia.* Clarendon, Oxford.

Alcorn, J.B. (1994) Development policy, forests, peasant farms: reflections on Huastec-managed forests. *Economic Botany* 38(4):389-406.

Alcorn, J.B. and C. Hernandez (1984) Plants of the Huastec region of Mexico. *Journal of Mayan Linguistics* 4:11-118.

Alvarez-Buylla Roces, M.E., E.L. Chavero, J.R. Garcia-Barrios (1989) Home gardens of a humid tropical region in southeast Mexico: An agroforestry cropping system in a recently established community. *Agroforestry Systems* 8:133-156.

Anderson, A.B. and E.M. Ioris (1992) The logic of extraction: resource management and income generation by extractive producers in the Amazon. In K.H. Redford and C. Padoch (eds) *Conservation of Tropical Forests: Working for Traditional Resource Use.* Columbia University Press.

Anderson, E. (1950) An Indian garden in Santa Lucia. *Ceiba* 1:97-103.

Anderson, E. (1954) *Man, Plants and Life.* Little Brown Co., Boston.

Asare, E.O., S. Oppong, K. Twum-Ampofo (1985) Home gardens in the humid tropics of Ghana. In K. Landauer and M. Brazil (eds) *Tropical Home Gardens.* UN University Press, Tokyo.

Ayres, W.S. and A.E. Haun (1983-1990) Prehistoric food production in Micronesia. In D.E. Yen and J.M. Mummery (eds) *Pacific Production Systems: Approaches to Economic Prehistory.* School Pacific Studies, ANU, Canberra.

Bavappa, K.V.A. and V.J. Jacob (1981) *A model for mixed cropping.* Ceres, May-June. FAO.

Bayliss-Smith, T. (1982) *Ecology of Agricultural Systems.* Cambridge University Press, UK.

Bayliss-Smith, T. (1986) Ontong Java Atoll: Population, Economy and Society, 1970-85. South Pacific Smallholder Project No. 9. University New England, Armidale, NSW.

Beccari, O. (1904-1986) *Wanderings in the Great Forests of Borneo.* Archibold Constable, London (1904), Oxford (1986).

Boonkird, S.A., E.C.M. Fernandes, P.K.K. Nair (1984) Forest villages: an agroforestry approach to rehabilitating forest land degraded by shifting cultivation in Thailand. *Agroforestry Systems* 2:87-102.

Bourke, R.M. (1983-90) Subsistence food production systems in Papua New Guinea: old changes and new changes. In D.E. Yen and J.M. Mummery (eds) *Pacific Production Systems: Approaches to Economic Prehistory.* School of Pacific Studies, ANU, Canberra.

Brierley, J. (1976) Kitchen Gardens in the West Indies with a contemporary study from Grenada. *Journal of Tropical Geography* 43:30-40.

Budowski, G. (1985) *Home gardens in Tropical America.* First International Workshop on Tropical Home gardens. Bandung, Indonesia.

Budowski, G. (1985) Home gardens in Tropical America. In K. Landauer and M. Brazil (eds) *Tropical Home Gardens.* UN University Press, Tokyo.

Chandran, S. and M. Gadgil (1993) State forestry and the decline of food resources in the tropical forests of Uttara Kanada, South India. *Tropical Forests, People and Food. Man and the Biosphere. MAB* (13). UNESCO.

Charon, C.M. (1995) Role of non-timber tree products in household food procurement strategies: profile of a Sri Lankan village. *Agroforestry Systems* 32:99-117.

Christanty, L. (1985) Home gardens in Asia with special reference to Indonesia. *Tropical Home Gardens.* UN University Press, Tokyo.

Christanty, L., O. Abdoella, G. Marten, J. Iskandar (1986) Traditional agroforestry in West Java: the pekarangan and kebun-talun cropping systems. In G.G. Marten (ed) *Traditional Agriculture in Southeast Asia, A Human Ecology Perspective.* Westview Press, Boulder, Colorado.

Clarke, W.C. and R.R. Thaman (1992) *Agroforestry in the Pacific Islands: Systems for Sustainability.* UN University Press, Tokyo/NY.

Clement, C.R. (1989) The potential use of pejibaye palm in agroforestry systems. *Agroforestry Systems* 7:201-212.

Dagar, J. and V. Kumar (1992) Agroforestry for the Bay Islands. *Indian Forester* June 1992.

Daw, M.E. (1991) *National Perspective of Pekarangan — Systems, Performance and Constraints.* FAO-AGSP, Rome.

De Forester, H. and G. Michon (1993) Creation and management of rural agroforests in Indonesia: potential applications for Africa. *Tropical Forests, People and Food. Man and Biosphere, MAB* (13). UNESCO.

De Jong, W. (1990) Recreating the forest: Successful examples of ethnoconservation among Dyak Groups in West Kalimantan. In O. Sandbukt (ed) *Management of Tropical Forests: Towards an Integrated Perspective.* University of Oslo.

Dury, S., L. Vilcosqui, F. Mary (1996) Durian trees (D. zibethinus) in Javanese home gardens: their importance in informal financial systems. *Agroforestry Systems* 33:215-230.

Ellen, R. (1988) Foraging, starch extraction and the sedentary lifestyle in the lowland rain forests of central seram. In T. Ingold, D. Riches, J. Woodburn (eds) *Hunters and Gatherers 1: History, Evolution and Social Change.* Berg, NY/Oxford.

Emperaire, L. and F. Pinton (1993) Ecological and socioeconomic aspects of extractivism on the Middle Rio Negro. In *Tropical Forests, People and Food, Man and the Biosphere, MAB* (3). UNESCO.

Fairbaine, I. (1971) Pacific Island Economies, *Journal of the Polynesian Society* 80(1):74-118.

Falanruw, M. (1985) Food Production of the Yap Islands. In K. Landauer and M. Brazil (eds) *Tropical Home Gardens.* UN University Press, Tokyo.

Felemi, I.M.N. (1991) *Evaluation of the Banana Export Scheme in Tonga, South Pacific Smallholder Project.* University of New England, Armidale, NSW.

Fernandes, E.C.M., A. Okting'ati, J. Maghembe (1984) The Chagga home gardens: a multistoreyed agroforestry cropping system on Mt. Kilimanjaro, Tanzania. *Agroforestry Systems* 2:73-86.

Flietner, D. (1985) *Don Ignacio's Home Garden.* Organization for Tropical Studies, Costa Rica.

Frazer, I. (1987) *Growth and Change in Village Agriculture: Manakwai, North Malaita (Solomons), South Pacific Smallholder Project.* University of New England, Armidale, NSW.

Fujisaka, S. and E. Wollenberg (1991) From forest to agroforest and logger to agroforester. *Agroforestry Systems* 14:113-129.

Gillespie, A.R., D.M. Knudson, F. Geilfus (1993) The Structure of four home gardens in Petan, Guatemala. *Agroforestry Systems* 24:157-170.

Gliessman, S.R. (1990) (ed) *Agroecology.* Springer-Verlag.

Gliessman, S.R., E. Garcia, M. Amadora (1981) Ecological basis for the application of traditional agricultural technology in the management of tropical agro ecosystems. *Agro Ecosystems* 7:173-185.

Godoy, R.A. and C.F. Tan (1991) Agricultural diversification among the smallholder rattan cultivators in Central Kalimantan. *Agroforestry Systems* 13:27-40.

Grove, A.T. (1986) State of Africa in the 1980s. *The Geographical Review* 152(2):193-203.

Guhardja, E. (1985) *Home-Gardening Activities at the Institute Pertanian Bogor.* Asian Vegetable Research Development Workshop, Bangkok.

Hall, D. (1969) *Free Jamaica 1838-1865.* Caribbean University Press, London.

Hocking, D. and K. Islam (1994) Trees in Bangladesh padi fields and homesteads. *Agroforestry Systems* 25:193-216.

Hocking, D., A. Hocking, K. Islam (1996) Trees on farms in Bangladesh. *Agroforestry Systems* 33:231-247.

Holden, S. and H. Hvoslef (1990) Transmigration settlements in Seberida: causes and consequences of the deterioration of farming systems in a rain forest environment. In

O. Sandbukt (ed) *Management of Tropical Forests: Towards an Integrated Perspective.* University of Oslo.

Horowitz, M. (1967) *Morne-Paysan: Peasant-Village in Martinique. Case Studies in Cultural Anthropology.* Holt, Rhinehart, Winston, New York.

Hutterer, K.L. (1984) Ecology and Evolution of Agriculture in SE Asia. In T.A. Rambo and P.E. Sajise (eds) *Introduction to Human Ecology Research on Agricultural Systems in South East Asia.* University of the Philippines Press, Los Banõs.

Innis, D.Q. (1961) The efficiency of Jamaican peasant land use. *Canadian Geographer* 5:19-23.

Isbozurike, M. (1971) Ecological balance of tropical agriculture. *Geographical Review* 61:518-529.

Jacob, V.J. and W.S. Alles (1987) The Kandyan gardens of Sri Lanka. *Agroforestry Systems* 5:123-127.

Jacobs, M. (1982) *The Tropical Rain forest: A First Encounter.* Springer-Verlag, Berlin.

Jambulingam, R. and E.C.M. Fernandes (1986) Multi-purpose trees and shrubs on farmlands in Tamil Nadu. *Agroforestry Systems* 4:17-23.

Jensen, M. (1993) Soil conditions, vegetation structure and biomass of a Javanese home garden. *Agroforestry Systems* 24:171-186.

Jensen, M. (1993a) Productivity and nutrient cycling of a Javanese home garden. *Agroforestry Systems* 24:187-201.

Jose, D. and N. Shanmugaratnam (1993) Traditional home gardens in Kerala: a sustainable human ecosystem. *Agroforestry Systems* 24:203-213.

Karyono (1981) *Structure of Home Gardens in the Rural Areas of Citirum Watershed.* Padjadjaran University, Bandung.

Kimber, C. (1966) Dooryard gardens of Martinique. *Yearbook of Pacific Coast Geographers* 28:97-118.

Kimber, C. (1973) Spatial patterning in the dooryard gardens of Puerto Rico. *Geographical Review* 63 (1):6-26.

Krishnarajah, P. (1982) *Soil Erosion and Conservation in the Upper Mahaweli Watershed.* Joachim Memorial Lecture Soil Science Society, Sri Lanka (Nov 10, 1982).

Kumar, B.M., S.J. George, S. Chinnamani (1994) Diversity, structure and standing stock of wood in the home gardens of Kerala in Peninsula India. *Agroforestry Systems* 25(3) 243-262.

Kunzel, W. (1989) *Agroforestry in Tonga.* South Pacific Smallholder Project No. 12. University of New England, Armidale, NSW.

Lagermann, J. (1977) *Traditional Farming Systems in Eastern Nigeria.* Weltforum-Verlag, Munich.

Lathrap, D. (1977) Our father the cayman, our mother the gourd: A unitary model for the emergence of agriculture in the New World. In C. Reed (ed) *Origins of Agriculture.* Mouton, The Hague.

Leach, H. (1984) *A Thousand Years of Gardening in New Zealand.* Reed and Reed, Wellington.

Lescure, J.P. and F. Pinton (1993) Extractivism: a controversial use of the tropical ecosystem. *Tropical Forests, People and Food, Man and the Biosphere MAB* (13). UNESCO.

Leuschner, W.A., and K. Khaleque (1989) Homestead agroforestry in Bangladesh. In P.K.K. Nair (ed) *Agroforestry Systems in the Tropics.* Kluwer Academic Publishers, Dordrecht, Neth.

McConnell, D.J. (1972) *Economic Analysis of the Samoa Banana Industry.* FAO Apia-Rome.

McConnell, D.J. (1986) *Economics of Some Tropical Crops – Trees and Vines* (Nipa palm for atap production in Johor; Sago palm in Johor; Bamboo in the Madiun Valley; Pandan in the Trenggalek area East Java). Midcoast, Coffs Harbour, Australia.

McConnell, D.J. (1989) *Project Proposal for Developing the Pekarangan Lands of Indonesia.* FAO-TCP Government of Indonesia Ministry of Plantation Industries.

McConnell, D.J. (1992) *The Forest Garden Farms of Kandy.* FAO, Rome.

McConnell, D.J. and K.A.E. Dharmapala (1973) *Small Forest Garden Farms in the Kandy District of Sri Lanka.* Farm Management Diversification Report No. 7. FAO-UNDP Perideniya, Sri Lanka.

McConnell, D.J. and W.J. Surjanto, Abu Yamin Nasir (1974) *Farming Systems and Land Use along a Traverse of the Solo Valley.* UNDP-FAO-AGL, Land Capability Appraisal Project, Bogor.

McConnell, D.J. and P.M. White, D. McCourtie (1982) *Diversification of Smallholder Agriculture in Malawi.* FAO-TCP Rome/Lilongwe.

Mendieta, R.M. and S. Del Amo (1981) *Plantas Medicinales del Estato de Yacatan.* INIRI-CECSA, Mexico.

Mergen, F. (1987) Research opportunities to improve the production of home gardens. *Agroforestry Systems* 5:57-67.

Michon, G. (1983) Village forest gardens in West Java. In P.A. Huxley (ed) *Plant Research and Agroforestry.* ICRAF, Nairobi.

Michon, G. and J. Bompard, P. Hucketsweiler, C. Ducatillion (1983) Tropical forest architectural analysis as applied to agroforests in the humid tropics: traditional village agroforests in West Java. *Agroforestry Systems* 1:117-129.

Michon, G. and F. Mary, J. Bompard (1986) Multistoreyed agroforestry systems in West Sumatra. *Agroforestry Systems* 4:315-338.

Michon, G. and F. Mary (1994) Conversion of traditional village gardens and new economic strategies of rural households in the area of Bogor, Indonesia. *Agroforestry Systems* 25:31-58.

Nair, M.A. and Sreedharan (1986) Agroforestry systems in the home gardens of Kerala. *Agroforestry Systems* 4:339-362.

Nair, P.K.R. (1993) *Introduction to Agroforestry.* Kluwer Academic Publishers/ICRAF, Dordrecht, Netherlands.

Ninez, V. (1985) Garden production in tropical America. *Tropical Home Gardens.* UN University Press, Tokyo.

Nunn, P.D. (1990) Recent environmental changes on Pacific islands. *The Geographical Journal* 156(2):125-137.

Okafor, J.C. and E.C.M. Fernandes (1987) The compound farms of Southeast Nigeria: a predominant agroforestry home garden system with crops and small livestock. *Agroforestry Systems* 5:153-168.

Orejuela, J.E. (1992) Traditonal productive systems of the Awa Indians of Southwestern Colombia and Ecuador. In K.H. Redford and C. Padoch (eds) *Conservation of Neotropical Forests: Working from Traditional Resource Use.* Columbia U. Press.

Padoch, C. and W. De Jong (1987) Traditional agroforestry practices of native and ribereño farmers in the lowland Peruvian Amazon. In H.L. Sholz (ed) *Agroforestry: Realities, Possibilities and Potentials.* Nijhoff/ICRAF Dordrecht, Netherlands.

Padoch, C. and W. De Jong (1991) The house gardens of Santa Rosa: diversity and variability in an Amazonian agricultural system. *Economic Botany* 45:166-175.

Padoch, C., J. Chota Inuma, W. De Jong, J. Unruh (1985) Amazonian agroforestry: a market-oriented system in Peru. *Agroforestry Systems* 3:47-58.

Pei, Sheng-ji (1985) Effects of the Dai Peoples cultural beliefs and practices upon the plant environment of Xishuanggbanna, Yunnan, Southwest China. In K.L. Hutterer, A.T. Rambo and G. Lovelace (eds) *Cultural Values and Human Ecology in SE Asia.* University of Michigan.

Pereira Dos Santos, H. and J.P Lescure (1993) Extractivism and agriculture: the choice of one community on the Rio Solimoes (Brazil). *Man and the Biosphere, MAB (*13). UNESCO.

Perera, A.H. (1988) *Role of Kandyan Forest Gardens in Germplasm Conservation.* Commonwealth Science Council UK – Natural Resources, Environment and Science Authority of Sri Lanka, Conference Proceedings, Paradeniya, Sri Lanka, May 1988.

Perera, A.H. and R.M.N. Rajapakse (1991) *A Baseline Study of Kandy Forest Gardens of Sri Lanka: Structure, Composition and Utilization.* Dept Crop Science, University of Peradeniya, Sri Lanka.

Pillot, D. (1980). Uses, species and cultural techniques of the dynamic culture of Haitian systems. *Journal of Traditional Agriculture and Applied Botany* 27(3-4):203-219.

Posey, D.A. (1985) Indigenous management of tropical forest ecosystems: the case of the Kayapo Indians in the Brazilian Amazon. *Agroforestry Systems* 3:138-158.

Primack, R.B. (1993) *Essentials of Conservation Biology.* Sinauer, Mass.

Prinz, D. and R. Rauch (1987) The Bamenda Model: development of a sustainable land use system in the Highlands of West Cameroon. *Agroforestry Systems* 5:463-474.

Probowo, D. and D.J. McConnell (1993) *Changes and Development in Solo Valley Farming Systems.* FAO-AGSP, Rome.

Raintree, J.B. and K. Warner (1986) Agroforestry pathways for the intensification of shifting agriculture. *Agroforestry Systems* 4:39-54.

Raynor, W.C. and J.H. Fownes (1991) Indigenous agroforestry of Pohnpei. *Agroforestry Systems* (16):139-157.

Raynor, W.C. and J.H. Fownes (1991a) Spatial and successional vegetation patterns. *Agroforestry Systems* (16):156-159.

Rico-Gray, V. and A. Chemas, S. Mandujano (1991) Use of tropical deciduous forest species by the Yucatan Maya. *Agroforestry Systems* 4:149-161.

Rico-Gray, V. and A. Gomez-Pompa, C. Chan (1985) Las Selvas Manejades por los Mayas de Yohalton, Campeche. *Biotica* 10:321-327.

Rico-Gray, V. and J.G. Garcia-Franco, A. Chemas, A. Puch, P. Sima (1990) Species composition, similarity and structure of Mayan home gardens in Tixpeual and Tixcucaltuyub, Yucatan. *Economic Botany* 44:(4).

Rugalema, S.H., A. Okting'ati, E.H. Johnsen (1994) The home garden agroforestry system of Bukoba District of NW Tanzania. *Agroforestry Systems* 26:53-64.

Saint-Pierre, C. (1991) Evolution of agroforestry in the Xishuangbanna Region of Tropical China. *Agroforestry Systems* 13:159-176.

Salafsky, N. (1994) Forest Gardens in the Gunung Palung Region of West Kalimantan: Defining a Locally Developed Market-oriented Agroforestry System. *Agroforestry Systems* 28:237-268.

Schultz, B., B. Becker, E. Gotsch (1994) Indigenous knowledge in a 'modern' sustainable agroforestry system — a case study from Eastern Brazil. *Agroforestry Systems* 25:59-69.

Senanayake, F.R. (1987) Analog forestry as a conservation tool. *Tiger Paper* 15:25-28.

Sholto Douglas, J. (1973) Forest Farming: an ecological approach to increasing nature's food productivity. *Impact of Science on Society* 23:17-32.

Soemarwoto, O. (1985) Constancy and change in agroecosystems. In K.L. Hutterer, A.T. Rambo, G. Lovelace (eds) *Cultural Values and Human Ecology in Southeast Asia*, Center for South and Southeast Asian Studies No 27. University of Michigan.

126   *The Forest Farms of Kandy – and other gardens of complete design*

Soemarwoto, O. and G.R. Conway (1991) The Javanese home garden. *Journal of Farming Systems, Research and Extension* 2(3):95-117.

Sommers, P. (1985) Mixed home gardening in the Pacific Islands: present status and future prospects. *Tropical Home Gardens.* UN University Press, Tokyo.

Souter, G. (1963) *New Guinea: The Last Unknown.* Angus and Robertson, Sydney.

Stavrakis, O. (1978) Ancient Maya agriculture and future development. *Culture and Agriculture* 5:1-8.

Tejwani, K.G. (1987) Agroforestry practices and research in India. In H. Gholz (ed) *Agroforestry: Realities, Possibilities and Potentials.* Martinus Nijhoff/ICRAF (Nairobi).

Terra, G.T.A. (1954) Mixed garden horticulture in Java. *Malaysian Journal of Tropical Geography* 4:33-43.

Thaman, R.R. (1983) *Food and National Development in the Pacific Islands.* University of South Pacific, Suva.

Thaman, R.R. (1985) Mixed home gardening in the Pacific Islands: status and prospects. *Tropical Home Gardens.* UN University Press, Tokyo.

Thien, I.B., A.S. Bradburn, A.L. Welden (1982) *The Woody Vegetation of Dzibilchaltun, a Maya Archaelogical Site in Northeast Yucatan.* Mari Paper 5. Tulane University, New Orleans.

Tickell, C. (1993) The human species: a suicidal success? *The Geographical Journal* 159(2):219-226.

Tricaud, P.M. (1989) *Urban Agriculture in Ibadan and Freetown.* UN University, Tokyo/Paris.

Wagachchi, H.R. and K.F. Wiersum (1997) Water management in agro-forestry systems: integrated buffalo ponds and forest gardens in Badulla District, Sri Lanka. *Agroforestry Systems* 35:291-302.

Wallace, A.R. (1869) *The Malay Archipelago: Land of the Orang-Hutan and Bird of Paradise.* Oxford University Press (1986).

Weirsum, F. (1982) Tree gardening and taungya on Java: examples of agroforestry techniques in the humid tropics. *Agroforestry Systems* 1:53-70.

Whistler, W.A. (1987) Ethnobotany of Tokelau: the plants, their names and uses. *Economic Botany* 42.

White, O. (1965) *Parliament of a Thousand Tribes: A Study of New Guinea.* Heinemann, UK.

Wilken, G. (1977) Integrating forest, and small scale farm systems in Middle America. *Agroforestry Systems* 3:291-303.

Xiong, W.H., S. Yang, Q. Tao (1995) Historial development of agroforestry in China. *Agroforestry Systems* 1995:277-287.

Yen, D.E. (1974) Aboriculture in the subsistence of Santa Cruz, Solomon Islands. *Economic Botany* 28:247-284.

Yen, D.E. (1983-90) Environment, culture and the colonization of the Pacific. In D.E. Yen and J.M. Mummery (eds) *Pacific Production Systems: Approaches to Economic Prehistory.* Department of Prehistory, School of Pacific Studies. Australian National University, Canberra.

Chapter 3
Kandy

with K.A.E. Dharmapala, G.K. Upawansa, S.R. Attanayake and Edmund Spencer

But were it not that Time their troubler is,
All that in this delightful Garden grows
Should happy be and have immortal bliss;
For here all plenty and all pleasure flows
*Without fell rancor or fond jealousy...*Spencer

Yes yes this is all very well, Edmund, but unless we also establish the economic credentials of the Garden these sentiments, and their other efficiencies according to Table 1.2 of Chapter 1, might count for little.

3.1   Background

The true origins of forest gardens in Sri Lanka will probably never be known.   From archaeological evidence in cave middens de Lanerolle (1985) puts the origins of earliest signs of 'agriculture' in Sri Lanka at 9-10 000 BC.   Significantly these findings have been in the forested hill areas, not on the plains;   and of equal significance they are tree food remains: the seeds of wild breadfruit, wild plantain and various nuts (Genesis).   It's an open question whether these first people the Yakkas and their successors the Veddhas were actually aboriculturists as well as hunter-gatherers (they were never cultivators).   But since they occupied the land for at least 10 000 years before the arrival of the modern Singhalese people as Indo-Aryan tribes from North India c 600 BC there was ample time for these prior Yakkas-Veddhas to have developed some form of aboriculture, at least of forest enrichment, though this is unproven and it would not have been as sophisticated as eg Posey was to find in the tree resource islands of Amazonia. (There are now only about 2-3 000 aboriginal Veddhas left.)

When the Singhalese ancestors arrived on this early Great Frontier of Elankai, later Lanka, before the time of Guatama the Buddha (d. 483 BC), they established themselves on the dry plains of the north central part of the island around what was to become their capital of Anuradhapura. Alexander's Greeks mapped what they saw as Thambapani later Taprobane, but didn't linger;   they had more ambitious Nation Building further up the Grand Trunk Road.   The ancestors developed a great agro-civilization based on irrigated rice, initially by ponding small monsoonal streams, these to develop into village 'tanks' then eventually into one of the world's great state planned irrigation systems, with networks of supply canals and even underground tunnels supplied by reservoirs...*so large the ancients called them Seas.* Incursions by the Tamil kingdoms of South India intensified over the centuries, culminating in the Singhalese shifting their capital further east to Polonnaruwa (c 1 070 AD) where a second great rice civilization was established, again based on massive public irrigation works financed by a rice tax.

This era of de Lanerolle's Ten Kings was terminated by continuing South Indian invasions, inter-marriage with Singhalese royalty, factional claimants for the throne. What followed is a common tale of foolishness of kings — intrigue, ambition, war, neglect of the irrigation works (sometimes their deliberate destruction), shrinking crop area and public tax base, siltation of the canals, stagnant water, mosquitoes, malaria, exodus...*And the people drifted off south and west away from the open spaces* (de Lanerolle). Or as Cook (1951) describes it...*The people were driven by degrees into the hill country and to the more moist and less healthy parts of the lowlands. And their morale was almost lost.* What they did there — their impact on the Great Forest of Sri Lanka's central hills and Wet zone, what it did to them — are almost blank pages. About all that's known with certainty is that the refugee Princes and the always contending princelings paused over the next four centuries to raise up other cities — Dambadeniya, Kurunegala, Gampola, Kotte and finally Kandy — but none so great as the old ones in the rice lands of the plain.

The scope of this new agriculture also expanded greatly. The forested hill country is a maze of small valleys in which irrigated rice cultivation was possible but only on a small and fragmented scale; and in these high-rainfall valleys there is neither scope nor need for 'great seas' as irrigation basis. Other forest species like cinnamon, cardamom and pepper and some of the palms were found or imported, and export trade in these through Arab merchants became the royal monopoly and main support of the reduced royal courts, in place of rice alone.

Renaissance Europe found the kingdom of Cotta/Kotte when Lorenzo de Almeida dropped anchor in 1505 and by 1619 all of the island, their *Celiao,* was in Portuguese hands, except the hill kingdom of Kandy which they never did subdue. The Portuguese are not remembered as farmers; but well before their ouster by the Dutch (in alliance with Kandyans, 1658) and through their manipulation of Singhalese feudalism, they had worked up what was probably the modern world's most comprehensive trade in extractive exotica (cardamom, gem stones, areca nut, pepper, ebony, pearls, leopard skins, ivory, live elephants, cinnamon).

The next wave of Nation Builders, the Dutch (1658-1796), were more serious. There are records of them introducing a few crops like coffee, and organizing the production of pepper and cinnamon on commercial plantations instead of simply trading for them with Singhalese farmers or forest collectors as in previous times. No doubt they introduced much more than coffee into their Zeilan (De Silva and Beumer 1988); but just *what* economic crops were present at the end of the 18th century — the indigenous stock plus imports by Tamils to their Elankai/Lanka, by Malay sea prowlers to the south, by Arabs to their Serendib – Resplendent Isle – by Portuguese to their Celiao, by Dutch to their Zeilan — is not too clear. Cordiner's (1807) advice probably still stands...*Any person desirous to be informed of the nature of the plants of Ceylon will find the greater part of them in the copious catalogue of Burman, Amsterdam, 1737, Thesaurus Zeylanicus Joannis Burmanni.*

. With acquisition by the British East India Company of their new asset, Ceylon in 1796, this uncertainty became irrelevant. John Company made a mess of it. According to history it 'so mistreated' the Ceylonese that their new property was quickly taken from them and governed directly as a British crown colony from 1802, the Kandyans eventually subdued in 1815. Now one begins to meet people like General MacDowell who before he left Colombo in 1804...*was in the habit of receiving boxes of trees and shrubs by almost every ship and in his garden an acre and a half was completely filled with peaches, apples, loquat, leechee and wampee. Melicocca of the West Indies thrives remarkably well and his mangosteen, nutmeg, clove, pimento, sapota, star apple will bear amply when of sufficient age* (Rev. James Cordiner 1807). Like the campesinos of Santa Rosa the general was an active seeker of plants. But Rev. Cordiner was no more impressed by the New Crops Development Program of the British than by those of the Portuguese and Dutch, in spite of these latter having been there 250 years...*Not one of the above crops had ever been brought into the island previous to the General's arrival, and it was his wish to have introduced a large quantity of nutmegs and cloves, but the scheme did not meet the approbation of Government.* Then one last sly jab at the old enemy...*Nutmegs were raised in considerable quantities many years ago under the Dutch, but the plants were afterward destroyed by order of the Government of Batavia.* Cordiner also suspected the China Tea might perhaps do well...

In a way the modern history of commercial agriculture in Sri Lanka, and the modern relevance of forest gardens, may be dated from 1869 when an outbreak of Hemilea vastatrix devastated the increasing number of colonial coffee estates which by then had been established, chiefly in the forested Kandy hills. Coffee was pulled out and replaced by tea on what was for those times a grand scale and at dramatic rate. (A little tea had been grown previously.) Within only six years 1 000 acres of this adjustment crop were in, then more...and more. The great tea boom was on. Within a few decades the tea fields covered the Kandy mid country (elevation c 400—1 000 m), then the high country and a good part of the low country too. And what wasn't occupied by tea was being cleared for rubber, cacao, cinchona... The necessary condition was clearing of the rain forests, also on grand scale (by elephants and cheap imported Tamil labor) that probably wasn't matched again in Asia until the contemporary Felda clearing-settlement schemes of Malaysia. There are also echoes here of Brazil's early land clearing for sugar; forest not taken for the monocrops was earmarked for the estates' boilers. This furious burst of agribusiness was to have three sets of contemporary consequences. First, well before 1960s 200 000 ha (500 000 acres) were under the tea monocrop in unbroken vista. It accounted for 15% of gross national product, 60% of all export earnings and provided 600 000 jobs or 16% of all national employment. In some years it returned 60% on tea estate investment. Tea and to lesser extent monocrop rubber and coconut were a huge economic success.

But second, this apparent success of collectively the estate sector was flawed. The national economy was over-exposed to fluctuations in the terms of trade (cost of imports relative to tea prices). The huge output of tea, second only to India in production and exports, was at the expense of food production. The sophisticated and admirable research on which this and other monocrops were based proceded with disinterest in the food or village or the non-monetized 'peoples' crops. The research ideology was as alien as the agro-technology itself. The resulting prosperity actually meant prosperity of the estates, their absentee European shareholders and the local mercantile class. In the political rhetoric of the 1970s these estates were easily and accurately denounced as 'islands of privilege and prosperity in a sea of peasant poverty'. The '600 000 jobs' actually meant high salaried positions for the foreign managerial class and minimum wages for the mass of field labor, most women and children (the bulk of the 600 000 jobs), often existing under appalling socio-economic conditions but kept docile by absence of alternatives and latent hostility of the rural Singhalese. By and large the local villagers declined to participate in these agribusiness adventures, preferring their small paddy fields and forest gardens.

Third, from an agroecological viewpoint, it was also apparent by the

1960s that large areas of this monocrop tea system were breaking down under the impact of accumulated soil erosion, especially in the steeper mid country and catchment of the country's major river system. So while the estate companies' books had run black with profits for almost a century this was paid for in the end by the Mahaveli Ganga running red with silt and top soil. In its downstream impact this soil erosion was especially alarming. Continuing population growth on the one hand and concentration of agricultural policy on export cash crops on the other meant that the country was no longer able to feed itself except by (a) ever greater cash crop exports for foreign exchange to import rice and flour, or (b) resuscitation of the ancient northern dry zone irrigation systems of the Ten Kings. But rebuilding these the way they had been was no longer practicable; the modern engineering alternative was to divert the Mahaveli Ganga to the northern rice lands, but this depended partly in improving the condition of the river, specifically in minimizing its silt load and reducing flood frequency through stabilizing hydrological condition of the catchment (most under tea).

These various economic, socio-political and engineering consequences of monocrop agribusiness converged in 1968 in the report of a commission of enquiry into the tea industry which *inter alia* recommended that 100 000 acres of tea and a smaller area of rubber be replaced by other more benign and socially acceptable forms of land use. It took about five years to decide just *what* these superior agroecosystems were, a story which can't be told here except that the best systems to meet these modern needs seemed to be the village forest gardens. These had been evolving in the Kandy hills — no more than tolerated as agro-technical anachronism — since those far times when the people, disheartened, had drifted away from the rice plains and broken promises of King Parakrama's Great Seas.

3.2 Forest Gardens

In Sri Lanka the forest gardens occur as a more-or-less continuous belt around the wet zone (Figure 3.1, map), wherever the original rain forest was not cleared for tea and rubber, elevation 200 – 1 000 meters. They also occur more generally in high rainfall areas, as agroforest village houselots, where the farm lands proper are devoted to pure stands of coconut, rice, cinnamon, citronella grass. Data discussed below are from several farm surveys through the 1970s and 1980s (References). Prices/costs will have changed but the farm structures and agrotechnical data hardly at all.

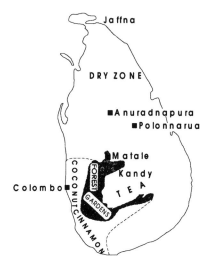

Figure 3.1 Sri Lanka Forest Gardens Zone

3.3 Botanical Composition, Crops

Those farms toward the drier north (Matale) tend to have more kapok (often with pepper vines attached); those on the high rainfall western escarpment have more palms and rubber (Kegalle). Over the years programs of 'village expansion' have taken in some estate tea land and where this has occurred the smallholders have inserted their tree mix into the matrix of an old tea field. This explains the presence of tea as an undercrop on the surveyed farms: otherwise it is seldom planted as a forest garden crop.

Species, crops   Main economic species on the surveyed farms are summarized in Table 3.1. Since the purpose of the surveys was to quantify the financial performance of these systems in comparison with tea and other plantation monocrops Table 3.1 does not include medicinals, ornamentals, timbers. However in a later survey of similar farms in the Kandy area Perera and Rajapakse (1991) compiled a list of all the main species present (economic and 'non-economic') and found 61 trees, 42 herbs, 12 shrubs and 10 major climbers and listed their uses (Annex 1).

Plant populations   The populations (Table 3.1) considerably understate the actual numbers present. Only plants in or nearing a production state are included. At the other end of the age scale those trees which had passed their productive phase are also not included; and except where the understorey plants are sufficiently numerous or important to be called a 'crop' these also are omitted. Considering these classes the total populations present of all significant plants would be up to twice the populations of Table 3.1 and as summarized in the distribution of these 30 farms shown in Table 3.2, and this also omits the common base crop tea (600—1 000 plants/acre). On some farms, as in a rain forest, there also dozens or hundreds of volunteer saplings from seedfall growing up under the canopy which will later be thinned out (but sometimes not, in which case these parts of the farm are best described as unproductive thicket). The big old senile trees and palms are often left as a standing store of fuel, timber and cash value depending on species, but felled if they pose a danger. As defined the modal population is 101-200 trees/palms/vines per acre (about 245-490/ha), compares with 400-600 *trees* of significant size in a rain forest.

Table 3.1   Distribution of 30 Farms by Plant Populations per Acre (0.40 ha)

No. trees/palms/vines	No. of farms	No. trees/palms/vines	No. of farms
below 50	4	301 - 400	3
51 - 100	6	601 - 700	2
101 - 200	12		30
201 – 300	3	187 ...	mean per acre

Plant populations and farm size   Lowest populations (less than 50/acre) are on four of the largest farms and the highest populations are on two of the smallest farms (10 and 24).   These latter are in excess of rain forest populations but accounted for mainly by smaller coffee trees and palms. This general inverse relationship between farm size and tree population is shown in Figure 3.2 and has been found in several other studies elsewhere (Reconnaissance).

Agronomic basis for high populations   On these farms one reason for high population levels relates to the presence of coffee, eg the highest population (and smallest) farms 10 and 24.   To maximize profits from these small one acre (0.4 ha) holdings their households have concentrated on coffee, this requires shade and microclimate, this is provided by inserting palms in gaps between the coffee, these require little ground space and so this coffee + areca + coconut + papaya combination (on farm 10) reaches an extremely high total plant population level (687 or 1 680 / ha *large* plants), as does

Table 3.2 Summary of Crops on Surveyed Farms

Farm no	(1)	(2)	(3)	(4)	(5)	(6)	(7)	(8)	(9)	(10)	(11)	(12)	(13)	(14)
Farm size (acres)	5.00	1 25	5.00	3.00	5.00	2.50	5.00	2.00	2.00	1.00	4.00	4 50	1.75	2.00
No. of crops	12	7	8	4	15	9	18	11	13	14	14	14	17	12
Trees, palms, vines on farm (1)	216	229	450	440	202	162	473	369	112	687	1 053	763	607	748
Trees, palms and vines per acre (1)	43	183	90	147	40	65	95	185	56	687	263	170	347	374
Areca (palms)	4	20		20	140	15	75	50	25	50	150	300	150	30
Jak (trees)	5	8	5	8		20	10	10	25	6	15	20	8	
Coffee (trees)		100	75		15	10	20		30	500	250	2	40	100
Pepper (vines)	15	77	250	400		50	50	200	6	15	200	200	150	60
Coconut (palms)	150		18		10	10	150	30	10	50	200	100	35	60
Banana (trees)	20			20	1	25	50	60	15	15	120	100	10	150
Tea (acres) (2)	5 00	1.25	5.00	3 00	2.50	2.50	4 50	2.00	2.00	1.00	4.00	4.00	0.50	2.00
Clove (trees)		22	20				4	6		6		40	100	150
Nutmeg (trees)			8				2		4			3	75	20
Citrus (trees)	7				8		3		2	5	25		5	15
Papaya (trees)	5				8	50	50	5	5	10		2		5
Vegetables (acres)					0 50		0.40			0.10	1.50	0 10		
Avocado (trees)	7				4		10	5	3	6				
Kitul (palms)	2	2			1	2					2	1	2	
Flowers (plants) (3)					(50)	(25)	(100)		(50)	(50)	(150)		(100)	
Mangosteen (trees)	1		2				12		2	5			5	
Cacao (trees)											50		15	
Breadfruit (trees)					3			2						
Yam (plants)							(few)				(400)	(600)		
Durian (trees)					2		2							
Paddy (acres)												(0.25	0.50	
Grass (acres) (4)	(2.00)						(1 00)							
Mango (trees)					2									
Cardamom (plants)														
Rubber (trees)												50		150
Pineapple (plants)														
Rambutan (trees)													few	
Passionfruit (vines)							25							
Ginger (acres)													0.50	

(1) Includes only pepper, omits other vines except one farm with commercial passionfruit.

(2) On all except (19) and (20) this is scattered tea of varying densities.

(3) Anthurium sometimes sold.

(4) Farm (8) has 2 milk cows

(15)	(16)	(17)	(18)	(19)	(20)	(21)	(22)	(23)	(24)	(25)	(26)	(27)	(28)	(29)	(30)	Farms with crops
4.50	1.00	1.00	1.50	3.00	1.00	3.00	3.00	1.50	1.00	1.00	5.00	1.00	2.50	1.50	2.00	
15	12	9	9	10	14	10	5	13	9	9	8	15	8	7	11	
447	107	296	225	78	223	522	320	145	615	194	186	111	362	237	629	
99	107	296	150	26	223	174	107	97	615	194	37	111	145	158	315	
125	25	100	30	12	80	200	5	20	50	50		10	20	50	150	28
8	3		4	4	7	5	5	5	2	100	6	6	20	0	20	26
70	15	50	30	35	60	50		6	500	12	65	30	200	10	200	26
20	1		2	15	12	100	300	15		10	100	10		100	2	26
	28	60	51		4	50		70	6	3	12	2	12	4	12	25
60	15	50	25	6	46	50	10		50			15	20	70	19	25
0.25				1.00	few	1.50	3.00				4.50	1.00		1.50	0.70	23
25	12	6			2	57		14	3	5		6	60	3	75	20
20	3	12	31		4	5			2	4	1				1	16
1	4	10	2						2			3	3			15
6		8		4	2			8				few				(15)
0.25	few	few		0.10				0.10	few		few	0.75			few	14
5					1			2				2			150	11
							5				10	2				10
	(3)										(60)					9
									2							7
100			50			3			1				30			7
			2	2					1							5
				(31)								(25)				5
1	1								1							5
											(0.50)					3
																2
6																2
					few								few			2
																2
												(25)				1
																1
																1
																1

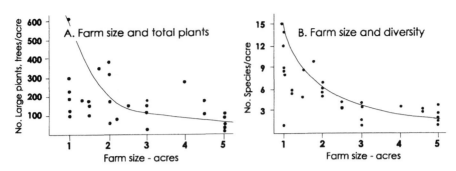

Figure 3.2   Farm Size, Large Plant Populations and Numbers of Species

farm 24 with coffee + areca + banana. (By way of comparison with monocrop estate conditions in the same area pure stands of coffee are 400-600/acre and of coconut 64/acre.) That is, farm 10 of Table 3.1 is roughly the equivalent of an acre each of coffee, areca/papaya, coconut and tea piled on top of each other, in addition to the other plants. These gap-filling palms are highly economic in their own right. The role of palms (and coffee) in contributing to high plant populations is perhaps better seen in the following comparison of highest v lowest population farms, Table 3.3.

Table 3.3   Plant Density in Relation to Coffee and Palms

Farm	Size (acres)	Trees/Vines	Coffee	Areca	Coconut	Banana	Kitul	Papya	Total palms
					Plants per acre				
Highest density farms									
10	1.00	687	500	50	50	15	0	10	125
24	1.00	615	500	50	6	0	0	0	56
14	2.00	374	50	15	60	150	0	5	230
13	1.75	347	25	150	35	10	2	0	197
30	1.00	330	200	86	12	19	0	0	117
						Palms, per farm mean:			145
Lowest density farms									
15	4.50	47	16	28	0	60	0	6	94
1	5.00	43	0	1	150	20	2	5	178
5	5.00	40	3	28	10	1	1	8	48
26	5.00	37	11	0	2	0	0	0	2
19	3.00	26	12	4	0	6	0	4	14
						Palms, per farm mean:			67

Diversity Species/crop diversity is also determined largely by farm size, increasing as size decreases (Figure 3.2B). The mean number of species/crops over all 30 farms is 11; 12 of 30 have between 6 and 10 crops, only 2 farms have less than 5 crops and at the other end of the diversity scale 2 farms each have 17-18 crops. Here 'diversity' means number of crop species and it excludes the non-economic species (Perera and Rajapakse, Annex 1). The other aspect is number of final *products* which can be obtained from the respective species (Nature, flexibility). Depending on the nature of these species it is possible that a farm with relatively few species might achieve a high level of economic or income diversity by processing these few primary outputs in a large number of different ways.

Crop yields Yield levels of all the species have been reported elsewhere (McConnell 1991); there are too many species and different quantitative units to discuss here — kg/tree, kg/ha, bunches, fruit numbers, bottles of kitul sap, sheets of rubber, boxes of mace etc. In comparison with performance of these same species when they are grown as conventional pure stands or monocrops some of these mixed crop yields appear to be low, typically 50-70% of the pure stands. This apparent yield inefficiency is the charge most often brought against forest gardens, especially by experts in one or other of the respective crops. It is valid if only criterion 1 of Table 1.2 (Chapter 1) is applied. Apart from this it is generally invalid for three sets of reasons.

First, the household operating objective is not money profits but household *security* and *sustainability* and this requires reference to the wider criteria of income stability, flexibility, risk avoidance, time dispersion etc (Table 1.2, Nature) which require a high species/crop/income diversity almost regardless of how any of the individual crop components might perform in comparison with some agronomic standard. Second, whole *farm* productivity is relevant, and this is almost always higher in the aggregate on these small mixed farms even if some individual crop components perform sub-optimally (Nature, productivity). Third, these conditions reflect in the following specific technical reasons for apparent low species yields:

Harvest v yield Much subsistence food produced especially perishable fruit is not harvested beyond family requirements (jak, breadfruit, mango...). This also occurs with some non-perishables (coffee, arecanut) where the amounts produced exceed household needs but are too small to warrant marketing. This un-harvested produce is not recorded as yield.

Harvest v collect On the less commercially oriented farms 'harvesting' is properly described as produce *collecting* when the need arises, rather than as one carefully timed step in a series of operations to maximize output/profit as on estates. (This does not apply to high value commercial crops.)

Several functions of trees The inclusion of old trees/palms well past their peak productivity ages lowers farm or species average yields. On estates these would be quickly removed; and on some forest gardens also where land is in extremely short supply (Kerala, West Java) the household must keep their trees in a permanently juvenile state of high individual productivity. But on Kandy farms where land is less of a constraint it might still be good management to retain old low yielding trees as long as they produce *something*, and as a standing store of fuel or marketable timber (especially the jak as a cabinet timber but also coconut and kitul palms).

Micro-environment Another reason for retaining old low yielding trees well past their horticultural use-by date is for their function of providing shade, humidity, wind protection etc for others (Energy). Around Matale (map) there are many big old rubber trees, some 50-60 years old. They are still tapped as rubber but their more important function is to provide micro-environment to cacao and coffee, and some are used also as live trellis for pepper. A rubber expert would shudder at this way of managing rubber and insist that they be pulled out every 25 years or so and replaced by whatever new miracle clone is coming on stream. A kapok export would say the same thing about that multi-purpose species. Forest farmers know better. The total economics of their mixed patch would be far superior to pure stands of these same species.

Agro-technology The short description of agro-technology applied on the wilder kind of forest gardens (but not the closely managed very small ones of Kerala) is that there isn't much. The thing looks after itself. The only species which is usually fertilized (and pruned) is tea. It is common for household waste to be dumped at the base of some nearby coconut, and an occasional handful of tea fertilizer might be thrown in the direction of a favorite jak tree, to the neglect of others. The bulk of farm production might come from these few pampered individuals. Average yields for the whole populations bear little relationship to potential yields which could be achieved if the need arose.

Table 3.4   General Architectural Arrangement of Species on Kandy Farms

Level 5 upper: 25-30m*
Durien	Durio zibethinus, tall open canopy, cash crop
Talipot palm	Corypha umbraculifera, leaves for thatching, handicrafts
Jak	Artocarpus heterophyllus, important food, timber, leaves livestock feed
Coconut palm	Cocos nucifera, subsistence food, toddy, timber, cash crop
Kapok	Ceiba pentandra, cash fiber and oil seed crop, live support
pepper	

Level 4 upper-middle: 15-25m
Bamboo	Several weaving and construction types home use, sale
Areca palm	Areca catechu, nuts for masticory/medicinal uses, sale
Nutmeg	Myristica fragrans, nuts and mace, mainly cash crop
Clove	Eugenia caryophyllata, cloves and parts, cash crop
Rubber	Usually old trees, tapped, also cacao and coffee shade
Wild breadfruit	Artocarpus nobilis, timber, forest remnant not deliberately planted
Kitul palm	Caryota urens, tapped/processed to toddy/sugar/honey etc
Mango	Many varieties, mainly fruits sold

Level 3 lower-middle: 10-15m
Pepper	Piper nigram, on live supports, important cash crop
Avocado	Several varieties, food and cash crop
Mangosteen	Seasonal and cash crop and food
Breadfruit	Artocarpus incisa, subsistence food staple, cash crop
Rambutan	Important seasonal fruit, food and cash crop
Citrus	Orange, pomelo, mainly cash crop
Papaya	Carica papaya, subsistence food, in places tapped for papain

Level 2 lower: 3-10m
Banana/plantain	Many types, food staple and sale
Cacao	Beans sundried, sold, low quality
Coffee	Subsistence and cash crop, low quality
Passionfruit	Minor crop, most consumed, some sold
Betel vine	On bamboo trellis sunny locations, family use, sale
Vanilla	Minor crop, most for medicinal uses, little sold

Level 1 ground-level
Tea	All sold as green-leaf to local factories
Cassava	usually only few plants for family use, not popular food
Ginger	Important cash crop under medium shade
Turmeric	Same as ginger
Anthurium	Few plants only under shade, flowers habitat, some sold
Pineapple	Minor food and cash crop, usually excessive shade
Chili peppers	Major vegetable condiment, cash crop
Vegetables	Okra, eggplant, beans, mainly family use

Yams   Minor shade-tolerant crop, food and sale
Grass/fodder   Only in clear spaces, for stall fed goats, cattle

*Top of canopy above ground-level; indicative. The agroeconomics of most of these as individual crops, in Sri Lanka, abstracted from the forest garden environment, have been discussed elsewhere (McConnell 1986).

Architecture Typical vertical arrangement of the 20 species is shown in Table 3.4. They fall into five storeys or layers, ranging eg from yams at/near ground level to the tallest emergents such as talipot palm, durien etc. The vertical distribution of total phytomass on these farms has important implications for their microclimatic conditions of productivity (Energy).

3.4 Economics

> *If a man has a dozen coconuts, a jak tree, a cow and part*
> *share in a paddy field he needs nothing else*
> ...Kandy saying, Williams 1950.

To introduce the economics of this system it's useful to outline the broad role of the most important crops in meeting household requirements. The following list is extended somewhat beyond Kandy to illustrate the self sufficiency of this farm type in the region generally. (A full list would run to hundreds of items.)

1. *Food – Bulk/energy/carbohydrates* Jak, breadfruit, plantains directly and sago and kitul palm flour, starch; and at ground level the yams, cassava, sweet potato depending on light conditions. In the pekarangan gardens on the slopes of Merapi (an active volcano in Central Java) the common bulk staple used to be root of the edible lily 'ganyong' (Canna edulis). But food also has a status dimension. When these villagers were surveyed after an interval of 14 years (Probowo and McConnell 1993) most of them were proud to claim they no longer ate ganyong but had advanced on to rice (paid for by clove sales). Rice is universally preferred, but in the Kandy survey only 3 or 30 farms had paddy fields; other households obtained rice by trading other produce for it.

2. *Vegetable proteins* There is no single major source except on those farms which have leguminous trees producing edible seed (tree beans, eg Malay 'puter') and no really satisfactory alternative to the soybean/peanut/gram field crops grown on conventional farms.

Obviously this lack is overcome on those farms which also have some area of clean-cultivated land. Most Kandy farm use cash from sale of pepper, cloves etc to buy these items (and the rice carbohydrate) or barter for them in the village.

3. *Vitamins/minerals/taste* Fruits (the species too numerous to list) and ground level vegetables; but usually these latter dominated by the red/green chilli pepper used probably as much to relieve the monotony of diet as for its vitamin content.

4. *Condiments/preservatives* Black pepper, cloves, nutmeg and mace, ginger, turmeric, coconut 'cream', coconut and kitul jaggery (unrefined sugar), cinnamon, cardamom, vanilla.

5. *Edible and cooking oils* Almost universally this is coconut oil but avocado is also important, although taken raw and not kitchen processed for oil extraction as is the coconut.

6. *Beverages/alcohol* Coffee, tea, cocoa, but these last two require too much effort to be quickly home processed and are sold or bartered for other items rather than used directly. In S-SE Asia the common refreshment is mildly alcoholic toddy tapped from the coconut or preferably the kitul palm, or after distillation the more potent arrack. In the dry zone of Sri Lanka and India the palmyrah palm (Borassus flabellifer) is tapped for these some purposes. (Some have described a leisurely progression through the toddy *kedais* under the palmyrah groves of the Batticaloa coast as one of the world's great journeys.) West of Surabaya on the Java Sea it's too dry for forest gardens; the village economies are structured around collecting salt, catching jellyfish and tapping the *siwalan* palm (B. sandaica); and if the economics of this odd suite are not appealing then the *toak* of Tuban would be sufficient reason for pilgrimage to that place.

7. *Narcotics* Nut of the areca palm chewed as a grated mix with lime and clove and wrapped in leaf of the betel vine (*sireh*) is near universal from Pakistan to Melanesia. (Extended to livestock the pigs of Bhutan are kept happy chewing bundles of fresh *ganja* (C. sativa) until summoned to their fate.)

8. *Medicinals* The oils of clove, nutmeg, cinnamon are most common in Sri Lanka; Jayaweera (1974) has reported the other medicinals (five volumes).

9. *Farm/household construction*
    (a) frame — coconut trunks, bamboo;
    (b) cladding — woven panels (*atap*) of nipa, sago, coconut leaf, split bamboo;
    (c) flooring — woven split bamboo (especially rural Indonesia);

    (d) furnishings — jakwood furniture, pandanus mats;

    (e) water piping — bamboo (especially Java, Bali);

    (f) implement handles, weapons, tea rollers — dense wood of the kitul palm;

    (g) buckets, containers — palm spathe, woven bamboo baskets;

    (h) all ropes, twine — coconut coir, kitul fiber.

10. *Household fuel* For cooking a small garden easily provides enough tree litter to meet requirements. For lighting the main traditional sources are coconut oil lamps (butter lamps in southern Himalayas). As late as 1980s the more isolated villages of the Solo Valley had street lighting systems of flickering coconut oil lamps maintained by each family in front of their house. Maybe they're still there. (In Peruvian Amazonia Gentry and others have described use of the seed of the *fevillea* vine for domestic lighting – Diversity.)

Basic model Planned as family survival units (ie *production* systems – Nature) these small forest farms of ¼ acre/1 000 sq m or so would probably be structured around 4-5 each of jak and breadfruit and 10 bananas/cooking plantains supplying bulk carbohydrate food in place of grain; 10 coconuts for oil/milk/flesh/husk fiber; half a dozen fruit species for vitamins, minerals, variety, barter; a clump of bamboo for tools and utensils; a few kitul palms for sugar and toddy; ground interspaces planted to chili and vegetables and 20-30 free ranging chickens for farm hygiene in place of agricides. A mature breadfruit (1st harvest in year 8) yields 500-700 fruit over 8 months annually x 1-1.5 kg/fruit and 70% of this is edible though 75% is water (ie 140 kg/tree dry carbohydrate food). A typical jak tree yields slightly more (fruits to 30 kg but fewer of them), 25% of useful fruit weight is seed but these have higher food value than the flesh and some varieties start producing within two years. (To illustrate the importance of jak food around Kandy 15 surveyed farmers there had an average of 15 trees, some intended primarily as timber.) If this subsistence farm was doubled to ½ acre and pepper, cloves, nutmeg, coffee, turmeric, more chili etc were added for cash income, and a sow for waste disposal (and itinerant courting boar conveyed around the village by wheelbarrow), such a household would be obscenely rich – at least in comparison with that 1/5th of global population who are either malnourished, chronically hungry or starving and for whom prospects of progress through Western grain-centred farm development models are a delusion.

Relative Importance of the Separate Crops

Table 3.3 summarizes the relative contributions of the crops to total farm income for the aggregated 30 farm sample, with 'income' consisting of actual crop sales plus imputed value of produce consumed or bartered.

Table 3.5   Relative Economic Importance of Crops and their Disposal

		Total income	Sold	Used/bartered	No. farms with crop
cloves		34.8	99%	1%	20
tea		16.0	100	0	23
pepper		12.4	96	4	26
coconuts		10.7	29	71	25
coffee		6.6	99	1	26
banana		4.7	87	13	25
nutmeg		3.1	100	0	16
areca		3.0	98	2	28
ginger		2.1	na	na	1
jak		1.4	7	93	26
paddy/rice		1.2	0	100	3
kitul		0.7	96	4	10
vegetables		0.7	88	12	14
anthuriums/flowers		0.7	97	3	9
avocado		0.6	88	12	11
cocoa beans		0.4	100	0	7
papaya		0.3	18	82	15
mangosteen		0.1	86	14	7
rubber		0.1	100	0	2
durien	below	0.1	60	40	5
breadfruit	"	0.1	83	17	5
citrus	"	0.1	62	38	15
rambutan	"	0.1	71	29	1
pineapple	"	0.1	92	8	1
yams	"	0.1	89	11	5
mango	"	0.1	95	5	2
passionfruit	"	0.1	90	10	1
grass/cow	"	0.1	10	90	3
cardamom	"	0.1	100	0	2

Species, crop and income diversity From the table these farms as a group are extremely income-diversified, possibly as much as any in the world apart from some farms in Peru (Reconnaissance); and as noted previously many of these crops yield several products and/or these are processed into several items. Also, the surveys were not particularly directed at diversity: if one set out to study highly diversified farms it would be easy to find units having two or three times the diversity levels discussed here. It is apparent that many of these crops are grown only on a few farms (last column). This indicates there is still much scope for further diversification. Eg all of these farms *could* grow also such crops as ginger, cardamom, rambutan, durien...without affecting productivity of the others, but only a few do so. This shortcoming — if such subjective judgment is allowed — is due partly to the lack of farmer perceived need for increased total farm production, but also partly to absence of research and extension effort to promote advancement of this system. As noted elsewhere about the only advice these forest gardeners have had over the years, here and throughout the tropical world, is to *reduce* the number of species, get serious, pull out all this heterogeneous rubbish and concentrate on proper farming...Which translates as those crops of interest to Government which further translates, as in King Parakrama's time, to developing that kind of agriculture which best supports the ambitions of the Prince.

Genetic diversity In addition to the large number of plant species (economic and non-economic from Annex 1), the number of their types/cultivars/landraces must also be large, but these await enumeration and evaluation. Most of the plants – including those of interest to the highly commercialized sector – have been evolving on these farms for centuries, refreshed as needed by seed or cuttings or exchange among neighbors and among villagers – or simply let run wild. This must be a genetic goldmine, a plant breeder's idea of heaven.

Farm Income Levels

Farm income levels are summarized in schedule A of the box (cash plus imputed value of consumption/use); the mean per farm was Rs 14 030 and per acre Rs 6 670. (At the time of the last survey Rupees 10 = US$1.) These monetary values are now out-dated but the per acre means were well above the income levels for paddy in the same area (as also reported in the Perera and Rajapakse 1991 survey) and this continues. A better comparison standard, one hardly affected by inflation over time, is daily wage rate of rural labor. This was Rs 8-10 per day, but even at this low rate ($0.80 -1.00/day) jobs were difficult to find for more than 200 days annually. The average household income then was about seven times higher than that of landless working households. (These differentials have undoubtedly increased due to increasing populations further out-stripping employment opportunities.) These gross farm income levels are also close to *net* farm income because there are very few purchased inputs and production costs (below).

Table 3.6   Distribution of Income per Farm and per Acre With and Without Dominant Crops

Income	A. Including all crops		B. Excluding dominant crops	
	Total farm income	Income per acre	Total farm income	Income per acre
(Rs)	Number of farms		Number of farms	
900 - 2 000	-	2	2	11
2 001 - 5 000	4	13	11	12
5 001 - 10 000	9	11	9	6
10 001 - 15 000	9	2	4	1
15 001 - 20 000	2	-	1	-
20 001 - 40 000	4	2	3	-
40 001 - 50 000	2	-	-	-
Mean income	Rs 14 030	Rs 6 670	Rs 7 930	Rs 3 650

Effect of dominant crops   It is sometimes objected that these small farms could not be economically viable without some particularly profitable dominant crop; and these 30 farms do in fact derive most of their income from a few crops – cloves, tea, coconut, pepper... This also implies considerable price and income risk. To test the effects of this dominance, part B of Table 3.6 repeats the income analysis with dominant crops entirely removed. These are defined as any crop on any farm which contributes 40% or more to that farm's income. With these income contributions omitted, mean farm income would be Rs 7 930 or Rs 3 650 per acre, which is still about four and a half times more than incomes of families dependent on wage labor (if they can find it).

Incomes of very small farms   For reasons of political threat – if not of equity and social justice – and given unemployment levels of 40, 50, 60...% which prevail in many developing countries (Nature, equity), it is highly desirable that whatever land exists be shared as widely as possible among the population. In structuring new farms in settlement projects, or restructuring old ones, this usually requires that these units be as numerous as possible by making them as small as possible subject to the requirement that they are economically viable and don't require continuing subsidy and become a drain on the public exchequer or on welfare system, if one exists.

   The following schedule shows farm and per acre incomes (with and without dominant crops) of only the smallest farms, those below 1.5 acres (0.6 ha). Part B of Table 3.7 then represents a 'minimalist' or worst case income scenario, per acre income of Rs 4 460, with dominant crops removed.

Table 3.7   Income Ranges for the 11 Smallest Farms, With and Without Dominant Crops

Annual farm income ranges (Rs)	A. Farm income including dominant crops		B. Farm income excluding dominant crops	
	Per farm	Per acre	Per farm	Per acre
	Number of farms		Number of farms	
2 000 - 3 000	-	-	2	3
3 000 - 6 000	4	6	5	5
6 000 - 9 000	5	4	4	3
9 000 - 12 000	1	-	-	-
12 000 - 15 000	1	1	-	-
Mean income	Rs 7 140	Rs 6 320	Rs 5 080	Rs 4 460

On this basis a household with, say, an acre of forest garden would still be twice as well off as would a landless laboring family. (In fact they would be much better off because this latter 'average' family might *not* be able to find employment and would probably have to pay accommodation rent, in cash or service to a landlord.)

Non-farm income   Only about a third of the households (11 of 30) supplement their farm income with off-farm activities. This occurs only on the smaller farms below 2.5 acres. Earnings are small, typically Rs 400-800 annually. Most jobs are in seasonal harvesting of paddy, cocoa, coffee or routinely tea on larger village farms. A main non-farm activity by women is operating a small roadside 'boutique', kedai, selling snacks, tea, jaggery, fruit...whatever an otherwise idle housewife has in surplus or can make. Except for plucking tea which is a regular and monotonous job, carrying disutility as the economists say, or if some special household purchase is intended, there is often only a vague element of economics in this 'employment'. It is pursued as much for opportunity to gossip or argue politics or simply to relieve the tedium of village life.

Farm Management, Inputs, Costs

Edmund...? *Ne needs there Gardener to set or sow, to plant or prune: for of their own accord all things as they created were do grow.* These earlier findings by Spencer (1552-1599) are verified by the following tables.

From Table 3.8 mean per farm operating costs, Rs 1 364, is very small in comparison with gross farm income (Rs 14 030). On the other hand, by far the main cost item is hired labor which introduces the paradox of employing outside labor while many of the household are apparently idle (below). These data are best examined in combination with those of Table 3.9 which shows how these various costs are distributed among the crops.

Table 3.8 Distribution of Farms according to their Total Production Costs

Farm operating costs	Number of farms with cost	Mean cost per farm	Cost item as % of total
hired labor	30	Rs 1 190	87.2
fertilizer	16	77	5.6
agricides	3	15	1.1
bags, boxes	9	6	0.4
transport	29	50	3.7
tools	30	20	1.5
marketing	12	4	0.3
seeds, plants	2	2	0.1
Totals		Rs 1 364	100.0

Most hired labor is used for tea, coconut, pepper, cloves. Apart from labor the other farm costs are trivial. Fertilizer is hardly used then primarily on tea (and paddy, but only 3 of the farms have a paddy field), and on coconut where this is a serious enterprise; the fertilizer used on ground level crops is usually a few handsful left over from the tea. The quasi forest supplies its own nutrients by litterfall and recycling (Energy, Nutrients). This also applies to agricides (3 farms only). Apart from tea and coconut plants and paddy seed all plants/seeds are retained or exchanged with neighbors, if the new plants are not transplanted from beneath a mother tree. Such

Table 3.9 Distribution of Inputs and Costs among the Various Crops*

Crop	Hired labor	Fert- ilizer	Chemicals	Boxes bags	Trans- port	Seeds plants	Market
Tea	61	41	62	81	67	48	-
Coconut	14	37	-	-	-	32	-
Pepper	9	-	-	-	10	-	11
Areca	2	-	-	14	12	-	21
Cloves	9	-	-	5	5	8	34
Coffee	2	-	-	-	5	-	34
Vegetables	1	9	38	-	1	2	-
Paddy	2	13	-	-	-	10	-
Total, %	100	100	100	100	100	100	100

Proportion of total variable cost on all farms absorbed by these eight crops (%):

%	92	100	100	70	81	100	73

*These 8 crops incur all or most of farm costs, as shown in the bottom line.

'costs' as transport and marketing, themselves extremely low, usually represent no more than a bus ride for a jolly day out in Kandy or Kegalle, perhaps taking a shoe box of mace or sack of arecanuts or coffee to some merchant, or for the more serious business of arranging a harvesting contract.

Tabasan and contract harvesting Both these previous tables considerably understate the real inputs of labor, marketing and transport as these items apply to cloves, nutmeg, pepper, (commercial) coconut and coffee. This occurs through the common system of selling forward these high priced crops, as *standing* crops, prior to their harvest. Typically the larger merchants, assemblers, maintain gangs of specialist harvesters who travel through the villages and do this work on their behalf, returning year after year. Usually the standing crop of pepper etc is sold to the merchant several months before it matures. This represents a cost to the farmer in the lower forward price paid than if farmers wait and harvest the crop themselves. In this case the merchant's crew are responsible for any agronomic maintenance and for guarding the crop if it is of high value and there is a hazard of theft (or against birds and animals in the case of rice, food crops). The longer this forward period is the lower the crop price paid. On the other hand farmers get cash money well ahead of when they otherwise would receive it and avoid all crop/price risk and also the bother of having to arrange their own harvesting, marketing.

The same system is widely used in Indonesia (*tabasan*). Here and in Sri Lanka, and in Malaysia etc it is impossible to generalize but a forward contract price to the farmer of 50-60% of the eventual commodity value at harvest is indicative. In some Javanese villages as much as 70% of the village rice crops are sold forward to traders under *tabasan* agreements (Probowo and McConnell 1993). The system is generally deplored as exploitive of ignorant villagers. But on the other hand where a formal banking system is lacking it is a better alternative than exposure to the rapacity of money lenders in the souk or debt bondage to a landlord (debt, below). With regard to the internationally traded commodities — coffee and the spices — there is no question that villagers are at great disadvantage in these *tabasan* agreements relative to merchants. These are in regular contact with their London, Hamburg, agents and know well in advance of farmers what their crop of pepper or cloves is going to be worth when it is harvested.

Apart from *tabasan* there are three other reasons for contract harvesting these crops by specialists, even though the farm family itself might have much idle labor (below). First, harvesting coconut (with a blade attached to a long bamboo pole) is hard work and dangerous; picking pepper on a fragile live support tree 20 or 30 feet above the ground equally so; and harvesting cloves requires deftness and skill. Second, harvesting was (and often still is) a caste occupation. Even without this ancient constraint, perceived family or personal status or the loss of it is at stake. (In Sri Lanka this is especially so with the cinnamon peelers of the South who have accumulated centuries of skill and experience.) It used also to extend to (Tamil) toddy tappers, especially of coconut and palmyrah palm, and since they have all the 'secret' ingredients and processes for making good toddy, and don't impart them, why try to compete...? There's a lot more to it than just the apparent comparative costs of these harvesting operations. Finally, in some societies (Java) there's still much perceived merit in sharing whatever work is available among as many as possible. For one to try to do such jobs as harvesting when they could be given to others with greater need would be judged as greed and conceit. Western farm economics — by compulsion of its personal profit maximizing model — would have to disagree.

3.5 The Households

Populations ... There was a time, ere England's griefs began, when every rood of ground maintained its man (Pope). Perhaps the main socio-economic characteristic of these farms is the high populations they support: on average 4.8 persons per farm or 1.85 persons per acre or 4.5 persons per hectare (but this ranging from one adult male only on one farm to 12 adults per *acre* or 29/hectare on another). Because of the very small amounts of supplementary non-farm income (above) these populations are almost entirely supported by the farm. (There is no relationship between farm size and household size.)

Labor and consumption Family/household populations have two economic dimensions: as consumption units (reflecting their needs) and as labor units (reflecting their potential labor capacity to supply those needs). Because of the heterogeneous nature of 'persons' — males and females of different ages and therefore of different consumption needs and work abilities — it is necessary to standardize these populations as 'adult male equivalents' or AME (consumption) and AME (work capacity). (The mechanics of such standardization need not concern us.) In most of S. Asia 'labor' has gender, age, task and often caste dimensions (McConnell 1991; McConnell and Dillon 1997). Distributions of these standardized farm populations as consumption and as potential work units are summarized in

Table 3.10. For the moment it's enough to note that *most* farms (15 of 30) are obliged to support 2-4 adult male equivalents (mean 1.45 per acre), and most farm households (23 of 30) had between 1 and 4 adult male person equivalents to do the necessary work (or 1 AME/acre).

Table 3.10   Distributions of Standardized Farm Populations

A. Consumption		B. Labor	
No. of farms	Household size consumption units	No. of farms	Household size labor units
2	0 - 1 AME	4	0 - 1 AME
3	1 - 2	11	1 - 2
15	2 - 4	12	2 - 4
6	4 - 6	2	4 - 6
3	6 - 8	1	6+
1	10+		
Mean per farm...	3.74 AME	...	2.5 AME
Mean per acre...	1.45 AME	...	1.0 AME

Farmers or Gatherers?

Table 3.11 shows how all utilized household labor is distributed among the respective crops according to task (plant, dig, weed etc). From this the system is primarily one of *collecting or gathering*, then processing, with these two tasks absorbing 57% of all farm labor used on all aggregated crops. This contrasts with conventional field crop agriculture where the main tasks are land preparation-planting-cultivation-weeding. This may be seen in the comparisons with two representative field crop systems at the foot of the table, commercial sweet potato and continuous vegetables. Not only are their total per acre labor requirements much higher (respectively 173 and 263 v 33 days/acre) but these are concentrated on land preparation-planting-subsequent crop maintenance. This has a social equity implication in farm planning.

These field crop tasks require strength and youth (digging or cultivating a hectare of wet paddy field several times by hand must be about the hardest work there is). In Western agronomic models applied to Eastern agriculture it is often forgotten that some considerable percentage (20, 30...%) of households do *not* have strong male persons to do this heavy work but consist of elderly people or female-only household heads with young children. In these cases it is especially important that the *kind* of tasks demanded by the agroecosystem be within the capacities of the elderly or the very young. A system which requires only or primarily

Table 3.11 Distribution of All Household Labor Used on All Farms by Crop and Task; and Comparison with Two Field Crop Systems (annual basis)

Crop or system	Plant	Dig land prep	Weed cult irrig spray	Pick h'vst collect	Process sort dry pack	Market	Total for each crop	As % of all crops
				Total labor days on 30 farms				
	(1)	(2)	(3)	(4)	(5)	(6)	(7)	(8)
vegetables	na	58	61	223		17	359	14.1
tea		8	327	21			347	13.7
nutmeg	3	11	9	141	98	45	307	12.1
banana	6	14		161		108	289	11.4
pepper	20	40	15	19	123	30	247	9.7
coffee	5	10	19	62	119	17	232	9.1
coconut	18	46	33	18		33	148	5.8
cloves	12	15	14		81	15	137	5.4
cacao		2		77	33	18	130	5.1
jak				109	16		125	4.9
flowers				54			54	2.1
papaya				18	14		32	1.3
yams		18		3		3	24	0.9
kitul				16		7	23	0.9
paddy	12		9				21	0.8
mangosteen				19		2	21	0.8
citrus		5		13			18	0.7
avocado				4		6	10	0.4
areca				4		6	10	0.4
other crops (9 of these)			2	2		2	6	0.2
								100

Comparisons Among Systems

(1) Kandy farms (the above)
Total days
all farms,
all crops: 76, 227, 489, 963 + 484, 303, 2,542
Task per cent: 3.0, 8.9, 19.3, (37.9 + 19.0), 11.9............... 100.0
Total days/acre: 1.0, 2.9, 6.4, 12.5 + 6.3, 3.9, 33.0

(2) Sweet potato (Java)*
Total days/acre: 32.7, 21.2, 96.3, 16.3 + 8.2, 8.0, 172.7

(3) Intensive relay vegetables (Java)*
Total days/acre: 40.8, 87.8, 98.0, 20.0 + 16.3, 0.0, 262.9

*Sweet potato labor requirements based on two such crops annually or their equivalent; intensive vegetable farms are relay planted continuously throughout the year.

Source: Probowo and McConnell 1983.

produce *gathering* and then subsequent kitchen processing is for this reason far superior to one which requires the impossibility of heavy field work, regardless of their theoretical comparative economics. This qualitative aspect of labor-task demand takes on additional importance in those many tropical countries where some large and increasing percentage of the nominal work-age population are in fact not available in the farming season because of more-or-less routine debilitation by malaria. Also and especially in central and southern Africa village work-age populations and their productivity, needed to support children and the elderly, are shrinking dramatically under the impact of AIDS. In brief, this qualitative dimension of village labor supply can be highly dynamic; but it is seldom if ever taken into account in agro-development planning.

The *collecting* nature of forest garden systems also has desirable environmental implications. Their minimal need for such normal production operations as 'dig/plant/cultivate/spray' means minimal disturbance of vegetation as fauna habitat and is conducive to these farms being a conservation refuge. There is possibly also a religious basis for Kandy gardens to be undisturbed habitats eschewing cultivation (not pursued in the farm surveys). Being traditional Buddhists these families would be averse to killing earthworms or other life that might venture into any hoe cultivation zone, especially in the wet season (when for similar reason good monks don't like to go on journeys). More materially, comparative absence of soil disturbance results in comparatively low emissions of soil/organic matter carbon dioxide from this pool, elsewhere a significant source of this Greenhouse gas (Energy, greenhouse), and in minimum soil erosion. And as for irrigation and the hydrological cycle...Edmund? *Ne do they need with water of the ford, Or of the clouds to moisten their roots dry; For in themselves eternal moisture they imply.* And what about inputs for pest control, Edmund? *Nor wicked beasts their tender buds do crop.*

The Paradox of Idle Labor

From the following distribution the farm populations (as AME labor units) have a labor supply far in excess of farm requirements. This is so regardless of farm size. On average 82% of potential available labor is apparently 'idle', only 18% of it is absorbed in farm work. Further, very little of this surplus labor is absorbed in off-farm jobs, mainly because of lack of opportunity. On the other hand from the previous discussion the main farm cost is in hiring outside labor. There are several reasons for curious situation — the type of crop, sale of standing crops, perceived loss of status attached to certain tasks (especially harvesting which has caste connotations), preference for calling in experts to pick the cloves or tap the palms...etc. Another explanation is that actual *farm* labor use is under-reported. In these systems it is difficult to distinguish *farm* from *household* chores. Especially on smaller farms such tasks as collecting produce, sorting, sundrying mace and pepper, smoke drying arecanut, maintaining the vegie patch....are done as part of the daily routine revolving around the kitchen which is also focal point for produce processing. On conventional farms these activities would be clearly distinguishable and counted as farm labor. A third explanation is that Kandyans put a high premium on what in the West would be called leisure. In the allocation of time, participation in social, political and religious affairs is commonly placed ahead of effort to increase income by working more days on the farm. In any case most of these crops by their nature are quickly labor-saturated.

Distribution of the Farms in relation to their Labor Supply Not Used

Farm size	% of idle labor	Farm size	% of idle labor
1.0 - 1.5 acres	84%	3.6 - 4.5	59%
1.6 - 2.5	75	4.6 - 5.0	72
2.6 - 3.5	75	Mean	82%

Calvin v the Prince Is this apparent idleness a good thing or bad? Should it be counted a strength of these systems and to be sought, or a weakness to be avoided? Probably the main thing from the viewpoint of governments is that a large number of people are maintained at more than reasonable level, even if they've nothing much to do. At least they're kept from further depressing the urban labor market. And maybe they'll seize this opportunity for cultural enrichment, as Kandyans commonly do...Or does the Devil still find work for idle hands?

Debt

It is difficult to over emphasize the importance of debt avoidance as a farm system attribute in most of S and SE Asia. As a system Kandy forest gardens are free of the worst kinds of debt but not of others (Table 3.12). These amounts of indebtedness are very small relative to farm incomes, except marginally on three farms where capital debt was incurred to buy a cow and equipment. In most farming systems in this region the main need for debt is in seasonal crop production 4. for plowing, seed, fertilizer, insecticide, extra harvest labor...especially on those farms which have adopted the Green Revolution high yielding varieties (HYV) package.

(The necessary debt to pay for this superior agronomy is a poisoned pill also slipped into the Package, surreptitiously by landlords: these have already discounted the Package's alleged farmer benefits by increasing tenants' land rents or fees for their oxen etc.) But with or without the GR Miracle Package the interest charge on this kind of interseasonal debt is usurious by Western standards - 50, 60, 70, 100% are common (Nature, equity).

Table 3.12   Debt Situation on 30 Surveyed Farms*

	No. of farms	Avg. amount
Households without debt	. . . 12	Rs 0
1. Capital: buy more land, equipment	2	1 000
2. Capital: buy livestock	1	1 500
3. Capital: develop land, crops	1	50
4. Crop production: short term, seasonal	9	176
5. Consumption and social	6	290
6. Emergency	-	-
7. Income supplemental	-	-
8. Family social development, education	-	-
Average per farm (30 farms)	. . .	Rs 100

*Average based on those farms which incurred such debt. These amounts can be related to farm annual income levels discussed previously.

Emergency (6) and income supplemental debt (7) are worse. Another feature of field crop agrosystems in S and SE Asia is their high level of risk — of drought, storm, flood, pests or tenancy insecurity. When the crop is seriously reduced or wiped out the farm families, especially tenants, must seek bridging loans of cash but usually in kind until the next planting season from whatever sources and under whatever conditions they can find. This frequently leads to perpetual indebtedness to landlord, storekeeper, produce buyer, money lender, goldsmith — sometimes with children placed in work bondage as security until the parents' debt is paid off (sometimes never) or until these chattels climb over the wall and flee to Karachi or Bombay from wherever it is that these agricultural advances are being recorded.

Because of the continuous flow of produce forthcoming throughout the year these kinds of debt (4, 6, 7) are usually not necessary on forest gardens (nor in such uniform-flow systems as tea, rubber, dairy cows, relay planted vegetables). The seasonal production patterns of some of the tree crops were recorded previously (Nature, time dispersion). More common is voluntary consumption and social debt, often entered into for frivolous purposes. This has little material basis, is outside the scope of farm-household economics and belongs under the heading of fashion or of Veblen's conspicuous consumption (weddings) or novelty.

3.6 Conclusions?

So these forest gardeners should happy be and have immortal bliss. It's only agrodevelopers they have to fear, and the Prince, and *wicked Time who with his scythe addrest, Does mow the flow'ring herbs and goodly things.* Are they much happier than other folks? Not really. They're remembered for great kindness and civility. Others, too critical, find them inclined to litigation – perhaps too ready to draw law, or blood, over the leaning of a coconut palm across a fence or the wanderings of a village rooster. Still, the conditions for paradise are there. If they want to screw it up that's their affair.

References

Cook, E.K. (1951) *Ceylon — Its Geography, Resources, People.* Macmillan, Basingstoke.

Cordiner, J. (1807) *Description of Ceylon* (2 vols) Reprinted Naurang Press, New Delhi (1983).

De Lanerolle, N. (1985) *The Reign of the Ten Kings: Sri Lanka 500 BC-1 000 AD.* Government Press, Colombo.

De Silva, R.K. and W.G.M. Beumer (1988) *Dutch Sri Lanka.* Serendib Publications, London.

Jayaweera, D.M.A. (1974) *Medicinal Plants (Indigenous and Exotic) Used in Sri Lanka* (5 vols). National Science Council of Sri Lanka.

McConnell, D.J. (1977) *Land Rehabilitation and Settlement in the Mid Country: Pilot Project in Nilambe-Atabage, Gurugoda-Ritigaha and Mahaoya-Kudaoya Catchments.* Ministry of Agriculture and Lands. UNDP, FAO, Kandy.

McConnell, D.J. (1986) *Economics of Tropical Crops: Mainly the Trees and Vines.* Midcoast, Coffs Harbour, Australia.

McConnell, D.J. (1992) *The Forest Garden Farms of Kandy.* FAO, Rome.

McConnell, D.J. and J.L. Dillon (1997) *Farm Management for Asia: a Systems Approach.* FAO, Rome.

Perera, A.H. and R.M.N. Rajapakse (1991) Baseline study of Kandyan forest gardens in Sri Lanka: structure, composition, utilization. *Forest Ecological Management* 45:269-280.

Probowo, D. and D.J. McConnell (1993) *Changes and Development in Solo Valley Farming Systems.* FAO, Rome.

Williams, H. (1950) *Ceylon: Pearl of the Orient.* Robt Hale Ltd, London.

Annex 1   Species Occurring in 50 Kandyan Forest Gardens (Perera and Rajapakse 1991)

Species	Family	Known uses[a]
(a) Trees		
1 Achras zapota	Sapotaceae	Fruit
2 Adenanthera pavonina	Leguminosae	Medicine
3 Albizia moluccana	Leguminosae	Timber, shade
4 Alstonia macrophylla	Apocynaceae	Timber, fuelwood
5 Anacardium occidentale	Anacardiaceae	Fruit, cash, medicine
6 Areca catechu	Palmae	Cash, polewood
7 Artocarpus heterophyllus	Moraceae	Food, cash, timber, fodder
8 Artocarpus nobilis	Moraceae	Food, cash
9 Azadirachta indica	Meliaceae	Timber, medicine
10 Berrya cordifolia	Tiliaceae	Timber
11 Bridelia retusa	Euphorbiaceae	No known use
12 Camellia sinensis	Theaceae	Cash
13 Carallia calycina	Rhizophoraceae	No known use
14 Carica papaya	Caricaceae	Fruit, cash
15 Caryota urens	Palmae	Cash, timber
16 Ceiba pentandra	Bombacaceae	Cash
17 Cinnamomum zeylanicum	Lauraceae	Cash
18 Cipadessa baccifera	Meliaceae	No known use
19 Citrus aurantifolia	Rutaceae	Fruit, cash, medicine
20 Citrus medica	Rutaceae	Medicine, cash
21 Cocos nucifera	Palmae	Food, cash, timber
22 Durio zibethinus	Bombacaceae	Fruit, cash
23 Eugenia corymbosa	Myrtaceae	No known use
24 Euphorbia tirucalli	Euphorbiaceae	Medicine
25 Euphoria longana	Sapindaceae	Fruit, cash
26 Feronia limonia	Rutaceae	Fruit, cash
27 Ficus altissima	Moraceae	No known use
28 Ficus asperrima	Moraceae	No known use
29 Flacourtia inermis	Flacourtiaceae	Fruit, cash
30 Garcinia cambogia	Clusiaceae	Medicine
31 Garcinia spicata	Clusiaceae	No known use
32 Gliricidia sepium	Leguminosae	Polewood, fodder, green manure
33 Hunteria zeylanica	Apocynaceae	Medicine
34 Hydnocarpus venenata	Flacourtiaceae	Medicine
35 Leea indica	Leeacae	Medicine
36 Litsea longifolia	Lauraceae	No known use
37 Macaranga peltata	Euphorbiaceae	Fuelwood
38 Maesa perrottetiana	Myrsinaceae	Medicine
39 Mangifera indica	Anacardiaceae	Fruit, cash
40 Melia dubia	Meliaceae	Timber
41 Michelia champaca	Magnoliaceae	Timber
42 Murraya koenigii	Rutaceae	Cash
43 Myristica fragans	Myristicaceae	Cash, medicine
44 Neolitsea cassia	Lauraceae	Polewood, fuelwood

45	Nerium oleander	Apocynaceae	No known use
46	Osmelia gardneri	Samydaceae	No known use
47	Persea gratissima	Lauraceae	Fruit, cash
48	Phyllanthus indicus	Euphorbiaceae	No known use
49	Polyalthia korinti	Annonaceae	No known use
50	Psidium guajava	Myrtaceae	Fruit
51	Punica granatum	Punicaceae	Fruit, medicine
52	Rejoua dichotoma	Apocynaceae	Medicine
53	Swietenia macrophylla	Meliaceae	Timber
54	Symplocos spicata	Symplocaceae	Medicine
55	Syzygium aromaticum	Myrtageae	Cash, medicine
56	Syzygium jambos	Myrtaceae	Fruit, cash
57	Syzygium operculatum	Myrtaceae	No known use
58	Tamarindus indica	Leguminosae	Food, cash, medicine
59	Tectona grandis	Verbanaceae	Timber
60	Terminalia bellirica	Combretaceae	Timber, medicine
61	Wendlandia notoniana	Rubiaceae	Polewood, fuelwood

(b) Herbs (including creepers)

1	Achyranthes aspera	Amaranthaceae	Green manure
2	Ageratum conyzoides	Compositae	No known use
3	Alocasia spp	Araceae	Food
4	Alternanthera sessilis	Amaranthaceae	Medicine, food
5	Alysicarpus vaginalis	Leguminosae	No known use
6	Amaranthus viridis	Amaranthaceae	Medicine, food
7	Amorphophallus campanulatus	Araceae	No known use
8	Anthuriam spp	Araceae	Flowers
9	Argyreia populifolia	Convolvulaceae	Medicine
10	Asparagus falcatus	Liliaceae	Medicine
11	Brachiaria mutica	Gramineae	Medicine
12	Brasica juncea	Cruciferae	Medicine
13	Brucea javanica	Simaroubaceae	Medicine
14	Carum petroselinum	Umbelliferae	Medicine
15	Centella asiatica	Umbelliferae	Medicine, food
16	Clidermia hirta	Melastomaceae	No known use
17	Commelina benghalensis	Commelinaceae	Medicine
18	Desmodium heterophyllum	Leguminosae	No known use
19	Dioscorea oppositifolia	Diocoreaceae	Medicine, food
20	Elephantopus scaber	Compositae	Medicine
21	Eupatorium odoratum	Compositae	No known use
22	Euphorbia hirta	Euphorbiaceae	Medicine
23	Gloriosa superba	Liliaceae	Medicine
24	Hibiscus abelmoschus	Malvaceae	Medicine
25	Hibiscus furcatus	Malvaceae	Medicine
26	Hydrocera triflora	Balsaminaceae	No known use
27	Ipomoea aquatica	Convolvulaceae	Medicine
28	Ipomoea batatas	Convolvulaceae	Food, cash
29	Ipomoea mauritiana	Convolvulaceae	Medicine
30	Mentha viridis	Labiatae	Medicine
31	Oxalis corniculata	Leguminosae	Medicine
32	Oxalis corniculata	Oxalidaceae	No known use

33	Phyllanthus urinaria	Euphorbiaceae	No known use
34	Rhinacanthus nasutus	Acanthaceae	No known use
35	Sida acuta	Halvaceae	Medicine
36	Stachytarpheta indica	Verbenaceae	No known use
37	Trichospus zeylanicus	Dioscoreaceae	No known use
38	Trigonella corniculata	Leguminosae	No known use
39	Urena lobata	Malvaceae	No known use
40	Vernonia cinerea	Compositae	No known use
41	Zingiber officinale	Zingiberaceae	Medicine
42	Zingiber zerumbet	Zingiberaceae	Medicine

(c)   Shrubs

1	Abutilon asiaticum	Malvaceae	No known use
2	Atlantia ceylanica	Rutaceae	Medicine
3	Bambusa vulgaris	Bambusaceae	Timber
4	Coffea arabica	Rubiaceae	Cash
5	Erythroxylum moonii	Erythroxylaceae	Medicine
6	Indigofera tinctoria	Leguminosae	Medicine
7	Leucaena leucocephala	Leguminosae	Fuelwood, polewood, food
8	Manihot esculenta	Euphobiaceae	Food, cash
9	Musa spp	Musaceae	Fruit, cash
10	Pavetta indica	Rubiaceae	Medicine
11	Pogostemon heyneanus	Labiatae	Medicine
12	Zizyphus oenoplia	Rhamnaceae	Medicine

(d)   Climbers

1	Cissampelos pareira	Menispermaceae	Medicine
2	Cyclea burmanni	Menispermaceae	Medicine
3	Derris scandens	Leguminosae	Medicine
4	Gleichenia linearis	Gleicheniaceae	No known use
5	Passiflora edulis	Passifloraceae	Fruit, cash
6	Piper betle	Piperaceae	Cash, medicine
7	Piper nigrum	Piperaceae	Cash, medicine
8	Pothos scandens	Araceae	No known use
9	Smilax zeylanica	Smilacaceae	Medicine
10	Vanilla planifolia	Orchidaceae	Cash, medicine

[a]Only the main recorded uses are listed.

Chapter 4
Diversity

*Long work it were, Here to account the endless progeny of all the weeds But so much as doth need, must needs be counted her...*Spencer.

There are thought to be 10-30 million species of plants and animals in the world (Wilson 1988, Ehrlich 1991); but this could be as high as 100 million, eg Erwin (1982) puts the arthropods or insects alone at around 30 million. The rain forests are by far the richest biome, occupying only 5-6% of world land area but having 50-80% of all known species, and the remnant rain forests of Australia occupying only 0.3% of that continent still have over 50% of its plants. More plant species have been reported on a single mountain in Philippines than there are in North America. These enumerations might obscure the close inter-dependence among species: the Amazon River system, artery of the greatest forest, has 3 000 fish species, more than the Atlantic Ocean and 75% of these depend seasonally on fallen forest fruit for their food.

4.1 Biodiversity Loss and Causes

Whatever the real global total now the maximum appears to have been reached about 30 000 years ago, well into the era of the first human agroecosystems but some 20 000 years before the origin of agriculture as granoculture (Origins, Genesis). Since then numbers have been declining, currently at galloping rate. Of the *known* high plants in all global biomes 33% are now endangered, 750 are recently extinct or thought to be and of all known trees 9 000 are critically endangered (IUCN Red Book of Threatened Plant 1998). Edward Wilson (1992), assuming a total number of five million in the rain forests, estimates their current extinction rate at 17 500 annually or more than a thousand times the normal background species extinction rate. Ehrlich's (1991) estimate is that the current loss in all biomes is 150 000 times the background rate, the highest it's been in the past 65 million years. Recent precise extinction causes globally can't be pursued here but taking continental Australia as a guide (Table 4.1) to those countries which have developed their agricultural sectors at rapid rate, clearing then operational agriculture are chiefly responsible for loss of both plants and animals.

Table 4.1   Comparative Causes of Species Extinctions, Endangerments

Plant extinctions and endangerment, Australia 1992

Cause	No. species	Cause	No. species
agriculture-grazing	105 (78) *	collecting	17
low populations	85	mining	11 (1)
roadworks	57 (1)	forestry	10
weed competition	57 (4)	pests and tree dieback	10
urban development	21 (3)	wild pigs, buffalo	3 (1)
fire frequency	17	All other causes	46 **

*(67) actual extinctions. ** drainage, erosion, salinity and 14 other causes.
Source: Castles/Leigh and Briggs 1992.

Causes of bird species endangerment Australia 1992

Habitat clearing/fragmentation	No. species	Other	No. species
agriculture	67	routine forestry	17
mining	9	fishing	3
monospecies forestry	5	pollution	7
urban development	8	disease	6
Subsequent land management		climate variation	11
grazing/trampling	51	species competition	24
soil erosion	2	predators	45
weeds	12	hunting, trapping	59
altered fire regimes	55		
altered hydrology	6		
loss of nesting sites, niches	24		

Overall total about 150 species endangered, several from more than one of these causes. Causes are often related and sequential eg predators (cats, dogs) now increase with urban sprawl; fire regimes change with climate variation.

Source: Glanznig 1995; data Australasian Ornithologists Union.

The clearing fires and implied biodiversity loss continue: through most of the 1990s around 5-600 000 ha annually were cleared for cattle and sheep in the state of Queensland alone (dry forests). Extending the Australian data globally, and taking habitat loss as a proxy measure of species loss, Table 4.2 summarizes this impact on five of the major biomes in a sample of 18 tropical countries.

Table 4.2 Percent of Habitat Lost to 1980s Since Agriculture Became Dominant Land Use

	Dry forest	Wet forest	Savanna grasslands	Wetlands	Man-groves
			% of biome/habitat lost		
Angola	45	48	17	na	50
Cameroon	69	56	72	80	40
Guinea	71	69	*	na	60
Liberia	20	87	*	na	70
Madagascar	62	84	78	na	40
Mozambique	57	*	20	10	60
Nigeria	70	91	80	80	na
Tanzania	39	80	49	na	60
Uganda	67	71	*	na	na
Zaire	54	57	30	*	50
Bangladesh	*	96	*	96	73
Cambodia	81	74	*	45	5
India	81	56	*	79	85
Indonesia	27	54	*	39	45
Malaysia	19	45	*	35	32
Thailand	78	57	*	96	87
Vietnam	68	79	*	100	62
Brazil	17	31	20	na	na

*Biome not present or na data not available.   These data now considerably underestimate real losses especially of wetlands and mangroves for aquaculture (SE Asia) and universally for urban/industrial expansion.
Source: World Resources Institute 1994-95; data from World Conservation Union, Woods Hole Center, World Conservation Monitoring Center.

Deforestation rates From official UN-FAO data the global clearing rate of all tropical forests in early-mid 1990s was around 15-16 million ha annually;   however these 'official' rates are based on what the various countries report to the chief responsible agency, FAO. Independent and more reliable estimates are much high than the official rates, sometimes to a bizarre degree: eg the official rates for Brazil, Myanmar, India, Philippines, Gabon were respectively 136, 10, 13, 9, 2 hundred sq km/year, while the unofficial estimates by experts with no political axe to grind were respectively 800, 68, 150, 21, 15 hundred sq km/year (Fearnside 1987; Whitmore and Sayer 1992; Mungall and McLaren 1990; Myers 1995).

Biological impacts This is for the forests as forests ie when loss can be considered in terms of the number of plant species it might have contained (seldom surveyed or enumerated before clearing) or in terms of its hydrological function as generator of the cooling towers (ITCZ) and as preventer of the soil erosion which in the wet tropics usually follows deforestation. In terms of zoospecies impact the situation is worse. To 'clearing' now has to be added the category of losses falling under the term 'wild forests brought under management' where such management can range from the benign to the terminal and is in some cases a euphemism for disturbance almost equivalent to clear felling in its biological impact; but because the forest continues to exist on maps and is presumed to eventually recover it is not included in the data for deforestation cited above.

Further, data relating to forest losses through road construction and settlement also grossly underestimate biological impact. This occurs first from the bisecting then quartering of the forests by roads built for whatever purpose into smaller and smaller compartments which at some point become too small to provide the necessary range of some animal species, eg one rule-of-thumb for the continuation of some mammals is a population of 500: any road or settlement plan which divides an initial range into compartments unable to support this number drives them to extinction...*One by one the isolated populations become locally extinct until eventually they disappear from large regions or vanish completely* (Leakey and Lewin 1996; this accurately describes the plight of the orang hutans, forest people, which used to occur up into South China but now only in Borneo and a few in Sumatra). Wilbanks (1994) offers maps of development roads in this compartmentalization process in Rondonia – the single trans-state highway of 1 400 km in 1979 mutated to 25 000 km of road grid network within just 11 years with corresponding fragmentization of the initial forest range.

Reinforcing compartmentalization is the 'edge effect' of further attrition as the roads expose non cleared areas to entry of settlers seeking supplementary income, food, building materials, curios, pets. This has been measured in Amazonia at 2.5: 15 000 sq km cleared there annually 1978-88 resulted in a further 38 000 sq km seriously degraded around these settlements and their road networks (Skole and Tucker 1993). First to go are usually the 'rare and beautiful' pretty much at a rate proportional to their now commercialized prices and ease of shipment out to markets. (The edge effect has also been measured in terms of hunting energy efficiency. In one Brazilian situation the food energy recovery per unit of energy expended was 9:1 in a new settlement but fallen to 2.5:1 in an older settlement; Moran/Vickers 1981.)

Deforestation and Agro Development

Myers (1995) offers a guide to the general causes of deforestation globally; it's now outdated as rates increase but the relative causes remain valid.

Table 4.3   Preliminary Guide to Recent Causes of Global Tropical Deforestation

Activity	Annual deforestation, 1989	
	Square kilometers	% of total
Destructive logging in SE Asia	30 000	21%
Clearing, ranching, Cent. America and Amazonia	15 000	11
Dams, mining, roads, worldwide	12 000	8
Plantations, tea-rubber-oil palm	8 000	6
Slash-and-burn farming, swiddening	77 000	54
Totals	142 000	100

From this the Shifters are the main culprits although their destruction is cyclical; only in the deep forest is there now pristine forest to approach. Nonetheless the 54% of all global deforestation attributed to these folk does suggest their stabilization on sedentary agroforestry farms, forest gardens or a stabilized farm/extractive range, would be a major contribution to reduction of deforestation. Clearing for 'ranching' in Latin America is due partly to agribusiness opportunism when this is by domestic or foreign land *concessionaires* attracted to cheap export hamburger meat; partly to pressures on national governments to repay 1st World debt; and partly to inherited cultural values which prize baronial land ownership for its own sake and for psycho-sociological reasons only Don Quixote in his saner moments could properly explain (but see also Fearnside 1987), with this essentially feudal model then adopted by a peasantry who scramble down from the Sierra to the new green El Dorado for their own place in the sun (Nutrients). The 'plantations' item is easier to approach.

Forests and settlement In recent decades clearing for this purpose in SE Asia has largely been by the Indonesian Transmigration schemes which attempt to divert surplus populations from Java-Madura-Bali to the 'under-developed' outer provinces of Sumatra-Kalimantan-Sulawesi-West Papua, and by Malaysia's Felda settlement-and-rural development schemes.

The transmigrasi settlements date from Dutch times, 1905, and the usual current target is around 50 000 households annually. Practically all are based on rain forest clearing. At 2½ ha per migrant family this would require some 200 000 ha of forest clearing annually for the actual farms, plus more for roads-towns-infrastructure plus the additional edge effects noted above. Apart from adverse socio-cultural impacts on indigenous populations whose ancestral lands are appropriated for the incoming migrants – and reaction by Achinese in Sumatra, Dyaks in Kalimantan, Papuans in that province and recently by East Timorese there – criticisms of the schemes by some supporting donor countries has come to focus on the ecological damage caused and the fact that the farm models for these schemes, lifted from Java, have been in many cases inappropriate for fragile rain forest environments, resulting in both great and unnecessary biological loss and not infrequently farm failure and abandonment (Holden and Hvoslef 1990).

The Javanese farm models emphasize settler food self-sufficiency in minimum time achieved through rice and other quick growing but erosive crops led by cassava and corn. Curiously the models for more sustainable and less destructive farming systems are also present on Javanese farms, in their pekerangan or household tree crops component (described previously, Reconnaissance) but these find little favor with settlement planners although they do often emerge spontaneously. In addition, at base of transmigrasi is the official presumption – codified in Indonesian land law – that the destination forests are...*empty, unowned, useless and abandoned*, as Anderson and Ioris (1992) put it in describing similar forest colonization in Brazil. In fact these destination forests are far from being empty, unowned and abandoned, and the forest agroecosystems of their traditional owners are on nearly all counts except ability to produce large volumes of carbohydrates usually far superior to the Javanese models with which they're replaced.

Let Thomas O'Neill (National Geographic 1994) summarize how he saw transmigrasi in West Papua: *The remains of the forests are burning. Everything looks raw – the ground gashed by bulldozer tracks and heaped with smoking tree trunks, rows of whitewashed cabins thrown together as if from a kit with wooden planks and sheets of zinc. A new Indonesian flag flaps to an anthem of chain saws and crackling flames.*

Felda Felda schemes (since 1956) also have been based on forest clearing at massive scale but the agro model then followed is somewhat different from Indonesia's, being the colonial era species rubber-coconut-oil palm estate, with this now transformed into a local Felda project to provide inputs/marketing facilities/infrastructure for the individual 'farmers' who comprise the scheme. Initially the schemes were modest, around 400 families on 1 800 ha; but the planning assumption that these would spontaneously develop into small urban centers and provide further secondary employment for rural Malays was not borne out; consequently the schemes were increased in size until they approach regional scale. By 1990 some 850 000 ha had actually been planted to the monocrops but settlements planned or in progress for the separate states will cover much more: 1.17 million ha in South Kelantan, 1.0 million in Pahang Tenggara, 0.4 million in Terengganu Tengah... On the ground these statistics lose their meaning. They're probably the largest contiguous clearing operations in the world. Destruction is total. Indeed only when all plant and animal life the forests contained is declared safely extinct are the settlements deemed fit for human habitation (and the contractors paid). *In little more than two decades vast areas have been replaced by uniform serried rows of plantation crops that march endlessly across the landscape. This owes little to the initiative of settlers themselves. It is the product of the managerial and technical skills of a giant authority* (Aiken and Leigh 1992, and one might add the product of World Bank funding of these schemes). Total environmental/ecological costs are probably beyond measurement. As in Indonesia there are many less destructive farm models the planners could have followed as alternatives, mostly diversified agroforestry systems which could be inserted into the standing forests, but these are seen as too 'primitive' to be worthy of consideration (Reconnaissance, Kandy, Genesis).

This time let Odoado Beccari (1904) offer his memories as summary: *On the hill where the former residence of Rajah Brooke used to stand, and in the park around his present residence, are grown most of the cultivated fruit-trees of Malaysia. The Rajah had also endeavored to introduce various kinds of plants which might if acclimatized have proved a source of wealth to the country...*Kuching 1865; and Cornelius and Stein, Tunku Allam and Sherif Ali and Jim and all those other Conrad ghosts who still prowl the wide brown rivers of Borneo would probably agree.

4.2 The Six New's – Some Lost Opportunities

Apart from ethical and perhaps theological considerations there are six main practical agricultural reasons for protecting both natural diversity and that which is a feature of many traditional agroecosystems:

1. *New crops, animal enterprises, foods* By far the bulk of these potential farm enterprises and global food resources are in the 3rd World rain forests, a major practical reason for salvaging what's left of them. Many indeed most of these 'new' resources come down to us

from ancient times. As foods they're as old as H. sapiens.

2. *New pharmaceuticals, drugs, fibers, industrial materials* Again these are primarily from the tropical forests and traditional agriculture, and few of them are really new except as raw material for new uses, particularly the medicinals; eg there are some 2 000 plants in the rain forests known to have anti cancer properties.

3. *New genes* This is the main concern of modern agriculture, particularly of Western field crop breeders. It applies both to the further development and defense of old crops and to potential new crops which require some gene pool from which domesticates can be bred. In the first group Prescott-Allen (1983) have listed the annually increasing number of the world's major food and industrial crops into which wild genes have been incorporated since 1870. A substantial number of these have tropical forest/swamp association, eg rice, sugar cane, cassava, oil palm, cacao, rubber. A sub-group are those for which the gene pool consists of semi-improved landraces now found on 'undeveloped' traditional farms rather than as truly wild forest/swamp resources. Taking Malaysian rain forest fruit tree species as representative of the first group, 24 of the wild species encountered by Saw and others (1991) in their Pasoh forest survey in West Malaysia were already important in cultivation. Counting these and wild species related to domesticated ones, 60 of the 76 total species present would be of value primarily for their genetic material rather than as new crops: *...The greatest significance of the wild fruit trees of Pasoh are their taxonomic proximity to cultivated crops.*

4. *New crop protection* Considered as service resources, the value of this kind of diversity lies primarily in those insect species which are the natural predators of crop pests and provide the basis for integrated pest management (IPM). The insects/arthropods comprise the largest of the animal kingdoms with about 900 000 species. The bulk of these are in the tropical forests. Their service role in crop protection is hardly explored.

5. *New management* This aspect of diversity exists in the different ways in which new plants grow themselves or are grown traditionally in different edaphic (soil, moisture) conditions and as members of associations. This is a pointer to the agronomic conditions they will require when brought under cultivation; it will also suggest the other plant (and animal) species it might be necessary to include as components of a crop-ecological package when the subject plant is domesticated.

6. *New knowledge* This concerns the range of known products and uses

of the subject new plant. These might govern the *ways* in which the species have been grown by traditional managers. This knowledge will be provided by ethnobotanists – but only if the source, the indigenous people, continue to exist, so it repeats Norman Myers' call for the conservation of traditional people and their folk science as well as their plants.

But agrotechnology and new crops development programs are always in a hurry and seem to prefer to spend large sums of money on developing a new agronomy to attend any new crop, knowledge which usually already exists in the indigenous societies from which the crop has been borrowed – or stolen.

4.3 Food Diversity – The Big Store

*We see in our own day barley replaced by wheat, maize preferred to buckwheat and many kinds of millet, while some vegetables and other cultivated plants fall into disrepute...*Alphonse de Candolle (1882).

Just how many potential food plants there are in the world is anyone's guess. Suzuki (1990) thinks possibly 70 000, the bulk of them with tropical forest associations. Some 7 000 globally are known to have been used significantly in the past; 4 000 have reported in Indonesia, 3 000 in southern Africa. In comparison with this potential for increasing global food supply only a handful are now in use and the number is probably shrinking with increasing global urbanization and adoption of simplified but intensified Western farming system models (though this is partly offset by 'new crops' development programs). In brief...*Modern agriculture is only a sliver of what it could be. Waiting in the wings are tens of thousands of unused plant species, many demonstrably superior to those in favor* (Edward Wilson 1992).

A typical Western supermarket offers 28 000 - 30 000 different food items and with migrations and broadening tastes this number increases by 700 - 1 000 items per year. With some exceptions (new foods from New Crops) this apparent cornucopia of diversity is misleading. Much of it consists of me-too items, products essentially the same but dressed differently by slightly differentiated color, taste, shape, size, package, usually to maintain sales of a line by cloning it when sales of a previous clone begin to falter. The number of basically different source crops and livestock types on which this diversity-as-marketing ploy is based have been steadily shrinking. About 20 crops now provide 80-90% of world food supply (Table 4.4) and concentration is particularly marked in the grains where only three crops, wheat-rice-corn, supply about half the world's food supply. It also applies to those relatively very few livestock species which have been extracted from the animal kingdom – 10 mammals (the hard hoofed Old Testament desertifying types) and fowl now supply 98% of all meat, milk, eggs etc with another 40 species making up the rest.

Table 4.4   Degree of Global Dependence on Crops

Food group	Crop contribution to food group	Food group	Crop contribution to food group
Cereals:		Vegetables, pulses:	
wheat	28%	tomato	13%
rice	25	cabbage	10
corn	25	melons	8
barley	11	onions	5
sorghum, oats		carrot, peppers,	
millet, rye	11	garlic, cucumbits	11
Root crops:		Fruit, nuts:	
potato	50%	grape	23%
cassava	23	banana	13
sw. potato	21	plantain	7
taro, yam	na	citrus	20
Oil:		apple	12
soybean	31%	mango	5
sunflower	13	dates, pineapple	3
oil palm	12	pears, stone etc	14
peanuts	8	all nuts	1
canola/rape	8	Sugar:	
coconut	8	cane	73%
olive, sesame,		beet	27
linseed, castor etc	10		

Source: Summarized from Prescott-Allen 1983.

Varieties, cultivars   Nearly 30 years ago Jack Harlan (1975) warned that the deadliest game in Global Village was not so much concentration on a narrowing crop base but, within these few crops, over reliance on a few superior cultivars (Table 4.5).   Harlan, himself a renowned breeder, referred to the rush towards increased concentration and reliance on the few varieties shown as 'these shocking statistics,' because they suggest that...*all of our cultivated crops have such a narrow genetic base as to be highly vulnerable to some new race of a current pathogen or some new biotype of insect pest.* Harlan's main concern was with the erosion and disappearance of genetic resources ('without variation the plant breeder can do little or nothing') and he attributed much of this to the success of modern plant breeding itself in the developing world (particularly the highly focused but ecologically asymmetrical kind which has characterized the Green Revolution). Of wheat as representative of these miracle cultivars and their impact on diversity, and ultimately on themselves...*The famous Mexican wheats have marched across Asia and much of Africa with amazing speed, wiping out old centers of diversity and replacing traditional mixed populations with uniform related varieties. The same thing was happening with rice and the pattern was set, indicating that all other crops would follow. There is no way out, no turning back ... There are two enormous dangers: the variable populations may be lost forever if not collected and conserved in time; and the replacement of adapted mixed landrace populations with uniform genetic materials is an invitation to disaster from epidemics of diseases and pests.*

Table 4.5 US Concentration/Dependence on Food Crop Varieties (1970s)

Food group	Crop	Total varieties	Main varieties	Dependence on main varieties
Cereals:	wheat	269	9	50%
	corn	197	6	71
	millet	na	3	100
	rice	14	4	72
Oils (etc):	soybeans	62	6	56
Root crops:	potato	82	4	72
	sw. potato	48	1	69
Legumes:	beans, dry	25	2	60
	green	70	3	76
	peas	52	2	96
Sugar:	beet	16	2	42

Dependence is in terms of percentage of each of the crop areas under the main varieties.
Source: US National Academy of Science 1972.

A more recent survey of genetic erosion by Hope Shand/RAFI (1997) re-issues Harlan's warning. Globally about 75% of the genetic diversity of (conventional) food crops has been lost in the past 100 years. Nearly all the 158 countries responding to a FAO survey ranked this as a serious problem. By 1990 half of global wheat lands and more than half of rice lands were dependent on a few Green Revolution cultivars. In 25 years (1950-1980s) the number of wheat varieties in China dropped by 90%. In Mexico, the home of maize, 80% of previous varieties are not now known. In Bangladesh a few Green Revolution wheats account for 94% of that crop and 64% of it is a single variety. In Philippines by 1985 90% of the rice lands are occupied by two 'miracle' International Rice Research Institute varieties. Her conclusion is the same as Harlan's, this is the most dangerous game in Global Village.

Livestock   These dangers extend to livestock and are perhaps amply illustrated by the outbreaks of BSE (bovine spongiform encephalopathy) in UK through early 1990s which decimated that industry then went on – as the human version of the disease CJD – to jump the species boundary and kill 69 people in UK by July 2000, and since it has a long incubation period possibly thousands more in future as projected by Wellcome Infections Disease Centre, Oxford. The UK foot and mouth outbreak of 2000 had comparable economic consequences there and extending into Europe. Malaysia's 1998 outbreak of Nipah disease cost over one million dead and sacrificed pigs and 111 dead people. Others like Newcastle disease in poultry occur so frequently around the world (eg Australia 1999) they're barely reported. A common thread running through these events, and dozens more which could be cited, is first the concentration on single species for their profit making potential (ie farming as chrematistics) then concentration of these populations in larger and larger production units to cut costs, factory farms, in environments of increasing artificiality. (And as in management of Nipah virus and pigs H. sapiens have apparently forgotten nothing and learned nothing: Australian pork suppliers were major beneficiaries of the Malaysian disaster, so much so that Australian producers exporting into SE Asia in place of previous Malaysian suppliers are now embarking on piggeries with up to 80 000 breeding sows, replicating precisely the conditions which devastated the Malaysian industry.)

Genetic engineering ('Tis Nature still but Nature methodiz'd)   The concentration of fewer and fewer species in the Big Store's shelves is further consolidated through agrotechnology as genetic engineering, although 'modification', GM, is a term found to be less publicly alarming. The industry's claims of potential benefits and critics' claims of negativities can't be pursued here. But most of the benefits claimed are based on the moral one of feeding the starving millions, with this GM industry viewpoint then amplified by such official agencies as UN

Development Program...*the GM technology can significantly reduce the malnutrition of 800 million people, especially poor farmers working marginal land in sub Sahara Africa* (UNDP July 2001). In the prosperous West those consumers in the Big Store worried about a possibly inherited proclivity to things like cancer and Alzheimer's will be able to forestall them by consuming the new nutriceuticals – strawberries, carrots, broccoli custom engineered for this purpose; and farmers too will come into fabulous fortunes if they subscribe to the nutriceuticals as New Crops soonest: *You could see one field of wheat or potatoes used to produce all the world's blood clotting agents to treat all the hemophiliacs and it would be worth $10 million at harvest* (Val Giddings, VP US Biotechnology Industry Association, Sydney, Aug 2000).

Critics base their case against the technology and support it with much evidence on the potential of these engineered organisms to actually reduce global food supply rather than increase it, by their erosion of existing biodiversity. If GM has any indispensable resource it must be a supply of the necessary genes with which to work. These are from three main sources: existing plants in mainstream agriculture; plants and all manner of organisms from the natural world; and plants, crops, landraces still existing in traditional 3rd World agroecosystems. The GM promise to 'feed the starving millions' is to be achieved largely by engineering crops to tolerate conditions presently unsuitable for farming (desert margins, acid soils, uncleared forests, places of excessive wetness etc). These places are also the repository of the largest gene pool left on earth. To the extent that the new miracle crops are actually introduced into them the clearing or draining of such areas to accommodate extension of agriculture must further contract that pool. Further, actual clearing of some bio-rich area is not a necessary condition for species loss through contamination. Taking corn/maize as example (and one of the world's main food crops, Tables 4.3, 4.4), the genetic bank for this crop still exists in the wilder parts of Oaxaca state in Mexico. To safeguard this resource Mexican government has banned the use of GM maize from 1998. Nonetheless by late 2000 it was found that native criollo maize had been contaminated with genetic material from GM maize, either from that previously grown 95 km away from the contaminated sites or introduced in US GM maize 'food aid' to this impoverished region (Nature Nov 2001).

4.4 Some Ancient Food Palms Revisited

Oakes Ames' observation (Genesis) that no new crops have been invented in the past 5 000 years seems to apply especially to the palms and medicinals; what's new is their rediscovery and new uses periodically thought up for them. The few palms approached here as food – starch, flour, sugar etc – also illustrate two other aspects of diversity: the multiplicity of their products as materials for village scale industry permitting diversity of employment and income, and the high level of biological diversity with which these palms are associated – the nipa and its tidal estuaries, the sago with its swamps, pandanus with the high moss rain forests. While these are human food directly their more important function is as pillars of these ecosystems – with these generating other foods from other food chains, of everything from fish and crocodiles to pigs and birds of paradise. Here it's possible only to sample a few of the palms from Asia and tropical America.

Some Asian and Tropical Palms

Kitul (Caryota urens) is a major multi purpose sugar palm throughout the Sri Lanka wet zone. It coincides with the forest garden zone and usually accompanied by coconut (or replacing coconut at higher elevations) it is an important forest farm component (eg around Kitulgula, Kandy). It's a relic of the rain forests cleared long ago for tea and rubber and wild kitul also survives in rocky wastelands and along streams. (Tappers insist it produces best with dry feet but to the music of running water.) It yields five main food products – Fresh sap tapped from the inflorescence stalk/spadix as *mira* (*neera* in Tamil for palm juice in India, a slightly alcoholic drink); this fermented for a few days as toddy, of slightly higher alcoholic content; mira distilled to *arrack* (brandy); 'honey' by slowly heating the mira until it coagulates; *jaggery* or brown sugar by further heating the honey and stirring until crystallization then molding this in half coconut shells; and trunk starch. Kitul mira/sap is popularly though superior to that of palmyrah and coconut for honey and sugar (verified by chemical analysis, Joachim and Kandiah, 1938); and Charavanapavan (1954) has outlined the methods of making good palm honey. Kitul sap contains more sugar (10-14%) than does palmyrah (7-10%) and yield over the palm's tapping life is 2 500 – 4 000 liters. To obtain these flows the inflorescences are pared each day and these fresh cuts are smeared with a concoction of chili, salt, wood ash, garlic…which means little because each tapper guards his exact formula jealously. (As with most palm tapping this is done by a specialist caste, seldom by the kitul owners.)

Tappers are expert at assessing whether a palm will yield much mira. If little it might be cut down for its trunk starch, and tapped palms at the end of their useful lives might also be felled for starch. By this combination of felling only the 'worst' and exhausted palms their starch yields are low, 5-15 kg, though it is prized and traded as fine cooking flour. Byproducts are the outer trunk for glossy and durable parquet flooring and implement handles and the pith for household fuel. Or alternatively – since most of Sri Lanka's working elephants are also housed in forest gardens – these inferior palms and their small amount of starch might be sold as elephant food or just left as a standing store of household fuel.

In Ceylon the fiber is prepared by a wild tribe of Rodyas and is brought down in the rough to Colombo (Madras Times 1891). And...*In this way the collectors of the Kittul kill a large number of palms yearly, to the great sorrow of the toddy and jaggery makers* (Ceylon Colonial Secretary 1890). The Secretary is referring to the network of black fibers, like horse hair and up to 5 meters long, which bind each leaf base or clasp to the palm trunk and fall naturally as the leaves do, to be collected, baled and exported as fibers for high quality water resistant brushes. With plastics this trade has nearly gone and its former suppliers, the hunter-gatherer Veddas, are nearly gone too. But it lingers. If harvested vigorously, as the 'Rodyas' did, the yield is about 1-1½ kg/palm/year. Commercially it is now collected as it accumulates around the base of the palm. A second fiber (village use only) are the ready made ropes about a meter long supporting the long strings of kitul fruit, used as lashing in house construction and fencing. From time to time agroeconomic budgets are advanced (Dharmapala and McConnell 1986) to discipline the kitul as a plantation/sugar New Crop; but there are too many loose planning parameters for these to be convincing. The kittul, having survived attack by that 'wild tribe of Rodyas,' is best left to forest gardeners.

Palmyrah (Borassus flabillifer) is the economic replacement of coconut/kitul in the northern and southern dry zones of Sri Lanka and dryer parts of South India. There are said to be seven Borassus species throughout the Old World, from the African fan palm of Sudan (B. aethiopum, *delieb*) to palmyrah (India to Thailand) to B. sundaica *(siwalan)* from Java through the Eastern Islands, others into Australia. (But Whitmore (1977) says this whole crowd are a single variable species.) In Sri Lanka palmyrah has a population of about 12 million, mainly in Jaffna (7 million) and on the Mannar coast (2 million), some around Batticaloa and fewer in the extreme south in Hambantota district. In Jaffna and Mannar it supports a large tapping-juice processing cottage industry. Ariyaratnam and McConnell (1986) list these main foods here as fresh fruit (also as cattle feed), tapped fresh juice (mira/neera); this fermented as low alcoholic toddy or this distilled to arrack (brandy); germinated seeds grown as a vegetable and boiled directly or grated into a flour. It also yields commercial/household twines and fibers from its leaves and seeds, and the leaves are indispensable as low cost housing panels (ataps), fencing and house yard windbreaks along that wind blown salty coast. Leaves drop naturally at about one per month. Life would be impossible without it.

In the Sri Lanka's northern dry zone there are so many of them in natural groves that only about one million are in fact regularly tapped and few if any are planted, except as vegetables. (Though the palms are individually owned tapping is limited by government licensing of the main palm sap outlets, the toddy shops.) Like talipot it is a late developer, taking 15-20 years to reach maturity but this offset by the 30-40 years it stays in production. (Tappers commonly might still be tapping some of the same palms their grandfathers brought into production.) Another attribute is that 4-6 inflorescences can be tapped simultaneously (compared with one or two for kitul and coconut).

Attempts at assessing the economics of palmyrah are frustrated by sap/sugar yields being a most rubbery parameter. On the Mannar coast female palms yield about 5 liters and male palms 3 liters per day, but with females tapped for 3 months and males for 5 months of the year (but these yields also depend on the number of inflorescences being tapped). Overall this is similar to India where average yields are believed to be around 150 liters per palm per year over a 30-40 years life. One tapper can handle 20-30 palms. Palm density in Jaffna is particularly high, 170-180/acre. For reasons noted only the higher yielding individuals are actually tapped. As with sago and nipa there have long been proposals to more fully exploit these resources. (They usually collapse because of unorganized collection, dubious product hygiene, lack of market development – and now in Jaffna with war.) For the Sri Lanka groves, assuming only one third of the palms are reasonably productive the arithmetic would be 60 palms (the best) x 200 liters/year or 12 000 liters/acre (30 000 liters/ha/year) and providing employment for 5 full time tappers per ha annually, plus downstream village jobs in jaggery making, fiber processing etc – ie about 10 jobs per hectare. There are very few forms of proper agriculture which could come close to matching the natural palmyrah groves in these socio-economic benefits (Ariyaratnam and McConnell).

The main type of trade fiber 'bassine' is extracted by pounding the base of the leaf petioles. (This industry in India has partly replaced Sri Lankan kitul fiber in international fiber trade; but in Sri Lanka bassine is only a household product.) One final gesture is that the palmyrah trunk makes an excellent sawn construction timber, improving in appearance and strength with age; and whole palm trunks as marine pilings are equally durable for more than 100 years in sea water. On the other hand in India (not Sri Lanka) these timber virtues, compounded by shortage of fuelwood, might well lead to the palmyrah's undoing. Davis and Johnson (1987) have described these fuel demands, especially for brick kiln fuel in Tamil Nadu state's 3 500 kilns each using around 700 tons of wood fuel annually, much of it palmyrah...*The numbers become staggering.* This is in addition to household fuel; the reliance of South India–Bangladesh on palms and timber from forest gardens is noted elsewhere (Kandy).

In India palmyrah's present and historical place has been reviewed by Davis and Johnson drawing on the Palmyrah Research Scheme at Tamil Nadu Agricultural University (Srivilliputtur). In South India it is the second most numerous large plant after coconut, about 40 million. Historically its leaflets provided the same service to literature as did talipot and this is said to have continued over 4 000 years, from the development of Sanskrit writing on them. (Until a few decades ago, with less exalted status these were used as stylus scribble pads in remote schools.) The term 'leaves' of a book has been offered as proof of antiquity and association. There are Indian references to over 800 palmyrah products and uses. It is possible that this is the longest used of all the palms. All the Borassus share the same preference for hot and seasonally dry places, the monsoon lands marginal to the present or recent rain forests. From this preference Whitmore speculates...*it may indicate an origin in the Jurassic when the world was rather dry.* So palmyrah or delieb or siwalan or whatever it calls itself has had more than ample time to accumulate 800 skills.

Sago (Metroxylon sagu) ...The Aru men have no staff of life but they have, however, many sorts of vegetables and raw sago (Wallace 1869). The sago palm is among the tropical world's important and largely unexploited resources. There are thought to be eight species occurring from West Malaysia (where it is not native) through Indonesia, Melanesia, well into the Pacific to the Carolines. Although it is sometimes incorporated into the forest garden plant *ensemble* it is more of a swamp forest agroecosystem in itself, natural or planted. Notable natural resources are the great sago swamps of West Papua (2.5 million ha) and Papua New Guinea (1.5 million ha). Its potential as a new crop has been reviewed by Flach and Schuiling (1989). Stanton and Flach (1980) approached it as a swamp resource. Barrau (1958, 1959) has described it as a food plant of the marsh dwellers of South Pacific islands (Genesis). There has been research on the wild resources at Salawati, New Guinea 'birdshead,' going back at least to 1950. Riezbos (1986) and Flach (1980) reported some sago-based cropping systems. FAO (1986) has discussed development of Indonesia's sago resources. McConnell (1986) described the industry, agroeconomics of smallholder production and mill processing in the Muar/Yong Peng/Batu Pahat area of Johor state West Malaysia where there are about 7 000 acres of planted sago, much of this replacing unsuccessful rubber which itself replaced swamp rain forest. (Rubber plantations on wet peat often get blown down by wind, sago doesn't.)

Sago agronomy is variable, its farm economics casual. Palm planting can be 20 to 30 feet square pattern, 280 plant/ha. However it is of untidy habit and develops into undisciplined clumps from suckers.   When managed (much is simply exploited) all but 6-8 of these stems are removed leaving this number to develop into trunks of a range of maturity stages, typically with the age distribution listed: and the whole clump occupying 9-10 sq meters. 1 - trunk mature; 2 - trunks half mature; 2 - stems emerging; 2-3 - replacement suckers; with the whole clump occupying 9-12 sq meters. These quickly shade out weeds but the leaves continue to fall and although these are cleaned up once or twice a year (the only maintenance, 10 days/ha/year) the whole thing is soon again in disarray. (These messy habits are one reason why sago is not welcomed in the *ensemble* of forest/kampung gardens; it's best left in the swamps.) Trunk starch accumulates from year 4 to a maximum of 200-300 kg/trunk at year 9-11 then declines sharply, so it is first harvested at this age and thereafter one trunk is harvested from each clump each 15-17 months, giving about 200 trunks x 200 kg or 40 000 kg/ha of dry starch on an annual basis…abstractions because many farmers sell a few trunks whenever they need money.

In Johor the small sago extraction mills are concentrated along a tributary of the Simpang Kanan river near Batu Pahat. For farm transport to them the trunks are cut into 40 inch lengths, wired together into strings and pulled to the mills by boats along the waterways which lace that country, and the sago logs are tethered to a mudbank until needed. For buoyancy these logs first have to be de-barked at the farms; but if transported by truck they are not, with the bonus that when de-barked at the mill this as fuel provides 10-15% of the heat to dry the recovered starch, or bake it to form the familiar 'pearls'. (All this, including harvesting, is commonly done or arranged by the purchasing mill so the farmer need not do anything much at all.) The outputs are pearled sago and fine flour, particularly valued for its non-stick properties in making Chinese noodles, *mie hun,* and the byproduct is residue pith as pig and pond fish feed.

Sago leaf produces the best quality ataps for roof and siding panels (among other things used for the classical rural houses of Malacca) and lasting 8-10 years. Palm spacing for atap production is much closer than it is for starch; leaf harvesting starts at 2-3 years and thereafter occurs each three months as new leaves emerge, this constant working keeps the grove neat and orderly until the palms grow too tall for leaf harvest and are abandoned to grow into starch.   About 8 000/acre whole leaves are harvested annually, these are woven into 140 atap panels (6 x 2 feet, of four leaves thickness), usually by Malay village housewives working between their other chores and doing 40-50 panels each per 5 hour day. It's a major cottage industry – low income, few alternatives. Marketingwise, as they say around Yong Peng, if ever a plant needed to seek legal advice it must be Metroxylon. Its name has been stolen, its recipes abused, its economic patrimony purloined by that awful pretender from Venezuela, *Manihot esculenta*, aka tapioca aka cassava, whose products thrust themselves forward among those 30 000 supermarket items as sago. 'The Aru men have no staff of life'…what a strange idea.

Nypa (fruticans) is another ancient food and general purpose palm and ecologically an especially important one. Its present distribution is from Bay of Bengal-Vietnam though the Eastern Islands to Australia and out to Solomons, Carolines; but there are fossil records of it from Brazil, Africa, even Europe, and it has recently been reintroduced (1970s) to Nigeria (Hamilton and Murphy 1988). Its habitat is a tidal zone between the seaward mangroves and the shoreward dry lands (or sago zone if this is present), particularly in river estuaries where it is both a foundation of littoral/marine ecosystems and, by entrapping silt and debris, an agent for raising the shore line. Pioneer farmers in Malaysia take advantage of this and plant nipa to reclaim mud banks along the littoral and 'dry' these new farms until they can plant other permanent salt tolerant species (starting with coconut); or alternatively they settle directly into an existing wild nipa grove and farm that, with or without additional plants, and subsist from nipa products and fishing. Its semi aquatic habitat is illustrated by optimum seed germination occurring when this is inundated by tides for four hours per day, but it can also be planted (as seed or seedlings or pieces of rhizome, like bananas) on land out of tidal reach. Nipa, bringer of fish and land creator.

Nipa throws its majestic leaves directly out of the mud to a height of around seven meters, with central fruit bearing stalks reaching one to two meters height. Both the leaves and the fruit stalks when tapped yield dozens of traditional products and many more have been commercially experimented with (eg the leaf and leaf spine/petiole for paper and particle board, the inflorescence stalk sap for industrial alcohol, vehicle fuel, the hardened fruit endosperms as vegetable ivory). The main actual village products are sap tapped for sugar, the small opaque fresh fruit (2-8 kg/palm/year), the leaves for *atap* panels, but these leaflets also woven to hats, raincoats and mats. Thick basal parts of the leaf petioles are also chopped and boiled to obtain culinary 'salt'. Fong (1992), Hamilton and Murphy (1987), Burkill (1935), Brown (1951), McConnell have outlined these uses. (Hamilton/Murphy offer extensive references for Bangladesh, India, Philippines, New Guinea.) On the east coast of Peninsula Malaysia in Pahang and Trengganu one village industry is cutting/processing the cylindrical 'swords' of emerging new leaves for *daun rokok* cigarette wrappers (these unopened leaves also used for storing cooked food); and on the west coast the main industry is weaving atap panels. In Malaysia fruit collection is only a minor household income supplement but in Thailand boatloads of it are brought down to the Bangkok markets.

Commercial efforts to exploit nipa, the wild groves or these replaced by plantations, have centered around its juice for sugar/alcohol. Like that for sago development much of this has been desultory, tentative, going back to the 1900s (Philippines), and in Sabah cars were run on nipa alcohol until WWII. (In the 1980s there were still two alcohol factories in Sarawak but no recent data. For vehicle fuel it can be added to standard gasoline/petrol to a ratio of 1:4 before carburetor modification is needed.)

Palm density in wild groves is very high, around 2 500/ha but if tapped only about 700 of these yield much sap. For commercial exploitation this wild population is reduced to about 500, and if planted this is reduced further to 400/ha. Whether wild or planted, commercialized nipa must be handled sternly – the tapping stalk kicked, punched, bent, rarely patted or stroked. Dennett (1927) and Hamilton/Murphy (1987) have described these various means of bullying the thing to flow to its maximum ('gonchanging'); they vary from India to New Guinea. (The 10 ha Sumatra nipa enterprise, below, was to have 5 of its 38 workers daily devoted to these assaults). They're quite specific, in New Guinea: bend the tapping stalk 12 times in one direction; slap it backward and forward 64 times; kick it 4 times; repeat these 4 times per week. Elsewhere this persuasion is less gentle – in India it's attacked with a mallet and in Philippines kicked every morning.

For alcohol as liquor, old boys around the Selangor Club still claim the first serious attempt in modern times to tame the wild nipa was by a distillery set up in that state in the 1920s. Like much else it never recovered from the war, and remembering Nipa Palm Scotch they admit this was not altogether a bad thing. On plantations for alcohol or sugar Dennett (1927) gives comparative per ha annual yield as:

nipa ...	15 600 liters	sweet potato	7-18 000	coconut	5 000
sugar cane	3-7 000	cassava	3-9 000		

For sugar production (14-17% sucrose) the comparison is 20 tons/ha/year from nipa and 8-9 tons from sugar cane (Java). In comparison with other palms one major advantage is that all tapping work is near ground level so that there is no need for skilled (and dangerous) climbing. From economic budgets for a small 10 ha model plantation in Sumatra, Hamilton and Murphy list nipa's other advantages over sugar cane as the greater per ha employment it would offer; this would be permanent year-round work in place of the seasonal and socially disruptive operations of a sugar estate; there would be no bagasse and its polluting byproducts to be disposed of; tapping the wild groves would require no agricultural land and this would save rain forest...perhaps. But commercial success of *planted* nipa might well threaten the natural stands and terminate their ecological role: the best place to clear and plant it would be where it plants itself. Fong has outlined the deterioration of wild groves in Malaysia in face of accelerated commercial pressures for their *atap* leaves. What was once a sustainably managed communal resource is now diminished by entrepreneurial over-harvesting. Similarly, technical success in developing paper and chip board would probably spell the end for the wild groves. Throughout the region there is already much pressure on the nipa lands (and mangroves) by clearing them for proper agriculture – and high pressure aquaculture ponds, seaside villas, yacht marinas and similar fatal idiocies.

Pandanus spp (screw pine) occurring through the Old World tropics from West Africa to Polynesia are now thought of mainly as an important weaving material (especially Polynesian soft mats) but it is also of continuing importance as a staple food in New Guinea (Hyndman 1984, Powell 1976, Barrau 1958). Worldwide there are 500 members of the Pandanaceae of which 66 occur in Papua New Guinea, from sea level to 3 000 m elevation. One area of concentration is the catchment of the upper Fly and Ok rivers, in the Star and Hindenburg mountains where there are thought to be seven species. This is in the territory of the Wopkaimin which ranges from lowland rain forests to high moss forests. Within a relatively small area studied by Hyndman the Wopkaimin themselves distinguish 41 pandan varietals. They have domesticated 30 of these for edible nuts and fat and plant them in groves, each having a mix of several types. The lowland forest groves also commonly contain sago and breadfruit but at higher elevations the groves contain only pandanus. These are small plantations of 100 sq meters with about 10 palms and each Wopkaimin man has several of them scattered through the forest.

Depending on variety the main food products are edible nuts or fat which is smeared over greens or taro. As well as being specific with respect to elevation and productivity (meaning yield, type of product, taste, seasonality) these many pandans are important and specific with respect to the kinds of game they attract – some are favored by large cassowaries, others by ground animals, parrots, birds of paradise. Pandans are acknowledged generally as the most important nut/fruit/fat source of the New Guinea highlands. (They are also associated archeologically with the rain forest origins of agriculture in Melanesia, and have been a pillar of agro survival systems throughout Oceania since ancient times (Genesis).)

Talipot (Corypha umbraculifera), claimed by Sri Lankans as the largest of the palms though not the tallest, is one of 6-8 species occurring from India to Australia. (Its size could be exceeded by Mauritia, below.) It takes 35-40 years to mature then throws up a huge inflorescence 30 feet across and 20 feet high, the largest of any plant and one of the great sights of the forest gardens in the Kandy valleys. It produces trunk starch at maturity similar in uses to that of kitul but taking so long to do it does not impress its owners, though it was an ancient food source. (According to Whitmore (1977) one of the other species C. elata in West Malaysia yields only 90 kg of this food.) In any case its modern use is for *tal-atta* umbrellas made from its leaves and traditionally carried by monks and tourists. Talipot is of greater cultural significance; the earliest extant Buddhist books (c 500 AD) were written on its leaflets with stylus. But monks tap their current epics on PCs. The talipot is out of vogue. In their 1970s surveys of Kandy farms Dharmapala and McConnell found farmers no longer bother to plant it, leaving this to bats.

South America ... More Palms and Food

From Clement's (1993) review of the tree food resources of Amazonia these could exceed 250 species when the epicenter of NW Amazonia is fully explored. He calls attention to the opinion of ethnobotanists that tree foods, with fishing and hunting, once allowed an Amazonian population of 5 million with little or no ecological cost to the forest – in contrast to the 15 million in much the same area who, with proper farming, can barely scrape a thin and temporary living by cutting it down. Kahn (1993) reports there are 39 genera of tropical American palms of about 200 species. Some 25-30 of these are known to produce a range of foods, but ethnobotanists could probably prove that a much larger number are used or have been used on substantial scale in the past, and that most if not all of these 200 have

Table 4.6 Habitats and Possible Management Systems for Some Amazonian Palms*

Species			Some products	Natural habitat	Possible system	Present commercial use
Acrocomia	latiospatha		fruit, oil	savanna	FAX	
Acrocomia	salerocarpa	(coco catarro)	yellow oil	dense groves	XFA	S
Acrocomia	totai	(parag cocopalm)	fruit, oil, fat	dry lands	PFA	S
Acrocomia	vinifera		food	dry lands	XF	
Arecastrum	roman		little oil	dry erlands	XF	
Aphandra	natalia		fibre, ivory	dry erlands	XF	
Astrocaryum	aculeatum		fruit	dry secondary forest	XF	
Astrocaryum	chambira		fibre, nuts (sim.coconut)	well drained	XFA	
Astrocaryum	jauari		fat, oil, hearts	flooded riverbanks	XF	
Astrocaryum	murumuru		fat, fruits	swamp forests	XF	S
Astrocaryum	vulgare	(tucuma)	fat, fruit, oil	savanna, sand	PAFX	S
Bactris	gasipaes	(peach)	hearts, fruit, starch, oil	dryer lands	PAFX	S
Batia	capitata	(cocos)	oil	NA	FXA	
Elaeis	oleifera	(am. oilpalm)	oil, butter	swamps	PAFX	S
Euterpe	oleraacea	(acai)	hearts	swamps, alluvial	XFA	S
Euterpe	precatoria	(acai)	hearts, leaflets	swamp forests	XFA	S
Jessenia	bataua	(milpesos)	'olive' oil, medicines	swamp forests	XFAP	
Jessenia	polycarpa		'olive' oil	moist forests	XFAP	
Manicaria	saccifera		'coconut' oil	swamps, rivers	XFAP	
Mauritiella	aculeata		fruit	swamp forests	XFAP	
Mauritia	flexuosa	(miriti)	oil, ivory, fruit, starch, wine	swamp sands	XFAP	S
Mauritia	vinifera		(same but more oil)	swamps	XFAP	
Maximiliana	maripa		fruit, oil	savannas	XFAP	S
Oenocarpus	bacaba		'olive' oil	dry primary forests	XFAP	
Oenocarpus	distichus		'olive' oil	dry primary forests	XFA	
Orbignya	martiana	(babassu)	indus. oil, animal feed, coke	humid rainforest	PAFX	S
Orbignya	speciosa	(babassu)	(same)	dry-to-wet	PAFX	S
Orbignya	oleifera	(babassu)	(same)	dry areas	PAFX	S
Orbignya	phalerata		oil, starch, animal feed	dry areas	PAFX	
Phytalephas	macrocarpa		fruit, ivory	swamps	XFA	
Sheelia	excelsa	(palm de vino)	'coconut' oil	riverbanks	XFA	
Sheelia	macrocarpa	(coroba)	oil	dry sites	FAX	

F - forest gardens; A - agroforestry on commercial small farms; P - monocrop plantations;
X - extractivism or small scale harvesting from wild stands. S - those species of which significant use is now made.
All of them have local indigenous use; all of them would have a place on local (mixed) forest gardens.
'Olive' and 'coconut' oil means having oil uses similar to these products.

Source: Compiled from Kahn 1993, Clement 1989, 1993, Clement and Urpi 1985, Balick 1979 and other sources.

yielded some range of products. A few with their products, habitats and possible management systems are listed in Table 4.6. This also suggests the appropriate development strategies for each species (table footnote). Some arid zone palms are noted elsewhere (Genesis).

Relatively few of the palms listed lend themselves to large scale commercial plantations (P), on edaphic grounds (preferred wet habitat, water logged soil) or economics (low yields, high costs, low prices). Most would be best exploited as natural stands (X) eg following the extractivist Combu Island model (Reconnaissance). Most also would best be approached with minimum ecological and social impact as components of mixed forest gardens (F). (A sample of American arid zone palms are later encountered in Genesis.)

Palm oils Balick (1979) has discussed the oil yields and production potential of 26 Amazonian palms (Table 4.6). With breeding and agronomic development some could significantly supplement production from the African oil palm, Elaeis guineensis, or perhaps in future partly replace it. When grown as monocultures (as also in Malaysia, Sumatra) this import is subject to pests and diseases. One of the Amazonian palms, Elaeis oleifera can hybridize with guineensis and its genetic material is of value for this reason, as well as being an important oil producer itself. Wilson (1992) offers one of the barbassu, Orbignya phalerata, as an example of potential oil production...*even though still harvested in the wild and semiwild states it gives the world's highest known yield of oil*, a stand of 500 trees producing about 125 barrels of oil annually (3 940 US gallons, 14 900 liters). Assuming 170 palms/ha as with Malaysian oil palm this would be a yield of 5 100 liters/ha/year, compared to 5 000, 1 600, 1 000 kg per hectare respectively for oil palm, coconut and sunflower. Of these and palms in general Balick notes that most populations are...*extremely variable. There are differences in individual fruit size, tree height, yield, susceptibility to predation...which are of vital importance for future breeding.* These are lost when only a few typical palms are spared in forest clearing, or when their seed is transferred to an *ex-situ* conservation facility, botanical garden.

Babassu The geography, products, promises and plight of the babassu (Orbignya martiana) have been reported by May and others (1985) and Unruh (1994). This is one of Brazil's main palms, occupying 200 000 sq km (1980s) and partly supporting 450 000 rural households. Its trunk and leaves have 24 non-food uses and 6 food-medicinal uses, its fruit 21, mainly as food. In spite of its vast area and the 2 million people who depend partly or mainly on its natural stands this traditional palm-based economy is under threat from two directions, government land development policy and inappropriate technology. (This oil industry based on *babassu* kernels is said to be the largest anywhere using wild resources.)

Subsidized land clearing for cattle ranching has destroyed large areas (one million hectares in 13 years in Maranhaõ alone), and land alienation on the babassu cattle frontier has made larger areas no longer accessible to traditional palm users. Development of industrial scale technology for cracking and separating the nut has altered the market from one for a multiplicity of kitchen products to one for the whole fruit for its oil kernels. Projecting the socioeconomic consequences of this latter, May et at believe that...*rural people will have to pay for industrialized substitutes of the goods that they previously obtained free.* The chief beneficiaries are the new class of commercially oriented palm-land owners. Fewer people will find employment in nut-only harvesting than previously in household processing. Men as paid agro-industrial labor displace women who drew a substantial equity from the previous cottage industry. Babassu is also said to be a rapid colonizer and restorer of degraded agricultural land (Anderson and Anderson 1983).

Palms of the aguajales Murrieta Ruiz and Ruiz (1993) have outlined the economics and potential importance of a group of palms (aguaje) occurring in the *aguajales* of Peru Amazonia, permanently flooded marshes along the Upper Amazon tributaries covering 6-8 million ha and about two million ha support more-or-less pure stands of these palms, ranging from 180 to 350/ha. Main species are Mauritia flexuosa, M. vinifera, Mauritiella (table), with also populations of most of the other palms listed for swamp habitats (Jessenia, Elaeis, Phytelephas etc). The *aguaje* palms proper (M. flexuosa, vinifera) produce 100-200 kg fruit/year containing an average of 24 kg oil. As an edible fruit this is one of the richest in the world (protein, oils, minerals vitamin A) and the edible oil is as good as soy, corn, cotton, olive, sesame. Another product, fruit paste mixed with water and sugar, is 'Amazon milk', as good as cows milk. Collecting-processing-selling this fruit is already an important industry in Peru Amazon towns (Padoch 1988).

Peruvian agencies (assisted by NY Botanic Garden, Orstom of France) have done much work towards finding economic but sustainable ways of exploiting these immense natural food resources, without land clearing and in country which often suffers food shortages and relies on imports. Apart from palm foods these aguajales are of great ecological importance, being the home of the black crocodile, jaguar, anaconda (and many poisonous snakes, another reason why they've not attracted development). Murrieta and Ruiz see one possibility as their transformation to extensive low cost aquaculture: fruit falling from the palms and other trees naturally supports large fish populations.

Mauritia flexuosa This palm of the *aquajales* in Peru is also abundant throughout Amazonia and because of this vast distribution, high density and economic uses Kahn (1993) ranks it the most economically important species. From the populations given previously (180-350/ha) its fruit yields have been measured at 6.5 tons/ha in Peru to 9.1 tons in Colombia. In addition to the products noted above the trunk contains up to 60% starch by dry weight. This makes it and the aguajales economically similar to the sago and its swamps in New Guinea. Unlike the sago, flexuosa is not yet similarly exploited but Kahn sees it as an important potential 'sago' resource. Flexuosa is claimed as the largest palm (to the chagrin of talipot), up to 30 meters tall and almost two meters circumference, with leaves three meters across and each a full load for a man – probably the reason Wallace compared its dense groves to...*a vast natural temple which does not yield in grandeur to those of Athens* (Kahn). But Clement (1993) adds this temple...*is being devastated around Iquitos by predatory fruit harvesting for market.* (Others have noted that, unlike SE Asia, many palm species are harvested by chopping them down.) Its other uses in Venezuela are noted below.

Açai The ecology and economics of these two palms have been described by Anderson (1988), Strudwick and Sobel (1988), De Castro (1993), Kahn (1993). Açai do Para (Euterpe oleracea) is a multistem palm concentrated in the Amazon estuary and Guianas (to Venezuela). Its wild stands are the major Brazilian source of canned palm heart (c. 60 000 tons 1980s). An important rural food/beverage use is its fruit pulp mixed with manioc/cassava flour and sugar as a staple food. The second açai, Euterpe precatoria with a single stem, has a wider range in central-western Amazonia to Peru and Bolivia. De Castro has described fruit collection from the wild stands along the middle Amazon, supplying Manaus markets. Here the dense palm stands of the seasonally flooded varzeas (swamp forests) contain around 180/ha adult fruit bearing palms, with about this number of infertile palms, thousands of seedlings and 30 other botanical families also present. Annual fruit yield (by climbing and cutting the bunches) is 2 000 kg/ha. Main commercial use of the fruits, with size and appearance of cherries, is their diluted pulp as non fermented açai wine (vinho do açai) sold by street vendors in Amazon towns. De Castro also notes that this açai is an integral part of forest gardens in this part of Amazonia...*açai is always planted in the polyspecific orchards which surround varzea houses in central Amazon.* Here it is more of a staple food, eaten with fish or mixed with manioc flour. Precatoria with only a single stem is not a suitable basis for a sustainable palm heart industry; nonetheless Kahn reports such use in Peru where a cannery severely depleted the wild groves. Here other uses are *chonta* salad made from the tender young leaflets, and the roots as medicine for relieving malarial fever. Both these açai fall within the palm group best left to managed extractivism, indigenous agroforestry systems, or mixed forest gardens (Table 4.6). They are also multi purpose palms – the inflorescens used as brooms, the trunks for house construction, the leaflets for ataps/thatch.

Jessenia Jessenia bataua is another of those palms in Table 4.6 which have high local food value but don't lend themselves to large scale commercial cultivation. (Its preferred habitat are swamps and seasonally flooded forests.) It provides important components of indigenous Indian diets in Amazonia as do others such as Manicaria and Astrocaryum, Oenocarpus. Balick and Gershoff (1981) analyzed its food products – fresh fruit, oil (to 50% of pulp), a milk from the pulp and the residual pulp as human food or pig/poultry feed. From this and feeding trials they concluded that the oil was comparable to fine olive oil, the milk a substitute for cow milk and the fruit protein 40% higher in food value than soybean. The oil is of such high quality that boutique specialty production from wild palms or small farms has been suggested.

Pejibaye Peach palm, Bactris gasipaes is one of the most important general purpose palms of tropical America. Its uses all have increasing modern relevance: food (cooked carbohydrate fruit), baking flour from its starch, vegetable oil, palm heart for canning/export, the fruit also as a widely used animal food. Its potential as a farm crop has been discussed by Clement (1993, 1989) and Clement and Urpi (1987) and its farm production economics in Costa Rica by Johannessen (1966). There is opinion that if as much research had been directed at pejibaye and barbassu as at the African oil palm these might become to tropical America what coconut is to Pacifica. Pejibaye's long association with H. sapiens is indicated by the fact that it has never been found in its wild state. It is cultivated widely on drier land, *terre firme*, so is one of the relatively few palms which could lend itself to large scale plantation development (P, Table 4.6).

From research at Yurimagues, Peru, Pedro Sanchez (1989) has

outlined one system for exploiting the pejibaye by interplanting it very densely (1½ x 1½ m) as seedlings in upland rice and harvesting it for palm heart after only one year for a yield of 1.5t/ha. This does not kill the *pejibaye;* it coppices, is thinned out and begins producing fruit when four years old. Crops of rice and cowpeas continue to be grown under the palms until they are shaded out at which time some legume ground cover crop (kudzu, centrosema) is planted. Palm fruit production reaches 10-15 t/ha/year and is equivalent to sweet potato as human food or animal feed. Production continues for 15 years then the palms can be cut down and used for parquet flooring.

From other research here and observation of Amerindian systems Sanchez has also pointed to good possibilities of finding lesser known species suitable for acid soils of the wet tropics. The well known acid tolerant species of mainstream agriculture include rubber, oil palm, coffee, cacao, but these face inelastic international demand. The lesser known species with high value/low volume – apart from pejibaye in its many management systems – include guarana (Paullinia cupana) for soft drinks, brazil nut (Bertholletia excelsa)...*and a myriad of tropical fruits for juice concentrates, ice cream and other uses* (Sanchez).

Panama hat palm   In their pursuit of the panama hat palm in Amazonian Ecuador (Carludovica palmata of three species and occurring from Mexico to Bolivia) Bennett, Alarcón and Cerón (1992) also list its many other products. Ecuador exports about a million hats a year, woven as an Indian household/cottage industry, but these authors estimate the palm's other products would have at least as great economic value though few of them except fine handicrafts baskets have yet entered the market. In Ecuador the palm is widely used traditionally or commercially by eight tribal/linguistic groups (including the Shuar or once notorious 'Jivaro'). Apart from fruit its main food products are the unopened leaf buds which have been compared to asparagus and often eaten as a salad. These authors see considerable commercial potential in this because among other things pulling out these young 'asparagus' leaves does not kill the plant, as does recovering cannery palm heart from other palm species. (Another use is oil extracted from the seeds by some Peruvian groups.) Although associated with lowland Amazonian rain forests C. palmata prefers disturbed open sites. Where it does not occur naturally it is sometimes planted and to this extent it's been domesticated. It also volunteers or is planted in managed rain forest swiddens (Shifters). On these and economic grounds it's probably best left on small forest gardens.

Colenda Feil (1996) in coastal Ecuador and Colombia found the palm Attalea colenda as forest clearing remnants and without any genetic improvement or management producing 18 kg of edible oil/palm/year from its seeds/kernels or 900 kg/ha at a natural population of 50 palms/ha. By comparison the African oil palm – but only after decades of breeding and agronomic research and with fertilizer and intensive management and a population of 170/ha – yields 3.5-7.5 tons oil/ha and no supplemental income from pasture. On a direct palm -v- palm comparison colenda thrives in dry areas with rainfall below 1 500 mm where the African oil palm doesn't. In Ecuador colenda is also grown as a component of mixed agroforestry systems (with coffee, bananas, papaya etc). Feil's reason for lack of interest in it is worth noting…*(farmer) unawareness that pastures may produce oil as well as cattle*, though this ignorance is almost universal and based on belief (as well as tunnel vision agronomy) that trees and 'crops' are competitive not complementary. Within wide bands nothing could be further than the truth.

Motacú Monica Moraes, Borchsenius and Blicher-Mathiesen (1996) chased the *motacú* palm (Attelea phalerata) through most of Bolivia into Peru and found it thriving everywhere – on islands in the wet savannas, flooded forests, deciduous forest, terre firme rain forests, a thousand meters up the Andes, like colenda somewhat subdued in pastures, producing fresh fruits, brooms, baskets, bract ashes for chewing with coca, long lasting thatch for sheds and Lonely Planet bus stops…but mainly oil which fends off time by preserving one's hair and preventing it from turning gray, though babies like it too for their skin and the roots when boiled are good for tuberculosis – if the generous motacú has not been early cut down for its palm heart salad. Another virtue is small size for most of its long life and ease of harvesting – first fruiting when less than a meter tall and the fruits hang to the ground, 3-4 meters when 25 years old, still producing well at 30 years (5 meters) before eventually reaching 10 meters. Most palms are exploited in dense natural stands. As a potential component of small mixed farms, forest gardens, the indicative oil yield is 3 liters of oil (with 2 liters from fruit mesocarp and 1 liter from kernel) per infructescence/fruit bunch of 500 fruits, with 2-3 infructescences/palm/year or 6-9 liters/palm. In Bolivia Moraes et al observed a few plantations with populations of around 250/ha; hand extraction methods (hammer-boil-skim) are still primitive but profitable (local oil price $15/liter).

Palms and Life – A Summary

Balick (1979) has also outlined the central place of 15 palms in the life of the Guahibo people of the savannas *(llanos)*, rivers and forest margins in Venezuela and Colombia. Some of these and a few of their products are listed in Table 4.5. Others include the natural 'styrofoam' of the petioles of Mauritia flexuosa used for fishing rafts. Because of this property and their vast numbers along the streams Balick suggests they would support a packaging materials industry (but this palm's main potential is as oil/food, Table 4.5).

More fundamental, Guahibo mastery of palm technology is ascribed to their previous nomadic hunter-gathering way of life on the *llanos*. Moving often they had to depend on the palms wherever they happened to be for just about everything from housing to clothing, food, oil, fat, medicinals, weapons, fish traps, alcohol... One notable dependence is on palm leaves for fabricating extremely fine or tight insect-proof housing, seasonally essential on the llanos. Similarly mastery of 'basketry' is ascribed to the need for many kinds of these, strong but light, in their wanderings. When the mosquitoes get too much another use of M. flexuosa is as a convenience liquor store. For this a palm is felled, holes are cut into the trunk and a liquid accumulates...*to a strength for rapidly getting drunk*; and if necessary the Guahibo make the rounds of several palms. Also, when a senile palm falls it rapidly fills with large Coleoptera larvae which are a prime protein source (about 500 per palm which Kahn says in Peru are now sold as a delicacy in Iquitos markets). Visitors have always been struck by the importance of this and other palms in Guahibo life. Balick cites Rivero (1956) who found this in a history of the Spanish missions in Bogota, apparently written by some wandering friar about 1730...*This (palm) is the earthly paradise of the Guahibos; this is their delight, their universal larder, their everything. This is the material of their conversation; about this they dream; and without this they could not enjoy life.*

4.5 Of Cures and Curanderos

While the Big Store shuts its doors to food diversity it eagerly holds them open to medicinal plants from the tropical world and to the village healers, shaman, obeah men who know how to use them. If these were stripped from the Big Store's shelves many would be empty. Wilson (1992) notes that even in the sophisticated USA 41% of pharmacy prescriptions are extracted from natural organisms – 25, 13, 3% from plants, microorganisms and animals respectively; and of the 119 pure pharmaceutical substances used globally 88 were discovered by leads from traditional medicine. Norman Myers (1997) estimated the global retail value of pharmaceuticals from rain forest plants at around $100 billion annually. In US-Canada (comparatively species-poor with only 10% of global flowering plants) Farnsworth and Soejarto (1985) estimated that with the then current plant loss rate its cost in terms of medical products sacrificed would be $3.25 billion annually by year 2000. (Incidentally these authorities note that globally only 5 000 plants had been *comprehensively* analysed by 1985 and that rejection of them and tens of thousands of others for specific purposes, 35 000 for cancer alone, does not rule them out for other future uses, if they're still extant.)

P.N. Suwal (1970) in his Medicinal Plants of Nepal reminds that medicinal knowledge of tropical (and Indo-Eurasian) plants goes back to the Rig Veda written down before c 1 600 BC with an eight section Ayurveda dealing with all aspects of healing (still flourishing in India, Nepal, Sri Lanka with departmental status). Suwal's Ayurvedic pharmacopoeia includes such mundane things as the large Himalaya cardamom spice (Amomum subulatum). Nim Dorji and McConnell (1989) pursued it through Bhutan monsoon forests as farm economics but quite missed its Vedic virtues as aphrodisiac, cure for gonorrhea, scorpion sting and snake bite, in any sequence. Currently one of the ancient plants of the Ayurvedic–Chinese stream, wormwood, is being described as 'the last shot in our locker' against emerging new strains of malaria. In recent historical times c 1660, Richard Grove (1992) has described how search for medical plants for their colonies led the Dutch to Malabar where they learned the local Ezhava caste had a more sophisticated medico-botanical classification system than did Europe and allowed one Dutch official in their then Cape Colony, a certain Meinheer Drakenstein, to have the Ezhava texts translated as the 12 volume Hortus Malabaricus and start promoting conservation of Cape Colony's own forest plant resources.

One evolutionary reason Wilson offers why plants contain these valuable substances is that they are themselves superb chemists...*Through millions of generations each kind of plant has experimented with substances to meet its special needs...the products have been tested by natural selection one generation at a time...*Once these active ingredients are identified and used as prototype for artificial synthesis they can give rise to new drugs for human use (the pan tropical 'weed' Madagascar periwinkle, Catharanthus roseus, with its alkaloids for cancer treatment, the meadowsweet Filipendula ulmaria and its salicylic acid as aspirin). Joining the curanderos and obeah men to scratch up some of this Vavilovian gold one learns that 80% of world population still relies largely or entirely on them and their plants – often unwillingly because 3rd World health budgets can't stretch to modern medicine, sometimes because science while not denying a malady exists can't handle it ('melancholia and sickening brought on by evil deeds'). One also learns from Coe and Anderson 1996, in Nicaragua but universally, that...*ethnobotanical knowledge is being lost even faster than species and habitat.* But much abides.

Habitats and chance   It's generally agreed that where both sources are present some two or three times more medicinal plants will be exploited from disturbed habitats than from primary forest, these being chiefly household/forest gardens, chacra swidden weeds, roadsides, secondary forests. One scientific reason (Voeks 1996, Abe and Higashi 1991) is that plants in both habitats have constituted their defenses (Wilson) differently...*It appears that pioneer and other fast growing species have more often evolved inexpensive (to the plant) lines of defense such as alkaloids and phenols whereas climax forest trees have tended to evolve more expensive structural defense systems such as cellulose and lignin.* From Piperno and Pearsall (1998) humans inadvertently created these disturbed habits in the forest by their first incursions there;   from Edgar Anderson (Genesis) the first useful 'crops' including medicinals occurred on garbage dumps round settlements as weeds where experiments were made with them;   and from Carl Sauer and Donald Lathrap (Genesis) the useful dyes, fibers, medicinals, magicals, from a forest would have been brought to the household for reason of collecting convenience, to which one might add minimizing danger to the collectors from the forest 'zone of sorcery', and worse.

Reliance on disturbed habitats as source, especially household/forest gardens, seems to increase along a gradient according to how much forest has been cleared in the vicinity;   but this also reflects historical chance events and the high levels of opportunism and flexibility-in-use of the plant growers/foragers in substituting plants for others no longer available, usually now because of forest clearing.   At one extreme might be the Yanomami Indians of NW Brazil (Millikan and Albert 1996), still forest dwellers and using 113 medicinal plant species and funghi but only 11 of these are cultivated around their villages.   Their recent 'chance events' include decimation by measles of the older generation who carried the medicinal knowledge, especially older women, and incursion by gold miners who brought malaria, the former resulting in plant lore being taken over by surviving men not able to pass it down through a female line of curanderas as the women had done;   but the latter leading fairly rapidly to Yanomami discovery of the necessary anti-malarial plants (though a general knowledge erosion now continues with availability of patent medicines).

In the village of San Andrés in Guatemala Simon Comerford (1996) found only a few old curanderas still carrying the knowledge with their Maya language, none with an apprentice, using around 80 plants of which 22 were grown in housegardens, 58 on roadsides, chacras, secondary forests with none now taken from the little remaining primary forest. (Here their adverse chance event had been programs by the 1980s military dictatorship to break up Mayan culture, including language and pharmacology, to make it more amenable to Nation Building.) Coe and Anderson (1996) found the Garífuna people of eastern Nicaragua using a total of 254 plant species for all purposes, 229 of these medicinals with many multipurpose (eg the annatto bush Bixa orellana as common red dye and condiment as well as medicinal). The Garífuna's historical event – in this case positive – had been their racial and cultural composition, based on initial Carib Indian and blended by chance with French and English colonial, slave African, with these adding their layers of plant lore now leavened by knowledge borrowed from their latest neighbours the Miskitu Indians.

Except in the deep forest, household gardens are central to these traditional healing systems and conservation of their plants. Coe and Anderson found 37, 34, 15 and 14% of these Garífuna curing plants respectively are herbs, trees, vines, shrubs and 74% were native to eastern Nicaragua. The health care network typically consists of 'wise old housewives', bone setters, midwives, general practitioners, shaman specialists, extended by obeah men into the spiritual realm to confront troublesome dead ancestors and evil, with these professions reflected in the composition of their garden pharmacy, or what they collect elsewhere...*Dooryard gardens owned by shamans have a species composition distinct from those owned by midwives and the general populace,* with the snake bite shaman growing his plants (Mikania cordifolia, bitta-wood Quassia amara etc) for this purpose, whereas in Maya Land Euphorbia lancifolia is valued by village pediatricians to stimulate the milk flow in nursing mothers. A broader function of these curers' gardens has been to provide ethnobotanical training for the young...*From a very early age children are encouraged to participate in care of dooryard gardens, assuring the passage of at least some ethnobotanical information from one generation to the next* (Coe and Anderson). Many of these garden plants are technically wild, being transplanted from the forest for reasons of convenience. On the other hand these authors found the Garífuna shaman must now make canoe journeys of

2-3 days looking for their special plants because of deforestation (though perhaps also because as happens elsewhere there's belief that wild plants are more potent than domesticated ones).

More disturbed habitats   In constructing his regional plant pharmacopoeia for Bahia, Brazil, R.A. Voeks (1996) found over 200 species used there which, after putting aside the magicals and staples of obeah men, left him with 100 species he lists together with their parts used (mainly leaves), modes of preparation and 24 afflictions treated, from syphilis to snake bite. In contrast to the Yanomami's reliance of primary forest plants. Voeks found 90% of his Bahia plants were from disturbed secondary forests, over 70% of them herbs and shrubs rather than rain forest trees and vines, and of the trees used about half are not native to Brazil and are cultivated. The 'chance historical events' which apparently led to reliance on plants from disturbed habitats, including their garden cultivation, are similar to the experience of the (initially Carib) Garífuna in Nicaragua. The initial Bahia pharmacopoeia was that of Tupi-speaking Indians (*with a cosmology and ehtnomedical system as rich as that of their Portuguese invaders* —Voeks), but seen as obstacles to Christian conversion and with the 'full fury of the Jesuits' directed against them the shamans and their ethnomedicine were both wiped aside within one generation, although not before the missionaries absorbed at least a smattering of this New World plant knowledge. It was replaced largely by the colonists' own *materia medica* brought from Portugal-Spain to their plantation 'kitchen gardens'. The next wave, slaves out of West Africa, carried an even richer layer of ethnomedicine to these gardens, though the plants were new, and being confined to their masters' sugar fields they had to apply the old knowledge to their weeds and whatever else was at hand in this by now most disturbed of disturbed habitats, never quite making contact with the Great Forest as the Tupi and the Yanomami had done.

Medicinals from the forest   Johnston and Calquhoun (1996) conducted such a survey of Kurupukari Indian lands in Guyana, prefacing their results with the note that of 122 Amerindian groups in Brazil the ethnobotanical resources of only 20 of them have been comprehensively studied, ie these forest resources remain largely unknown except to indigenous users. This group of 65 forest Kurupukari were using 130 different plants, most multipurpose, for a total of 246 uses (ie number of species used for all of six use categories), led by medicinals which accounted for 40% of all plant uses. In terms of plant species, again allowing for multiple uses, about 64 of these 130 species had one or several medical uses, 13 of these plants specifically and solely for malaria. None had yet been screened, nor had any of the other plants been scientifically tested for their uses in skin irritation, snake and insect bites, ulcers, sore eyes etc. Johnston and Calquhoun also found a wide range of fruits and handicraft species which might be developed commercially, the basic idea being to harvest these together with timber as a permanent sustainable agroforestry system.

There are practical development problems, if the planning strategy is to 'pick winners', ie attempt to identify those plants with greatest commercial prospects, greatest expectations. First, until the medicinals and fruits are screened the Kurupukari wouldn't know which ones to concentrate on. (This doesn't apply to the timbers which commercially are well known or the handicrafts, mainly palms, which have screened themselves over millennia.) Second, small output volumes discourage commercialization. But third, if the Kurupukari and others like them were persuaded into specialist production (of medicinals, fruits) to boost supply there's no guarantee this demand would last or that these 65 would benefit from it (the dynamics of pharmaceuticals, the fickleness of consumers in the Big Store). These practicalities and risks considered, the best strategy would seem to be – for the Kurupukari and ourselves – to put least trust in Princes, and most trust in continuing diversity as modified and managed extractivism, Kurupukari Gardens of Complete Design (Reconnaissance).

Areca ... Friend, these are the thirteen qualities of tambula not available even in heaven. Areca catechu is probably the most numerous palm of all under cultivation and to betel chewers—when the nut is chipped, leavened with a sprinkle of lime, garnished with a clove or two, perhaps a pinch of country tobacco and wrapped in a leaf from the betel vine (no relation)—certainly the most indispensable. Bulman (1988) has charted the distribution of these lotus eaters from Natal in the west through all of S-SE Asia to Taiwan then south through Ocean's gates at Santa Cruz to Fiji. This is about where Carl Sauer's ancient fisherfolk wandered and invented agriculture (Genesis), so areca would have been among their earliest crop suite. But that was in pious times. The betel trail which Bulman follows goes back only to 13th century BC, then later to Chinese rumors of 'black tooth people in distant lands'. These days it scrapes in under the heading of food in that its unopened leaf cabbage is still appreciated as a vegetable or its cabbage sap made as 'tea' in parts of Indonesia, but only if it has to be cut down for some other purpose, usually construction poles. It belongs more firmly with the drugs and medicinals: cure for asthma, anaemia, indigestion, eructations, constipation, diarroea, dysentery, worms, sleeplessness, bad breath, sexual disinterest...and to mid 19th century, curing nervousness in Chinese prisoners awaiting beheading (Bulman citing Morarjee and others). In India the Bearer of the Betel Bag was an important royal functionary and in battle the betel carriers, often women, doubled as field surgeons.

In his survey and review Bulman found 76 Areca species with 24 of these occurring in Sulawesi-Borneo-Malaysia (but only one species cultivated), indicting this area as the likely origin. Use-concentration and economic importance as a farm crop are greatest in S. India and Sri Lanka. In Kerala it's both a small mono-stand plantation crop and mixed with other trees-palms-vines on forest gardens (Kandy, reconnaissance). In Kerala some 28% of garden area is devoted to areca (Bavappa, Nair, Kimar 1982). Similarly in Sri Lanka Dharmapala and McConnell found most Bearers of the Betal Bag these days to be the forest gardeners (Kandy). Only 5-6% Kandyan farms *don't* grow areca + betal vine; usually they have 60-70 palms/household which yield 4-500 nuts/palm/year from year 8 with their economic life continuing for 40-50 years, but this can reach 100 years. One important use is palm shade and microenvironment for coffee and cacao, with the palms' own direct outputs as nuts, leaf spathes for food containers/wrappings, papaya-papain latex tapping receptacles, cafechu extracted from nut for hide tanning, poles for construction. Common nut storage methods on farms to counter price fluctuations are smoke drying then storing in sacks over the hearth fire, or covering with water in an earthen jug and burying it in some moist place. Black tooth people in distant lands.

Fevillea – things that burn with a clear blue flame ... In view of the inappropriateness of temperate-zone agricultural systems in the poor-soil lowland tropics it is especially critical to find forest products that can be harvested on a sustained yield basis without destroying the forest (Gentry and Wettach 1986). They investigated oil from Fevillea vine fruit (F. cordifolia and pedatifolia) in Peru Amazonia as one of these possibilities. The actual number of species (8-10?), their geographical distributiona and uses are not well known but traditional medical uses of the oil include 'curing leprosy', snakebite, a general poison antidote (F. cordifolia, the antidote vine in West Indies), rheumatism, jaundice, purgative and there are several minor actual and major potential industrial uses. The vine fruits are remarkable for their large size (to 14 cm diameter), large seeds (to 6 cm or about 9 grams) and their very high seed oil content (about 55%)...*or on an oil weight per fruit basis among the highest ever recorded for any plant.* Extracted and dried the whole seeds 'burn slowly with a clear blue flame' and are used for house lighting by the Campa Indians in place of kerosene lamps. Similar reported use is to cut the seeds into cubes and skewer them on thin sharp sticks as candles. In this part of Peru Gentry and Wettach found F. cordifolia's habitat to be the seasonally flooded forests where it drapes itself over trees while pedatifolia apparently tolerates poor laterite soil, suggesting that varieties/cultivars could be developed for a range of edaphic and farm conditions. If planted in a forest – from observation of how it plants itself – indicative production would be: 70C vines/ha x 100 fruits/vine x 15 seeds/fruit x 0.75 gram oil/seed = 790 kg oil/ha, or this multiplied several times with development *(without cutting down a single tree)*. Alternatively it could be added at almost zero cost as an element in many agroforestry systems. (We'll take some over to Otto Soemarwoto in Java. He'll want them for his Garden of Complete Design – Reconnaissance.)

Neem The Big Store might also want the neem tree for its pharmaceutical and agricide departments. In Africa and S-SE Asia crop losses to weevils, insects, vermin, fungus in farm storage commonly range 30-40% of the harvest (and incidentally one reason why a continuous flow of produce from mixed tree crop systems is superior to a seasonal grain crop which requires such storage). Globally there are around 20 000 crop and storage pest species, mainly in the humid tropics, and around 1 600 plants/products are used traditionally to counter them. Representative and most celebrated in India-Pakistan are the neem tree's several products (Azadirachta indica) also used as human medicinals since Vedic times (antiseptic, hepatitis, rheumatism, leprosy, syphilis, ulcers...and now commercial toothpaste and cosmetics). Its agricultural uses (Saleem Ahmed and Michael Grainge 1986) start with natural seedfall of 30-50 kg/tree/year which is depulped then crushed yielding seed oil and cake residue, this latter spread over a field as fertilizer and to discourage pests in the subsequent crop. Water from this process can be sprayed as a natural pesticide on the growing crop. After harvest the grain in farm storage (usually a mud bin or silo) is protected by mixing 2-5 kg neem leaves in with each 100 kg of grain. In addition leaves are ground to a paste and this mixed with the mud at time of silo construction. For grain storage in sacks these are soaked in neem leaf water overnight before filling. Neem oil sprayed on crops has proved effective against locust attack. By use of these several methods 2-3 neem trees are enough to protect one ha of grain as crop and stored product.

The long living neem (200 years) is also esteemed as a timber tree, for land reclamation in semi arid areas (Africa to Americas, Philippines), urban and livestock shade (Middle East), its cake residue as cattle feed and its pulp as feedstock in farm/village scale methane gas generating plants.

Unruh (1994) has described how the neem spread throughout Sahel, its seed carried by birds and bats, since introductions were made in Senegal-Mali-Sudan early in the 20th century.

Parting salutations   It's well established in Maya medical circles that the most effective curing days are Thursdays and Fridays (Comerford); that for each illness there's a specific prayer to be recited as the plants are chopped and mixed; that a little aspirin or Vicks Vaporub (or axle grease in Nicaragua) can often fortify the mixture; that the curanderos mixture is always better at fixing snake bite and pregnancy than is modern medicine but for this to be effective the recuperating patients should not push their luck by eating fruit from long creeping vines... (We also learned from Milliken and Albert the best way of tricking Yanomami babies into taking The Mixture is to sprinkle it over sticky banana juice smeared around their mother's nipple and let them go for it.) But it's also well known in corporate pharmaceutical circles (Coe and Anderson citing extensive research) that 70% of the 229 Garífuna medicinal plants do have one or more bioactive substances. Samoa has around 900 vascular plants and George Uhe (1974) found 150 of these used in traditional medicine, formerly by priests, many aimed at countering daemons but many also efficacious at a physical level. So putting aside the more fanciful aspects of shamanism and the distraction of lonely planet trippies in their restless quest for the magicals an immense body of medicinal lore remains. On a global basis statistically this is about twice as likely to be 'right' than 'wrong' in offering pharmaceutical and medical research an economical guide to plant substances which might be efficacious for particular ailments.

But with the forests disappearing and the curandera ladies getting older and some missionary zealots confusing the obeah men with devil worshippers and the sorcerer's apprentice no longer willing to follow the obeah man for 20 years to learn his trade, and Nation Building generals breaking up the curanderos' culture to grab their land, this favorable probability is disappearing too...before Meinheer Hendrik Adrian van Rheede Tot Drakenstein can add it to his Hortus Malabaricus.

4.6 Biodiversity Conservation Strategies

Most conservation strategies consist of one or a combination of those listed. Underlying these are two general conditions: (a) a change in social economic values to those more conducive to conservation by whatever means, and (b) for in situ conservation a greater more meaningful role of indigenous peoples in the conservation process.

Table 4.7 Strategies for Tropical Plant Conservation

A. Ex situ

Intensive preservation	...	seed banks, tissue collections
Standing collections	...	botanical gardens, aboria, zoos
New Crops	...	dissemination of old crops to new owners

B. In situ

Public specific	...	scientific reserves
Public general	...	parks, biospheres
Sponsored parks, reserves	...	anticipating biodiversity values
Carbon forestry	...	diversity and Greenhouse
New/old industries	...	requiring ecosystem preservation
Vignette farms	...	the showcase Ark
Traditional crops	...	specific races, cultivars, species
Traditional farming systems, forest gardens	...	mixed plant/animal associations
Traditional extensive land use	...	shifters, extractivism

Ex Situ Conservation ... Preserving Much, Losing Much, Using Little (Hope Shand)

From FAO (1996) surveys there are about 1 300 national and regional *ex situ* crop plant genetic resource storages of significant size in the world, consisting of seedbanks (90% of accessions), standing plant collections in 1 300–1 400 aboreta and botanical gardens (8%) and *in vitro* tissue collections (1%).   These facilities contain about 6 million accessions. However only about 400 of the 1 300 *ex situ* facilities are secure medium- and-long term facilities meeting international standards; only 35 countries operate adequate long term storages; 45% of all accessions are held by 12 countries and only three of these are in the developing 2nd and 3rd Worlds (Brazil, China, India). By far the bulk of this *ex situ* material is from the 3rd World.

This geographical imbalance is paralleled by the kind of materials held (Shand/RAFI 1997):   40   and   15%   respectively   are   cereals   and pulses/legumes. Wheat alone accounts for 14% of all accessions. While this is reasonable in that 'most of the world' now relies on these few crops (previous tables) the thrust of global *ex situ* conservation of largely 3rd World materials is towards those species of prime importance to the urbanized 1st World, and in this towards those offering the greatest commercial potential to seed companies. On the other hand those crops of interest to the tropical 3rd World are not well served by *ex situ* conservation with eg only 0.5% of accessions being cassava, 0.18% various yams and 0.16 plantains/bananas...*Many of the plant species most vital to subsistence farmers and the household food security of millions of poor people in the South, including non domesticated species, are grossly under represented in genebank collections* – Shand. The comparative advantages /disadvantages of *ex situ* conservation have been canvassed by Prescott-Allen (1983), Altieri and Merrick (1987), Gradwohl and Greenberg (1988) among others. Seed storage is possible for up to 100 years but facilities are expensive and consistent high management required (reasons why *ex situ* conservation is concentrated in a few First World countries). Even here security is not guaranteed. Interest might wane with changing food fashions, farmer needs, government budgets. Seed collections must be rejuvenated periodically by growing them out to whole plants and in the larger facilities like Russia's Vavilov Institute this a hugely expensive exercise. The accidents of power failure, natural disasters, war, pose other threats to *ex situ* security.

The ex situ alternative for those species which can't be stored as seed (plantains, roots, breadfruit etc) is storage as standing crops; but it's likely that the enviro-climatic conditions at any site will suit only relatively few species, cultivars, types. The others will hang in there, as in a museum, but they might not thrive. At more fundamental level Altieri and Merrick find weakness in *ex situ* conservation in that storage of seed, tissue, standing crop...*freezes the evolutionary process by preventing new types or levels of adaptions or resistance to evolve because plants are not allowed to respond to the selective pressures of the environment.* There are increasing objections at political level as the 3rd World sees their plants taken off for storage to the First for the benefit of science. Science often turns out to be the bottom line of a multinational's balance sheet. They can have its plants back at any time and now much better than they were before, as HYVs or miracle cultivars (now fitted out with terminator genes), and while these might now need new supports to prop them up (artificial fertilizers, agricides, irrigation and advanced technology generally), fortunately the multinationals are in a position to help with these inputs too. Which leads the campesinos to a dilemma: if they purchase all this as the Package they often find in toting up its full costs that the net benefits are illusory; if they don't they'll be chastised for this by their Prince for being backward and ignorant. They'll survive well enough with their old landraces but their role in the 3rd World is of conservator of resources for eventual use by plant breeders of the 1st. The 1st World has not been much interested in these problems; its lawyers have been too busy filing for patent right on the campesinos property.

New Crops and the new dispersals   The New Crops development industry is also a form of ex situ conservation. Some species are threatened by land clearing in their places of origin, notably for fruits the epicenters of Borneo-Malaysia and western Amazonia and part of this treasure trove is being relocated to Australia, Florida, Israel, South Africa, Mediterranean countries, Oceania, by official agricultural development programs there. (Eg one list of so-called New Crops for possible use in Australia runs to 4 236 species, most from the tropics – Fletcher 1993.) In addition from 1950s many newly independent states began attempts to broaden their agro economic and domestic food supply base away from monospecies colonial agriculture – Fiji and Samoa from coconut, Sri Lanka from tea and rubber etc (Kandy). Taking the example of new fruits into Australia, plant importers there report about 200 of these being currently in the pipeline from epicenter to farm but only about 30 of these are truly rare and fewer will eventually make their way into the Big Store to compete with the apples-oranges-bananas triad. The rest won't be discarded. They'll be adopted by plant *aficionados* whose purpose might be to set up a few exotics on their lawn as conversation piece – or no particular purpose at all – and include attempts to collect all the main tropical fruit trees of the world (7 000 are vaguely mentioned).

This group also collect, exchange and import their own, sometimes without much formality and in such novel forms as cuttings disguised as coathangers when the need arises. Through their enthusiasm they might be a decade or two ahead of the New Crops development departments. These are those compulsive 'keen seekers of plants' we've met before – soulmates of Posey's Indians setting up their shopping malls in Amazonia; General McDowell unloading his boxes of wampee on the Colombo docks (Kandy), the old Rajahs Brooke in Kuching, the *ribereños* of Santa Rosa.

In Situ Conservation ... Leave it Where it is

The general advantages of conservation by this broad strategy can quickly be summarized: Plants left to fend for themselves on reserves, parks, biospheres or grown under adverse conditions on traditional farms have a resiliency to pests, diseases, low fertility, aridity, properties they can develop and maintain only in their testing environment. This is particularly important when what is of value is not simply the possession by the plant of genes for use for achieving higher crop yield, but ability to withstand those pests and pathogens which coevolved with the plant as parts of its inheritance. There can be thousands of these separate packages for landraces, notionally one for each combination of soil type, moisture regime, associated microfaunal set etc, all of these differing sufficiently to give to each *in situ* conserved population some degree of uniqueness. In situ conservation in the tropical world can procede along a maze of strategies, most complementary.

Parks, reserves, biospheres   This is the ecological ideal, often also the economic ideal when these upstream biospheres serve the pragmatic purpose of protecting downstream coral reefs and fishing industries (Sian Ka'an reserve in Yucatan) or rice irrigation water supply (Dumona Bone, Sulawesi). It's also mandatory in that discussion so far has ignored the bulk of the world's species, the insects, earthworms, algae, fungi etc without whose services in soil formation, nutrient conversion and pest control the higher plants as agriculture or potential agriculture could hardly exist. With few exceptions (insect houses at high cost) these service biota are not amenable to *ex situ* conservation. Their prospects for *in situ* conservation on modern high pressure farms are not much better.

Globally there are about 30 000 publicly dedicated bio-protection areas of all types covering 13.2 million sq km (UN List 1998), ranging from smallest local reserves for a single species to the large UNESCO sponsored biospheres, about 200 of these, the largest said to be Manu in Peruvian Amazonia of 1.5 million hectares screened by another 0.3 million hectares in which traditional land use is permitted. Another new one in Surinam covers 1.62 million ha of mainly rain forest in that country. However only about 4, 2 and 6% of the rain forests of Africa, tropical America and Asia respectively are reserved (Wilson and Peter 1988); and of such reserves in general, as Gradwohl and Greenberg (1988) put it, *far more tropical forest is preserved on maps than on the ground.* A related estimate is that about 6% of the world's potentially useful land could be devoted to bio-reserves before these became directly competitive with agricultural production. This is with a global population of 6 billion (to probably double within 80 years), most in the 3rd World demanding land, so prospects for diversity conservation through these formal reserves on the scale required are dim.

There are then three main potential strategies (they're partial and complementary): (a) continue with attempts at park/biosphere expansion until the competitive land limit is reached; (b) intensify the best existing productive agriculture, which can be made sustainable (but most of this is in the 1st World so it also requires changes to global food distribution systems and equity considerations); and (c) fundamental changes in the peasant farming systems for which most tropical land is cleared, and incorporation of bioconservation functions within these systems. Most of these latter are low input agroforestry systems. Under (a), limited global conservation budgets are best spent on areas which have both large number of species *and* highest endemism, ie species unique to such areas. Norman Myers and colleagues have recently identified 25 of these 'hot spots' around the world. They cover just 1.4% of earth's surface and would protect 44% of its plants and 35% of animals – ie they're a sort of minimalist fall-back position as far as H. sapiens' future is concerned. By far the bulk of these 25 areas are in the humid tropics and 14 of them have already been stripped of at least 90% of their cover, usually for unsustainable peasant plow farming. Formal protection of these 25 core areas, the ideal, would cost $480 million annually.

The world record in paper parks is probably held by Indonesian planners in West Papua who have devoted one sixth of that province to them (including some real ones). The largest, 1.5 million ha Lorentz, with its 16 000 foot Puncak Jaya, tropical glaciers, lowland forests and Arafura coastal reefs must be among the most valuable scientific real estate on earth; but with mining in the area park boundaries have proved most fluid. Ignoring paper parks the real ones can have several practical limitations. Dedicated with fanfare they often become resented as a drain on the public purse, then with insufficient maintenance budget become a haven for feral animals and threat to nearby agriculture, dangerous stores of combustible materials in the wildfire season – Hazards valid or fanciful but easily stimulated as urban sprawl encroaches on them and realtors contemplate their higher use. There's always pressure on them – for their land by hungry campesinos, their extractive fauna and flora by unemployed villagers, their timber by corrupt administrators and GDP-deprived governments. Destructive logging in Indonesian national parks by Suharto's cronies would be a case in point; and now with post-Suharto democratization the pillage is greater. But it's unfair to select these examples from the 3rd World. In 1998 Australia's government warmly endorsed uranium mining in that country's UNESCO approved World Heritage Kakadu National Park on grounds that this would provide jobs, increase exports and boost the GDP. Or taking that country's most famous and largest park, the Great Barrier Reef marine park, these can be undermined internally by porous Thou Shalt Nots (trawl fishing through the coral) or externally by agrochemical pollution runoff from uncontrollable sources, often agriculture, outside the park.

Sponsored parks These have begun emerging in various forms. In one case (Wilson 1992) Merck, world's biggest pharmaceutical company, paid Costa Rica $1 million in exchange for screening its biodiversity and rights to develop substances found, plus royalties on these, the money earmarked for biodiversity conservation in that country. It is anticipated other corporations will follow, sponsoring conservation in specific areas even without such *quid pro quo* – led by those with environmental image problems. While these sponsorships or subsidies can make all the difference to a poor country they're peanuts to an image-conscious multinational (and money is better spent this way than on court cases which, won or lost, only call attention to the corporation's alleged villainy).

Parks, reforestation, carbon forestry and conservation This path is an adjunct to carbon forestry and carbon emissions offset projects. In poor countries they're one means of cross-subsidizing diversity restoration with financial returns from the carbon absorption component. Some carbon forestry projects are positive; others based on quick-growing carbon gulping forest monospecies (conifers, eucalypts) will be negative. Much also depends on whether preservation of *existing* forests will be granted carbon forestry status under the Kyoto protocols, as distinct from carbon credits earned by planting new ones.

Ecosystem conservation through new/old industries Again this is largely incidental, resulting in conservation of whole ecosystems because one of their components takes on some particular commercial value. On the Australian NW coast there's a thriving tourist industry based partly on fine seafood the jewel of which is large mudcrab. These things are normally caught and boiled and that's the end of it. However Aboriginal people, also now engaged in this 'bush food' homestay business, snap the main claws from the crabs then release them in the mangroves to grow more, with the whole littoral-mangrove ecosystem now valued and protected because of this single renewable product. Vietmeyer (1996) offers a similar example from crocodile farming in New Guinea where previously the croc was hunted nearly to extinction. Now farmed for boutique meat as well as skins (as also in Africa, Australia, SE Asia) the young reptiles for this are caught in swamps, with these and all else in them now valued because of this one product. Use it or lose it.

Vignette farms These blur the line between in/ex situ conservation. They're typically a small working farm where local plants/crops/animals are concentrated, aimed at metro or 1st World visitors, a small room in ecotourism by no means confined to the 3rd World. Others collect plants/animals regardless of origin or contemporary practical use. They're even found in the Kandy hills, Sri Lanka, performing three conservation roles. First is species/cultivar collection. They're mini agro-botanical gardens and like formal ones the more diversified the more successful they are. The second is conservation and showcasing of the folk technology in processing these products. (Watching cinnamon being peeled is far more interesting than watching it grow and takes less time.) Since the trippers' time slot is inflexible another important function is condensed instruction in what diversity looks like and why it could be a good thing. The trippers – willingly paying much more than necessary for a vignette pack of spices on their way back to the bus to support the vignette farm – are a good thing too. They're economics and goodwill, more powerful than scientific pleading.

Commerce and organic markets This also comes under the heading of new industries, owes little to government action and taps the world wide trend towards organically grown produce. This doesn't automatically support diversity (it might mean organically grown monocrops of cabbages), but it's more closely associated with traditional farms including highly mixed ones like forest gardens than with undiversified systems. Organic produce *per se* is now big business worth $7 billion annually in Europe alone and in North America it's growing at 25% annually. Some forms of this trade tap an association with the rain forests ('new fruits from Brazil'); and one export firm in Sri Lanka obtains its produce exclusively from forest gardens there (Forest Garden Products), its overseas marketing structured around 'farms which save the rain forests' as well as new crops and their clean green rain forest image. Again this is a reminder of the power of commerce in bringing tropical diversity conservation within reach of distant urban consumers, especially when supported by ecotourism and hands-on vignette farms.

Diversity as traditional crops *'There are about three billion farming people in the world. They have almost infinite capacity, experience and application to select and maintain crop germplasm'*...David Wood, plant collector. About 60% of the world's agricultural lands are farmed by traditional subsistence households; their contribution to genetic and species conservation is immense. The amount of external official attention, money and scientific expertise directed at *in situ* conservation in this 3rd World is very small in comparison with that directed at *ex situ* conservation in the First. An emerging exception has been through the efforts of non government organizations (NGOs) to assist some 3rd World governments and their farmers in this kind of peoples/village conservation. Since 1988 the Biodiversity Institute of Ethiopia supported by Canadian NGOs has provided a model for farmer involvement in agro-resource conservation in Africa (the Seeds for Survival Program). Since 1993 this has expanded globally as the Community Biodiversity Development and Conservation Program. (Not surprisingly this impetus came in Ethiopia, Vavilov's 8th center – Origins.)

By this approach traditional farmers would be encouraged and if necessary subsidized by the 1st World for preserving diversity. So far this has been defined narrowly and technically as cultivars, types of some important crops (potatoes, cassava, rice, barley...) especially those important to the 1st World. The several conditions for success (Altieri and Merrick 1987, 1988) are that farmers understand *what* they are to conserve, *why* they are doing it and be involved in planning *how* at farm level it can best be done. To these must be added necessary farmer *incentive* and intelligible policy and administrative structure before farmers are approached. The first three can usually be assumed: traditional farmers have known all about the what, why, how for millennia. The problem is in educating the developers not the farmers. In experience with about a hundred official agro development projects through the 3rd World the author can't remember a single one of them that did not have the sweeping away of old crops, old cultivars, old socio-agricultural systems, old technologies in the shortest possible time as its main objective, specified or implied. Almost by definition these projects have little or no incentive to preserve the traditional methods/crops which they're set up to replace. This need not be so.

Diversity and agricultural development projects   Bio/agroconservation could with only slight technocratic pain be specified as an objective in most rural development projects, especially the more comprehensive integrated rural development projects, IRDPs, which such agencies as IFAD, UNDP, FAO, World Bank, bilateral donors, have been running for decades. It would require that development be redefined to include deliberate non-development of targeted landraces, cultivars and their associated agrotechnology and folk knowledge, non-development as a shadow activity. This is easily represented as neocolonialism, a call to freeze traditional land managers along with their plants in some sort of agro-ethnological zoo; and in some SE Asian countries it's resented as interference with what governments see as their civilizing mission among their more 'backward' indigenous subjects. And who will pay the subsidy, the zoo's upkeep? Why *should* plant breeders of the 1st World pay for diversity when these simple campesinos might maintain it of their own volition? In fact for the basic food crops there's a fair chance they will: *As rural populations become increasingly impoverished a sizable portion of the peasantry are renewing use of the traditional varieties and low-input management practices* (Altieri and Anderson).

This hunt for traditional diversity also finds much in the past. Two recent titles by a prestigious authority convey its flavor: *Lost Crops of the Incas* and *Lost Crops of Africa* (US National Research Council). The first suite contain some 80 Andean tubers, grains, vegetables, fruits (Srivastava et al 1996). Outside this region the high altitude root crops are practically unknown; the only ones to be universalized have been a few of the potatoes (some others of the root crops are listed elsewhere – Origins). The African suite south of Sahara contain over 2 000 food plants; less than 20 are well known globally.

Traditional farming systems This is similar to the previous strategy except that preservation of whole agroecosystems, related plant associations, blocks of mutually dependent diversity is the objective rather than some single crop element. Representative of these traditional farms are forest gardens (the wilder the better), the more sustainable forms of shifting and forest extractivism. The basic condition is that these be recognized and promoted as legitimate agriculture, for their biodiversity value as well as their other attributes. There are four main factors working in their favor as conservation vehicle, according to Table 1.2 (Nature).

Diversity purposes Diversity conservation already exists in the multiplicity of species and cultivars kept and reasons for doing it. There's no need to repeat these lectures to the forest gardeners. These farm plant inventories have been compiled by many workers around the world (Kandy, Reconnaissance). In addition to the economic plants a theme running through these inventories is the large proportion kept for no particular reason, some spiritual or half forgotten cultural connotation or something as vague as habitat. This also implies presence of a wide range of microfauna dependent on these many plants, and more diversity in their higher faunal consumers. So this repository, this last genetic bank which Michon describes in Indonesia, and Pereira and Rajapakse in Sri Lanka (Kandy) has many rooms. One of these is the household pharmacopoeia – drugs, medicinals and magicals; and with many of these now included in the IUCN Red Book list of endangered plants (particularly in Africa) this 6 000 year old chapter from the Rig Veda is also drawing to a close. Concern is enough to cause an international conference (Bangalore Feb 1998) to urge botanical gardens and community groups around the world to set up medicinal plant conservation programs. Janet Alcorn (1984) has similarly called attention to the conservation function of the highly diversified Huastec farms of Mexico (Reconnaissance).

Economics and costs Another of these rooms is farm economics. Some traditional forest farming systems are at least as profitable as any alternative. On the other hand most strategies targeted specifically at biodiversity conservation (parks, reserves, botanical gardens…) incur costs in administration, wages and land use productivity foregone. Species and genetic conservation on traditional farms are already a 'profitable farm activity' otherwise the forest gardeners wouldn't, incidentally, be doing it.

Domestication, breeding Forest gardens and most other 'underdeveloped' agroecosystems have been involved in these activities since ancient times. In this they go far beyond both *in situ* conservation on bio-reserves and *ex situ* storage in seed boxes. It's no accident that the people of Pohnpei in Micronesia have 155 varietals of breadfruit, 50 of taro. This diversity extends to the shifters, recalling that Katherine Warner found the Tagbanwa in Philippines had bred and were using up to 20 varieties of rice, these in particular socio-ecological niches. Even those troublesome hillbillies, the Mountain Ok of New Guinea, have selected and cultivate 41 varietals of pandanus, again according to ecological zones. And Rico-Grey says of the Maya farms of Tixpeutal, Mexico…*Before the Spanish conquest 80% of the garden trees and shrubs were a product of selection from the native elements of the flora by Maya people.* That is, these gardens were centers of domestication and breeding long before Western plant science was invented (Genesis). That these plants are sitting in an ethnological museum waiting for Science to come and clean them up as New Crops, new genes, more bars of Vavilovian green gold, might strike some as cultural chauvinism of some degree. Experiments have long been made with them by 'obscure and unknown men', as Candolle has said.

Security There are two aspects. Conservation reserves unless well policed are always exposed to attrition and often illegal occupation, locally rationalized on grounds that no one except professors in distant universities knows what their purpose is. Formal biospheres try to avoid this by integrating people with plants, permitting some non conforming traditional land uses on pragmatic grounds that these offer a screen to the core of the reserve against more rapacious intruders. With traditional agroecosystems of the more developed kind, as informal biospheres (Kandy, Java, Kerala) this security screen is present and has legal weight through land title, though in the wilder kind their traditional titles are usually not recognized and these – both titles and the people – are more easily brushed aside.

Another security threat comes from pressures for their redevelopment as agrosystems, the 'eco' bit is discounted. Students of them in their conservation mode express warm admiration, but then putting on a second hat as agronomist or forester or farm economist they find them backward and sorely in need of improvement. Mergen (1987) concludes of the Chagga in Tanzania...*Though these gardens represent an ecologically sustainable land use their productivity needs to be increased to support a larger population*, and follows with instructions on how to do this: more fertilizer, pollard and discipline the trees, optimize plant population densities, reduce competition, improve genetic quality...with trained foresters and agriculturists called in to elevate the Chagga in this agronomy. This must be sound advice if the object is to replicate a Kent apple orchard or maximize productivity or human population. But it's counter to the basic concept of this kind of agroecosystem, antithetical to its diversity conservation function and recipe for its destruction. In any case it would be futile. More productivity and income would only result in more people and their greater demands, these in more external inputs applied to fewer and cash generating species to pay for the inputs, these in faltering productivity but still more people...until biodiversity collapses into just another optimized but disease ridden banana patch and the young people drift off to Dar es Salaam.

4.7 Summary

From Table 1.2 of Chapter 1 one item on that checklist is how well or badly a farming system contributes to preservation of biodiversity. In the wet tropics most traditional agroforestry systems score highly in this regard. It seems apparent that their social role as *conservator*, rough and incidental as this might be, is far more valuable globally than is their local economic role as *farmer*. Their prosperity as farmer, where this exists, is only relative. Few of them are rich Kandyans with clove trees or Weinstock's Dyaks who got obscenely rich on rotan. They're still poor people in poor lands. If their role as conservator really is so valuable, how best to compensate them for this service?

References

Abe, T. and M. Higashi (1991) Cellulose centered perspective on terrestrial structure. *Oikos* 60:127-133.

Ahmed, S. and M. Grainge (1986) Potential of the neem tree (Azadirachta indica) for pest control and rural development. *Economic Botany* 40(2):201-209.

Aiken, S.R. and C.H. Leigh (1992) *Vanishing Rain Forests: the Ecological Transition in Malaysia.* Clarendon Press, Oxford UK.

Alcorn, J.B. (1984) Development policy, forests and peasant farms: reflections on Huastec-managed forests' contributions to commercial production and resource conservation. *Economic Botany* 38:389-406.

Altieri, M.A. and L.C. Merrick (1987) In situ conservation of crop genetic resources through maintenance of traditional farming systems. *Economic Botany* 41(1):86-96.

Altieri, M.A. and L.C. Merrick (1988) Agroecology and *in situ* conservation of native crop diversity in the Third World. In E.O. Wilson and F.M. Peter (eds) *Biodiversity.* National Academy Press, Washington DC.

Anderson, A.B. (1988) Use and management of native forests dominated by açai palm (Euterpe oleracea) in the Amazon estuary. *Advances in Economic Botany* 6:144-154. NY Botanical Garden, NY.

Anderson, A.B. and E.S. Anderson (1983) *People and the palm forest: biology and utilization of Babassu forests in Maranhoa.* Botany Department, University of Florida, Gainesville.

Anderson, A.B. and E.M. Ioris (1992) The logic of extraction: resource management and income generation by extractive producers in the Amazon. In K.H. Redford and C. Padoch (eds) *Conservation of Neotropical Forests: Working from Traditional Resource Use.* Columbia University Press, NY.

Ariyaratnam, A.E. and D.J. McConnell (1986) Palmyra in the northern dry zone of Sri Lanka. In D.J. McConnell (ed) *Economics of Tropical Crops: Some Trees and Vines.* Midcoast, Coffs Harbour, Australia.

Balick, M.J. (1979a) Amazonian oil palms of promise: a survey. *Economic Botany* 33(1):11-28.

Balick, M.J. (1979b) Economic botany of the Guahibo: I. Palmea. *Economic Botany* 33(4):361-376.

Balick, M.J. and S.N. Gershoff (1981) Nutritional evaluation of Jessenia bataua palm: source of high quality protein and oil from tropical America. *Economic Botany* 35(3):261-271.

Barrau, J. (1958) *Subsistence Agriculture in Melanesia.* Bishop Museum Bulletin 219, Honolulu.

Barrau, J. (1959) Sago palms and other food plants of the marsh dwellers of South Pacific Islands. *Economic Botany* 8:155 following.

Bavappa, K.V.A., M.K. Nair, T.P. Kimar (1982) *The Arecanut Palm.* Kerala, India.

Beccari, O. (1904-1986) *Wanderings in the Great Forests of Borneo.* Constable, London (1904), Oxford (1986).

Bennett, B.C., R. Alcarón, C. Cerón (1992) The ethnobotany of Carludovica palmata in Amazonian Ecuador. *Economic Botany* 46(3):233-240.

Brown, W.H. (1951) Useful Plants of the Philippines Vol I. Department of Agriculture and Natural Resources, Manila.

Bulman, T.L. (1988) The origin and diffusion of Areca catechu, the betel nut palm. *Malaysian Journal of Tropical Geography* 17 (June 1988)11-17.

Burkill, I.H. (1935) *In Dictionary of Economic Products of the Malay Peninsula Vol II.* Crown Agents for the Colonies, London.

Castles, I. (1992) *Australia's Environment – Issues and Fact.* Australian Bureau of Statistics, Canberra.

Charavanapavan, C. (1954) Improved methods of producing cane and palm jaggery. *Tropical Agriculturist* (Apr 1954). Ceylon Dept Agriculture, Perideniya.

Clement, C.R. (1989) Potential use of the pejibaye palm in agroforestry systems. *Agroforestry Systems* 7:201-212.

Clement, C.R. (1993) Native Amazonian fruits and nuts: composition, production and potential use for sustainable development. *Tropical Forests, People and Food, Man and Biosphere* 13. UNESCO.

Clement, C.R. and J.E. Mora Urpi (1985) Pejibaye palm (B. gasipaes: multi-use potential for the lowland humid tropics. *Economic Botany* 41(2)302-311.

Coe, F.G. and G.J. Anderson (1996) Ethnobotany of the Garífuna of Eastern Nicaragua. *Economic Botany* 50(1):71-107.

Comerford, S.C. (1996) Medicinal plants of two Maya healers from San Andrés, Petén, Guatemala. *Economic Botany* 50(3):327-336.

Davis, T.A. and D.V. Johnson (1987) Current development and further utilization of the palmyrah palm (Borassus flabellifer) in Tamil Nadu State. *Economic Botany* 41(2):247-266.

De Candolle, A. (1882-1959) *Origin of Cultivated Plants.* Hafner Publishing Company, New York.

De Castro, A. (1993) Extractive exploitation of the açai (Euterpe precatoria) near Manaus, Brazil. *Tropical Forests, People and Food. Man and the Biosphere, MAB* 13. UNESCO.

Dennett, J.H. (1927a) Final observations on the nipah palm as a source of alcohol. *Malayan Agriculture Journal* 15:420-432.

Dennett, J.H. (1927b) Alcohol fuel and the nipah palm, a popular outline. *Malayan Agriculture Journal* 15:433-445.

Dharmapala, K.A.E. and D.J. McConnell (1986a) Kitul in Sri Lanka. In D.J. McConnell (ed), *Economics of Tropical Crops: Some Trees and Vines.* Midcoast, Coffs Harbour, Australia.

Dharmapala, K.A.E. and D.J. McConnell (1986b) Arecanut in the forest garden farms of Kandy. In D.J. McConnell (ed), *Economics of Tropical Crops: Some Trees and Vines.* Midcoast, Coffs Harbour, Australia.

Dorji, N. and D.J. McConnell (1989) Cardamom in Bhutan. In *Crops, Livestock and Farming Systems of Bhutan.* FAO Commodities Division, Rome/Thimpu.

Ehrlich, P. (1995) The scale of the human enterprise and biodiversity loss. In J.H. Lawton and R.M. May (eds) *Extinction Rates.* Oxford University Press.

Ehrlich, P. and A. Ehrlich (1991) *Healing the Planet: Strategies for Resolving the Environmental Crisis.* Surrey Beatty Sons, Sydney.

Erwin, T. (1982) Tropical forests: their richness in Coleoptera and other arthropod species. *Coleopterists' Bulletin* 36(1):74-75.

FAO/BPPT (1986) *Development of the Sago Palm and its Products.* FAO Consultation Report (255 pp). Jakarta Jan 1984.

FAO (1996) *State of the World's Plant Genetic Resources for Food and Agriculture.* FAO Rome.

Farnsworth, N.R. and D.D. Soejarto (1985) Potential consequence of plant extinction in the United States on the current and future availability of prescription drugs. *Economic Botany* 39(3):231-240.

Fearnside, P.M. (1987) Causes of deforestation in the Brazilian Amazon. In J.S. Levine (ed) *The Geophysiology of Amazonia: Vegetation and Climate Interacting.* Wiley, NY.

Feil, J.P. (1996) Fruit production of Attalea colenda in coastal Ecuador – an alternative oil resource? *Economic Botany* 50 (3):300-309.

Flach, M. (1986) Agronomy of sago based cropping systems. In FAO/BPPT 1986, Rome.

Flach, M. and D.L. Schuiling (1989) Revival of an ancient starch crop: review of the agronomy of the sago palm. *Agroforestry Systems* 7:259-280 (Kluwer, Netherlands).

Fletcher, R. (1993) *Listing of Potential New Crops for Australia.* Department of Plant Production, University of Queensland, Gatton, Queensland.

Fong, F.W. (1992) Perspectives for sustainable resource utilization and management of nipa vegetation. *Economic Botany* 46(1):45-54.

Gentry, A.H. and R.H. Wettach (1986) Fevillea – a new oil seed from Amazonian Peru. *Economic Botany* 40(2):177-185.

Glanznig, A. (1995) *Native Vegetation Clearance: Habitat Loss and Biodiversity Decline.* Biodiversity Paper 6, Australian Department of Environment, Canberra.

Gradwohl, J. and R. Greenberg (1988) *Saving the Tropical Forests.* Island Press, Washington DC.

Grove, R.H. (1992) Origins of Western environmentalism. *Scientific American*, July:22-27.

Hamilton, L.S. and D.H. Murphy (1988) Use and management of nipa palm (Nypa fruticans). *Economic Botany* 42(2):206-213.

Harlan, J.R. (1975) *Crops and Man.* American Society of Agronomy. Madison, Wisconsin.

Holden, S. and H. Hvoslef (1990) Transmigration settlements in Seberida: causes and consequences of deterioration of farming systems in a rain forest environment. In O. Sandbukt (ed) *Management of Tropical Forests: towards an Integrated Perspective.* University of Oslo, Norway.

Hyndman, D.C. (1984) Ethnobotany of Wopkaimin pandanus: a significant Papua New Guinea plant resource. *Economic Botany* 38(3):287-303.

Joachim, A.W.R. and S. Kandiah (1938) Analysis of Ceylon foodstuffs. *Tropical Agriculturist* (Sri Lanka) Jan. Ceylon Dept Agriculture, Perideniya, Sri Lanka.

Johannessen, C.L. (1966) Pejibaye palm: yields, prices and labor costs. *Economic Botany* 20.

Johnston, M. and A. Calquhoun (1996) Preliminary ethnobotanical survey of Kurupukari: an Amerindian settlement of central Guyana. *Economic Botany* 50(2):182-194.

Kahn, F. (1993). Amazonian palms: food resources for the management of forest ecosystems. In *Tropical Forests, People and Food. Man and the Biosphere* 13. UNESCO, Paris.

Leakey, R. and R. Lewin (1996) *The Sixth Extinction: Biodiversity and its Survival.* Weidenfeld and Nicholson, London.

May, P.H., A.B. Anderson, M.J. Balick, J.M. Frazão (1985) Subsistence benefits from the babassu palm Orbignya martiana. *Economic Botany* 39(2):113-129.

McConnell, D.J. (1986a) The sago palm in Johor. In *Economics of Tropical Crops: Trees and Vines.* Midcoast, Coffs Harbour, Australia.

McConnell, D.J. (1986b) The nipah palm for *atap* production in Johor. In *Economics of Tropical Crops: Trees and Vines.* Midcoast, Coffs Harbour, Australia.

Mergen, F. (1987) Research opportunities to improve the production of home gardens. *Agroforestry Systems* 5:57-67.

Milliken, W. and B. Albert (1996) The use of medicinal plants by the Yanomami Indians of Brazil. *Economic Botany* 50(1):10-25.

Moraes, M.R. F. Borchsenius, U. Blicher-Mathiesen (1996) Notes on the biology and uses of the Motacú palm (Attalea phalerata) from Bolivia. *Economic Botany* 50(4):423-428.

Moran, E. (1981) *Developing the Amazon.* Indiana University Press.

Mungall, C. and D.J. McLaren (1990) *Planet under Stress.* Royal Society of Canada, Oxford University Press.

Murrieta Ruiz, J. and J. Levistre Ruiz (1993) Aquajales: forest fruit extraction in the Perusivan Amazon. In *Tropical Forests, People and Food. Man and the Biosphere. MAB* 13. UNESCO.

Myers, N. (1997) *The Sinking Ark: a New Look at the Problem of Disappearing Species.* Permagon, Oxford, UK.

Padoch, C. (1988) Aquaje (Mauritia flexuosa) in the economy of Iquitos, Peru. *Advances in Economic Botany* 6:214-224. NY Botanical Garden, NY.

Piperno, D.R. and D.M. Pearsall (1998) *Origins of Agriculture in the Lowland Neotropics.* Academic Press, USA.

Powell, J. (1976) Ethnobotany. In K. Paijmans (ed) *New Guinea Vegetation.* Australian National University Press, Canberra.

Prescott-Allen, R. and C. (1983) *Genes from the Wild: Using Wild Genetic Resources for Food and Raw Materials.* Earthscan, London.

Riezebos, E.P. (1986) *The Economics of Two Sago Based Cropping Systems.* FAO/BPPT Report, Jakarta/Rome.

Rivero, J.P. S.J. (1956) *History of the Missions of the Llanos of Casanare and of the Orinoco and Meta Rivers.* National Publications Press, Bogota.

Sanchez, P.A. (1989) Soils. In H. Leith and M.J.A. Werger (eds) *Ecosystems of the World.* Elsevier, Amsterdam.

Saw, L.G., J.V. La Frankie, K.M. Kochummen, S.Y. Yap (1991) Fruit trees in a Malaysian rain forest. *Economic Botany* 45(1):120-136.

Shand, H. (1997) *Human Nature: Agricultural Biodiversity and Farm Based Food Security.* Rural Advancement Foundation International (RAFI, Canada).

Skole, D. and C. Tucker (1993) Tropical deforestation and habitat fragmentation in the Amazon: satellite data 1978-88. *Science* 260:1905-1909.

Stanton, W.R. and M. Flach (1980) (eds). *Sago: the Equatorial Swamp as a Natural Resource,* Proceeding Second International Sago Symposium, Kuala Lumpur, Nijhoff, Hague.

Strudwick, J. and G.L. Sobel (1988) Uses of Euterpe oleracea in the Amazon estuary Brazil. *Advances in Economic Botany* 6:225-253. NY Botanical Garden, NY.

Suwal, P.N. (1970) *Medicinal Plants of Nepal Bulletin No. 3,* H.M. Government Department of Medicinal Plants, Kathmandu, Nepal.

Uhe, G. (1974) Preliminary survey of the use of plants for medicinal purposes in Samoa. *Economic Botany* (1967) 28:1-30.

Unruh, J.D. (1994) Role of land use pattern and process in the diffusion of valuable tree species. *Journal of Biogeography* 21:283-295.

Vietmeyer, N. (1996) Harmonizing biodiversity conservation and agricultural development. In J.P. Srivastava, N.J.H. Smith, D.A. Forno (eds) *Biodiversity and Agricultural Intensification.* World Bank, Washington DC.

Voeks, R.A. (1996) Tropical forest healers and habitat preference. *Economic Botany* 50(4):381-400.

Wallace, A.R. (1869-1986) *The Malay Archipelago: the Land of the Orang-Utan and the Bird of Paradise.* Oxford University Press (Reprints), Singapore.

Whitmore, T.C. (1977) *Palms of Malaya.* Oxford University Press.

Whitmore, T.C. (1984) *Tropical Rain Forests of the Far East.* Clarendon, Oxford.

Whitmore, T.C. and J.A. Sayer (1992) *Tropical Deforestation and Species Extinction.* Chapman and Hall, London.

Wilbanks, T.J. (1994) Sustainable Development in Geographic Perspective. Presidential Address, Association of American Geographers 48(4):541-556.

Wilson, E.O. (1992) *The Diversity of Life.* Belknap-Harvard University Press, Cambridge, Mass.

Wilson, E.O. and F.M. Peter (eds, 1988) *Biodiversity.* National Academic Press, Washington DC.

World Resources Institute (1992-3, 1994-5) *WRI-UNEP-UNDP.* Cambridge University Press, UK.

Chapter 5
Nutrients

This chapter goes looking for reasons why rain forests and forest gardens - respectively the world's richest biome and agroecosystem - can thrive on some of the world's worst soils.

From the efficiency criteria discussed in Chapter 1 (Table 1.2) and especially in the 3rd World humid tropics a farming system which needs no artificial fertilizer is better than one which needs much, for the reasons listed below, though it's also recognized that in some cases, (5) and (7), artificial fertilizers can be used positively to reduce deforestation and rehabilitate land. In the context of forest gardens this requires brief review of the nutritional arrangements of farming systems and of their parent, the Great Forest.

Some Undesirable Impacts of Fertilizer Use

1. Cost to farmers and foreign exchange drain on governments
2. Runoff pollution (of soils, aquifers, surface waters, biosystems)
3. Induced plant nutrient imbalance, acidification
4. Propensity to accompany fertilizer use with agricides (the Package)
5. Induced land clearing which would not be feasible without fertilizer
6. Atmospheric pollution in fertilizer manufacture and use as greenhouse gases
7. Induced soil erosion by farming beyond a sustainable level
8. Postponement of development of permanently sustainable agroecosystems.

5.1 Plant Nutrient Supplies

A popular and sufficient view of the nutritional arrangements of conventional crops is that they are vegetative mechanisms with root systems which suck up their nutrients in water solution from the soil and any deficiency in this is met by ladling out more N, P, K...to them from a bag. Other things equal (temperature and moisture regime, soil acidity etc) the more of these nutrients available in balance in the root zone the more productive the plants will be. This also defines good agricultural land. The arrangements of rain forests and forest gardens are much different. A starting point is with the approximate chemical composition of the leaf fraction of their phytomass (Table 5.1), which implies the elements which must be provided for its continuance.

Table 5.1   General Chemical Composition of Most Plants

	% of live plant	Volumetric concentration in atmosphere
Water (H$_2$O)	90%	(varies with humidity)
	% of dry weight	
C - carbon	45%	(0.03)
O - oxygen	45	(21.00)
H - hydrogen	5	(0.01)
N - nitrogen	1 - 3	(78.00)
Other	1 - 5	(traces)*

*(P - phosphorus, K - potassium, Ca - calcium, Mg - magnesium) S - sulphur: plus 7 micronutrients zinc, copper etc which are considered essential.

The bulk of these are sourced from their atmospheric reservoirs: water (in the hydrological cycle), carbon, oxygen, hydrogen, nitrogen.   Except locally for water these are in unlimited supply and in this sense are 'free goods.'   The others are vital for plant production but are quantitatively minor.   Some are usually in deficient supply and to the extent that it costs money to make this up they are *not* free goods. Agricultural plant nutrition is concerned with these latter nutrients;   but Walker (1992) argues that these are more in the nature of *condiments*, becoming part of the plant but not contributing energy, eg in comparison with the bulk 'nutrient' of carbon which comprises nearly half the plant.

Carbon

Plants receive their carbon directly from the atmosphere as carbon dioxide entering (mainly) the leaf stomata. Combined in leaf cells with water and in the necessary presence of chlorophyll and light energy, carbon-rich compounds are produced (starch, sugar, cellulose) which become plant organic matter via equation 1 (Energy). In conventional field crop farming the atmospheric supply of CO$_2$ is given, it's assumed not a critical productivity constraint and there's no interest in it from the viewpoint of possibly 'managing' it. On the other hand there is ample evidence that increased CO$_2$ supply to plants stimulates their 'growth', and this fertilizer effect via equation 1 has been proposed as one of the few offsetting favorable consequences of CO$_2$ increase leading to global warming. For several reasons the CO$_2$ supply to forests and those agroecosystems which mimic them structurally is more complex and here there could be some possibility of CO_2 management through deliberative architectural structuring (Energy, architecture).

First, the very large number of species/crops which comprise these systems handle their CO_2 assimilation and starch manufacture in different ways. (In this regard Richards 1952 has grouped the rain forest plants into four behavioral groups and given examples.) Second, in addition to the general atmospheric CO_2 'pool' from which they all draw their supplies, they themselves produce CO_2 at different rates by their respirations ($C_6H_{12}O_6 + O_2 \rightarrow CO_2 + H_2O$ + usable energy). Third, CO_2 is also being continuously emitted by decaying biomass (litter and logs) on the forest floor and from soil organic matter beneath it. Fourth, CO_2 emitted by these several sources (including zoo species) at several levels within the forest or farm is being distributed vertically to the plants which use it by wind zephyrs; consequently CO_2 production, distribution and use are determined substantially by the forest/farm species present and their architecture. Finally, production-distribution-use (at each level) are continuously changing through the day-night cycle. Kira and Yoda (1989) measured these diurnal changes and elevation differences as CO_2 concentrations over a day-night cycle down through a Malaysian rain forest, Figure 5.1 (shown for two levels only). The internal box summarizes the mean vertical gradient. This consists of many processes operating differently at specific levels: at and near ground level CO_2 concentrations are always high and fairly uniform. Here production is occurring day-and-night from decomposing biomass, soil organic matter and root respiration, but comparatively little of this is used locally in photosynthesis during the daytime by the small amounts of L0-L1 vegetation there; and especially at night there is little mixing wind (the 'stagnation of atmosphere' Allee described in Panama). At higher (30 meter) levels – approximately the main L4 canopy – the photosynthesis workshop starts to shut down after 4 pm, with the departing sun, and CO_2 accumulates overnight, then rapidly falls after the workshop opens up again at dawn. At highest level in and just above the emergents (at 55 meters, not shown), the situation is similar to that at 30 meters except photosynthesis and CO_2 drawdown starts earlier in the morning and its concentration never falls far because it is most easily replenished by free atmospheric CO_2 (and probably more of this is imported by the higher winds at this exposed level).

Relevance The variable CO_2 resource is apparently a 'production factor' of practical significance: its different concentrations down through a forest or forest garden and its changing daily regime probably represent a set of niches for the plant synusiae and associated zoo species at the different levels. On farms, it is hardly likely that forest gardeners set out to 'manage their CO_2 resource' in the sense of deliberately apportioning it optimally among the different synusiae as crops – much to the yams and coconuts etc. On the other hand such incidental or accidental apportioning, efficient or not, results from their choice and arrangement of species and individuals and the ways in which these produce and use CO_2 and distribute it among themselves by wind and sunlight governed by their architecture. Walsh (1996) has pointed to the relevance of these regimes in forests...*Carbon dioxide concentrations and their dynamics above and within the rain forest are important because at times they become a limiting factor for photosynthesis and carbon dioxide measurements...can be used in estimating the productivity of the rain forest and its components.*

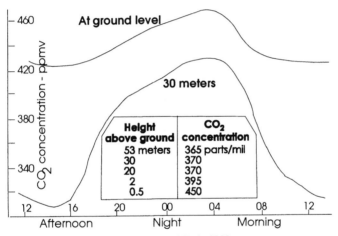

Source: Drawn from data by Kira and Yoda 1989.

Figure 5.1 Daily March of CO_2 Concentration in a Rain Forest

Just what this productivity effect is – either on total forest or farm productivity or at different points through the ecosystem – remains unknown. But that it exists seems somehow more appealing than does the idea of CO_2 regimes signifying nothing more than accident or chaos...And one would expect no less from any Garden of Complete Design.

Oxygen

Oxygen is also obtained directly from the atmosphere. Like carbon it's not a conventional plant nutrient but necessary for plant metabolism (except for the micro anaerobic plants which can get along without it). Just how oxygen first got into the atmosphere is conjecture – geologists suggest by general degassing of the early earth's interior, perhaps later supplemented by photosynthesis processes involving blue-green algae. Its contemporary source is as a by-product of the above conversion of carbon dioxide and water into plant matter. In this for each molecule of carbon dioxide absorbed to become organic matter one molecule of oxygen is released. Belief is that all contemporary atmospheric oxygen has been and continues to be produced in this way by plants which is another reason for keeping them around, especially the forests and their agro-analogues. Custodians of arcane statistics reckon that if all the oxygen-generating plants were to disappear the existing oxygen in atmospheric reservoir (21% by volume) would still be enough to last another 2 000 years. Others, not entirely convinced, plant trees as oxygen factories (operating via the above equation for respiration).

Nitrogen and the Nitrogen Cycle

Comprising 78% of this atmosphere this is the freest of the free goods; another useless statistic is that there are about 39 000 kg of nitrogen in atmospheric suspension over every hectare of farm land on earth. Sadly this can't be used by most plants directly and must be routed through the nitrogen cycle Figure 5.2 where it's converted into useful forms, ammonia and nitrate. There are three main routes (1, 2, 3) by which nitrogen can enter the plant systems of a forest or agroforest. On Route 1 some originate from volcanic eruptions and reactivation of 'fossil' nitrogen from rock weathering via the three equations at the top of the sketch where elemental N_2 is combined with oxygen by lightning to yield nitrate which enters the system in dust and rain. One estimate of lightning-fixed nitrogen arriving on this route is one kg/ha/year of earth surface. Combined with this is other nitrogen in dust, rain, stolen from some other agroecosystem/biome as wind erosion. One estimate of these combined 'new' and 'old' nitrogen inputs on Route 1 is 6-9 kg/ha/year globally but much more than this where sources are plentiful and wind erosion and/or transport are significant.

A second and more important external source is Route 2 where atmospheric nitrogen is fixed into useful forms by one set (A) of bacteria, fungi, algae, in association with plant roots. On farms this is usually captured by plowing leguminous crops under (or direct sod seeding into them). This can be a very important source; some typical nitrogen capture rates for sub/tropical conditions are summarized, one rule-of-thumb being the following amounts of nitrogen added to soil at the rate of 3% of the biomass of N-fixing field crops in this type.

cowpeas (6 months crop)   150 kg   D. lablab (2 years)   250 kg/ha
soybeans (7 months)   240 kg   alfalfa (6 years)   140 kg/ha/year

In forests and forest gardens this source depends largely on the populations of tree species present (leguminosae) which have this fixing capacity. The third source, Route 3, is nitrogen already in the soil (its 'fertility') and parent rock beneath it, and on the capacity of the root systems to exploit it, determined largely by root system architecture (but this root architecture also depends on nutrient availability).

Internal cycling   Once within the system the nitrogen cycle is as sketched in Figure 5.2, from plant growth through animal, litter, detritus to plant feeder roots, being transformed by differnet sets of bacteria (B, C, D) into useful nitrates and ammonia along the way to arrive at the plant via Routes 4 and 5. Under some conditions eg low oxygen levels the reverse process of denitrification can occur, involving another set of bacteria (E) in which useful nitrate is degraded (lower equations). Such denitrification is not always present. Where it is it would eventually, theoretically, remove all of the useful nitrogen from the system. But at this point the molecular nitrogen N_2 removed in denitrification is returned to the atmospheric pool to there supplement the N_2 which continues to arrive from the atmospheric sources, Routes 1 and 2.

Some economic and ecological implications   Taking Figure 5.2 as also broadly representative of other nutrient cycles (P, K, CA, Mg etc), and insofar as external inputs (Routes 1, 2, 3) might be very small, the critical factor becomes how to maintain nutrients within the system. The three main conditions are: (a) nutrients taken in harvested produce must not exceed some limit (ie if the system is a managed one there's a limit to the intensity with which it should be operated); (b) if the several bacteria and microorganism sets (A–D) are to operate effectively their necessary microenvironmental conditions of heat, moisture, light/shade, humidity...must be present; and (c) nutrient losses in surface erosion and leaching must be held to a minimum. These conditions are generally met in undisturbed natural systems and their farm analogues. Of (a) in forest gardens of the wilder kind, harvested produce is small relative to total biomass (Kandy); most production stays within the system as litter. Of (b), these necessary conditions are provided by the farm plants themselves, as they are in a rain forest (Energy; architecture, microclimate). Of (c), erosion losses are minimal but for those which do occur...*no amount of recycling can make up for what is not there* (Young 1989).

When a forest or other natural biome is cleared for modern field crop farming these three conditions are generally and quickly violated and two strategies can be substituted. The first tries to adhere at least partly to the principles of Figure 5.2, substituting leguminous or soilage crops of the types noted above for the original biomass and retaining *some* of the faunal components. The second – high pressure commercial farming led by

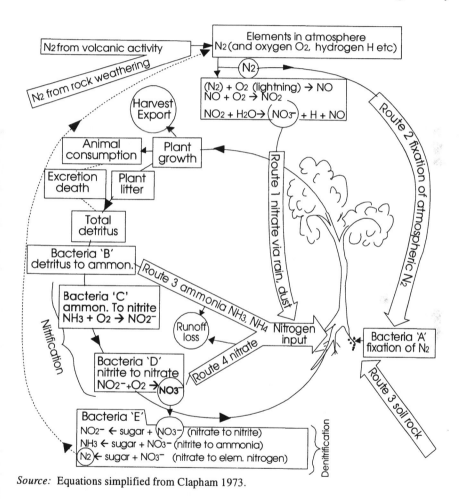

Source: Equations simplified from Clapham 1973.

Figure 5.2 Overview of Nitrogen Cycle in a Forest or Forest Garden

agribusiness – dispenses with Figure 5.2 altogether, gets its nutrients in on a truck and declares the fauna redundant. Globally other consequences have followed. Manufactured fertilizers require much energy which requires much fossil fuel which emits much carbon dioxide; and at point of farm use some of these fertilizers (and land disturbed for their use) are also major greenhouse gases. Some also (ammonia, nitrates) are highly soluble, not easily attached to soil particles and are carried by water percolation into and through the soil to aquifers and streams. By some estimates about one half of all nitrogen fertilizer applied to US crops is not actually used by them. If flow routings for artificial nitrogen (and phosphates) were added to Figure 5.2 they would meander through poisoned aquifers, undrinkable wells, entrophic lakes, meadows of blue-green algae, red tides, fish kills and degraded river basins (supplemented along the way with more unwanted nutrients from manure when the crops are fed to livestock) before they again emerge in sediments cleaned up by geochemical time. Data for one British agro-catchment suggest how important the farm fertilizer nitrogen pathway to river and ocean sink has become, increasing fivefold in 40 years.

Table 5.2   Increase in Nitrogen River Load and in Probable Causes

	Annual N load from all sources	Increase in contributing components – kg/ha*			
		Sewage	Rainfall	Plowing grass	Farm fertilizers
1940	5.50 kg/ha*	1.70 kg	1.80 kg	4.80 kg	15.10 kg
1950	8.00	1.70	1.90	0.01	22.30
1960	7.50	1.80	1.80	1.30	45.30
1970	12.60	2.30	1.60	0.50	63.00
1980	17.40	2.70	1.80	0.50	74.80

Severn-Trent catchment 580 sq km. *kg per ha of catchment. The four sources shown are independent series and don't total to the N shown as river load. Pollution rates for many UK rivers have now been reduced, stabilized.
Source: UK Min. Ag. for Fish., cited by Tivy 1990.

External nutrient inputs   In the absence of data relating to external inputs specifically on forest gardens one obtains a guide from forests.

Route 1 from atmosphere   By this route...*tree crowns act as dust traps* (Szott et al 1991). From Table 5.3 its importance is variable, depending on element, presence of a dust source (usually desert) and prevailing wind as transport mechanism. These latter seem to be illustrated by the high dust/nutrient loads into forests in Ghana and Ivory Coast, presumably from Sahara-Sahel dust. This source is probably far more important near centers of volcanic activty, eg Indonesian-Melanesian (New Britain) volcanoes. After all, as Wallace said of Java...*Java contains more volcanoes, active and extinct, than any other district of equal extent. They are about forty-*

five in number...and the noble and fertile island owes its existence to the same intense volcanic activity which still occasionally devastates its surface. Now Java's forests are nearly gone but her volcanoes are still there, Merapi, Bromo, Gunung Kelud and the others, daily streaming out their dust in which there is undoubtedly much nutriment.

Table 5.3 Atmospheric Inputs of Elements, kg/ha/year

Wet tropics	N	P	K	CA	Mg
Ghana (Nye 1961)	4	0.3	14	10	9
Ivory Coast (Szott)	21	2.3	6	30	7
Costa Rica	5	0.2	3	1	1
Brazil	10	0.3	-	4	3
Malaysia	14	-	6	4	1
Wet tropics means:	11	1	8	9	3
Arid (Niger)	(trace)	(trace)	1	3	1

Source: All except Nye were compiled by Szott, Fernandes, Sanchez 1991 for 17 sites for which only representative data are shown, but means refer to all 17 wet tropical sites.

Route 2, fixation by soil bacteria This applies to nitrogen fixed by free-living soil bacteria (non-symbiotic fixation) and by symbiotic fixation by other bacteria in association with the forests' feeder roots. Symbiotic fixation occurs with both leguminous and non-leguminous tree species. As a guide to nitrogen supply from this source Young (1989) suggests about 16 kg/ha/year in a primary forest, increasing to 40-100 kg in an immature secondary forest. Greenland and Kowal (1960) found in a 40 year old secondary forest in Ghana that only about 8% of the small trees and 4% the larger trees present were leguminous and concluded that while the total rate of soil nitrogen accumulation in the forest biomass had apparently been more than 40 kg annually it was unlikely that much of this could be attributed to symbiotic fixation. But this parameter varies widely among forests. In Guiana Richards (1952) emphasizes its general importance: *Since the soils are so poor in nutrients the possession of root nodules containing nitrogen-fixing organisms is probably of considerable advantage to rain forest plants...* He quotes percentages of leguminosae populations in five forests there (39, 33, 15, 14, 53%) to illustrate the probable importance of this nitrogen source, but also its high variability among forests. In the Amazonia forests they surveyed Klinge et al (1975) found that 20% of all trees in the upper two storeys were leguminosae and that this family was also the most numerous at lower storeys (contributing 25% of total forest phytomass). There are no comparable data for forest gardens; but Jensen (1993) notes in passing that on one West Java farm he analysed some 18% of the fine roots were colonized by Mycorrhiza.

Route 3, from rootzone   Unlike field agrosystems in which the amounts of N-P-K-Ca-Mg extracted from the rootzone are considerable and predictable the amounts extracted by forests and forest gardens from this soil source vary from most to very little. This depends on the nutrient status and depth of the potential rootzone, the existence and mineral composition of underlying parent rock to which deep roots might penetrate, the root architecture of the subject tree species (ie whether or not they *have* such deep roots or the incentive to develop them), and the ability of the species to make alternative nutritional arrangements.   This latter applies particularly to the mechanically dependent plants – mosses, epiphytes, parasites and to plants of the separate principalities of the aerial marshes and to quite large trees which start life in a pocket of debris high in the canopy and perhaps never reach the ground. These plants with their own 'rootzones' can easily account for 15% of the forest biomass.

These aside, estimates of their nutrient requirements which rain forests extract from the rootzone soil range 20–70% depending on the factors listed. Forests and forest gardens flourish in conditions ranging from some of the world's richest soils (though as on the volcanics of Java this is why there are few forests left) to some of the worst, or where the concepts of *rootzone* and soil *quality* hardly apply. Examples are those forests and gardens on the karst (limestone) hills of Thailand, Malaysia, NE India, Borneo; the forest gardens on similar thin dry soils over limestone rock along the south Java coast; the forest farms below Sri Lanka's SE escarpment (with trees restricted to any spaces they can find among the boulders); the rain forests on Upolu, Samoa, entirely on volcanic *scoria* (clinker) rocks where there is practically no soil at all, only pockets of debris. Great swamp forests exist in the seasonally flooded lands of SE Irian Jaya, Kalimantan, Amazonia. Littoral forests and gardens mingle with mangroves and *nipa* thickets throughout Indonesia-Melanesia where the 'rootzone' is a tidal mudbank of indeterminate depth and uncertain geological parentage, with no meaningful soil profile and a microenvironment of extreme salinity.   Richards (1952) lists similar examples of fine rain forests on impossible soil conditions: the highly acidic (pH3.0) swamps and peats (100 meters deep) of Malaysia and Sumatra, the bleached white sands of South America and Sarawak. Newman (1990) describes the rain forest soils of tropical America generally as *exceedingly thin and fragile* with a topsoil (A horizon) of not much more than 5 cm (2 inches) over infertile laterite. (The best use of this material when it's stripped of vegetation and hardens under the tropical sun is for building development roads and pyramids, as every Prince will verify.)

Root Systems and Filter Mats

Rain forests on poor soils are characterized by shallow root systems but these are about three times more dense than in temperate forests. The depth of soil in the rootzone (as well as its quality) provide limits to the amounts of nutrients which can be extracted from this source (route 3); and the configuration and position of the rootmass are governing factors in the efficiency with which soil nitrogen-fixing bacteria and mycorrhizal fungi operate. These shallow rootzones are illustrated in the sketch for a Ghana forest (Greenland and Kowal 1960).

Depth			
Cm	Inches	Each sub-zone	Cumulative
0	-		
2	1	23.8%	(23.8%)
7	3	24.9	(48.7)
12	5	16.0	(64.7)
17	7	8.7	(73.4)
29	12	12.1	(85.5)
59	24	9.0	(94.5)
88	36	2.9	(97.4)
118	48	1.6	(99.0)
147	60	0.5	(100.0)

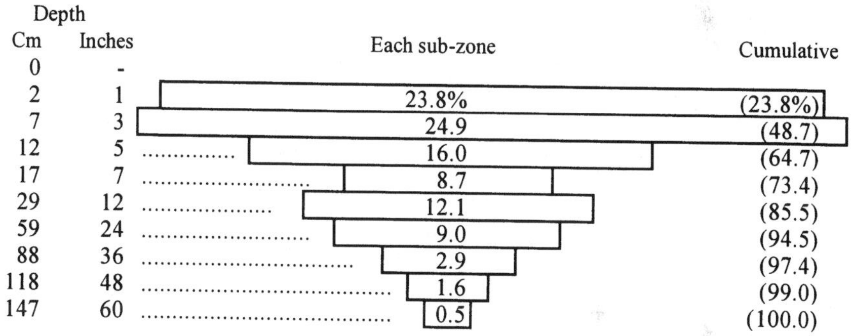

Length of bars show % of total at each soil depth layer.
Source: Sketched from data by Greenland and Kowal 1960;
a secondary forest 40-50 years age.

Figure 5.3   Percentage Distribution of a Rain Forest Rootsystem by Depth

Here nearly half the roots were above 7 cm depth and 86% were less than 29 cm below the surface. These were big trees in 40-50 year old secondary forest yet their root system configuration would be little deeper than that of a backyard apple tree. The soil nutrient supply which could be extracted from such a shallow depth of poor soil must be extremely limited, the recycling capacity great. Rootzone distribution also depends on soil fertility at the several levels down through the soil profile. Huttel (1975) mapped tree root system distributions under three forests in Ivory Coast: in the most fertile of these soils (3) half the roots were concentrated in the top 10 cm, compared with their necessary deeper location in the poorer soils (2) and (1).

Location of Main Rootmass in Relation to Different Soils Qualities

Forest	(1)	(2)	(3)
Fertility (EC) of soil at 10-20 cm depth:	0.40	0.15	0.87
50% of tree roots located above soil depth of:	30 cm	15 cm	10 cm
EC, exchangeable cations (me/100g soil)			

Filter mats   On poor soils in Amazonia Moran (1981) and Sioli (1985) have described forest root systems as gigantic absorption mats or filters, working in two directions.   Working 'down' they capture whatever nutrients are available from the rootzone soil in the usual way.   working 'above' they capture incoming nutrients in falling litter as soon as these arrive or are mineralized by decomposing bacteria; and they also filter out nutrients from incoming rain wash. Being far from industrial centers (or volcanoes) this incoming rain is itself nutrient poor but it absorbs some nutrients by canopy leaching and as trunk trickle flow and as previously deposited dust. These nutrients are more or less immediately absorbed from above by the filter mats; and when the water passes through them to become ground water it again is very nutrient poor, ie very little escapes through the mats. Sioli makes the further point that this filtering process extends to nutrients washed down as manure from the canopy where most rain forest animal mass concentrates; so these root mats are also a linking mechanism between faunal concentration providing nutrient *output* and plant needs for nutrient *inputs. (Observe how system into system runs...)*

Role of the fungi Incoming nutrient supplies in litter are relatively secure while they remain in organic form but less so after their mineralization. Again to minimize possible nutrient escape, research in Amazonia has shown that by association of the mat feeding roots with root fungi (Mycorrhiza) the rootlets can also feed directly from the decomposing surface litter without waiting for the mineralization process to be completed and these nutrients to infiltrate into the soil proper.   Sioli has described...*fungi with one end of their hyphae inside a decomposing litter leaf and the other inside the living root.* In other mutualistic relationships (Newman 1990) some fungi, in exchange for sugars, directly inject nitrogen they glean from the soil into the tree roots. In other associations insoluble phosphorus and zinc are also supplied directly to the roots. The combined effects of these several strategies are to keep losses of all incoming litter nutrients to only 10-12%. At a directly practical (and economic) level these fungi-root associations are also mechanisms which enable a forest to *rapidly* take up nutrients from an apparently infertile environment and so enable its *rapid* recovery after serious disturbance, as when it's cut down by shifting cultivators. If the fungi (and decomposers) have been 'cooked' in the clearing fires the forest regeneration process will be impeded or prevented (Shifters, regeneration). There's also an increasing likelihood that they'll be 'cooked' by radiation increase (Energy, ozone depletion; Shifters, forest regeneration).

Nutrient losses from soil surface and rootzone   These root system mechanisms and rapid action by the decomposers and converters result in nutrient losses which are very low in forests and on most forest gardens and might otherwise occur through runoff and leaching.   Szott et al (1991) provide a guide to these combined nutrient hydrological losses in a sample of forests. No comparable nutrient loss data are available for forest gardens but these authors believe they would be within the range shown for the natural forests.

Table 5.4   Nutrient Losses in Erosion under Forests

	Nutrients - kg/ha/year				
	N	P	K	CA	Mg
Moderately fertile					
Costa Rica	19	-	4	6	7
New Guinea	-	-	15	25	51
Oxisols/Utisols					
Venezuela	-	30	5	4	1
Brazil	(trace)	(trace)	1	(trace)	(trace)
Mountain forests					
Venezuela	5	(trace)	2	2	1

Source: Szott, Fernandes, Sanchez 1991; original source attributed to Vitousek and Sanford (uncited).

Phytomass as soil fertility criterion   Emilio Moran (1981) brings a human dimension to all this in his discussion of how two groups of forest settlers selected their land on the Xinghu River, Brazil.   The size of forest vegetative mass and of its individual trees are often mistaken as evidence of fertility of the soil beneath them. The first settler group, newcomers to the rain forest, used this criterion. The second group, local *caboclos*, used the presence of certain tree and vine species only. From later tests the promised high soil fertility under the more luxuriant forest, particularly under large buttress trees with shallow root systems, proved an illusion: their soils were poor in comparison with those chosen by *caboclos*, as shown in Table 5.5.

Table 5.5   Comparative Soil Qualities of Lands Selected by Different Settler Groups in Brazilian Forests

		Newcomes (poor soils)	Caboclos (good soils)
acidity	(pH)	4.7	5.5
soil organic matter	(%)	2.9	2.9
phosphorus, P	(ppm)	6.0	20.0
potassium, K	(mE/100g)	0.2	1.6
cal-mag,CA-Mg	(mE/100g)	2.0	7.0
aluminium, AL*	(%)	35.0	5.0

*low level desirable.
Source: Moran 1981.

Summary

1.   The standing biomass in forests and most forest gardens is very large.
2.   To maintain this biomass and productivity from it large amounts of nutrients are needed.
3.   The amounts provided by incoming rain and dust are marginal. Amounts routinely extracted from the rootzone may be small or large depending on its fertility: nutrients which might otherwise be tapped by deeper roots are often not present.
4.   Nutrient needs must often be met largely by the forest or farm re-cycling whatever nutrients are in the system from its litter. This recycling must occur rapidly, intensively. Losses to erosion and leaching must be held to a minimum and on farms the amount of product taken as percentage of NPP must be small.
5.   The condition for rapid recycling is the presence of litter decomposers and mineralizing bacteria. These require the appropriate micro-environmental conditions shade/light, heat, moisture, humidity, absence of wind, conditions which must be provided by the architecture of the forest or farm itself (Energy, micro climate).

5.2   Internal Nutrient Cycling in Forests and Forest Gardens

There are four main pathways by which rain forests cycle nutrients once they're in the system and more on forest garden farms: fall of fine litter (leaves, fruit, flowers, bark); fall or decay in place of large branches and tree trunks; leaching of nutrients in the canopy by rain throughfall; and similar leaching from branches and trunks by rain as trickle flow. On farms the decay of large trees is seldom a source; these are usually felled and their nutrients leave the system a household fuelwood and sold timber. (Ash from hearth fires is retained but applied only to vegetables and flowers.) On farms another source can be household waste, but often applied to only a few 'favorite' trees eg a coconut or jak tree. Another can be farm fishpond mud, with or without enrichment by excreta. This fishpond mechanism for recycling nutrients is complicated by the type of water source used for these ponds. This might be water runoff from the farm itself in which case pond nutrients are internal. But in hilly high rainfall areas like West Java it might be a small trickle passing through the village, to or from some paddy field, and both delivering nutrients to and taking them from any fishpond it feeds along the way (Jensen's Farm). Animal manure is also internally cycled but to the extent it might originate as grass cut off the farm it's not an internal nutrient.

Sanchez (1989) offers a guide to nutrient amounts supplied by canopy and trickle flow leaching in forests. Depending on phytomass amounts these would be similar on farms. The amounts are small, nonetheless they augment the nutrient content of the incoming rain itself by a factor of 2-6. (As this is an island forest there would be comparatively little dust.)

Table 5.6   External Nutrient Inputs to a Forest from Rain, Throughfall and Stem Flow

	parts per million*				
Source	NO_3	K+Na	C_A	Mg	SO_4
arriving in rain	0.1	3.6	0.9	0.7	1.1
added by throughfall	0.6	14.1	2.1	1.2	3.0
added by stem flow	1.7	4.7	1.7	0.9	0.1

*from a Puerto Rico forest.

Nutrient Content of a Rain Forest

For present purposes it's hardly necessary to go beyond the classic studies of Greenland and Kowal (1960), Nye (1961) and Nye and Greenland (1964) of a 40-50 year old secondary forest in Ghana, Table 5.7. The forest, previously subject to shifting agriculture, had accumulated 358 tonnes of total above-and-below ground phytomass containing the nutrients shown, and these in relation to nutrients in soil storage to a depth of 30 cm beneath the forest. (Total soil reserves would be much greater, extending down to a meter or so according to rootzone configurations described above.) As a first approximation, if the forest was cut down and burned it would theoretically yield 2 026 kg of N, 135 kg of P etc. However in a shifting system only the litter, vines, smaller branches of biomass are burned, the large trees and stumps left (Shifters). In this case the nutrients recovered for cropping would amount to around 0.4 tonne of superphosphate, 0.2 tonne of magnesium carbonate, a tonne of limestone etc or their other fertilizer equivalents (Greenland and Kowal). Some of these would then be lost in the burning, amounting to 10-40% of the nitrogen and smaller amounts of the others (Szott et al 1991); and much also as ash might be blown away by the wind (Shifters).

Table 5.7   Phytomass Fractions and Nutrient Content of a Secondary Rain Forest, Ghana

A.	The forest	Kg dry weight per hectare				
	stems and large live wood ...	171 500 kg	leaves and twigs ...	25 300 kg		
	large dead wood	71 200	vines	14 300		
	stumps	48 500	litter	2 200		
	roots	24 500	Total	... 357 500		
B.	Nutrients in phytomass	Kg per hectare				
		N	P	K	CA	Mg
	stems and large live wood	723	44	420	996	165
	large dead wood	228	189	36	519	51
	stumps	184	20	92	184	34
	roots	212	11	87	144	43
	leaves and twigs	475	32	192	481	65
	vines	168	8	62	278	21
	litter	34	1	10	44	6
	Totals in phytomass ...	2 026	134	898	2 646	386
C.	Nutrients in soil to depth of 30 cm (1 foot), kg	4 550	13	644	2 553	366
D.	Total nutrients in system, kg (to depth of only 30 cm)	6 576	147	1 542	5 199	752
E.	Nutrients in phytomass as % of total, % (to depth of only 30 cm)	31	91	58	51	51

Source: Greenland and Kowal 1960.

Nutrients in Other Plants, Synusiae

To the extent that only the larger and independent plants are usually reported the total nutrient contents of forest or farm phytomass are often considerably underestimated and the picture incomplete. Also present are the epiphytes, parasites, fungi and the semi autonomous principalities of the aerial marshes (discussed elsewhere – Energy, synusiae, principalities). In total these can account for 5-15% of the forest biomass, up to 30% of individual trees. They have their own separate nutritional arrangements: epiphytes draw some of theirs from rain and litter; carnivorous plants from trapped insects and small animals; the parasites from the host plant and to this extent are ground-independent; the fungi from decaying matter. Some micro fungi are also parasites and draw on live animals as well as live plants. The nutrient cycles within plants-animal systems of the aerial ponds and marshes are hardly known (Energy, principalities).

Nutrients Cycled Annually

Nye (1961) also measured the amounts of nutrients in the forest of Table 5.7 which cycled annually. Some of these data are extracted in Table 5.8. The forest produced 10 400, 11 100 and 2 550 kg of litter, tree fall and decomposing roots respectively and these returned 242 kg of N, 11 kg potassium, 76 kg of potash etc to the forest. In addition, the incoming 1 745 mm of rain brought 2.2 kg of N, 0.4 of P, 15.6 kg of K etc *to* the forest

Table 5.8 **Nutrient Amounts Cycled Annually, and Added (kg/ha/year)**

	*Material amounts					
Source	(dry weight)	N	P	K	CA	Mg
from litter fall	(10 400 kg)	198	7	68	202	44
from tree fall	(11 100 kg)	36	3	6	81	8
from root decomposition	(2 550 kg)	8	1	2	19	2
Totals	...	242	11	76	302	54
from rain fall **						
in incoming rain	(1 745 mm)	2.2	0.4	15.6	11.3	10.1
leached from canopy	(1 476 mm)	8.2	3.3	196.3	25.8	15.9

*Amounts produced by forest annually. **Incoming annual rain 1 745 mm (72.7 inches); reaching ground 1 476 mm; absorbed by canopy 269 mm; this through-fall rain leaches the nutrients shown from the canopy and adds these to nutrient supplies at ground level.
Source: Nye 1961.

(an external input) and as it passed down through the canopy this rain also leached 8.2 kg of N, 3.3 of P and 196.3 kg of K etc *from* the canopy and delivered it to the rootzone. This leached fraction can be high, even in comparison with the main internal cycling path, that of the litter; here nearly three times as much potash K is returned via rain canopy-leaching than comes back through litter decomposition. Each rain drop passes over and leaches many leaves, so this source depends somewhat on tree/forest/farm architecture (Energy) which is, at least in principle on forest gardens, amenable to management control.   In this example and considering only the internally cycled elements (ie ignoring the external input from incoming rain) the nutrient turn-over rates for N-P-K-CA-Mg are 20-10-30-12-18% annually, on basis of nutrients stored in the forest of Table 5.8.

Litter Cycling Rates

As noted previously one important requirement for continuation of a forest, especially on poor soil, is its ability to internally recycle its nutrients. This is also the main requirement for sustainability on farms without resort to external nutrient inputs, or without growing crops specifically for this purpose. From Table 5.8 the most important route is cycling of the litter fraction. Nye estimated the litter cycling rate in this Ghana forest at (10 400 kg x 100)/2 200 kg or 470%, where there was 2 200 kg litter on the ground at any one time and 10 400 kg of it was produced annually, ie any layer of litter would decompose and be replenished by fresh material 4.7 times annually. This compares with a cycling rate of only 6-12% in high latitude temperate forests. In Amazonian forests investigated by Sanchez about half of any litterfall is mineralized (becomes available for plant food) within 10-12 weeks.   Proctor (1984) compiled a data set for litter production rates in 198 tropical/subtropical forests around the world: the usual range is 6-11 tonnes/ha/year but many forests in Malaysia, Thailand, Colombia, Brazil have rates of 12-14 tonnes (the highest recorded was 23.3 tonnes in a South Thailand forest). These yields don't include large fallen branches and trunks; their nutrient N, P, K... compositions don't vary markedly from that in Nye's forest Table 5.8.

Effect of Climate and Comparative Litter Depletion Rates

Using a similar measure of litter depletion rate (D, or per cent per year depletion of a given litter pool) Esser and Lieth (1989) reported depletion rates for 45 different kinds of tropical forests and savannas and related them to temperature and rainfall at each site. A few of these measurements are extracted in Table 5.9 to illustrate the wide range of this process (last column). From these, highest litter cycling rates are associated with high temperature but only moderate rainfall, eg Thailand and Ghana but not Colombia. Savannas also can have high cycling rates where these climate conditions prevail (Nigeria). The relationship is more clearly illustrated in Figure 5.4. Mean temperature ranges of the three groups of forests are respectively 0-10, 10-20 and 20-30 C degrees (the first representing high mountain forests). Within these three temperature groups maximum litter depletion occurs with approximately 1 000, 1 500 and 2 000 mm annual rainfall respectively. One possible reason for there being an optimum probably related to termites.

Table 5.9   Litter Decomposition Rates under a Range of Temperature and Rainfall Regimes

	Annual temp C	Annual rain mm	Depletion rate % D
Grass/savanna			
Central Queensland	21	500	50
Nigerian savanna	28	1 270	220
Forests			
primeval rain forest Nigeria	26	1 230	273
Thailand rain forest	26	1 410	866
Ghana rain forest (Nye)	26	1 820	465
Very high rainfall forests			
Colombia	27	9 120	169
New Guinea	13	4 000	48
P. Rico	20	3 000	60

Source: Esser and Lieth 1989.

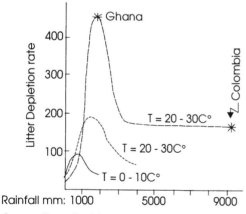

Source: Generalized from Esser and Lieth 1989.

Figure 5.4   Effects of Rainfall and Temperature on Litter Depletion Rate

In Praise of Termites

Although their several roles have not yet been satisfactorily quantified termites are a very important component of rain forests and commonly this extends to forest gardens. These roles are summarized; most relate to recycling and translocation of plant nutrients:

1. direct atmospheric nitrogen fixation (amount not known);
2. direct consumption of organic matter, usually dead biomass but with some species also extending to live matter;
3. breakdown and increase in surface area of larger chunks of litter so that their nutrients are more easily mineralized by smaller organisms;
4. falling of even large trees, live as well as dead, by bringing fungal pathogens into tree trunks; these large trees then begin their process of decomposition on the forest floor;
5. shifting litter about in the forest for nest construction etc which results in some redistribution of its nutrients (some slash-and-burn farming systems take advantage of this – Shifters);
6. concentrating litter nutrients in nests or mounds (on trees or ground) which especially after their vacation by the termites provide a rich habitat for seedling establishment and so are related to forest regeneration (and perhaps to tree species diversity);
7. collection and relocation of seeds which is one factor in determining the spatial distribution of plant species in the forest;
8. live termites are also an important link in the faunal food chain, attracting some birds and mammals which themselves subsequently perform some of these forest modifying tasks, especially importing seeds.

Termite biomass Collins (1989) has consolidated data which indicate the number of termite species increase from 4-5 at temperate latitudes (37 degrees N) to over 60 in some equatorial forests. Within this equatorial belt (the three Great Forests) typical termite biomass levels are 70-100 kg/ha. This compares with about 2-8 kg for mammals (mainly monkeys) and 1-2 kg/ha for understorey birds, making termites usually by far the largest forest zoomass except for soil micro-organisms. However both the numbers of their species and their total biomass fall off sharply with increasing elevation, eg to only 7 kg in one Sarawak forest at 1 300 - 1 800 meters elevation and none in a second forest above 2 000 meters. Apart from declining temperature the reason for this seems to be increasing rain at these higher elevations. Here the ground dwellers can't tolerate water-saturated conditions, their foraging columns would get drowned, while the tree-dwellers would be exposed to more or less continuous tree stem trickle flow. From Figure 5.4 (Esser and Lieth) litter decomposition rates which decline markedly with increasing rainfall and decreasing temperatures probably reflect changing termite activity.

Litter consumption Collins also notes that termites (and cockroaches) are the only insects known to fix atmospheric nitrogen, but the amount is not known. No doubt more important are the other nutrients which they make available through litter consumption. These amounts are also not known but are implied in the termites high litter consumption rates, reaching 1 650 kg/ha/year in one Malaysian forest and consuming 15% of its litter production (Table 5.10). Estimates of termite consumption in savannas are much higher, in some cases exceeding the per hectare vegetation consumption of all large grazing animals. At global level the main downside with termites, so to speak, is their inordinate capacity to generate methane, a major greenhouse/global warming gas. In this their mounds are the second most important *natural* source of emissions (after swamps and wetlands). They've been accused of producing anything from 10 to 100 million tonnes of methane annually; no one is too sure.

Table 5.10 Consumption of Organic Matter by Termites in Representative Rain Forests

Forest	Organic matter consumption kg/ha/year	% of annual litter production consumed
Sarawak	200	2
West Malaysia	1 650	15
Venezuela Rio Negro	70	5

Source: Collins 1989.

Litter and Moisture in a Fragile Forest Environment

At Cherrapunji in NE India Kliewtam and Ramakrishnan (1993) studied litter nutrient production under a much different kind of forest. This is a sacred grove of stunted vegetation on poor highly leached soil over porous karst (limestone) at 1 300m elevation under extremely high rainfall (annual mean 10 400 but increasing to 24 600 mm, 967 inches!). Total litterfall is 11 780 kg/ha annually and both the seasonal distribution and its varying nutrient content, because of leaching, is correlated with seasonal rainfall.

The fine feeder rootmass is 14 000 kg/ha, which these authors also describe as 'mopping up' the litter nutrients. Here again most of the rootmass is in the top 10 cm and 6% is *above* the soil. Further, the configuration and vertical position of this fine rootmass varies seasonally, apparently in some relationship with litter-fall and the seasonally varying amounts of nutrients in this available for capture. In spite of the high rainfall there are dry months and under such porous soil/low rainfall conditions the second function of litter is moisture conservation for the fine root mass...*This (litter nutrient cycling and moisture storage) probably explains why the sacred grove (the forest) stands alone as an island in a vast expanse of degraded grasslands.* These grasslands were once forests and are the result of shifting cultivation 'long ago'...*Degraded grasslands such as these are not able to revert back to the climax forest* (Shifters, regeneration). One reason for considering Khiewtam and Ramakrishnan on this is that forest gardens also exist under such adverse conditions. (A second reason is that these impossible forests once supported Hoabinhian, but that also was long ago – Genesis.)

5.3 Nutrient Cycling in a Javanese Forest Garden

Jensen's Farm

Jensen (1993, 1993a) measured nutrient flows into, from and within a typical forest garden in Bandung area, West Java, of 1 200 sq meters of which 280 sq m were occupied by house and *baruan* (cleared social area) and 920 sq m by a mix of trees, bushes, vines, grasses of 60 plant species dominated by those listed.

Some Species on Farm (main only)

Trees—

coconut	rose apple	banana	guava
cloves	mango	locus bean	salak fruit
jakfruit	jakfruit	coffee	soursop

Shrubs, vines, grasses—

pineapple	papaya	taro	turmeric
lemongrass	yam	sweetpotato	cassava
betel vine	sugar cane	chayote	hyacinth bean

Of these, 39% supplied some economic product (food, fiber, timber, fuel, condiments, drugs, beverage). Total trees numbered 169 corresponding to a large-plant population of 1 835/ha, similar to a rain forest, but a highly modified one in that three economically important species (coconut, clove, langsat fruit) comprised 95% of the biomass, with the other plants inter- and under-planted. Apart from plants the farm's other enterprises were free ranging poultry and a small fishpond, with all these inter-connected by their nutrient exchanges. The 'useless' trees/plants/weeds as well as the economic ones contribute nutrients as litter as well as microclimate for the decomposers and soil organisms by providing 99% ground cover (or in Tao-Buddhist terms, the utility of the useless). The pond is continuously fed by a small stream passing through the village and evacuated by an outgoing channel, bringing nutrients to and from the pond, and the pond mud is used as fertilizer on some of the plants.

Whole-system nutrient stores and flows Total plant biomass is 126.4 tonnes (dryweight) and its fractions and nutrient compositions are summarized in Table 5.11. Nutrient flows to, from and within the system

Table 5.11 Nutrients in Phytomass and Soil of a West Java Forest Garden

		Total Phytomass (dry) tonnes/ha	Nutrients, kg/ha N	P	K	CA	Mg
Trees (stems), leaves, roots, fruit etc)		116.4 t	907	72	648	642	145
Shrubs		2.1	21	2	43	13	4
Other:	fine roots	1.9	13	2	-	-	-
	weeds	1.6	22	6	53	10	6
	ground litter	4.4	32	3	7	25	11
Total		(126.4t)	995	85	751	690	165

are sketched in Figure 5.5 using Jensen's data. *External inputs* to the system from four sources are shown in the top section - from rain, fishpond stream inflow and soil nitrogen fixation in the amounts shown. Nutrients from the soil might be regarded as a 'general reserve' to be drawn upon to make up any deficiency in the ability of the other sources to keep the system sustainable. These external nutrient resources enter the farm's 'nutrient pool,' a conceptual device. External *outputs* or losses from the whole-system nutrient pool are shown on the left of the model in the amounts circled; there were losses in fishpond outflow and soil leaching but little in surface erosion. In addition there were exports as fruit/vegetables/spices sold, fish sold, in excess of these products consumed

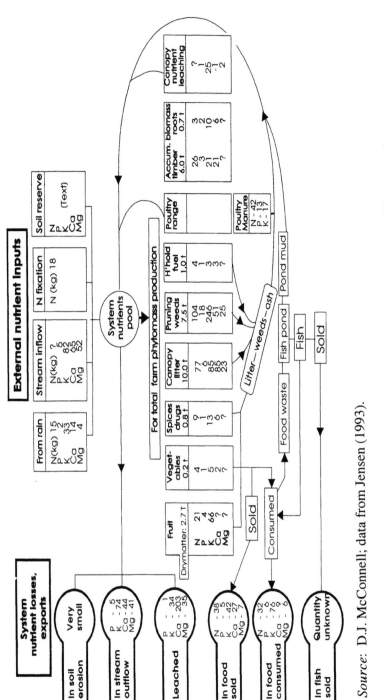

Source: D.J. McConnell; data from Jensen (1993).

Figure 5.5 Nutrient Flows Into-Within-From a Javanese Forest Garden (of 920 sq m; kg of elements per year per ha)

and going into the sink of household maintenance/reproduction (supplemented by purchased food from sale proceeds). Nutrients from the 'pool' go to produce the economic items shown (fruit, vegetables, spices, drugs) and the other biomass outputs of tree litter, prunings, weeds, wood as household fuel, poultry range (weeds, insects) and to accumulate tree/plant biomass as timber and rootmass. Nutrients from these outputs and/or byproducts cycle back into the pool, supplemented by more nutrients leached from the farm phytomass.

Summary – system nutrient balance   As so far depicted this farm of Figure 5.5 would have a negative balance, lines (i) -v- (ii) of the summary Table 5.12. However if food consumed is regarded as staying within the farm-household system, inflow and outflow nearly balance, lines (i) -v- (iii). The deficits are very small as a percentage of nutreints available from the soil reserves, line (iv), except for potash (0.7%). They're actually less than this: it was noted previously (Reconnaissance) that the traditional practice in this part of Java is to locate the household toilet over these fishponds, so food waste (Figure 5.5) would cycle back to the fish, not directly as feed but to fertilize microorganisms on which fish feed. It's also been the practice to allow, indeed encourage, non household members to use these pond toilets (providing a fifth external nutrient source not shown in Figure 5.5). With these adjustments the system would be well in balance, achieved by more plant growth from enriched fish pond mud fertilizer – or more visitors.

Table 5.12   Summary of Nutrients Inflow – Outflow – Balance (kg/ha/year)

		N	P	K	CA	Mg
1. Inflow (excluding N in streamflow)	kg:	33	7	115	79	56
2. Outflow (including food consumed)	kg:	70	17	226	283	89
3. Outflow (excluding food consumed)	kg:	38	11	150	274	83
4. Negative balance as % of soil reserves	%:	0.1	0.1	0.7	0.4	0.4
5. Nutrients internally cycled	kg:	223	38	373	135	50
6. Cycled as % of nutrients in vegetation	%:	22	44	50	20	30

Ignoring domestic waste cycling the mounts of nutrients cycling within the system (line v) are high relative to those imported or exported and amount to 22, 44, 50...% of N, P, K... contained in the farm's vegetation. Although this farm is highly productive in subsistence and commercial produce from the crops listed previously, 61% of the farm plants are 'non economic'. This always leads to pressures to simplify, develop and 'improve' these farms, suggestions which usually ignore the finely tuned nutrient balances Jensen measured, and the probability that these higher economic outputs from simplification and substitution of high yielding varieties would also need higher nutrient levels beyond the system's ability to provide them internally. If left alone Jensen's farm could go on forever. A Garden of Complete Design.

Effects of litter removal   Jensen also measured some of the effects on soil quality of regular litter removal. On Javanese and Malay farms some small area immediately around the house, the *baruan* is kept clean for social purposes and crop sun-drying by daily sweeping it of litter from the surrounding trees. The cumulative effects of this over a long period of 50, 60...years in comparison with the undisturbed part of the farm are indicated, a general deterioration of all the qualitative parameters.

Accumulated Effect of Regular Litter Removal on Soil Condition

Parameter	0-10 cm zone of soil profile	
	tree/garden part	baruan
texture: sand %	35	47
clay %	33	23
pH	6.6	6.3
humus %	2.4	1.6
$C_A\ CO_2$ %	0.5	0.1
N (total)	166	96

Reasons for litter removal   Sometimes much of the litter on these small farms is removed. This happens on those forest gardens growing cloves around Bogor in West Java where there are clove leaf oil distilleries and a good part of these farms' litter as naturally fallen clove leaves is sold for this purpose. The peak leaf-fall months are the dry season August-October but some fall continues year-round, a mature tree yielding around 150 kg of (wet) leaf annually. To ensure good quality oil only fresh leaves are picked up (with a pointed stick), the other litter remains. Nonetheless on clove-dominant farms more than half their litter would not be available for nutrient re-cycling within the system (Nasir and McConnell 1986). More general reasons relate to farm hygiene. Particularly in the wet season the litter on much larger areas than the *baruan* (or its equivalent elsewhere) is swept up and burned in small smouldering fires as an anti-mosquito measure, a village evening ritual. Another reason is to discourage serpents, particularly the small deadly kraits which can easily hide under a few leaves. So while maximum nutrient cycling through litter management is an admirable concept there are sometimes more pressing concerns.

Nutrient Flows Extended ... Through Cows and Fish and Lotus Ponds

Many kinds of aquatic systems can be combined with the plants of forest gardens - tilapia fish fed on enriched micro-organisms as on Javanese farms; grass carp fed on cut roadside grass or napier/para grass or cassava leaves (if these crops can be grown under the tree canopy); fish combined with water hyacinth (the latter boiled and fed to pigs, their manure returned to fertilize aquatic organisms for the fish...). Most enchanting and profitable of these are *lee koh* fish in lotus ponds. A useful by-product of animal manure when it's used to generate household methane gas is its residual as a fertilizer. Among other uses this can be applied to fertilize the aquatic organisms on which fish feed as well as to the lotus plants. There's a symbiotic relationship between *lee koh* fish and lotus: the plants' large leaves shade and protect the fish from predators; the fish keeps the lotus clean of insects and aquatic pests and fertilize the plant by its excrement and does not eat it. (Other fishes like the grass carp might be more productive but they're partial to lotus leaves.)

The full set of nutrient relationships might start with the cow who has provided the manure for the methane tank digester and the tank residue as fertilizer to establish the newly planted lotus, and the methane to later cook some of the fish. They extend also to the farm household who eat some of the fish but sell most of it along with the lotus products to purchase food, excrete the residue of this which often goes into the pond as additional fertilizer – not to forget the other plants which are fed with pond mud when it's periodically cleared, as Jensen does it on the farm. The system looks about as sketched in Figure 5.6, on a one acre (0.4 ha) basis annually (but most farm ponds are only one tenth of this size except in places like Ipoh, Malaysia, where there are many larger lotus fish farms in old tin mines).

In Malaysia and Cambodia the main lotus product is the root (yields are shown) used in cooking pork and chicken soup. The root nodes are often separated out, stored under water in a clay jar and used as needed to make medicinal 'tea' for sore throats. The seeds are valued for filling mooncakes for that festival and used routinely in steaming duck. The leaves are a by-product, left for children to collect and sell as wrappers for hot food (as teak leaves are used in Java). They retain heat better than do plastic bags and add a desirable flavor to the food. Only a few are harvested intermittently so as not to disturb the *lee koh's* habitat.

The cow thinks the lotus pond celebrates ecology but she's wrong. Its operations are scheduled around high *lee koh* and lotus market prices at

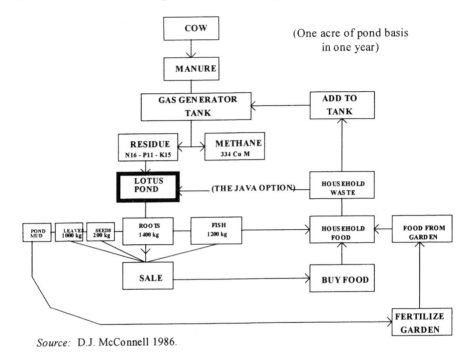

Source: D.J. McConnell 1986.

Figure 5.6   Nutrient Flows from Cow Manure Through Methane to Fish and Lotus Products

lunar New Year and Mooncake time. Long ago, even before the incident of the Three Rivers, a princess took bad medicine and was captured by heaven. To get her back her husband shot arrows at the moon. When this failed he offered it cakes made of the best ingredients; the finest he could find were lotus seeds...But then again, in early Manchu times the people were so oppressed that each 10 families were permitted only a single knife among them. They organized against the Manchu by distributing more knives and secret messages from house to house in mooncakes. The lotus seeds became a symbol of these mooncake messages. So the basis of the pond is not ecology at all or economics but liberation of Han from Manchu...Or perhaps only a foolish story of a man shooting arrows at the moon.

5.4   Forest Agro-development Strategies – Other Voices, Other Views

Most of the world's untapped tropical soil resources including those under rain forests are latisols (oxisols) and red-yellow podzolics (ultisols). Pedro Sanchez (1989, 1989a) and Szott et al (1991) have listed their agricultural limitations in order of importance as low nutrient reserves, aluminium toxicity, acidity, high phosphorus fixation. Concensus among optimists is that most of these chemical constraints and some of the physical constraints can be removed by such agro-technical practices as liming, appropriate fertilizer regimes, formulating crop combinations/rotations for both field crops and tree crops, and more recently crop genetic modification is touted as a strategy for overcoming acidity and aluminium toxicity. With these technical prospects for overcoming the constraints, government development agencies in Brazil have estimated the amount of their potentially 'good farm land' now under forests as 8-10% of it. (Independent authorities think that 4% closer to the mark.) These good soils include the *varzea* riverine lands regularly flooded for several months annually. In this regular replenishment of nutrients by silt and debris and in their sustainable agricultural capacity they have been compared to the Nile (Newman, 1990). In addition and based substantially on agronomic work reported by Sanchez and others in Peru, estimates of the area of Amazonia in which economic agriculture *could* be feasible are as high as 80%. These soils are described as being generally comparable to those of Southeast USA (though given the agro-social history of that region this might seem hardly encouraging).

This claimed potential, based mainly on North American type field crop rotation and fertilizer trials in Peru and Brazil will be good news for those who view the rain forests as corn-peanut-hamburger potential, or a calamity for those who don't. Agrotechnology sees no limits to what it can achieve in bringing the tropical forests into revenue...*Sound soil management technologies exist for overcoming these (chemical) constraints and permit sustained production of annual crops, pastures and perennial crops. The judicious use of this technology will promote a balance between conservation and development in the humid tropics* (Sanchez 1989a).

Agronomic research aimed at finding suitable modern cash crop rotations and fertilizer regimes, specifically for Yurimaguas, Peru, but representative of similar rain forest environments is summarized in Figure 5.7 (Jordan 1987). It says much for this approach to tropical land development that here in Peruvian Amazonia – close to what Gentry and others have measured as the richest concentration of botanical resources on earth – that the agro-ideological basis of this research could not extend itself beyond the North American cliché of rice-soybeans-corn, propped up by the usual package of lime, NPK and six micronutrients. (With no intended irony these forest clearing/wetland draining schemes throughout the developing world often fall under the heading of Diversification Projects, and what this involves is sacrifice of great natural biodiversity in these biomes for increased national economic diversity by adding corn or cabbages to the national exports list – Reconnaissance, schemes.)

From Figure 5.7A in the first three years after forest clearing the yields of six successive rice crops declined rapidly, whether or not these crops received a complete dose of lime + NPK (the top curve), or a compensating dose (middle), or none (the bottom curve). All that fertilizer did was delay the inevitable. In sketch B from 1975 on this situation was salvaged by increasing the dose of NPK on the heavily fertilized plot and adding the micronutrients of zinc, boron, molybdenum etc. (These yield curves of (B) refer to a rotation of corn-soybean-rice.) Without fertilizer the yields of this rotation quickly went to the near zero and stayed there. In other words it is technically possible to develop and farm these soils but only with a research station's comparatively unlimited budget for fertilizer and expertise.

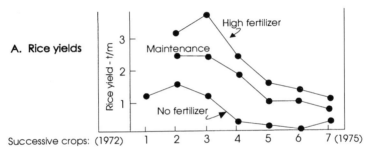

A. Rice yields

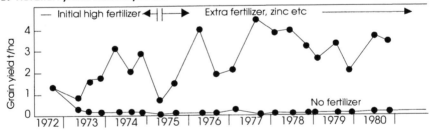

B. Rotation yields corn-soybeans-rice

Source: Modified from Jordan 1987.

Figure 5.7   Rice and Crop Rotation Yields With and Without Fertilizer, Yurimaguas, Peru

Theory

In face of criticism that this type of research must result in greater deforestation pressure it has mounted three main defences. First, forest loss will be minimized if per ha crop yields on existing cleared lands are maximized. This requires soils-agronomic research. Second, there exist vast areas of less biologically valuable secondary or degraded forest which if converted to proper farming would save the primary forests, again requiring research. Third, there is the problem of somehow stabilizing the shifters on productive farms, saving both primary and secondary forests. Some will find this logic appealing, development research is necessary for biological and diversity conservation. Others will find it sophistry.

Reality

The fine print of the thing must be scrutinized. First, it's a truism that whatever promising results are obtained under ideal conditions on research stations are seldom if ever fully attained when this same technology is moved onto farms. (In Australia eg with comparatively progressive and wealthy farmers the research discount rule-of-thumb is a minimum of 33%; ie new technology on farms is only two thirds as effective as might be projected by research.) The effective adoption rate is then further reduced by attrition as farmers later abandon key parts or all of the new technology for irrational as well as rational reasons. Second, the *possible* initial technology adoption rate is a formidable parameter: adoption of the advanced agronomy by unskilled impoverished campesinos and immigrants would require money, accessible credit, supplies of the new inputs, social justice in prices for outputs, dedicated intense extension effort and faith. Of these in places like the upper Rio Negro and Kalimantan there are essentially none. Third, to the extent that farm reality falls short of the Great Expectations of research – for the reasons listed and a dozen more – farm budgets based on the higher costs or management or labor effort will not be met, settlement projects based on these will falter, *caboclos* will revert to their old low-input land use habits, the cycle of forest clearing - farm degeneration - abandonment will be picked up again but now since more people will have been enticed into it this will be on a larger scale than previously.

Sanguine views from the research station seems so far removed from socio-economic reality that others must be considered. One of these is that of Jacobs (1982) whose *vignette* is of a family of hungry *campesinos* who have managed to scrape up ten dollars for the bus trip, then ride down from the poverty of the Sierra along some new mining road to their new green Ed Dorado. There they half-clear 50 hectares, plant grass, buy cattle on credit, overstock. The pasture lasts five years before it wears out, the gaunt cattle are eaten or disposed of, along with the *rancho*, to some more substantial *patron*. Then the settlers disappear with what they can salvage back into the Sierra. Jacobs is describing spontaneous migration into Amazonia through the 1970s and 1980s of several hundred thousand such families. For most of these in the end nothing has changed; they remain as poverty stricken as they ever were. Only the forest has changed. From the air it now takes on the aspect of a rat-chewed carpet. For every   2 000 settlers moving in, whether they persist or retreat, 100 000 hectares of the treasure house shut down forever.

Decades ago Robert Pendleton (1950) professor of tropical soils and agronomy at Johns Hopkins University was substantiating how erroneous these technocentric research beliefs were, particularly that SE USA crop technology could be translocated to Amazonia and Far East to produce cattle, abaca, rubber, sugar, without massive and continuing imported agronomic inputs... *On the whole the soils in humid tropical lowlands have distressingly limited possibilities for plant (crop) production.* Records of Belgian attempts to graft European agriculture onto Congo forest ecosystems and social systems in the 1920s had earlier illustrated this futility (Shifters).

Alternatives

More ecologically sensitive approaches to land management and fertility maintenance include Erwin's work in a swamp forest in Peru (Newman 1990). In this application of Mayan technology a moat is excavated for drainage and the small enclosed area (400 square meters) is filled with rich swamp forest muck. With multiple and relay cropping this intensive food garden is claimed capable of feeding four families. As with the conventional field crop research noted above the ecological *rationale* of this system lies in the alternative it offers to extensive farming which by definition requires deforestation. In the Mexico-Guatemala Maya homeland itself in what is now the Sian Ka'an Biosphere Reserve in Quintana Roo, Gradwohl and Greenberg (1988) report attempts to reestablish these classical Maya systems and to determine the ecological impact on the forest of appropriating its litter and humus for crop fertilizer. (Similar forest materials are used in Central Java and the same question is unanswered – Nature, range.)

Around the town of Tome Accú near Belem in Brazil there is much scientific interest in learning how Japanese settlers there have made their poor soils so productive and sustainable. Jordan and Uhl (1987) have described the principles:

1. Utilize trees and select these economic species in which the biomass harvested (fruit, latex) is only a small proportion of the total plant biomass, thereby minimizing nutrient drain from the system.
2. Aim for maximum species diversity to fully exploit soil nutrients and microenvironment niches, eg sunlight, and to inhibit disease and insects (Energy, architecture).
3. Maintain ground cover to maximum extent to minimize erosion and deterioration of soil physical properties (aggregation, infiltration, moisture storage etc).
4. Recycle all animal and vegetal organic matter as fertilizer.
5. For annual crops, use only high value species and plant them only once or twice; then follow with long life species of possibly lower economic value but also lower nutrient demands.

These are substantially the principles of forest gardens elsewehre though few of these except perhaps in Kerala would be as closely and frugally managed as the farms of Tome Accú. Their farm structure has been evolving since 1930s, after initial attempts at monocropping cacao and then pepper failed because of disease. With continuing plant imports and experimentation these farms have gone back to what they were – multi-storey forest but now agroforest as is apparent from some of the species mixes grown: rubber, oil palm, local palms for palm heart, juice and pig feed, peach palm (often seeded by macaws), cacao, coconut as an overstorey, pepper and vanilla as vines often attached to live supports, shade tolerant ginger as an under-storey crop, many fruit trees... Which nearly brings us back to Kandy, and a long way from the troubles of that corn-soybean-hog proper agriculture which collapsed in Figure 5.7.

References

Clapham, W.B. (1973) *Natural Ecosystems*. Macmillan, Basingstoke.

Collins, N.M. (1989) Termites. In H. Lieth and M.J.A. Werger (eds) *Tropical Rain Forest Ecosystems. Biogeographical and Ecological Studies*. Elsevier, Amsterdam.

Esser, G. and H. Lieth (1989) Decomposition in tropical rain forests compared with other parts of the world. In H. Lieth and M.J.A. Werger (eds) *Tropical Rain Forest Ecosystems: Biogeographical and Ecological Studies*. Elsevier, Amsterdam.

Gradwohl, J. and R. Greenberg (1988) *Saving the Tropical Forests*. Island Press, Washington DC.

Greenland, D.J. and J.M. Kowal (1960) Nutrient content of a moist tropical forest in Ghana. *Plant and Soil* 12:154-174.

Huttel, C. (1975) Root distribution and biomass in three Ivory Coast rain forest plots. In F.B. Golley and E. Medina (eds) *Tropical Ecological Systems: Trends in Terrestrial and Aquatic Research*. Springer-Verlag, Berlin.

Jacobs, M. (1982) *The Tropical Rain Forest: A First Encounter.* Springer-Verlag, Berlin.

Jensen, M. (1989a) Productivity and Nutrient Cycling of a Javanese homegarden. *Agroforestry Systems* 24:187-201.

Jensen, M. (1993) Soil conditions, vegetation structure and biomass of a Javanese homegarden. *Agroforestry Systems* 24:171-186.

Jordan, C.F. ed (1987) *Amazonian Rain Forests: Ecosystem Disturbance and Recovery.* Springer-Verlag Berlin.

Kira, T. and K. Yoda (1989) Vertical stratification in microclimate. In H. Lieth and M.J.A. Werger (eds) *Tropical Rain Forest Ecosystems: Biogeographical and Ecological Studies.* Elsevier, Amsterdam.

Kliewtam, R.S. and P.S. Ramakrishnan (1993) Litter and fine root dynamics of a relic sacred grove forest at Cherrapunji and NE India. *Forest Ecology and Management* 60:327-344.

Klinge, H., W.A. Rodrigues, E. Brunig, E.J. Fittkau (1975) Biomass and structure in a central Amazonia rain forest. In F.B. Golley and F. Medina (eds) *Tropical Ecological Systems: Trends in Terrestrial and Aquatic Research.* Springer-Verlag, Berlin.

McConnell, D.J. (1986) *Economics of Some Tropical Crops: Lotus in fish farm systems in West Malaysia.* Midcoast, Coffs Harbour, Australia.

Moran, E. (1981) *Developing the Amazon.* Indiana U. Press, Bloomington.

Nasir, Abu Yamin and D.J. McConnell (1986) Distilling clove leaf oil in Java. In D.J. McConnell (ed), *Economics of Some Tropical Crops.* Midcoast, Coffs Harbour, Australia.

Newman, A. (1990) Tropical Rain Forest. Facts on File. NY.

Nye, P.H. (1961) Organic matter and nutrient cycles under moist tropical forest. *Plant and Soil* 13:333-346.

Nye, P.H. and Greenland (1964) Changes in the soil after clearing tropical forest. *Plant and Soil* 21:101-112.

Pendleton, R.L. (1950) Agricultural and forestry potentialities in the tropics. (US) *Agronomy Journal* March 115-123.

Proctor, J. (1984) Tropical forest litterfall II: the data set. In A.C. Chadwick and S.L. Sutton (eds). *Rain forest - the Leeds Symposium.* Leeds Philosophy and Literary Society, UK.

Richards, P.W. (1952, 1996) *The Tropical Rain Forest: an Ecological Study.* Cambridge.

Sanchez, P.A. (1989) *Properties and Management of Soils in the Tropics.* Wiley, New York.

Sanchez, P.A. (1989a) Soils. In H. Lieth and M.J.A. Werger (eds) *Ecosystems of the World.* Elsevier, Amsterdam.

Sioli, H. (1985) The effects of deforestation in Amazonia. *The Geographical Journal* 151(2)197-203.

Szott, L.T., E.C.M. Fernandes, P.A. Sanchez (1991) Soil - plant interactions in agroforestry systems. *Forest Ecology and Management* 45:127-152.

Tivy, J. (1990) *Agroecology.* Longman, UK.

Walsh, R.P.D. (1996) Microclimate and hydrology. In P.W. Richards (2nd ed) The Tropical Rain Forest: an Ecological Study. Cambridge University Press, UK.

Young, A. (1989) *Agroforestry for Soil Conservation.* CAB International, Wallingford, UK.

Chapter 6
Energy ... or Prometheus Extended

All farming systems depend variously on four categories or stages of solar energy as incoming radiation or insolation. First stage energy is direct insolation as this is received in specific wavelengths (the electromagnetic spectrum) and modified by earth's atmosphere and operating most notably through plant photosynthesis for growth and production (equation 1 below). Second stage energy consists of insolation rearranged at macro level as climate and weather. At third stage this may to a minor or major degree be further arranged as micro climate by the plants themselves on each farm. The micro parameters of heat, light, temperature, humidity etc are discussed below. The possibilities for making this further refinement vary according to type of farming system. They are great in polycultures, especially in diverse mixes of tree-palm-bush-vine species as on forest gardens but negligible in mono-field crop systems (architecture below).

Fourth stage energy consists of past insolation which has resulted in accumulated biomass, now recovered as fossil fuels (coal, oil, gas) and which can't be used at farm level until it's routed through industrial processes and delivered transformed as artificial farm inputs both upstream and downstream from the farm of everything from tractors and their fuel to produce refrigeration. The energy chains of stages 1, 2 and 3 are short; that of the stage 4 is very long and historically has been increasing; and since each energy transformation and application of it to do work along this extended chain results in some loss (to entropy, 2nd law of thermodynamics) these systems become increasingly energy-inefficient (but with this negated as a farm productivity/economic constraint by energy pricing policies which treat fossil fuel energy as largely a free good except for the cost of exploiting and marketing it).

From an energy viewpoint four main types of farming systems can be distinguished. The first, represented by forest gardens and forest extractivists, relies almost entirely on stage 1-2-3 energy: direct insolation + climate 4 biomass, with this biomass then (a) nutritionally maintaining itself by litter-fall (processed to the nutrient energy of N, P, K, Mg etc by the energy of micro zoosystems engaged in breaking and decomposing), and (b) also yielding the farm's economic product. The second farm-type is similar except that pulses rather than continuous streams of nutrient energy are extracted from the biomass periodically each 5, 10… 20 years or so by the combustion energy of fire or the bio-energy of termites. This describes sedentary swiddeners and the Shifters. In the third and classical farm-type situation nutrient energy is produced for the farm by 2nd stage energy as climate/rain, impacting through soil particle detachment in the soil erosion process (below) then carried to the farm by the energy of flowing water, first as overland flow then stream flow – Central Africa and Ethiopia's bare hills feeding Egypt, Anatolia subsidizing Babylon… or the 2nd stage energy of wind detaching and transporting nutrient energy of loess from central Asia to Man farming Woman weaving along China's Three Rivers, where some think it might have all begun (Origins).

In the fourth situation modern agriculture dispenses with the natural energy of stages 1-2-3 to the maximum extent it can and relies instead on fossil energy, once lost now found – what Marcel Proust's melancholia might want to call energie perdu.

Its logical conclusion is to bring the farm ever closer to the factory supplying its inputs or processing its outputs until the farm becomes just another branch of the factory (factory farms);   or enviro-controlled greenhouses;   or hydroponics in which regiments of cucumbers slurp their energy in synthetic chemical solutions from plastic gutters. According to criterion 30 of Table 1.2, Chapter 1, the wise use of energy is an important attribute of any farming system. Economics has been of little use as a guide to achieving it. On the contrary: through its distorted market pricing mechanism it demands of farming systems that they use more energy not less and applauds this as economic development;   and in its discounting of the social externalities from such profligate use it encourages more of them. But according to Aeschylus this is hardly new. ManWoman's relationship with energy has always been at best ambiguous, initially as its beneficiaries then its victims;   though that Greek was always an awful liar.

6.1 The Story So Far…

Heaven and Earth, Uranus and Gaea, have twelve children, the Titans, and their grandsons from these include Prometheus (Forethought) and Epimetheus (Afterthought). All these are gods but one cousin, Zeus, shows a distinct drift towards a monopolistic monotheism as God, stimulated by anthropomorphic acclamation. Contemptuous of Zeus' delusions and hoping to curry favor with the Humans, Prometheus steals fire from Zeus' heaven and brings it down to them as their greatest gift after life itself. With this harnessed as energy their potential power should be unlimited, perhaps some day allowing them to dispense with Zeus himself. Almost as reprehensible and to wring more kudos from the Humans, Prometheus by trickery diverts an increasing percentage of earth's fruits, its net physical product or NPP, from Zeus or God to the Humans' economy, now called their GNP, leaving Zeus with only a few bones camouflaged with a wrapping of fat.   At first this theft of the NPP goes unnoticed but it increases over the centuries;   and when Paul Ehrlich later measures it (1988, 1991) he finds that it's increased to 40% of all terrestrial net NPP, most as those biomes cleared for farms by the Humans or for fiber, timber, livestock food and much of it wasted.   Paul Ehrlich gets angry...*There is no way that co-option by one species of almost two fifths of earth's terrestrial food production could be considered reasonable, in the sense of maintaining stability of life on this planet.*   Zeus gets angry too and banishes Prometheus to a gulag in the Caucasus where he's bound to a rock and his liver is to be pecked at by eagles for all eternity.   Or by ravens evermore.

But God relents.   Shaken but not contrite Prometheus makes his way back to Athens only to find his brother Afterthought has married Pandora, a silly cow who brings a mysterious box with her.   Forbidden to open it she promptly does and before she can get the lid shut releases all the demons which have afflicted the Humans ever since.   Prometheus watches in horror as he recognizes most of these as offspring of the fire energy he stole from Zeus, and counts four score of these imps as they scuttle out the door – biome destruction and species loss (including those which Zeus had been reserving for the Humans as New Crops when the need arose);   soil erosion and acidification;   ozone holes and cancer;   sulphur and acid rain;   irrigation demand and wars over water;   global warming, sea level rise, eco refugees and boat people;   unreasonable Arabs, fuel prices and road rage;   tanker spills and Enron scams...   All these devils escaped Pandora's box and the Virtues which might have contained them vanished too...except for Hope.

Full of guilt and puzzlement, because the potential for good in Promethean fire seems so easily subverted to evil by the Humans, Prometheus flees to Glasgow, an ice-bound colony in the North where he's befriended by a cheerful moral philosopher, Adam Smith who, learning that his new friend is in the energy supply business, confides to him that he, Adam Smith, is conducting an enquiry into the nature and causes of the wealth of nations.   His results prove conclusively that if the Humans are to find true Economic Fulfillment each must be free to maximize their own self interest.   Only then, in this innocent condition of *laissez faire* would the Humans be led as if by an invisible hand to orbit in a natural harmony with Zeus.

Prometheus is skeptical. If greed is to be the new protocol governing the Humans there won't be enough energy to go round – to make all the pins and broadcloth and shoe buckles and things, not to mention the cannons and round shot needed to open up the new markets for Adam Smith's scheme in India and the inter tropical convergence zone. However these concerns vanish when Adam Smith takes him over to inspect James Watt's new infernal machine: here is a new source of power undreamed of. With this, and unlimited quantities of Promethean fire energy which he promises by digging up a little energie perdu and chopping down a few forests, and Adam Smith's laissez faire protocol to guide them, the Humans should be able to conquer any recalcitrant remaining Titans then even Zeus when the thing gets fully operational and the rough bits get polished of the *laissez faire* protocol by a more circumspect economics, say by 21st century.

His theft now gloriously vindicated Prometheus sets off on a triumphant return to Athens;   but dark doubts resurface when passing through Paris he runs into Sadi Carnot (1796-1832), a theoretical engineer who has spent most of his life inventing a perpetual motion machine. It was to run on hot air using Promethean fire as heat source. Now Sadi admits it won't work and is inventing the laws of thermodynamics instead to prove why such a machine is impossible: (1) Conservation – The energy of the universe is constant, it can't be created nor destroyed;   and (2) Entropy – Conversion from one form of energy to others to do work always involves some loss to useless heat, friction, noise and is not recoverable. Further, entropy as the unavailability of energy tends always to a maximum which will be reached when all matter is at the same lowest temperature and there are no energy gradients left to guide the arrangement of matter and it wanders helter skelter in chaos before coming to a final rest. These laws and not Zeus' tantrums, says Sadi, will define the Humans' use-by date, though he does concede this is still a fair way off.

Worse is to come. Taking a short-cut home through the Harvard Yard, Prometheus runs into the Romanian-American economist, Nicholas Georgescu-Roegen (1971). To Sadi's three laws of thermodynamics (the 3rd is irrelevant) GR is now adding a 4th. This consists essentially of a dual or the dimension of matter to supplement Sadi's laws concerning energy: (a) energy can't be applied in any economic or natural process without some matter to work on or through;   and (b) for such application there's no such thing as a perfect material which generates no waste matter...*The economic process is entropic: it neither creates nor destroys matter or energy but only transforms low into high entropy, ie into irrevocable waste, or with a topical (1971) term, into pollution.*

Prometheus now has to confront reality. The end product of all economic systems is to be exhaustion of energy if one follows Sadi Carnot, and a world drowned in its own pollution if one follows Georgescu-Roegen, though these time rates of depletion and pollution can to a significant extent and for a time be relaxed if the Humans are wise enough to read Adam Smith with more discrimination and, in the present context, to search out and preserve those farming systems which are most energy-efficient and sustainable, for a few centuries at least. It will do no good to run weeping to Zeus.

Waiting until dark Prometheus slinks back to his brother's house. Pandora has had a mid life crisis and run off with a younger god but her box is still there. Afterthought lets him in. They sit in the uncertain candlelight talking quietly of winter in the Caucasus and the bird life there, the price of fish in Glasgow, what a bitch Pandora was and wondering if her box is really empty. Prometheus wants to rummage through it to see if small Hope might still be there. Afterthought thinks perhaps it's better not to know... Now read on.

6.2 Stage 1 Energy – Insolation

All bodies emit radiant energy at some level depending on their temperature. The hotter the body the greater the radiation necessity and to achieve this the shorter the radiation wavelength. While the sun has a surface temperature of about 6 000 C degrees its radiation originates at different depths within its atmosphere, consequently it radiates on all wavelengths from extremely short high intensity gamma and x-rays at one end of the spectrum (part B of Figure 6.1) to extremely long radio waves at the other. From part A of Figure 6.1 (sun) the bulk of solar radiation is in the 'short' wave region of the spectrum and of this some 43% is in the visible part of the spectrum as light, 49% is in the near infrared (IR), 7% is in the ultraviolet (UV) part (further broken down into UV-A, B, C in part D of the graph) and less than 1% is at the very short and very long ends of the spectrum. From the distribution of radiation from sun (part A), most received radiation to earth is on comparatively short wavelengths of between 0.2-4.0 micrometer (millionth of meter), the maximum amount occurring at about 0.5 micrometer in the visible light range of the spectrum (part B).

Solar Energy, Reflectivity, Distribution

Energy emitted by the sun is about 6 300 000 watts per square meter of its surface.   Intensity/amount diminishes with distance from this source in inverse proportion to the square of distance (149 600 000 km); consequently the energy reaching the outer envelope of earth's atmospheric system as incoming solar radiation, insolation, is only about 1 350 - 1 370 watts per square meter (the solar constant) However this is measured on a flat surface perpendicular to the sun directly, and since most of earth's globular surface does not face the sun directly and about half of it is turned away in local night, insolation to each average square meter of earth surface is only about 342 W (Figure 6.2).

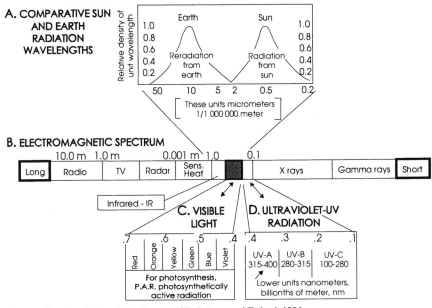

Source: Based on Walker 1991, Bryant 1997, Lutgens and Tarbuck 1986, Ramanathan et al 1989.

Figure 6.1   Electromagnetic Spectrum, Comparative Sun/Earth Radiation, Photosynthesis and UV Radiation

Also from Figure 6.2 (left side) this 342 watts is again a gross incoming amount: about 31% of it is reflected by earth's albido components without it being available to drive earth's systems – 30 watts (9%) reflected back by earth surface, 52 watts (15%) by clouds and 25 watts (7%) by atmospheric molecules, leaving 235 watts (69%).

Figure 6.2 shows how the net 235 W of incoming solar energy is apportioned and disposed of. On the incoming (shortwave) side, 168 W are absorbed at earth surface and 67 W are absorbed by the atmosphere (to the right).   Other inputs to absorption by atmosphere are 24 W from earth surface carried up in air thermal currents, and 78 W carried up from evapotranspiration. Still on the righthand side, once within the atmospheric system most movement or exchange is between land surface (outgoing 390 W) to the near atmosphere to return as back radiation (324 W), with this turnover occurring at an average of seven times before it finally exits the system to space as degraded heat (Bryant 1997).

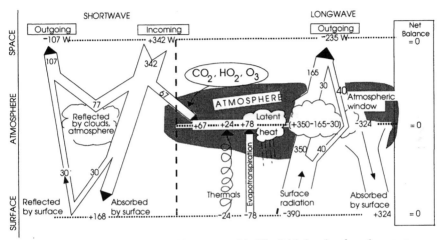

Source: Redrawn, modified from IPCC 1995. Measurements might differ slightly from those from other sources, Houghton 1997, Bryant 1997, Wuebbles and Edmonds 1991, Lutgens and Tarbuck 1986 - References.

Figure 6.2 Earth's Radiation and Energy Balance – Watts per Square Meter of Average Earth Surface

At what farmers will insist is a more practical level the useful tasks which insolation does while within earth's systems are summarized:

Table 6.1 Task Distribution of Insolation

heating the atmosphere, earth surface and oceans	... 42% (of 342 W)
evaporating water or condensing I for the hydrological cycle	23
driving winds, ocean currents, waves, clouds	1
plant photosynthesis and farming	0.023
(albido, reflected to space by earth)	... (31)
(uncertainty and variability margin)	... (3)

Only a very small amount, 0.023%, is used by all plants for their growth through photosynthesis via equation 1 and of this the percentage used by farming proper is minute while—by one arcane parameter or guess—two thirds of the 0.023% is commandeered by the more industrious algae.

Photosynthesis (1) $6CO_2 + 6H_2O + hv\ 4\ 6C_6H_{12}O_6 + 6O_2$ (Clapham 1973)

Here hv is light energy is as defined in Figure 6.1 part C, the other inputs are atmospheric carbon dioxide and water and the output is plant growth (which as food is ultimately the basis of all human systems) and the byproduct is oxygen.

Safeguards, Ozone

The benefits of the first and most direct category of natural energy as insolation have been praised, at least probably to Prometheus' satisfaction, and it only remains to sketch safeguards against excesses. Like Alice one must start at the beginning. According to the canonical model of the evolution of life (eg Boyden 1992, Levine 1991) the first life forms arose by impact of high-energy incoming (UV) solar radiation on a 'watery soup' of pre-biotic gases to build up chemical compounds, about 4 billion years ago. From these amino acids then proteins the earliest simple life forms emerged about 2.5 billion years ago and fed on other compounds which also had developed. High energy radiation was then the condition for life's emergence; but on the other hand this same radiation was to be a main constraint to life's evolution (and eventually, as ozone depletion under H. sapiens' management, a main threat to its continuation).

Some of the earliest life forms developed chlorophyll and used this, with radiation as light, as the basis for a more advanced economic system, photosynthetic capacity (via equation 1). The atmosphere had evolved to a condition in which the necessary carbon dioxide and water were available and the photosynthesis process generated more oxygen, as well as food for the single cell plants/bacteria. By about one billion years ago some of these plant-like cells had evolved into multicellular organisms, enabling reproduction by sexual cell union rather than simple cell division. In turn this required deoxyribonucleic acid (DNA) so that hereditary would be possible, a new organism could function in the pattern of its parents, chaos avoided. One other requirement was some mechanism for discouraging foreign organisms from colonizing its tissues; these became the immune systems of higher animals.

Some of these earliest forms developed the specialization of respiration using oxygen, by then abundant and supplemented by oxygen *from* algae/plant photosynthesis; and in exchange they produced more carbon dioxide *for* the plants, via equation 2.

Respiration (2)   sugars + oxygen 4 carbon dioxide + water + energy, or
$C_6H_{12}O_6 + 6O_2$ 4 $6CO_2 + 6H_2O$ + usable energy
(Clapham 1973)

With much simplification those early organisms which obtained sustenance through equation 1 became the plants; those which consumed the plants through equation 2 became the animals.

Life multiplied and diversified but for 80% of the time since its emergence it was limited to the 'soup' or what became the ocean, confined there by the same high energy UV radiation which had brought it forth from the compounds. Its affinity with ocean has always been closer than with land. About half of all annual global net primary production (NPP) still occurs there and from other estimates up to 70% of all animal life still exists there, from zooplankton to whales.

This oceanic confinement continued until 450-550 million years ago until the UV radiation threat had been removed by accumulation of enough ozone in the upper atmosphere (stratosphere) to filter and block out the most harmful of the UV radiation. This was and is produced in the stratosphere (above the equator where radiation is greatest, at about 350 tonnes per second) and disperses around the globe to provide the ozone shield, the ozonosphere, 90% of it at 12-40 km height, via equation 3:

Ozone generation (3a) ... O_2 + hv 4 O + O (Fraser 1989; O'Hare and
Wilby 1995)
(3b) ... O_2 + O + M 4 O_3 (ozone) + M

In this, photodissociation of oxygen is followed by recombination of the atomic oxygen with ordinary oxygen, where M is any third molecule and hv is solar radiation at a specific wavelength less than 242 nm on the solar radiation spectrum (Figure 6.1, part B).

With this *production* process established it was then 'necessary' that an ozone destruction process also be present to keep ozone at an equilibrium life-optimum level. One of these correcting processes is

Ozone destruction (4) ... $2O_3$ + hv 4 $3O_2$ (Bryant 1997)

where radiation hv is on wavelengths less than 1.14nm and in which atomic oxygen reacts with ozone to again produce ordinary oxygen. Other normal and benign ozone destruction processes involve naturally generated oxides of nitrogen and hydroxyl.

The net result of this ozone generation process (equation 3) and the several destruction processes was establishment of earth's ozone shield against harmful UV radiation. Life could then, 450-550 million years, come out from the protective ocean onto land. In the new terrestrial environment it thrived and further diversified. One strand struggled on to emerge as H. sapiens. Another, the sea plants of the littoral, made the transition to become the land plants, to culminate in the Great Forest and its remnants in the forest gardens. Associated steps were in the increasingly complex and accelerated feedbacks between life and the atmosphere through respiration, nitrification, denitrification, excretion, methane emission, decomposition…and eventually through the increased incidence of fire. As Levine (1991) has summarized the relationship between life and the atmospheric gases…*Almost every atmospheric gas is (now) cycled through or produced by microscopic and macroscopic plants and animals – nitrogen, oxygen, carbon dioxide and monoxide, methane, hydrocarbons, hydrogen, nitrous oxide, nitric acid, ammonia, hydrogen sulphide.*

6.3 Stage 2 Energy – Redistribution of Insolation Energy

The second category of insolation energy is that involving its redistribution among earth's systems. According to IPCC (1995) and Trenberth and Solomon (1994), incoming solar radiation and its blessings do not fall alike upon the just and unjust but are distributed according to Figure 6.3. Incoming (daytime) radiation is incident to only about half the earth while balancing outgoing radiation (the small arrows) occurs over all of it, day and night. On an average annual basis this results (from the scale Part A) in an 'excess' of net insolation between latitudes 36 degrees north and south (amounting to 68 watts/sq meter at the equator) and a deficit N and S of these latitudes reaching a maximum of 125 and 100 w/sq m at these poles. These excesses/deficits exist in the sense that if heat was not redistributed the equatorial and high latitude zones would get progressively hotter and colder. A modicum of equity in radiation as heat – but not as much as Sudanese and Siberians would like – is achieved through the two large arrows (figure) redistributing it north and south from the equator.

There are two sets of mechanisms for achieving this, earth's great circulation systems winds and ocean currents. (These also govern ancillary heat redistribution processes such as drifting ice.) Globally winds and currents – defined to include vertical up/down air drafts and up/downwelling of oceans as well as horizontal movements – are about of equal importance as heat (re)distribution mechanisms. Their component subsystems are all closely or eventually linked, and they operate on time scales which vary from days (the jet streams) to centuries. All these atmospheric and ocean systems are probably linked to accelerated climate change (global warming, atmospheric pollution), as subjects but also as initiators of further change in yet other systems far removed, by teleconnections.

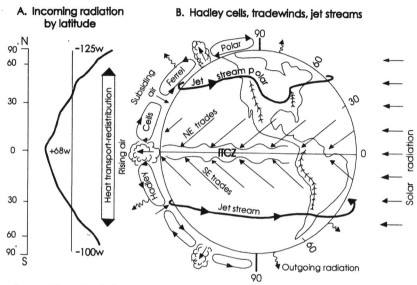

A. Incoming radiation by latitude

B. Hadley cells, tradewinds, jet streams

Source: 'Excess' radiation at equator is 68 W/sq m; deficits at poles are 125 W and 100 W. Redrawn-extended from IPCC/Trenberth and Solomon 1994.

Figure 6.3   Incoming Radiation in Relation to Some Atmospheric Systems

A few of the major wind systems are sketched in Figure 6.3B, including the uprising and descending legs of those systems which operate longitudinally, the Hadley, Ferrel and Polar cells. For present purposes the Hadley cells are especially relevant because their up-rising legs and high rainfall from these define the moist and monsoonal forests and the forest garden zone while their descending legs (air empty of moisture) define the world's main deserts. The trade winds shown periodically falter and sometimes reverse to a W-to-E course as part of the El Niño phenomenon (below).

The ITCZ

As the term suggests, many of the relevant systems converge to form the inter tropical convergence zone (Figure 6.4) notably the trade winds and subsequently the N/S or longitudinal wind systems, Hadley cells, which distribute heat and moisture from this zone. It lies slightly north of the equator though with some branches (because the longer and stronger SE trades prevail over the NE trades) and is a zone of concentrated cloud formation and thunderstorms, with their moisture carried into the zone by wind systems crossing the oceans and over land from surface evaporation and plant transpiration.

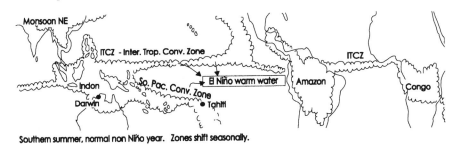

Southern summer, normal non Niño year. Zones shift seasonally.

Figure 6.4   Cloud Belts as Indicator of Trade Wind Convergence and Convection Zones

From Figure 6.4 based on satellite survey by far the bulk of the ITCZ over land is located in the three Great Forest basins of Amazonia, SE Asia-Melanesia and Central-West Africa which correspond also to the forest garden zone. As the warm moisture-saturated air rises in convection it cools, is unable to contain its moisture which precipitates out locally as rain: the day's expenditure of solar energy in surface moisture evaporation and plant evapotranspiration becomes the evening thunderstorm. To the extent that this arises from forests or forest gardens these systems irrigate themselves; eg in a normal non-Niño year about 60% of the rain in Amazon basin arises within that basin and the other 40% is imported as evaporation from over the Atlantic. Perhaps more important globally is that the thunder storm cloud masses arising from all the Great Forests act as cooling towers or heat stratospheric fountains to counter global warming by disposing of heat from biosphere to atmosphere and stratosphere, as well as feeding this north and south to heat deficit regions according to the needs for this sketched in Figure 6.3A. On the other hand these cooling towers also carry large volumes of those pollutant gases which might be present in the ITCZ, notably from the fires of burning forests and the seasonal operations of the Shifters (biomass burning and ozone, below).

But at present clearing rates the Amazon forest will be gone within 50 years, that of Indonesia-Melanesia sooner. The role they've always played in the global hydrological cycle and through this in heat energy redistribution will be terminated. The forests are typically replaced by attempts at modern agriculture (Nutrients) having very little biomass and therefore of evapotranspiration capacity (and CO_2 storage and almost zero biological conservation capacity). For the specific *local* impacts of change in the hydrological cycle through deforestation in Amazonia see especially Sioli (1985). It might seem strange that development agencies and banks are amongst the keenest supporters of this transition. Conversion to highly diversified forest gardens would be preferable on all three counts but even these are a poor substitute for the standing forest.

Jet Streams

These are tunnels or ribbons of wind which mix and transport very large quantities of air and the heat it contains—and pollutants—rapidly around the globe. They commonly braid and wander. The best known polar jet operates at about 10 km height with a speed of 200-300 km/hr; another the polar night jet at 50 km and 600 km/hr; another is the Krakatoa easterly so named because it carried huge quantities of volcanic dust quickly around the world when that Indonesian island erupted in 1883 and among other things this reduced incoming solar radiation to West Europe by 10-20% for three years (Bryant 1997).

From a human viewpoint the jets are benign and in any case they're beyond management control when they do wander from their 'normal' paths and this causes mischief. This errantry is alleged to be responsible for or closely associated with late arrival of India's main SW monsoon rains, droughts in the US Midwest, and also proposed as a trigger for initiating an El Niño event (below). On the other hand as the Krakatoa experience illustrates they might not now be so benign as transporting agents of whatever pollutants enter their path (ozone holes, below).

The Trade Winds and El Niño

By one description the El Niño/La Niña phenomenon is an integrated winds and ocean currents with its memory in the deep ocean and which in years of occurrence affects around 60% of the global population and their economy, either directly or through relationships with other removed weather systems, teleconnections. The specific El Niño mechanics are sketched in Figure 6.5 which is essentially an extension of Figure 6.3A (the excess radiation heat incident to the tropics and which must be redistributed N and S from there), and of Figure 6.3B (the Hadley cells which operate in a longitudinal direction to perform this task together with the Ferrel and

Polar cells, and the east-to-west trade winds).

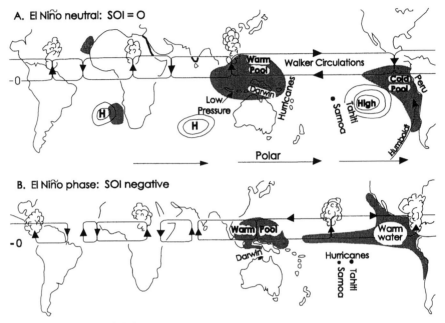

Walker circulations in vertical plane
Source: Nicholls 1987, Lockwood 1985, Philander 1990, Allen 1996.

Figure 6.5   El Niño: Walker Circulations, Warm and Cold Pools

Two functions of the trade winds are relevant here. First, globally they pile up warm water to an additional depth in the western margins of the tropical oceans.   To counter this there are upwellings of cold water in the eastern margins of the oceans. In the west Pacific this pile-up occurs in the Philippines Sea to form a slightly elevated Warm Pool to an additional depth of 10-20 cm and the main offsetting cold upwelling occurs NW of Peru where the thermocline or interface of warm surface and cold deep water is often at the surface. The cold water extending out from this Peruvian upwelling (reinforced by the cold Humboldt current) forms a large Cold Pool (Figure 6.5), accompanied by normally high atmospheric pressures (for convenience measured at Tahiti for purpose of comparison with the normally low pressures around the Warm Pool measured at Darwin, with this differential a proxy measure of the strength or absence of an El Niño event).

The second role of the trade winds is to absorb moisture from the warm ocean then when saturated this air rises in convection, precipitates this as rain then the dry air returns eastward to descend and repeat the process in a series of Walker circulations (Figure 6.5), the uprising legs defining areas of high and the descending legs areas of low rainfall.

This applies to 'normal' non Niño years. About every 4-5 years and perhaps with an increasing frequency these normal mechanics go awry for reasons still speculative. The trade winds slacken and sometimes reverse blowing west-to-east; the East Pacific cold water upwelling is overwhelmed by warm water coming from the west which then extends as a tongue far out to the central Pacific, and Tahiti air pressures drop. The Walker circulations are geographically dislocated (Figure 6.5B) to deliver rain to normally dry or even desert areas (NE Brazil, Peru) and drought to the normally moist tropical forests and farms (Indonesia), with these disruptions also extending far beyond the tropics through disruption also of those weather systems linked to the tropical systems, by 'teleconnections' (Europe). With the impounding force of the trade winds removed or reversed water in the elevated West Pacific cold pool begins to shift to the east in a series of giant but non visible Kelvin waves, smothering the eastern cold upwelling and raising sea levels along the American coasts (eg by 30 cm in California in the 1982-83 El Niño and bringing tropical conditions – and intense rains and mudslides (Malibu 1997) – to the west coast extending from Canada to Chile (a sea level rise of 60 cm and increase in sea temperature of 10C degrees in 1982-83).

Whether, how and to what extent El Niño the Child is himself related to recent global warming and climate change is speculative. There are geological records of El Niño episodes at least 10 000 years ago (Noller 1993); and Quinn and Neal (1992) have put together El Niño's record since 1525 in South America – 115 El Niño's over 1525-1987, a recurrence interval of four years and of very strong El Niño's each nine years.

On the other hand El Niño dryness is certainly conducive to global warming by anthropomorphic extension. In the tropics (and elsewhere through teleconnections) El Niño dry years are adverse for agricultural production and sever El Niño's have historically resulted in famine (the terrestrial equivalent of periodic smothering of the East Pacific cold upwelling then cessation of the current which normally brings fish to Peruvian-Equadorian coastal populations). Some 9-13 million famine deaths are attributed to the 1877-88 El Niño in China (but this also stimulated by landlordism and civil war) and 4 million to the 1943 El Niño in Bengal (but this amplified by bureaucratic bungling). On the other hand these Niño dry years provide ideal conditions for clearing normally moist forests for agricultural expansion.

The 'wildfires' which swept SE Asia-Melanesia, parts of South America, Southern Mexico, East Florida, Greece, Cyprus, Spain in the 1997-98 El Niño event were fueled by drought. In the 1982-83 event similar fires ranged into Manchurian forests. The 1997-98 Indonesian fires alone, 90% of which were deliberately lit by loggers and plantation companies to reduce their development costs, put as much CO_2 into the atmosphere as European industry does in six months. An alternative guesstimate is one billion tonnes added to earth's CO_2 greenhouse load. (It's probably much more. One feature of the Borneo and Sumatra forests is that many are on deep peat or over coal seams and when these are lit by whatever agency they smolder for decades, even centuries, emitting both heat and CO_2 to atmosphere. In 2001 700 of these usually non visible fires were located in East Kalimantan alone.)

El Niño costs too can only be roughly estimated, $80 billion in econoic loss and damage globally for the 1997-98 event in addition to the incalculable social costs of lost forests and their plant-animal species (notably the orang hutans, literally 'forest people', in Kalimantan). And since a dry El Niño is typically followed by an abnormally wet phase of the cycle, La Niña, these also result in costs of similar magnitude eg from the 1998 Niña: 100 million displaced in Yangtse valley, inundations of Bangladesh and Assam, torrential rain and mudslides terminating severe El Niño drought around San Antonio and Los Angeles, typhoon Yanni in Korea and hurricane Georges in Caribbean-Florida, rouding off the year with hurricane Mitch in Central America from which it will take countries like Honduras 30 years to recover. (The 2002-El Niño is running true to pattern: drought throughout most of inland Australia; severe floods in NW South America but drought in eight US states and massive wildfires (Colorado, Arizona); unusual floods in N. and S. China, southern Russia...)

The child and the sunbelt, the greenbelt, the mob and their realtor
Agricultural developers and nation builders are not alone in their fascination with the potential for boosting El Niño effects with a little Promethean fire. Being more candid the Italian press reported their 1997-98 event (torrential Niña rain and landslides at Sarno and hundreds dead following Niño drought) not under climate but under crime:

Rome May 9 1998: The reason so many landslides take place in Campania and Italy has nothing to do with rain or soil...Across southern Europe (Portugal to Turkey) vested interests have combined to eliminate natural methods of disposing of heavy rain. The first step involves clearing of land designated 'green belt'; the easiest way is a forest fire. The growing incidence of such fires around the Mediterranean is not coincidental; large numbers are lit by developers to ensure their targeted areas lose their natural beauty. One side effect is to loosen soil. Then comes development. In the past three years 205 000 houses in Italy have been built without permission, many without proper foundations or on river beds that seem empty...In Naples area the process reaches its delirious apogee, exacerbated by the intimidatory powers of the Camorra. Il Messaggero described this kind of development as 'collective suicide'. Interior Minister Giorgio Napolitano said of Sarno, 'I am witnessing scenes unworthy of a civilized country.'...(UK) Guardian

The South Polar Vortex and Ozone Destruction

In the relatively unindustrialized tropics most pollutants which enter the lower atmosphere (troposphere, to 5-10 km) then the stratosphere (10-50 km, varies) are carried there by thermals associated with the tropical thunderstorms or cooling towers or stratospheric fountains of the inter tropical convergence zone and here most of these pollutants arise in forest burning for permanent agricultural expansion and in the cyclical operations of the Shifters in forests and savannas. This also includes most nominally sedentary smallholder agriculture at the margins of the true wet forests; eg in sub Sahara Africa any traverse of that region say Kenya to South Africa before the rains arrive will be through a continuous pall of smoke as crop trash and bush fallow is burned in preparation for the coming season.

Biomass burning A starting point might be with Joel Levine's (1991) warning: *It may turn out to be that the destiny of our planet is also bound to global biomass burning through its atmospheric, climatic and biospheric implications.* Global totals of biomass burned and its carbon released from the main sources are summarized from Andreae (1991) in Table 6.2.

These data for global carbon release (mainly as carbon dioxide, monoxide and methane) are 'gross' in that some of the release in these categories will subsequently be offset by carbon dioxide recovery in forest regrowth (if this occurs) and, in the case of crop wastes, in the next crop cycle. Most interesting are the immense carbon emissions from savanna burning, about three times more than from the rain forests. For permanent agriculture the forests are cleared and burned once and that's the end of it. The savannas are burned every year or every 4, 5... 10 years or whenever they have recovered enough biomass to be worth burning for their nutrients for the next crop (Shifters).

Table 6.2 Types of Biomass Burned and Annual Carbon Emissions

Biome or source	Biomass burned (Mt, dm/year)	Carbon released (Mt, C/year)
tropical forests	1 260	570
tropical savannas	3 690	1 660
boreal and temperate forests	280	130
world fuel wood	1 430	640
world charcoal	21	30
world farm crop wastes	2 020	910
World	8 700	3 940

Mt, million tonnes of dry biomass and of its carbon content. Other estimates of annual total biomass burned range around 6 000 – 7 000 million tonnes in a 1990s average year.
Source: Andreae 1991.

From Table 6.3 about 85-90% of all global biomass burning of all types is occurring in the tropics, 7 580 of the global total of 8 700 million tonnes. Andreae from whom these estimates are taken calls attention to the great importance of global savanna burning particularly in Africa where this single cause contributes about one third to global biomass burning/carbon release. (Other estimates are that around 750 million hectares of the 1 530 million hectares of the world's savannas are burned annually.)

Table 6.3 Geographical Distribution of Biomass Burning

Region	Forests	Savanna	Fuel wood	Crop wastes	Total biomass	Total carbon million tonnes C
		million tonnes biomass				
Americas	590	770	170	200	1 730	780
Africa	390	2 430	240	160	3 210	1 450
Asia	280	70	850	990	2 190	980
Oceania	(?)	420	8	17	450	200
Total tropics:	1 260	3 690	1 260	1 360	7 580	3 410

Forest burning Permanent tropical forest clearing/burning through the 1980s was concentrated in nine Brazilian states mainly to expand cattle pastures; and Brazil, Colombia and Indonesia were responsible for about half of all that carbon release through deforestation and these together with Peru, Ecuador, Mexico, Congo, Ivory Coast, Thailand, Philippines were responsible for about three quarters of carbon release through deforestation globally (Detwiler and Hall 1988). Fearnside (1991) has estimated carbon released by all forest burning in Brazil (and then further release from the cleared soils) at 0.27 billion tonnes annually in 1980s and 1990s, and that if all Brazilian forest were burned—as they rapidly are being for agricultural expansion/settlement—this would add 51.1 billion tonnes carbon to earth's atmospheric load (less a small amount recaptured in crops, pastures or other sink).

Greenhouse gas yields from biomass burning Table 6.4 (summarized from Andreae 1991) offers a guide to the types and amounts of the major of 16 greenhouse gases/aerosols released globally from all types of biomass burning and their relative contributions to total global output from all sources. From this 26% of all global CO_2 output is from biomass burning, 10% of methane, 21% of nitrogen oxides etc. Some of these gases (methane - methyl chloride - nitrous oxide) are also involved in ozone depletion (below) and the sulphur gases and nitrous oxides are also the major cause of acid rain.

Table 6.4   Global Gas Emissions from Biomass Burning

	World emissions from all sources	From all biomass burning World	From all biomass burning Tropics	From world biomass as % of world total
		Million tonnes of elements per year		
Carbon burned in biomass	-	3 940	3 410	-
CO_2 carbon dioxide	8 700	3 460	3 000	26%
CO carbon monoxide	1 100	350	300	32
CH_4 methane	380	38	33	10
N_2O nitrous oxide	13	0.8	0.7	6
NO_x nitrogen oxides (NO, NO_2)	40	8.5	7.4	21
SO_x sulphur gases	150	2.8	2.4	2
$CH_3 Cl$ methyl chloride	23	0.5	0.4	22
O_3 ozone	1 100	420	360	38
Aerosols	1 530	104	90	7

Altogether 16 gas species and types of particles are emitted.
Source: Andreae 1991, from many primary sources.

Stratospheric ozone depletion The presence of some necessary level of ozone in the atmosphere as a condition for the initial emergence of terrestrial life and as a necessary continuing screen against harmful UV radiation, its natural formation and natural destruction processes were noted previously. Some 90% of total atmospheric ozone is in the higher atmosphere or stratosphere and 10% of it in the lower troposphere. Concentration in a normal atmosphere is very small, only 10 to 100 parts per billion by volume.

As distinct from its benign role as the main UV screen ozone is associated with three types of atmospheric pollution problems: (a) *depletion* of higher altitude ozone which erodes this screen and exposes biosystems to increasing radiation especially of UV-B;   (b) *increasing* concentration of ozone in the troposphere as an important greenhouse/global warming gas;   and (c) *increasing* ozone in the lower troposphere/biosphere as an important contact plant pollutant.   Here it is possible only to sketch the destruction process as it relates to (a) where this occurs in its most dramatic form over Antarctica.

The gases involved destroy ozone through their chlorine, bromine, fluorine or iodine contents; eg via one of these, CFC-11:

Chlorine release (5) $CFCl_3 + hv \ 4 \ 3Cl +$ products (Fishman and
Kalish 1990)
Ozone destruction (6) $Cl + O_3 \ 4 \ ClO + O_2$ (Walker 1992)

The worst of these chlorine and bromine gases have been or are being phased out of use in developed countries under the 1987 Montreal Protocol but this still permits their usage in the developing countries up until 2010-2015 depending on the gas species. (And in any case while the substitute gases being phased in have no chlorine or bromine content and therefore no ozone destroying capacity these also are global warming gases.)

Although the bulk of the ozone destroying gases are from industrial or urban sources a substantial volume of them are from either biomass burning or from routine agriculture or that part of industry which is involved in producing farm inputs or processing-storing-transporting agricultural products in modern agrosystems. Eg one is methyl chlorine which is the second largest source of atmospheric chlorine and about 22% of it is from farming or biomass burning for farm expansion (Table 6.4). Another is methyl bromide, a wide spectrum agent used to fumigate soils and grain handling shipping facilities. These are in addition to a large range of gases used at various points along the farm inputs-production-outputs chain which, while being phased out in developed countries will be used increasingly in the developing countries as their agriculture adopts the 1st world's high energy Euro-North American model.

Destruction processes Mechanics of the Antarctica (and now Arctic) ozone-deficit areas or holes have been known in principle since 1974 (Molina and Rowland 1974; NASA 1988; Gribbin 1988; Fishman and Kalish 1990; Fishman et al 1990; Booth 1994; Fraser 1997; Rycroft 1990). They involve: (a) a giant cold temperature chemical processor which is provided by the south polar vortex over Antarctica, sealed off from warmer air by its swirling winds; (b) ice crystals in the processor on the very cold surfaces of which the ozone-destroying chemical reactions can occur; (c) solar hv radiation at specific wavelengths and (d) a supply of the necessary chlorines.

Winter temperatures within the Antarctic vortex drop to very low levels of about –90C at 20 km altitude. Polar stratospheric ice clouds or PSCs form at about 18-20 km consisting of water and nitric acid. Chemical reactions then occur in winter on the cold surfaces of the PSCs. These reactions release active chlorine from CFCs or similar sources which have been led into the vortex, then when the August-September spring sun returns to provide the necessary hv for this last reaction this chlorine reacts with and destroys ozone. As a result ozone above and around Antarctica seasonally falls to very low levels. These deficiencies have increased since measurements began...*essentially all the ozone in the middle layer (15-20 km) has been destroyed each spring in the 1990s* (Fraser 1997). Measured in terms of geographical area of deficiency this increased from 2 million sq km in 1980 to 26 million in 1998, with much year-to-year variation. Measured in terms of time during which the seasonal deficiency or hole persists this now sometimes continues into the southern summer.

The consequence is that as the low-ozone air migrates north it brings increased radiation exposure to the biosystems of southern countries South America, southern Africa, Australia, New Zealand. Recently similar but smaller ozone deficiencies have begun appearing in proximity to north polar regions. In 1995 ozone levels over North America-Europe were 10-15% below those of 1980s. Here the process is relatively constrained by an atmospheric vortex which is much weaker than the southern, winds are more variable and allow entry and mixing of more warm air to limit the ice cloud/chemical reaction process.

There are still many uncertainties, and global land management and the types of farming systems which this includes are involved in some of them. One is the suite of chemicals which are actually entering the vortex and the possible reactions among them in this cold processor and while they're in high level transit to it, with the thermals or stratospheric fountains then the jet streams as their transporting agents. Chemicals fed into the fountains in the inter tropical convergence zone from extreme episodes of biomass burning could be of special significance. The exceptional El Niño fires of 1997-98 in Indonesia were noted; but in nearly all years over the past two decades deforestation there has been 1.0 – 1.3 million ha/year. In the Brazilian states of Acré and Rondonia 5.5 million ha of forest were burned in one year alone, 1987; and according to Newman (1990) one of these Rondonia fires of 1 400 sq km was the single biggest planned fire in human history (although this and others *did* allow a 30 fold increase in cattle numbers there in 11 years and a 10 fold increase in humans to tend them). In one Australian state alone, Queensland, the annual forest clearing/burning rate also for cattle has *officially* been 0.4-0.6 million ha annually since early 1990s (with government monitoring agencies conceding privately this has been closer to 0.9 million ha).

Another uncertainty is the rate of increase in concentration of stratospheric water and nitric acid with which the ice clouds are formed for the first stage of the ozone destruction process. By some estimates (Booth 1994) stratospheric water has increased by around 30% in the past 40 years. The several possible causes include formation of 'new' atmospheric water from the greater levels of methane now entering the atmosphere (and which is also a major global warming gas in the troposphere) via equation (7):

New strat. water (7) $CH_4 + 3O_2\ 4\ CO_2 + 2H_2O$ (Walker 1992)

and some 25-40% of annual increase in the necessary atmospheric methane for this is from routine agriculture (mainly farm animals and irrigated crops) and from biomass burning for agricultural expansion (Table 6.4).

Also unknown is the actual linkage between ozone destruction and general global warming, but one scenario runs as follows: Since incoming solar radiation is constant at the top of earth's atmosphere except in periods of solar flare activity (Figures 6.1, 6.2) and the lower atmosphere or troposphere is warming, the stratosphere must be cooling to maintain this overall sun-earth heat balance, and these further lowered temperatures there would be conducive to increase in formation of the vortex ice clouds.

Some consequences   A first direct consequence of depleted ozone/increased UV radiation (especially UV-B, see Figure 6.1B) is increasing incidence of skin cancer and a range of other maladies (Australia National Health and Medical Council 1989). On a global average basis for all latitudes a 1% decrease in ozone would result in a 2% increase in UV radiation (Australia Environmental Council 1989); then eg in the tropical state of Queensland this in a 2.4% increase in nonmelanocytic skin cancer, which in this state would translate to 2 500 new cases of NMSC annually. These general relationships also apply more fatally to malignant melanomas.

A second and perhaps eventually more serious concern is impact on those micro organisms which under-pin agriculture and pollination. Incremental amounts of atmospheric CO_2 should themselves have a growth-stimulating effect (via equation 1), at least to some point but economically this will probably be negated by stimulation of weeds/costs too. The effect of depleted high-atmosphere ozone (through increased radiation exposure) is likely to be negative; and increasing low level ozone as a contact plant pollutant is known to be highly negative. Krupa and Jager (1996) offer a (partial) list of 43 common crops adversely affected by increased UV-B exposure. These include the world's main food and fiber crops (listed in Diversity). One way of mitigating this effect, not really possible with field crops in monospecies stands (wheat, cotton, corn etc), is to arrange the architecture of a farm or its biomass stratification to screen out the UV radiation. Such stratification (see Figure 6.7 and architectural-production factors) is a prominent feature of forest gardens and similar quasi-forest systems. (To a marked extent it also governs the supply of CO_2 throughout the profiles of such farms/forests – Kira and Yoda 1989; Walsh 1996; Nutrients.) Though they don't address forest gardens specifically but polycultures in general Kruper and Jager and Runeckles (1992) have outlined in principle these management possibilities as they relate to UV-B and ozone: *With both these stresses (UV-B and ozone), plant spacing of monocultures and mixtures will influence the exposures received. With UV-B this will result from canopy density and shading; with ozone, canopy structure will determine the flux of gas into the foliage as well as dictating the microclimatic conditions* (Kruper and Jager; Runeckles 1992). In summary: to be really complete any Garden of Complete Design must now add the possibilities of UV-B reduction to its attributes.

The Conveyor Belt

Another piece of heat energy redistribution apparatus representing the deep currents is the conveyor belt, Figure 6.6. This girdles the globe, picks up cold salty and therefore heavy downwelling water in the North Atlantic and delivers it via the Indian and Southern Oceans to become upwelling water in the North Pacific. In exchange, warm tropical water flows back to the North Atlantic in a network of surface currents. The circuit takes up to a thousand years. One of its functions is to distribute equatorial heat to the

North Atlantic. Climate modeling (Geophysics Fluid Dynamics, described by Houghton 1997) has shown that if increased North Atlantic rainfall was to occur, as by greenhouse warming impact on the global hydrological cycle, this could decrease ocean salinity there, impede the North Atlantic downwelling of now heavy highly saline water, this would interfere with the global conveyor belt, reduce the return of warm tropical water and this would bring large climatic changes to North Europe, ie regional cooling in a world that generally is warming.

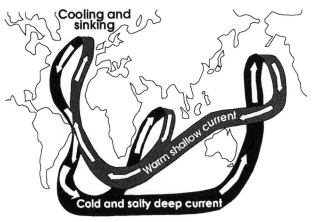

Source: Australian Meteorological Bureau.

Figure 6.6 The Conveyor Belt

There's precedent for such interference during past Ice Ages when severe glacials locked up enough water in the ice sheets to lower sea levels by 100-130 meters, blocking the conveyor belt's return through-flow via the Indonesian island channels (Figure 6.6). This shut down the conveyor belt as a normal North Atlantic warming mechanism, with this becoming in effect a feed-back mechanism to prolong or intensify the northern glacial in countries around the North Atlantic. As far as is known there is no direct connection between the conveyor belt and the (Pacific) El Niño phenomenon. But a possible indirect connection exists through the massive wildfires which El Niño forest dryness promote, and by their anthropomorphic extension, and which provide the necessary greenhouse gases (along with those from other sources) which, through climate change and alteration of global geographical rainfall patterns could adversely affect operations of the conveyor belt, as projected by the Fluid Dynamics model.

There are further implications. In a world which is warming due to build up of the greenhouse gases in the atmosphere, mainly carbon dioxide, it is essential to both reduce CO_2 build up and find safe sinks for that which exists. CO_2 is essentially indestructible; it cycles through earth's systems at time rates ranging from a few months to millennia – from atmosphere to plant to animal to soil to atmosphere... One major way in which global CO_2 is disposed of is through absorption at the atmosphere-ocean interface then burial in the deep ocean by the downwelling phase of the conveyor belt. And since these disposal processes operate on a time scale of centuries it will take a similar time to ascertain if and to what extent the disposal processes of the conveyor belt and similar systems are impeded by climate change. Or perhaps more to the point it would take similar magnitudes of time, great leaps of it, to rehabilitate them... If the Humans are still there and with more vigorous scholarship can figure out a technical fix. And if Zeus has not lost interest. (Epimetheus, is Hope still there...?)

6.4 Stage 3 Energy – Further Refinement at Micro or Farm Level

Accepting insolation and its redistribution by those macro systems which govern climate as given, there is then the potential for refinement of this energy supply on farms—and indeed down to the level of an individual plant—by selecting appropriate plants, crops, animals and integrating these into systems which mimic most closely natural plant-animal communities. These opportunities are greatest in the tropics, especially in the inter tropical convergence zone where insolation is greatest, rainfall and humidity high and there are the greatest number of species to manipulate (but they are by no means confined to the tropics).

Comparative Biomass and Productivity

One starts with a comparison of the biomass and its annual NPP of a few natural biomes (table). By far the largest standing biomass (practically equivalent to plant mass) and annual net productivity in terms of net physical product NPP occurs in the rain forests (and some tropical wetlands). This suggests that those farming systems which closely mimic these forests—agroforestry especially forest gardens—will also be superior in these attributes. It doesn't necessarily follow that these latter systems will be superior in terms of their total *economic* output, when prices are attached to physical output, but high farm biomass and annual output are the basis for such superiority. Sanchez (1989) has listed phytomass of some rain forests usually ranging from 270-400 t/ha (dryweight);

Fearnside has estimated biomass of the main types of forests in Amazonia; Nye and Kowal measured biomass by its fract ons of stems, leaves, roots etc in a Ghana forest (Nutrients); and Klinge et al 1975 offer a very detailed analysis of all parts of a primary rain forest in Amazonia, summarized in Table 6.6. The biomass fractions are also shown for the respective vertical strata, ie down though the forest mass according to above-ground elevation. (This also provides a basis for discussing the supply and/or management of energy at these various levels, below.)

Table 6.5   Biomass and Annual Net Physical Product (NPP) of Major Biomass*

Biome	Standing biomass (dry) tonnes/ha	Annual NPP (dry) tonnes/ha
Tropical rain forest	450	22
Tropical seasonal forest	350	16
Temperate forest	300	12
Boreal forest	200	8
Savanna	40	9
Temperate grassland	16	6
Semi-desert scrub	7	1
Cultivated land	10	7

*global averages.
Source: Whittaker and Woodwell 1971. Woodwell 1985.

The dead matter (D) is a snapshot of the forest feeding itself by internal cycling (Nutrients). The vertical distribution (E) represents a large number of niches in terms of light/shade, temperature, humidity, moisture, wind exposure, carbon dioxide supply (discussed below). The very large number of plants – 93 780/ha of all types – implies high species diversity of both plants and their associated zoosystems. Even without the forest's other principalities of aerial gardens and 'swamps' in the canopy Klinge's bare numerals are describing 'a thing of miracle and wonder'.

Biomass on Farms

Apart from Jensen's (1993) studies of a small West Java farm (Nutrients) there have been few analyses of total phytomas and its fractional composition on forest gardens. These also vary greatly according to degree of 'wildness'/development and farm size. Proxy measurements of these biomass variations are provided by José and Shanmugratnam's studies of 80 forest gardens in Kerala (Reconnaissance). This was distributed according to the following subjective judgments, Table 6.7.

Table 6.6 Phytomass of One Hectare of Amazonia Rain Forest by Vegetation Fraction and Elevation Stratum (Klinge et al 1975)

A. Above-ground fresh weight

		Tonnes/ha		%	
Leaf matter:					
Trees and palms above 1.5m height		17.5			
All plants below 1.5m		0.6	18.1	...	2.5
Branches and twigs:					
Trees above 1.5m height		202.2			
Plants below 1.5m height		0.2	202.4	...	27.5%
Stems:					
Trees and palms above 1.5m height		467.6			
All plants below 1.5m		0.6	468.2	...	63.6%
Other plants (total phytomass):					
Lianas		46.0		...	(6.3%)
Epiphytes		0.1			
Parasites		0.1	46.2	...	6.4
Total aerial phytomass		734.9 tonnes live			100.0%

B. Below-ground dry weight

Rootmass:			
Fine roots		49.0	
Other roots		206.0	255.0

C. Total live phytomass   989.9 tonnes (450-500 tonnes dry weight)

D. Dead wood, ground litter, soil organic matter

Large stems and branches	25.8 tonnes/ha	
Fine leaf litter	4.0	
Fine wood litter	3.0	
Flowers and fruit	0.2	33.0
Organic matter in soil beneath forest	250.0	
Total dead matter of forest	283.0	

E. Vertical distribution of tree and palm live phytomass, section A above:

Stratum	Trees	Individuals/ha Other	Palms	Leaves	Aerial phytomass (tonnes/ha) omits lianas etc Branches + twigs	Stems	Total	Total %
L5 — A	50	0		2.3	48.7	139.2	190.2	27.6
L4 — B	315	0		7.1	123.1	269.3	399.5	58.0
L3 — C (1)	760		15	3.9	26.1	47.3	77.3	11.2
L3 — C (2)	2 765		155	2.0	3.6	10.0	15.6	2.3
L2 — D	5 265		805	2.2	0.7	1.8	4.7	0.7
L1 — E		83 650	0.6	0.2	0.6	1.4	0.2	
Total		93 780		18.1t	202.4t	468.2t		100%

Table 6.7   Tree Populations as Proxy Biomass Measure on 80 Small Farms in Kerala

Farm size class	mean trees/ha	over-crowded	% of farms in each class		
			crowded	optimum	sparse
below 0.2ha	891	46%	35%	13%	6%
0.2 – 0.4	639	13	56	25	6
0.4 – 1.0	613	22	33	44	-
1.0 – 2.0	251	-	29	-	71

The very small farms had around 890 trees/ha and on agronomic grounds (ie optimal spacing for yield maximization) 81% of these were judged crowded or over-crowded while only 29% of the largest farms fell into these categories. Such evidence of over-crowding (according to agronomic criteria) is often used as rationale for reducing farm plant populations and biomass but it overlooks the role which the 'excessive' biomass might play in farm microclimate and nutrient cycling (Nutrients, Jensen's farm).

Synusiae in Forests and Farms

In addition to segregating/classifying rain forest and farm phytomass according to its botanical identity and strata or elevation position (below), Richards (1952, 1996) offers the broader *synusiae* scheme which groups plants having similar life form and occupying a similar niche within the forest or farm; a third group C are here added for the aerial principalities, marshes and some parallels on forest gardens are noted. Importance of the mechanical dependent plants is suggested by Table 6.6 where Klinge found they accounted for 6.3% of above-ground phytomass of one Amazonia rain forest. This fraction can be much higher, especially in cloud and moss forests. Newman (1990) notes these dependent synusiae can account for 30% of the total phytomass of individual trees. The liana equivalents on some forest gardens – the pepper-on-kapok systems around Matale Sri Lanka – consist of multiple vines (2-3 per tree) which grow as tall as their live kapok tree trellis (30-35m) and the vine mass can easily account for 45% of this combined biomass (McConnell 1986). One of the most important economic lianas is the rattan, both on farms and forests (Shifters). Apart from heading for the sunlight then draping themselves across the forest canopy these have no fixed abode or vertical level (L1-L5 below). Burkhill (1935) in West Malaysia found one of these wanderers 126 meters (410 feet) long but it must have been more because it had been broken off by elephants; there are other reports of them exceeding 200 meters. The fungi have an important economic farm parallel in commercial mushrooms, eg high value Japanese *shiitake* now grown under the rain forests in Bhutan but in these forests many kinds of wild fungi have always been collected as an important subsistence food item.

Table 6.8   Forest Plants Arranged as Synusiae and Some Forest Garden Equivalents

Forest synusiae   Some farm equivalents
A. Autotrophic plants (chlorophyll)
   (1) mechanically independent plants
      (a) trees, saplings .... (all trees and palms)
      (b) herbs.. (citronella grass, fodder)
   (2) mechanically dependent plants
      (a) climbers, lianas................................... (passionfruit, pepper, rattan)
      (b) stranglers... (no equivalent)
      (c) epiphytes .... (vanilla, tree orchids)
B. Heterotrophic (no chlorophyll)
   (1) saphrophytes (fungi) .... (tree/ground mushrooms)
   (2) parasites (fungi) .... (fungi)
C. Aquatic (aerial ponds, marshes)(fishponds: algae, lotus)

Principalities ... Where there are Symbiotic Bargains to be Struck (Wilson)

Table 6.6 (Klinge) portrays a kingdom which often also contains many semi autonomous principalities, notably the aerial dryland 'farms' in the forest canopy and its aerial marshes, often found in and around the epiphytes and parasites. Longman and Jenik (1974) describe one of these dry aerial 'farms' in which soil is carried into the canopy by ants and here seeds also carried up by ants or birds or rodents establish themselves to form aerial gardens having no contact with the ground. These gardens might then also be tapped by certain otherwise usually ground-rooted trees which send out aerial roots to the garden for this purpose. Some rain forest tree roots...*are not always subterranean organs* and can grow in any direction including upwards. Newman (1990) has described aerial ranches operated by Aztec ants in American forests. These colonize the trumpet tree (Cecropia spp) where they herd mealybugs, corral them in tree hollows and milk them for their nectar. The ants are very fierce, drive off intruders, and when not milking their herd keep themselves busy around the ranch trimming back vegetation. Their economy also extends to a little forest extractivism in the form of glycogen taken directly from leaf nodules.

    Wilson (1992) has described another of these arboreal principalities, an epiphyte garden in the Monteverde Cloud Forest of Costa Rica (explored by N. Nadkarni and others) as possibly the 'most complex arboreal ecosystem in the world' and compares the profusion of microplant species on a single tree limb to a thicket of miniature woodland, with the epiphytes themselves supporting small trees, the trees supporting smaller plants growing on their leaves, insects browsing on these and bacteria living in the tissue of the insects. In the comparatively poor subtropical rain forests of Dorrigo, Australia, single acacia trees supporting 5 - 10 000 miniature orchids are common.

More luxuriant are the aerial marshes held in the rosette crowns of certain trees also described by Newman of up to 110 liters capacity and sustaining an aquatic world of algae, small plants, frogs, fish, crabs, floating insect larvae and their larger resident or visiting predators. Some nutrient input to these biospheres is from incoming rain water, some from incoming forest litter, and the muck at the bottom of the marsh also nourishes the supporting tree. Compared to these natural micro principalities modern agriculture which bills itself as 'sophisticated' might be seen as the crudest of crude artifacts.

Forest and Farm Plant Architecture

Architectural factors as determinants of farm energy distribution Here architecture is used to describe a plant's spatial form or these collectively as the configuration of a forest or forest garden:

Architectural Factors
M – mass or total quantity of the phytomass
L – the way this total is organized into vertical strata or layers
S – the shapes of its individual trees
D – their unit density
B – their bulk or total space occupied

From an energy viewpoint the function of forest or farm plant architecture is to rearrange at farm or micro level incoming solar radiation directly (light/shade), or as this has been previously modified as climate and weather in order to best provide the necessary conditions for all the species at all vertical levels which comprise the system. Since there are a very large number of these in a forest or forest garden a large number of niches are implied. While the object of each species is to maximize its own welfare, meaning growth and reproductivity, it can do this if its niche conditions are provided by others while in turn it creates the necessary niches *for* others. This may be contrasted with the relationships among individuals when these are organized as monocrops in which no mutual support is possible, no competition tolerated by management (weeds, pests) and the only available niche is occupied by the same species—or potential niches left empty, the main reason why monocultures are in total less productive than polycultures. As Thomas Berry says of human society, it is literally true that a forest or forest garden is a 'community of subjects and not a collection of objects.' (Species or crop combinations are also common even in highly commercialized tropical farming systems eg cattle under coconuts, pepper vines on kapok trees...but in the interests of management convenience these are usually restricted to two or three species/crops.)

Architectural factors as determinants of production factors   Given the initial conditions of local insolation and climate, productivities of the respective species and the forest/farm as a whole can be managed or optimized through farm architecture as this is determined by species selection and combinations of them.   (For the typical architectural configuration of many tree species in isolation see Hallé and Oldeman 1970; Hallé 1974; Hallé, Oldeman, Tomlinson 1978; Tomlinson 1983.) The most important productivity parameters are listed in Table 6.9.

Table 6.9 Productivity and Environmental Niche Factors

P	-	the specific plant species present	ET	-	evapotranspiration
G	-	light/shade intensity, regimes	Rn	-	net rain received
A	-	air temperature	N	-	nutrient supply
H	-	humidity			- via roots
W	-	wind exposure			- via plant absorption
E	-	free surface evaporation	C	-	carbon dioxide supply
					(see Nutrients)

(Identification symbols are arbitrary. NR that part of gross incoming rain or mist less losses to runoff, deep percolation, evaporation.)

These parameters are specific to each set of plants which occupy each elevation level vertically down through a forest or farm.   Following Richards (1952, 1996) there's general agreement that there are five of these levels or strata in both rain forests and the 'wilder' kind of forest gardens (see also Klinge for forests and Kandy for forest gardens).

But on gardens this architecture also depends on many other factors; eg Michon et al (1983) in West Java found it varied along a traverse because specific plants are grown at specific distances from the house for reasons of security (theft), aesthetics, utility (daily food gathering), safety (falling coconuts).

Table 6.10   Layering – General Comparisons of Forests and Forest Gardens

	Rain forest			Forest garden		
L5	30 – 42 m	woody trees		20 – 30 m	(durien, jak, coconut...)
L4	20 – 27 m	woody trees		15 – 25 m	(areca, clove...)
L3	10 – 16 m	woody trees		10 – 15 m	(avocado, rambutan...)
L2	3 – 10 m	herbaceous shrubs		3 – 10 m	(coffee, bananas...)
L1	0.5 – 2 m	treelets, herbs		0.5 – 2 m	(tea, chili...)
L0	0 – 0.2	ground litter		0 – 0.2	(litter, vegetables)
L-1	(to – 2 m)	root mass		– 2 m	(root mass)

The strata elevation ranges are for forests in Guiana, Borneo, Nigeria; in Amazonia some forests reach 90 meters. The forest garden is typical of Kandy.

In the production factors listed above mutuality and circularity are rampant: eg temperature and wind determine humidity; gross incoming rain (not shown) is partly generated by the forest/farm's own evapotranspiration;   light/shade in any niche determine what plants are feasible there then these partly determine light/shade conditions for lower ones; net rain available to a given plant is partly determined by the amount of rain intercepted and absorbed by those whose canopies are above it; all drop litter at different rates to become nutrients which are used by others at their different rates...   Further, all these parameters and mutual dependencies are highly dynamic as trees grow, mature, die (or are removed) and replaced by others. The architectures of individual plants also change with their development, perhaps growing up as spears to more easily penetrate the canopy then flattening out to table-shapes to dominate those then below them. This makes such a forest probably the most complex of ecosystems and the forest gardens which mimic them the most complex of agroecosystems – repeating Tajwani's observation that no one fully understands them.

Most of the production factors exist as gradients down through the forest or farm biomass: light, temperature, wind exposure, evaporation decrease; humidity increases; incoming rainfall is redistributed and carbon dioxide supply changes. These varying conditions or range of vertical niches favor some plants (and associated zoospecies) more than others – some needing the open exposed conditions of the high canopy others humid darkness of the forest floor. In forests such a system manages the production parameters itself, and this is also nearly the case in the wilder kind of forest gardens. (For these vertical gradients see eg Allee (1926) on light in Panama; Gillespie, Knudson, Geilfus (1993) on forest gardens in Guatemala; on temperature Brown (1919) in Philippines; Evans (1956) in Nigeria; on wind Allee (1926); on rain distribution Walsh (1996), McLean (1919) in Brazil; Freise (1936); evaporation and transpiration Walsh (1996); on humidity Shultz (1960) in Surinam; on CO_2 see Nutrients.)

Summary model of a forest garden The flow chart of Figure 6.7 attempts to summarize directions of dependence in a forest garden. (To avoid clutter only the production factors of G-light, A-temperature, H-humidity, W-wind and Rn- net rain at each level are included and levels 4, 3 and 1 are bypassed.) Briefly, the productivity of trees at any level depend initially on the species and their populations present, their supply of nutrients and the five factors listed. 'Productivity' is taken as the NPP; part of this might be harvested as 'economic product' (but as noted previously much potential product is *not* harvested, and becomes litter then nutrients for the system as a whole). In forests it will be harvested by birds, animals, and the balance again discarded, and these animals themselves become litter.

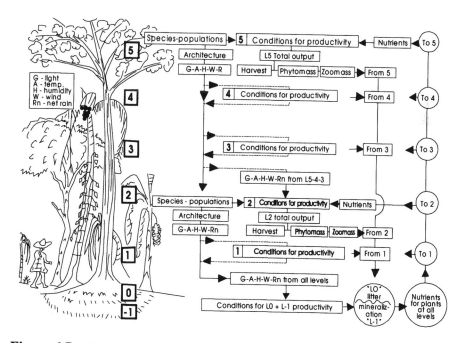

Figure 6.7 Architecture, Levels and Microclimate Factors in a Forest Garden

Taking for example L2 plants as a group, the amounts of light, heat, humidity etc available to them or to which they're subject are determined by the species, populations and the architecture of all trees at all levels L5, 4, 3 above them, with the architectural effects of these on L2 (positive or negative) accumulating downward. On the other hand, all plants at all levels contribute variously to the common litter pool and when this is converted to nutrients they share variously from it. A more complete scheme would have a similar common pool for soil moisture, with incoming rain diverted/absorbed/evaporated by the species/layers (variously) as it passes down through them, then the common pool of soil moisture exploited by them according to species needs, abilities, competition. Especially on poor soils these systems can exist only as long as the litter conversion-nutrient return process operates, and here also the optimum conditions required by the microorganisms involved (heat, shade, humidity…) are provided by the architecture of the plants above them. Again this forest system in its farm analogue may be compared with 'scientific' agriculture where mutuality in species support and niche creation hardly exists.

Edmund, can you help us out with a summary? *There was a pleasant arbor not by art, But of the trees' own inclination made…*

6.5 Stage 4 Energy – Fossil Fuels

While this fourth energy category has several sources (Table 6.11) these are dominated by the fossil fuels, Marcel's energie perdu, contributing around 77% of the global total of all harnessed energy, 900 billion watts of the 12 TW total (TW = 10^{12} watts). But this immense amount from the fossil fuels is still only around 1/15 000th of contemporary incoming solar radiation ie of 1st stage insolation energy supply via Figure 6.1 (Houghton 1997).

This great reliance on the fossil fuels and the cheap energy they have provided, even when discounted by possibilities of substituting some of the other energy sources listed, generates four main concerns. First is depletion and therefore increasing cost of known fossil fuel reserves now guessed to be around 400-years supply globally. Second concern is atmospheric pollution from use of the fossil fuels directly as this impacts through global warming and climate change, led by emissions of CO_2 from these fuels, the main greenhouse gas. Third is similar pollution when energy—whether itself 'clean' (eg geothermal, wind) or dirty (led by coal)—is applied to materials to do work in industrial or agricultural or urban/domestic uses (Georgescu-Roegen's 4th law). In the present farming systems context some notable examples of this latter are the leakage of CFCs to atmosphere to destroy the ozone shield when they're used as refrigerants for farm produce; leakage of nitrous oxide as a greenhouse gas from both the cultivation of and nitrogen fertilizing of soils; agricide contamination of aquifers etc, regardless of whether these materials are made or applied with clean or dirty energy.

Table 6.11   Sources of Harnessed Global Energy

	Actual – 1990 % of total supply, use	Possible – 2020 % of total
Commercial scale		
fossil fuels (coal, oil gas)	77.0%	(?)%
nuclear	5.0	(?)
hydro (large dams)	5.3	5.8
biomass (eg from sugar cane, Brazil)	1.4	5.0
geothermal	0.1	0.8
oceanic (tides, waves)	0.0*	0.5
Mainly local, village scale, farm		
biomass (direct heat, biogas)	10.6	9.3
small scale hydro (running water)	0.2	0.6
solar	0.1	3.1
wind (windmills, generators)	0.0*	1.9

*Exists but is not yet significant globally.
Source: Adjusted from World Energy Council/Houghton 1997.

The fourth concern is impact of energy use generally through species extinctions and biodiversity loss as popularized by Paul Ehrlich's $I = P \times A \times T$ equation where adverse impact is the result of human population level times their consumption level times the environmental damage caused by the particular type/intensity of technology used to satisfy these consumption needs (or wants). Ehrlich later (1995) consolidated this as $I = P \times E$, population times their level of energy use. Since recognizable farming began some 10 000 years ago (Origins, but see also Genesis) estimated global energy use has increased some 13 000 times until 1990. Biodiversity has eroded, but at an unknown past rate, and species loss rate now is around 10 000 per million of those existing. Through history, and from archeological evidence through pre history most of this biodiversity loss has been through clearing biomes and habitat for agricultural expansion, initially by direct use of Promethean fire as the energy source for clearing (Shifters) then this more recently converted to chainsaws and bulldozers.

Farm Energy Efficiencies

Energy efficiency analysis of economic and farming systems is invariably directed at the first concern noted above, that aspect of energy as a scarce resource and in this taking little or no account of the three other concerns. For this narrow approach to farming system energy efficiencies E, these are expressed as ratio of energy content of outputs to energy expenditures in producing them, E = Outputs/Inputs. By now inputs, outputs and E ratios have been determined for most modern agrosystems and some traditional systems (Odum 1989, Leach 1976, Pimentel 1980, 1989, 1990, Tivy and O'Hare 1981, Slesser 1975). There are important differences between modern and traditional farm types. While most farm food outputs (cassava, corn, meat...) are easily measured as calories or joules of nutrition, with the shifters/forest gardeners/collectors/semi-nomads this still leaves a large range of outputs – condiments, medicines, magicals, skins, fibers, wild vegetables – for which energy content is unknown or irrelevant, though it can often be brought to a common energy basis through market price equivalence with rice or cassava or whatever the common staple of the country. On the other hand the difficulty of applying energy analysis to a heterogeneous range of produce has often meant that only the recognizable items are included, the rest ignored and the E ratios then considerably understate the real outputs and efficiencies of these traditional highly diversified systems. Nonetheless the results will usually be close enough to allow energy comparisons among most systems.

On the energy inputs side most of these also have been standardized as (a) direct cultural inputs (labor, manure, ox power, retained seed, tractor fuel... or (b), more completely, these plus the energy incorporated in whatever farm capital is used. Traditional systems have little of this – an axe, hoe, buffalo plow eg energy-rated at 15 000 kcals of energy input per kg of steel in their manufacture; and in these systems most other farm capital represents only human energy required in its fabrication, eg at 2 400 kcals per labor day. Energy (like nutrients) circles within an almost closed system; energy exchanges are between farmer and nature – so much labor energy (at eg 10 MJ per day) for so much food at 7 MJ energy per kg of eggs etc. (They're not quite closed because entropy loss still occurs.)

On the other hand in modern systems the energy exchanges are with markets – so much corn exchanged (via money) for so much fertilizer. Energy flows across the farm as an open system. Their input chains are very long. These again start with on-farm cultural inputs (of which human labor energy is here minimal), wind upstream to where the farm capital is fabricated (tractors, combines, irrigation pipe, steel sheds), further up to their plate mills and rubber factories, on up to the ore mines and rubber estates, with this energy as farm capital tied together by the energy expenditures of transport networks and power supply companies, and lubricated by still more energy expended by banks, marketeers, imagemakers and drug store clerks dispensing pills to ulcerous salesmen. Going downstream from the farm a similar energy chain winds off through processing plants, cold stores, wholesalers, Madison Avenue, stress counselors...to MacDonalds. Such chains or webs, on basis of their food energy delivered to final consumers per unit of energy expended (the ratio E), must be among the least efficient mechanisms yet dreamed up by H. sapiens, easily rivaling Concorde transport. Purely on-farm (cultural) energy becomes almost marginal.

Energy dimension of agricultural development Black's (1971) longer view of energy efficiency takes in the whole sweep of progress from gorillas in trees through Kalahari gatherer-hunters and shifters to Texans on tractors (Table 6.12). From this a chimp economy is well left behind; these apparently spend 70% of the food energy they gather in looking for more food (the price they paid for staying in their 'green caves'). The Kalahari Kung do better with an output/input ratio of 3.0 (and have to work only three days a week to get it). Moving up the scale, most shifters do still better, especially when they grow a bulk energy food like cassava which yields an E of 60-70 (Shifters), though some in Gambia (E = 3) may as well have stayed as gatherers. Climbing further up, by 1960s the mechanized farming systems were about 200 times more efficient than are most of the shifters (for US rice E = 3 700). There has obviously been an enormous increase in energy efficiency when this is defined in terms of direct crop cultural labor; but when other inputs are included (as machine fuel, fertilizers, pesticides, energy embodied in farm equipment) these mechanized systems become only half as efficient as the shifters, 9 -v- 17, or much less, 9 -v- 34 for rice in Sarawak (Shifters). Or in the case of corn, in Guatemala human energy is applied with a hoe, a little more is applied as seed, energy as sun and rain do the rest and tortillas are consumed. When this hoe is replaced by a tractor (US corn), and the Indian lady's food replaced by diesel fuel and the tractor is followed back to its stable where the energy in its steel, rubber, copper are assembled and all this toted up, corn's efficiency has fallen from 16-40 to 2.5. Then when the 2nd law of thermodynamics pursues this corn downstream through the cow (US feedlots) the best she can do is an efficiency of 0.07 – 0.10; ten times more energy has gone in than comes out. European heated greenhouses with efficiencies of around 0.002 suggest further development possibilities.

Table 6.12   Comparative Energy Efficiencies of Farming Systems Around the World

	Labor only	All inputs
Primate collectors (apes, gorillas) ... E =	1.4	
Kung gatherer-hunters	3.0	
Shifters (mean of 13 systems)	17.0	
Draft animal systems		
Mexico corn	26.0	6.0
Vietnam rice	18.0	14.0
Mechanized systems	Labor only	
USA rice	3 700	9.0
USA oats	4 580	29.0
USA corn (1983)		2.5
UK wheat		3.35
UK wheat-as-bread		0.52
Settled Peasants		
Guatemala, corn with hoe ... E =	16	
Chinese gardens	42	
Range		
US beef	1.8 – 2.5	
Intensive		
US feedlot beef	0.07 – 0.10	
UK glasshouses	0.002	
UK broilers	0.10	
UK eggs	0.14	
Europe ocean fish	0.004 – 0.05	

Source: Compiled from Black 1971; Leach 1976; Tivy and O'Hare; Pimentel and Dazhong 1990.

No similar data are available for forest gardens. From a previous chapter (Kandy) they need practically no external inputs and little labor, being on the 'wilder' farms almost gathering systems, but have high output so their Es are probably around 70-80. This is suggested by the Chinese gardens which have very high labor (and recycled waste) inputs but still manage efficiencies of 40 or 50. This is accompanied by their high *per unit area* productivities (*...the intensively cultivated garden is one of the most productive creations of mankind* – Leach 1976 quoting Kropotkin). Leach also cites the case of London backyard gardens being much more energy efficient (E = 1.3) than British agriculture overall (E = 0.37), a fact he attributes not surprisingly to their use of polycultures, the way most shifters do it; and one might recall also the household gardens on former USSR state farms which always outperformed the official farms, in both energy and per unit area productivity, to embarrassing degree. But wait, there's more...

Historical trends David Pimentel and Wen Dazhong (1990) taking US corn with its increasing yields but also its increasing energy inputs as representative of modern agriculture have chartered how total energy efficiency has historically been declining (Table 6.13). Energy definition is here expanded to include that embodied in *all* inputs of the kinds shown.

From this the E ratio of 10.5 in USA in 1700 would represent a technology comparable to that used by Mexican and Vietnamese animal-plow farmers today (Table 6.12) and by most farmers in the world. From that time the news has been progressively better or worse, depending on one's viewpoint. Sanguine spectators will have their eye on the 340% yield increase made possible by the greatly increased energy inputs, even if it did drive the US corn E ratio (2.5 in 1983) below what a shifter achieves in Gambia in a bad year. (This 2.5 E ratio is also close to that of high input Green Revolution rice technology in Asia.) Pessimists worried about depletion of fossil fuels – especially as the developing world also adopts this high-input US model – will have their fears confirmed by the steady deterioration in energy efficiency since mechanized farming started after about 1910.

Table 6.13   Changes in Inputs and Energy Efficiency Ratio for US Corn 1700-1983

	1 000 kcals per hectare – US averages*				
Year:	1700	1910	1950	1970	1983
labor	650	65	24	12	6
machinery	20	280	560	910	1 020
fuel	0	0	1 630	2 110	1 260
fertilizer	0	0	300	2 050	3 910
agricides	0	0	10	210	600
irrigation	0	0	130	1 130	2 250
drying	0	0	10	380	660
Total energy inputs	720	1 300	3 110	7 540	10 540
Yield/ha (kg):	1 880	1 880	2 380	5 080	6 500
Yield 1 000 kcal:	7 520	7 520	9 530	20 320	26 000
Energy efficiency ratio E:	10.5	5.8	3.1	2.7	2.5

*Data rounded, several individual items not shown.
Source: Pimentel and Wen Dazhong 1990.

Really dedicated Cassandrists will worry their way through this issue of energy inefficiency to get at the more fundamental one of Greenhouse gas emissions and climate change, these atmospheric pollutants arising in the manufacture of all the farm inputs listed (except labor), and from their application at farm level – carbon dioxide, nitrous oxide, methane (in accordance with Georgescu-Roegen's 4th law of thermodynamics). They'll find nitrous oxide especially promising, arising both from nitrogen fertilizers and tearing up Mother Earth with tractors while the Congo Medje, those pioneers of minimum tillage and energy efficiency, only prod her gently with a stick (Shifters).

A summary   From Georgescu-Roegen's supplementary 4th law of thermodynamics energy can't be applied to materials to do useful work without also producing some waste as pollution and the more energy applied the more pollution is produced.   In economics terminology any economic production-distribution-consumption process concerns both an intended or planned output (which broadly can be assumed to contribute to human welfare) and a set of unintended by-products or externalities. These can be positive in that they operate as feed-backs to enhance the output or efficiency of the production process (eg accumulating technical knowledge), but by far the bulk of them are negative in that they can reduce input-output efficiency of the given process itself or corrode general welfare in unforeseen ways.   To the extent that these externalities are not known or are camouflaged by a market pricing system which does not recognize them, the given production process (farm, firm, industry sector) appears far more efficient and therefore socially desirable than it actually is.   The major externalities of agriculture are listed in Table 6.14; while some (soil erosion, salinization, the trans species diseases) are as old as agriculture itself most are increasingly attributable to global adoption of Western high-energy and specialized 'scientific' farming models.

Table 6.14 Some Agricultural Externalities

Atmosphere: pollution, as manifest and measured in
1.   enhanced Greenhouse, global warming
2.   sea level rise, inundations
3.   climate change and shifts in biome boundaries
4.   increasing frequencies of extreme weather events
5.   depletion of stratospheric ozone, increased radiation exposure
6.   acid rain and contact pollution of biosystems.

Lithosphere:
7. water/wind soil erosion
8. soil/water salinization, acidification, agrochemical pollution
9. aquifer and surface water depletion.
Biosphere:
10. habitat destruction, species and genetic diversity loss.
Oceanosphere:
11. run-off pollution of the littoral and reefs (and their economics).
Econosphere:
12. contraction of the crops/livestock human food base; disease risks.
Urbanosphere:
13. the unintended, unplanned and largely ungovernable megacities, in which the owners of the urbano-industrial enterprise barricade themselves in ghettos, surrendering the nights to street sleepers whom the enterprise cannot accommodate and to Fagan's children.

Only a few of these can be pursued in following sections – 1, 5, 6, 7, 8; and 10 and 12 were discussed in the previous chapter Diversity.

6.6 Population and Pressures on Biosphere, Especially Tropical Forests

Global population now around 6.1 billion is projected to increase to 9.0 billion within 50 years (Un Pop. Division), with developing countries contributing at least 80% of this increase. There's much debate whether the number who are chronically malnourished (or starving) are increasing or decreasing: they have hovered at slightly less than one billion 1969-96, but in any case from UN data cited in World Bank annual reports about 3 billion of the present 6.1 billion are living on less than $2/day and 1.2 billion are existing (or not) on less than $1/day. And since the bulk of these are farmers—or used to be or would be if they could find land—these data seem not a warm endorsement of global land management practices or the farming systems which now predominate. The almost universal model proposed to relieve this situation (especially by economists with their eye on the previous experiences of countries like Japan, Singapore, Korea, Taiwan) is industrialization and/or modernization of agriculture. However this is a catch 22 situation since it would require greatly increased inputs of fossil fuel energy in the developing world and this would only further increase environmental destruction there and—through greenhouse— globally. In the moist tropics ie in and at the margins of the ITCZ the most direct impact of population growth has and will continue to be destruction of the rain and monsoon forests (Table 6.15).

Table 6.15   Population Growth and Pressure on Agricultural Lands and Forests in Tropical Forest Countries

	Popluation million		Annual growth rate	Persons per ha of crop & pasture land	
	1990	2025	1985-90	1990	2025
World	5 292 m	8 467 m	0.02	1.10	1.76
Africa (WT=22)	648	1 581	3.00	0.60	1.47
Ivory Coast	13	40	4.10	0.78	2.40
Madagascar	12	33	3.20	0.32	0.90
Nigeria	113	300	3.40	1.59	4.21
Uganda	18	55	3.50	2.12	6.47
America N & Cent (WT=14)	427	595	1.30	0.63	0.93
Cuba	10	12	0.70	1.59	2.00
El Salvador	5	11	1.90	3.85	8.46
Mexico	89	150	2.20	0.90	1.50
America South (WT=9)	297	498	2.10	0.48	0.80
Brazil	150	246	2.10	0.61	1.88
Equador	11	28	2.30	0.41	3.59
Venezuela	20	38	2.60	0.94	1.78

There's a high correlation between these as cause and effect; Table 6.15 summarizes data for 60 tropical countries, shows regional totals and a few representative observations. In some countries notably Bangladesh, China, India, Mauritius the population increase/forest land clearing correlation is weak because there's not much left to deforest (and in parts of some countries, Philippines and Indonesia, the pressures are now being directed also at other biomes, notably the littoral and off-shore coral reefs). Viet Nam with a population growth rate of 2.2% and deforestation rate of 28.8% represents those countries in full economic development flight, converting biocapital to industrial Kapital soonest, and in this substituting energie perdu for traditional forms of energy, the Western model most warmly recommended by IMF and—now with some hesitation—World Bank.

6.7 Contraction of the Global Soils Resource Base

Some recent estimates of global lands adversely affected (including lost) and the main forms of degradation are summarized in Table 6.16. Some 2 billion ha were so affected by all the causes shown and water + wind erosion accounted for 1.642 billion ha. (There is some double counting; much degradation has more than one cause.)

Table 6.16 Global Lands Deterioration 1945-93 (million ha)

World Total	Soil Erosion water	wind	Soil compaction	Drainage, subsidence	Nutrient loss	Salt	Chemical pollution	Acid
1964	1 094	548	68	16	135	76	22	6

Soil Erosion

Accelerated soil erosion is activated by the 2nd stage of solar energy as wind or flowing water as this is enabled by the expenditure of human mechanical energy or 4th stage energy in clearing land of its protective cover. Table 6.17 breaks the soil erosion component of Table 6.16 down according to degree of severity and region. Globally 'strong' erosion is concentrated on small farms in the 3rd World (S-SE Asia, Africa, Central America) where there is usually insufficient farm (or national) income to take expensive preventative/rehabilitation works, or insufficient effort at relaxing intensity of land use pressure and solving this problem by biological methods. Globally the area of 'strong' erosion is about 246 million ha and much of this is on its way to 'extreme' status within 20-30 years as population pressures mount, adding to those imposed by governments themselves to find the capital for industrial development or increasing national exports to repay debts.

Table 6.17 Severity of Soil Erosion by Region, Water and Wind

	World	Africa	America N & C	America S	Asia	Europe	Oceania
Total, million ha:	1 642	414	145	165	663	157	99
	%	%	%	%	%	%	%
1. light	37	35	12	44	39	16	96
2. moderate	48	38	71	49	48	76	4
3. strong	15	26	17	7	13	8	*
4. extreme	9	*	*	*	*	*	*

*exists but too small to record on million ha scale.
Source: WRI, UNEP, UNPD 1991-92, 1994-95 Reports.

Abandoning the 9% (148 m ha) of extreme erosion which by now are beyond recall (except by afforestation), if it is assumed that limited available conservation resources are directed to where they would have greatest cost effectiveness, and further assumed that farmers themselves could reverse 'light' erosion by adjustments to farming practices, the soil conservation effort would be concentrated in areas of so far 'moderate' erosion.

Soil Conservation Needs

On this minimal basis at least 700 million ha would have to be treated or somehow stabilized over say the next 20-30 years, 527 m ha of this against water erosion. If an engineering methodology were used this task would be far beyond the capacity of all the soil conservation services of the world. Reliance must be largely on alternative and/or supporting biological methods. Fortunately in the high rainfall tropics where most water erosion occurs these biological methods (or combinations of biological-engineering methods) are economically feasible and they are available, largely from agroforestry and traditional indigenous farming systems. The solution would require substantially abandoning Western models of field crop farming which are increasingly imposed on the 3rd World in the guise of development or modernization, or to recoup 1st World debt. The most profitable crops for this latter purpose are often export cotton, corn, peanuts, tobacco and other modernized row crops which are the worst from a soil erosion viewpoint.

6.8 Soil Conservation Enabling Conditions, Strategies, Methods

There are probably eight main sets of conditions for a successful soil conservation program, or negatively, reasons why these programs are not implemented or can fail.

1. Strategic policy This requires that the public authority (state, society) limit its demands on the agroecosystem to its sustainable capacity. This is the most fundamental of the conditions. But very few governments (North Europe and Japan) make realistic assessment of this capacity and shape their agricultural demands and land use policies accordingly. In the developing world excessive pressures are the norm not the exception, and this applies also to the agricultural exporting countries of the 1st World.

2. Debt management The first condition of awareness, policy sensitivity, extends to international banks, development partners, the IMF and again it is seldom met in the developing world. First concern of these financial agencies is with things like debt repayment and correcting the balance of payments, typically achieved by increasing the pressures on farmers and their land and not reducing them.

3. Economics, perceived benefits v costs　This condition requires that perceived benefits from conservation be greater than perceived costs. The qualification is important because especially in non-monetized societies both benefits and costs are largely subjective — what farmers think they are, as distinct from what soil conservation technicians calculate them to be. On the benefits side possible measures to improve the benefits/costs ratio include increasing the prices of less erosive crops/systems relative to prices of the more erosive (pasture v row crops). In practice this is difficult to do because it can require substantial adjustments to the agricultural economy and its crop mix, and many prices are set beyond a country's boundary. On the costs or inputs side measures include low interest conservation loans, free technical advice, inputs subsidies. But conservation is also a long term investment which must be evaluated (discounted) over time. The critical factor is length of farmers' planning horizons. For hungry campesinos it will be very short — today, tomorrow. With insecurity of tenure or a bullying landlord it might be weeks, a few months. So improving the benefits/costs ratio, or increasing perceived net benefits, must often take the conservation planner along any of those strange paths which lead to expanded farmer planning horizons and lower rates at which they subjectively discount the future — improved tenure security, reduced price risk, guaranteed markets, poverty amelioration. Few go down these paths because they are too numerous and devious, and in the absence of conditions (1) and (2) the effort would be futile. By-and-large economics offers far more powerful tools to achieve conservation than does technology, though they are seldom used unambiguously for this purpose.

4. Knowledge transfer　This condition is broadly defined to include transfer of *awareness* that a problem exists, of the *will* or incentive to fix it, then transfer of the necessary *technology*. This might loosely be called extension. At its best this means objective assessment and positive assistance; at its worst it degenerates to exhortation — empty patriotic rhetoric urging farmers to repent, reform, conserve, while at the same time denying them the means to do so.

The direction of attempted transfer is also critical. Until recent decades modern Western technology assumed this direction was downward, the knowing passing instruction to the ignorant, mainly because it does in fact have a large fund of knowledge to impart. It is now realized, but not always so, that there is at least as much informal traditional soil and water conservation technology in the villages. This is often more relevant, appropriate. It calls for lateral transfer among villages, regions. If modern scientific conservation in the tropical 3rd World is going anywhere at all this seems to be towards discovery, verification, absorption and lateral transfer of this traditional stream (FAO 1965, 1977; Roose 1988, 1966; Hudson 1977, 1987, 1991; Nair 1993; Young 1986, 1987, 1989; Konig 1992; Lal 1976; El-Swaify et al 1982; Seager et al 1965).

5. Simplicity, resiliency This requires that whatever technology is applied it should be of maximum simplicity for installation and maintenance, and resilient in that it can withstand the shocks of abuse and neglect. This applies particularly to engineering structures. As a first principle the more elegant or complex the technology is the less its chance of survival. A corollary is that if such a system breaks (a series of terraces, diversion channel, dam wall) the resulting damage through concentration of energy is often greater than with no system at all.

6. Compatibility This condition recognizes that erosion from running water occurs on a catchment basis but typically its impact, the incentive to reduce this and take counter action occur on pieces of the catchment, individual farms. This introduces the need for spatial compatibility: all farmers should arrange their farming programs so that these do not impact adversely on others eg through runoff diversion, and each technical component (location, size, function) should be integrated with others. This cliché is often difficult to achieve, especially with complex engineering systems superimposed on clusters of traditional farms, and again especially in those ex-colonial countries whose governments inherited a partiality to rectilinear survey.

   But in much of the tropical world with less individualistic notions of land ownership (and more fluid concepts of land survey) this particular problem might not arise. In rural Java for example field land (not pekarangan land) is to varying extent subject to village decision making. But even here there might be five villages in a catchment, five hamlets (sub-villages) around a village, with 20, 30 40... land 'owners' along a proposed terrace or ditch with this number of different ideas regarding its location, size, desirability, and nearly as many degrees of enthusiasm regarding its maintenance or dismantling — especially if it serves no visible local purpose but benefits mainly other farms, villages, elsewhere in the catchment. This diversity of opinion usually does not extend to irrigation supply channels the value of which is commonly acknowledged. In this there is something to be said for packaging soil conservation engineering works as water management (as in fact it is). *Temporal* compatibility requires that whatever technology is used it must be compatible with farming operations through the full seasonal cycle. A system which is optimal for the cropping phase might well be impossible during the subsequent grazing phase when the village cattle are turned loose on the stubble to trample earthworks, perhaps create break points in the banks, or if vegetative controls are part of the system chew down the trees. (In Africa there have been many large scale soil conservation projects of admirable technical design which failed through the impact of uncontrolled cattle, or through preoccupation with technology to the neglect of sociology.)

7. Technology A — engineering, mechanical controls As a condition the availability of this kind of technology is partly optional, it can often be replaced by biological/vegetative methods. Nonetheless it is taken very seriously, to the extent that modern soil conservation *per se* early came to be defined in terms of engineering methodology, and because this was easier to apply to running water than to falling rain, concentration of effort was on processes (3) and (4) below, which was a good thing because engineering structures are often necessary. But also not so good. For 50 years this official type of conservation has planned its campaigns from the ramparts of applied hydrology, its heavy ordinance Manning's formula for flow velocity, calibrated to shoot in various directions (largely according to the field manuals left by Horton 1937, 1938, 1945) supported by the small bore fire of slope and length exponents, roughness coefficients, Reynolds numbers for laminar flow and half the Greek alphabet to screen basic process behind the camouflage of arcane equation — an expenditure of ammunition balanced by paucity of result (Tables 6.16, 6.17 and following). The main objection to all this noise and smoke is that it distracts attention from the main game, conditions (1), (2), (6) and (8). Globally it is now generally acknowledged that this 'official' type of soil conservation has failed, not because of its inadequacies as technology but because it is overwhelmed by sociological and political conditions.

Essentially this Technology A amounts to application of engineering methods to reduce the amount/depth, turbulence and velocity of surface flow (by checks, rough tillage, absorption, diversion). In this soil conservation can hardly be distinguished from water management and the best technology aims at conservation of both. In the West it has been most comprehensively developed and applied in USA as the Universal Soil Loss Equation, followed by Southern Africa, Australia, Europe, and used in support of field crop farming, especially erosive row crops. In the East its ancestor is wet rice terracing which achieves soil conservation through earthworks for crop water management (Java, Philippines, Sri Lanka, China). There is a paradox here. Taking Java as example her rice lands are as carefully conserved as anywhere in the world, even where they have been extended up into extremely steep slopes of 40, 50...% and consequently the *sawahs* (ponded fields) at the top are as small as one square meter. But where water for rice is not available, regardless of soil quality, Java's *tegal* or unirrigated hill lands are as badly managed as anywhere. One concludes that in the West engineering conservation technology is constrained mainly by its cost relative to perceived financial benefit whereas in Java it is constrained mainly by lack of irrigation water. It is also possible to read a cultural dimension into this because many of these neglected uplands are, or were, quite good. The fact that they are so abused seems due to their inability (through lack of water) to produce the cultural staple rice. Another curiosity is that Java's population has grown immensely since a century ago but land management ability seems not to have evolved with it, apart from intensification of the rice lands. The *availability* of conservation technology, even when this is supported by favorable benefit/cost ratios and stimulated by growing land scarcity and food demand, is no guarantee that it will be applied. Again one returns to condition (1).

8. Technology B — biological controls This condition requires recognition of the role of biomass (plants and soil fauna) in erosion prevention and control. To a large extent it is possible to substitute these biological controls for the mechanical controls of Technology A. The emphasis is on processes (1) and (2) (below). The immediate economic advantage of this approach is that trees, bushes, grass etc usually yield some economic product while mechanical structures yield only costs, although their space may still be utilized. The first technical advantage of biomass controls is that they can counter the two main types of erosion, raindrop impact and overland flow whereas mechanical structures can counter only the latter, and that partly. A second technical advantage is that control by biomass is not limited by such factors as degree and length of slope, soil depth and rainfall intensity, all of which restrict use of engineering structures. Conventional slope limits for a range of mechanical structures are noted below. In practice beyond these slope limits (10, 30, 40...%) permanent vegetative cover is the only feasible technology, and at still higher slopes (or with thinner soils, higher rainfall etc) only permanent forest or quasi forest are feasible. There are always exceptions, these technical conventions are only guides. The terraced rice fields of SE Asia procede almost regardless of slope as long as soil depth and water are available and they have been there for centuries, but only at great cost of labor in their maintenance, and great heartache in their reconstruction when they periodically fail.

Alternative Conservation Strategies, I and II

With these two technologies available there are broadly two conservation strategies. Strategy I is to accept some more-or-less erosive farming system and attempt to plug up its deficiencies using Technology A (earthwork) or Technology B (plants, biological controls) or A and B in combination and mutual support. In reality most official conservation programs are usually built around earthworks of one kind or other (Technology A). The limitations, requirements and costs of a range of these are discussed below. Strategy II is to select a farming system which by its nature does not need either earthworks (A) or plants used in a specific soil conservation mode (B). In effect this says that the best kind of soil conservation system is none at all. It uses plants/animals but these are parts of the normal farming system. There are very few agroecosystems in which this is possible and these are close analogues of natural ecosystems – wet terraced fields for taro, rice, which are analogue swamps; some grass-livestock systems as analogue savanna; some agroforest systems as analogue rain forests, and the closest of these are forest extractivism, some palm tapping systems (Diversity) and forest gardens of the wilder kind. Or put more negatively, the further a farming or agroecosystem departs from ecosystem the more it must be propped up by Technology A or B. Some of the extremely low soil erosion rates of forests and agroforests, in comparison with field crop farming, are listed in a following table.

The Soil Erosion Process

In sheet-rill erosion there are four main subprocesses:

1. Soil particle detachment by raindrop impact energy;
2. Splash erosion by shifting these particles up or down slope from point of impact (Figure 6.8);
3. Down-slope transportation of 1 and/or 2 by running water as overland

flow (Figure 6.9);

4. Further particle detachment as sheer erosion in the course of overland flow. (Another sub-process might also be present as soil tunnel erosion caused by either site soil characteristics, 'tunneling soils,' compounded or not by tunneling along dead root cavities in ex-forest land or following animal burrows.)

These processes are listed and discussed separately but they operate jointly, more or less simultaneously. For example, particle detachment and displacement in 1 and 2 are determined primarily by the velocity/kinetic energy of incoming rain, while processes 3 and 4 are determined mainly by the velocity/energy of subsequent running water. But 3 and 4 are also determined by the degree of turbulence of surface flow and this is also caused partly by rain impact on the sheet of running water.

Subprocess 1, Soil Particle Detachment ... Raindrops Keep Falling

This is governed by the energy of falling rain (its amount and intensity), the resistance of soil to this energy impact (which depends largely on soil quality characteristics eg organic matter status), and the degree of soil exposure (ie the extent to which vegetative cover is present and its location and type). The first of these must be accepted for any given site. The second can be manipulated and process 1 curtailed by some mechanical practices (minimum, zero, rough tillage), and by some chemical practices (eg gypsum to improve aggregation), but primarily by soil/crop biological practices which increase the resistance of exposed soil particles to rain impact dispersion. The third, degree of exposure, can be modified only through vegetative cover, crop selection (or vegetative matter used in a mechanical role in the case of cut-and-placed mulch). For a given soil and vegetative cover (if any) the extent of particle detachment is a function of storm energy.

Wischmeier and Smith and Uhland (1958, 1978) developed an equation for determining the kinetic energy of a storm from its intensity as $E=916+331 \log_{10}I$, where E is kinetic energy in foot-tonnes per acre inch of rain and I is intensity in inches per hour: eg a storm of 2 inches/hour delivers a kinetic energy of 1 016 foot-tonnes per acre inch of rain. The metric equivalent is $E=206+87 \log_{10}i$ where E is energy in joules per centimeter of rain on a sq meter surface at an intensity of centimeters per hour (FAO 1965). (These storm energy values are the basis of one of the erosion predicting factors R in the Universal Soil Loss Equation, below.) Examples of energy equivalents of rainfall of five intensities are:

Intensity, inches/hour:	0.1	0.5	1.0	2.0	5.0
Foot-tonnes per acre inch:	585	816	916	1 016	1 147

Subprocess 2, Particle Downslope Displacement – More About Raindrops

On the basis that drop velocity is determined by drop size and this by rain intensity (inches/hour), Ekern (1954) expressed the erosive effect of rain in process 2 directly in terms of intensity and degree of landslope: Erosion = $SI^{1.46}$. This would apply to Figure 6.8 where for S, the slope shown, the amount of dirt shifted downhill depends on storm intensity.

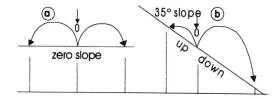

Figure 6.8   Effect of Land Slope on Splash Displacement

Other things equal (wind, angle of incoming rain etc), downhill displacement erosion increases relative to upslope displacement as slope increases, Figure 6.8(a)-v-(b). Displacement occurs in both situations but in (a) there's no net displacement, outgoing particles are replaced by incoming ones. This displacement distance caused by the impact of single rain drops can be large, easily 1.5 meters on a field of zero slope as in Figure 6.8(a) (Schwab et al 1977) and correspondingly more as slope increases. In terms of relative soil amounts Ellison (1945) offers the example of downslope displacement exceeding upslope displacement by a factor of 3 on a 10% slope. Ekern (1954) offers a general guide for estimating upslope v downslope soil displacement as a function of slope only, eg on a 20% landslope:

% of dirt moved downslope	= 50 + slope %	(eg 50 + 20%	= 70%)
% of dirt moved upslope	= 50 – slope %	(50 – 20	= 30%)
net loss at any point	= 2 x slope %	(2 x 20	= 40%)

Most engineering controls (terraces, checks etc) aim at curtailing erosion in its overland flow stage, subprocesses 3 and 4. But Schwab et al point out that regardless of how close together these might be they are ineffectual in preventing downslope splash erosion. On narrow steep fields most erosion is of the downslope splash type. Apart from the largely biological counter measures of increasing ground cover and soil resistance to splash dispersion the only mechanical strategy is to decrease land slope, ideally to zero by bench terracing, a very expensive alternative except when terracing is needed anyway for agronomic reasons. This direct measure of erosion has increasing modern relevance: as populations increase farmers are always moving up to steeper land (increasing S); and there is substantial opinion that rainfall/storm intensity profiles are changing towards storms of greater intensity as a result of greenhouse and climate change.

Crop selection, vegetative cover and splash erosion   As a first approximation subprocesses 1 and 2 are more or less erosive according to how much vegetative mass is on and above the ground and its timing. Clean tilled field crops represent a worst case scenario. Typically they are planted in bare ground at the start of the wet season, just when the rain impact erosion hazard is greatest, and subsequently they provide comparatively little bulk protection against rain. The role of architecture (species, mass, bulk, shape, density) in rearranging incoming rain was discussed previously. In terms of erosion subprocesses 1 and 2 this plant architectural rearrangement is both good news and bad. Rain energy impact is partly a function of drop size (Ellison above). The first effect of canopy interception—in those agrosystems where significant canopy exists—is to break up these drops to smaller size and reduce their velocity. But—depending on architecture—the second effect might be to reassemble them again as drops of crown drip and throughfall, larger and more erosive than they initially were. There is substantial evidence that any intercepted reassembled drop falling from the canopy quickly reaches terminal velocity on ground impact, ie its effect is as great as if no trees were present. Its effect might even be greater because reassembled drops as crown drip from tall trees are usually bigger, though there are fewer of them. Nair (1993) suggests this critical height for terminal velocity at impact at about 20 meters; Hamilton (1990) cites experimental evidence that drops can reach 90% of terminal velocity in a fall of only 10 meters.

Partly offsetting this negative effect of canopy on splash erosion is some good news that in a forest or agroforest a considerable part of incoming rain is diverted to arrive at ground level as trunk trickle flow, some is diverted to bark absorption, some to in-transit evaporation. Once on the ground, trickle flow itself can cause erosion but not of the splash impact kind. On balance there is little doubt that the presence of a forest or quasi forest is beneficial in terms of sub-processes 1 and 2.

Ground litter   The really good news is that most agroforestry systems, especially forest gardens, have a second line of defense — ground litter. Whatever of those initial or rearranged drops do reach the ground their erosive potential is largely dissipated by this barrier. In general it is much more important than is the interception effect of the live phytomass. Another important aspect is the timing of litter supply for this mechanism to function. Usually, even in so-called evergreen forests, most leaf fall occurs in the dry season as a mechanism for relieving plant water stress. This is exactly the time when leaf fall-as-soil protective litter is most needed for accumulation in preparation for the next (erosive) wet season rains. And village lore has it that (a) the longer the dry season the greater the litter fall, and (b) also the longer this dry season the more likely it is to end with unusually intense erosive storms. If this is about correct it is almost as if the forest or quasi forest, this particular bit of Gaea, is solving its own forthcoming storm season rain impact-erosion problem by thinking 3, 4, 5... months ahead. Further, it is also likely that the natural removal/depletion of fallen litter for transformation to plant nutrients by microorganisms is at a slower rate in this dry season than in the wet, because absolute humidity is lower. One could claim that such a forest as evolved natural system, in solving the problems of subprocesses 1 and 2, is far ahead of the most sophisticated artificial technology if the forest is converted to farm ... And that forest gardeners, insofar as they mimic the forest in this regard really don't have much to learn from modern soil conservation.

Subprocesses 3 and 4 ... the Ballad of Overland Flow

Overland flow starts when incoming rain intensity exceeds outgoing soil infiltration and continues for as long as is necessary to evacuate this excess from the field. For a given soil this is possible within wide limits by selecting a farming system which is conducive to high infiltration. The management objective at this point is to reduce the amount of runoff forming the overland flow by increasing infiltration amount to a maximum and this can be done by increasing the time rate of infiltration (mm/hr) or ponding it on the field for a maximum time. These can be achieved respectively by selecting a crop or land use system which results in many vertical micro channels down which infiltration can occur and a crop/system which leaves the soil surface in rough condition with many micro storage ponds. Both of these occur in natural systems, especially forests; they usually don't occur in the compacted and smoothed soils of commercialized agriculture (but other methods of aggregating soils to improve infiltration are possible, eg use of gypsum).

Infiltration under forests and forest gardens – in praise of small things
One reason why infiltration rates under forests and forest gardens are high is that a large percentage of their total biomass is rootmass; eg in the Amazonian forest surveyed by Klinge total roots amounted to 255 t/ha, 80% of these were *large* roots. These are always in process of decay/regrowth, leaving networks of infiltration channels. When these root systems are removed in clearing rain forests for farming, Roose (1996) in Ivory Coast has described the consequence for accelerated erosion, though he attributes their function to binding the soil together and not specifically to their infiltration role: *...There are slopes of over 65% protected by dense forest. If (this) is cleared manually without destroying the root network...the soil can resist the aggressiveness of rainfall for one or two years. However if forest or savanna is cleared mechanically, scouring the fertile topsoil, erosion and runoff assume catastrophic proportions, further aggravated on steep slopes.*

Roose also has pointed to the role of the fauna channels in infiltration, particularly that of termites and earth worms by measuring infiltration only 5-12mm/hour on crusted bare soil, this increasing to 60mm where termites were present, and to 90mm when the sealed surface crust was also not present (as when the soil surface is also protected by litter). The holes created by these and other microfauna can be 4mm diameter. (Downward flow through them is given as flow $= \text{diameter}^5$, Roose.) Small soil fauna are notoriously absent under intensive modern cropping and 'improved' pasture where this is routinely topdressed with superphosphate or unbalanced urea, or where it is only a short phase of a crop rotation (grain, cotton etc) which leaves residual agricides. The first hydrological-erosion control function of such pasture, that of mitigating overland flow turbulence (below) is still performed but its second role of increasing infiltration via the fauna channels is curtailed.

Infiltration and land management practices   Mechanical land clearing generally has three adverse effects of decreasing infiltration rate and, through compaction and reduction in pore space, of decreasing permeability and soil moisture storage capacity. Large scale land clearing (Felda type projects in SE Asia) is done with bulldozers operating at various levels of intensity or number of required passes over a field. These methods range from simple pushing and heaping to pulling a 100m long ship's chain through the forest attached to two D8 machines followed by blade heaping or raking etc. Barber and Romero (1994) found that in clearing rain forest in Bolivia there was little difference among the four most common clearing methods or levels of machine intensity and all caused adverse hydrological damage to soil down to a depth of about 30cm (one foot) compared with undisturbed forest: average infiltration rate decreased about 3½ times, soil density increased by 11% (and the land surface was lowered 27mm by machinery compaction).

Effects of Machine Use Intensity on Infiltration Rate and Soil Bulk Density

Machine intensity	Cumulative infiltration (30 minutes)		Soil bulk density
1. (highest)	3.6 mm		1.38
2.	7.1		1.33
3.	2.6		1.34
4. (lowest)	4.2		1.29
Mean, all clearing methods ...	4.4	...	1.34
Undisturbed forest	15.1		1.19

Source: Barber and Romero 1994.

Effects of farm operating intensity on infiltration Once land is cleared the adverse hydrological impacts of reduced infiltration rate, pore space and moisture/rainfall holding capacity continue through soil compression by tractors and tillage/harvesting equipment or by hard-hoofed cattle/sheep/goats in the case of grazing land. The adverse impact of farm machinery on soil hydrology in row crop systems can be dramatic. Ankeny et al (1995) compared infiltration rates under a range of crop rotations on different soils, locations, across five US Midwest states and found that normal machinery traffic could reduce infiltration rates by up to 95% (191 v 9), even where the land had been chisel plowed. No-till (or minimum tillage) cropping technology is conducive to soil conservation in that by leaving crop trash and avoiding soil disturbance they reduce both raindrop impact and overland flow erosion, and these conditions also increase pore space and soil fauna channels; but as the data show (134 v 15) these beneficial effects can be offset by compaction by machinery traffic for other than initial tillage operations. These authors conclude that close control of machinery is essential in this type of farming if infiltration rates are to be maximized. More generally one could conclude, from this and the previous data for land clearing, that the best way to maintain the initial hydrological condition of soils under forests is to leave them alone, ie 'develop' them – if that is necessary – as forest analogue agroecosystem.

Subprocesses 3 and 4, Particle Transport and Shear Erosion

When overland flow starts it does two things (Figure 6.9). First, in

Effects of Farm Machinery Traffic on Infiltration Rate

System Chisel (plowed)	Relative infiltration rate	System no-till (not plowed)	Relative infiltration rate
no traffic	191	no traffic	134
traffic	9	traffic	15

An Iowa soil.
Source: Ankeny 1995.

subprocess 3 it picks up and brings into suspension particles which have been detached by raindrop impact, subprocesses 1 and 2. The distance these are transported downslope — ie the effectiveness of overland flow in transport mode — depends on the rate at which the natural tendency of these particles to settle is countered by the upward turbulent thrust within the running water. Whether pickup-transport predominates over settling depends on the size/density of the particles (hence the rapid removal of fine v coarse particles) and the velocity and turbulence of flow. If this flow

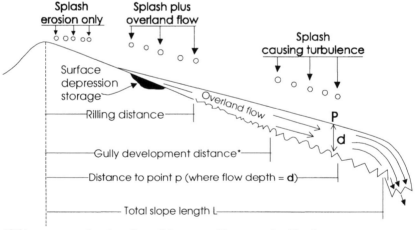

*Rill becomes gully when it can't be erased by normal cultivation.

Figure 6.9 Profile of Overland Flow

contains sufficient energy it can also transport large coarse particles by a rolling action. Second, in direct erosion mode as subprocess 4 this flow might detach, pick up, roll, additional particles by applying a shear stress, depending on whether this stress is greater than shear resistance of the soil over which the flow passes. This process also depends on particle detachability v resistance, velocity and turbulence of overland flow. As Figure 6.9 indicates, different *types* of erosion occur down a slope. At the top of the slope only splash erosion occurs (subprocesses 1 and 2); lower down this becomes splash-plus-turbulence; then still further down erosion becomes primarily due to turbulence of overland flow. At this lowest section of the slope rain impact energy is still expended but the deepening body of moving water itself acts as a cushion, so that rain impact energy here operates through increasing the (erosive) turbulence of overland flow, ie subprocesses 3 and 4 operate. Also at some point down the slope, rills form (the rilling distance) and increase in depth and if the slope is long enough these might fully develop into gullies. The main purpose of mechanical checks/diversions/terraces is to break up a long slope into shorter segments before this can happen. But another (biological) strategy, mainly through choice of land cover crops, is to change the nature of this overland flow to one of minimum (erosive) turbulence.

Flow turbulence and land cover – the story of m The factor *m* is a measure of turbulence of flow. In laminar flow the water moves downslope in parallel layers with no crosscurrents in a rolling action. In this, downslope velocity of the flow is greatest in the surface layers and decreases with depth. Velocity is relatively small at the bottom which is in contact with the soil surface. Through this lower velocity it exerts a relatively smaller shear stress on this contact soil and so detaches relatively little of it for subprocess 4 erosion. By contrast in turbulent flow the water has many crosscurrents and eddies. These mix the several water layers and their velocities; there is less difference between the velocities of the upper and lower parts of the flow. In laminar flow most of the water energy is devoted to moving it forward downhill. In turbulent flow a good part of the energy is dissipated in crosscurrents and eddies, then in these exerting shear stress on the containing soil for detachment of its particles. To this extent laminar flow is benign in the erosion process while turbulent flow is not so in that it contributes relatively more to subprocess 3 by preventing particle settlement, and to subprocess 4 by detaching more soil particles in transit. As indicated in Figure 6.9 the impact of falling rain is cushioned by the body of water in overland flow so no or little erosion by subprocesses 1 and 2 occurs; but this rain impact now causes turbulence and through this increases in subprocesses 3 and 4. As Ekern (1954) states it...*The kinetic energy of falling raindrops greatly exceeds the energy of shallow flow of runoff waters. The drops striking into the shallow flow add vertical currents irrespective of the initial character of the flow and drastically alter the nature of detaching capacity of the water...*

Fully developed turbulent flow has an *m* value of about 1.67 and laminar flow 3.0. The objective of overland flow management is to make it as 'laminar' as possible. Within wide limits this is possible by land surface management and selection of the appropriate vegetative cover. Figure 6.10

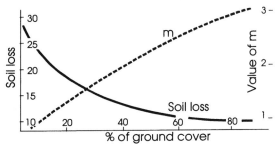

Source: Schiff 1951.

**Figure 6.10   Effect of Ground Cover on Flow Turbulence
and Soil Loss**

(Schiff 1951) shows how the value of *m*, and its resulting soil loss, can be manipulated by increasing the density of vegetative ground cover. Here by increasing ground cover from 40 to 80%, the type of overland flow passing over this field would change from near turbulent (*m* = 2.0) to laminar (*m* = 2.9) and resulting soil loss would decrease from about 13 pounds to 10, or by the same proportion on a per hectare basis.

Consolidation: the Universal Soil Loss Equation

Previous description of the several subprocesses has been necessary in order to construct a technical basis for considering the costs, limits and efficiencies of alternative conservation approaches, especially Strategy I -v- II, and the relative advantages of forest gardens and similar quasi forest systems which usually require no specific soil conservation investment. Fortunately, at the practical level of farm conservation planning most of this can be by-passed by resorting directly to the Universal Soil Loss Equation, USLE, which digests these subprocesses – rainfall impact, infiltration rate, effect of ground cover, overland flow etc – and allows feasible combinations of engineering and biological conservation methods to be rapidly determined.

This is 'universal' to the extent that the physical erosion processes which it summarizes are about the same everywhere, and it is directly applicable with little modification in much of the world outside North

America (southern Africa, Australia, Europe, much of South America). On the other hand it is not universal in that it was formulated with North American farming conditions specifically in mind: climate and rainfall regime, large scale mechanized production, small range of field/ground crops (corn, wheat, soybeans, sunflower...), deep soils, moderate degree of slope and long lengths of slope for machinery operation. These conditions and the type of erosion controls appropriate to them (long earth terraces, strip cropping, etc) are not found in much of the tropical world, especially on small mixed farms where the number and types of crops, species, are immense, they are often mixed, trees and livestock are mixed with the crops, soils can be shallow and/or steep (either one preventing excavated terraces and other mechanical controls) — and probably most limiting of all — where land 'ownership' and a multiplicity of decision makers in land management often prevent conservation planning on the required hydrological or catchment basis. Nonetheless for all of its limitations USLE (and its variants) usually provides the most useful point of departure for direct field application. An example is shown below.

This particular combination of the USLE factors (defined) would result in an average annual soil loss of 36 t/acre/year. This would be generally far too high for any site. The planner/farmer would then re-work his plan eg reduce LS by installing terraces/diversions to reduce length of slope or decrease the intensity of the planned rotation and increase C or reduce the factor P by installing a range of farming practices until a feasible plan is determined, ie that which is most profitable/productive while staying within limits of the allowable soil loss. Several limitations emerge when this procedure is applied to small farms in the high-rainfall tropics ie where most soil erosion is occurring.

USLE: A = R   x K   x LS x C x P
(Example: 36 = 1 000 x 0.20 x 3.0 x 0.2 x 0.3)
where A is annual soil loss in tonnes/acre and:
R = local rainfall erosion index   (1/2 mean annual rain, mm*)
K = soil erodibility factor
LS = length and degree of slope topographical factor (sketch 8)
C = effectiveness of ground cover factor   (0 – to – 1.0)
P = supporting management practices factor   (0 – to – 1.0)
*can be used if no local rainfall erosion index is available.

Allowable Soil Loss

In the context of USLE this concept of there being some quantity (tonnes, mm) of soil to lose in farming is a compromise, recognition that *some* loss will occur under any farming system and that it is more or less important according to how much of it there is to lose, its depth. (It is based partly on the comparatively deep soils of North America.) Some quantitative conversions are:

Soil bulk density	=	1.4 grams/cubic cm	
1 mm depth	=	14 tonnes/ha	= 5.7 tonnes/acre
1 tonne/ha	=	0.007 mm	
1 mm	=	10 cu meters/ha	= 4.1 cu m/acre

Ideally quantitative soil loss rate would be held at or close to the rate of new soil formation. For purposes of applying the USLE under North American conditions allowable loss is set at 3-10 t/ha (1-5 t/acre) depending on initial depth, formation rate and erodability (K below). Time rates of natural soil formation by weathering have been a contentious issue, except there's general agreement that loss rates are measured in tonnes/mm per year while formation rates are measured in terms of 100s or 1 000s of years. Formation rate depends on climate as well as type of underlying rock. David Pimentel's (1993) estimates of formation rates are listed; and his conclusions regarding these in relation to erosion rates are also worth noting: *Soil loss rates on cultivated lands that range from 10 to 100 tons per ha are exceeding soil formation rates by at least ten times in most of the world's agricultural nations.*

Guide to Soil Formation Rates

	Tropics	Temperate
Years taken to form 1 inch/25 mm or 340 tonnes	200	1 000
Depth of soil formed in 1 year, mm	0.120	0.025
Tonnes of soil formed per year per ha	1.700	0.350

These formation rates are consistent with other estimates (Edwards (1991) which are around 30 mm per 1 000 years (or 0.4t/ha/year). The effect of parent material on formation rate is also great, eg 2 mm per 1 000 years on hard rock to 200 mm on soluble limestone (Edwards). To put this into context, soil *loss* under row crops in the Sri Lanka hill country has been *30 mm in 30 years* (Wickramasinghe 1989). In summary, even under conservation-conscious conditions in the temperate zone where a loss limit might be put on soil use, this is likely to be 10-20 times higher than the rate of soil formation. Estimates of actual average soil loss rates for the US range around 10-15 tonnes/ha/year or 20, 30... times higher than formation rate. Assuming an optimistic formation rate of one tonne/ha/year, it would be at least theoretically possible to bring soil loss rate down to something approaching this by use of the best available agrotechnology (which is the broad thrust of modern conservation services).

But consider the high rainfall tropics: here soil loss rates if approached on a farm-by-farm, catchment-by-catchment basis can be not 5 tonnes/ha/year but higher by a factor of 10, 30, 50... Some observations for specific sites, farming systems are:

Burkina Faso	35 t/ha/year	India	40-100
Kenya	72	Brazil	10-20
Lesotho	250	Jamaica	36-800
Nigeria	10-220	Guatemala	200-3 600
China	10-250	Kalimantan	200
Indonesia	30-380	Sri Lanka	18-190

These are by no means exceptionally high or selected loss rates; they're a fair sample of what's been reported (eg by Roose 1996, De Graff 1993, Holden and Hvoslef 1990, Krishnarajah 1982, Wickramasinghe 1989). Further, these localized loss rates are accelerating because they commonly occur on small farms on steep lands. To the extent that these farms are not sustainable and population pressure is increasing, these factors will drive them further uphill to farm even steeper land with at least equal intensity, resulting in even higher erosion rates. These specific loss rates, local downward spirals of sustainability, are hardly noticeable when they're pooled as a national average of eg '40 tonnes in Madagascar' etc when the real situation is that 80% of its uplands are approaching collapse; and in Somalia where Seager et al (1965) reported that country having only 300 sq miles of arable land proper and 90% of this was subject to or threatened by sheet and gully erosion. It would be churlish to find fault with USLE and its variants. These must be acknowledged as among the best (largely US) land management technology. Nonetheless there is some feeling that by their existence (and misuse) they hold out unwarranted hopes that an engineering erosion control solution is possible. The fact is that in the tropics with soil loss rates of 100, 200, 500... tonnes/ha there is no mechanical technology known to man which, under orthodox field crop farming, can hold soil loss within the allowable limits of soil formation, only reforestation or some agroforestry systems. Under adverse tropical conditions of high rainfall intensity (R) and steep slopes the proper role of mechanical structures (ditches, diversions) is to stabilize a site until vegetative controls can become established.

Qualitative soil loss Especially in a rain forest or ex-forest environment the main problem with the quantitative allowable soil loss concept is its assumption of constant proportionality between soil-as-dirt and soil-as-productivity, or that any centimeter down the profile is of equal productivity value as others... *The aims of erosion control should be reformulated with more emphasis on productivity decline. The loss of soil volume or thickness only becomes serious when erosion has proceded to an advanced stage. Long before this is reached, serious losses of production occur, with consequent decline in soil physical properties and nutrients* (Young 1989). In this environment there is often not much soil to begin with, biomass nutrients must be cycled continuously at rapid rate, this process occurs in the top few centimeters (often in the litter zone *above* the soil - Nutrients). Once this essentially biological process is terminated by erosion of this top thin layer the productivity of the site can drop dramatically. In this case the allowable soil loss (more appropriately defined as biologically produced nutrient supply) would have to be set at that amount which the farming system or crop can continuously re-generate. There are very few farming systems which can do this and these are close analogues of natural systems. Granatstein and Bezdicek (1992) have also discussed the need for a qualitative soil loss index and difficulties in constructing it. (Such indices are in fact available but not widely used.) There are a dozen or so qualitative soil properties, most inter-dependent. Consensus is that organic matter content is most important and representative. Loss of nutrient status (kg of NPK...per hectare) is of direct economic importance and many soil erosion studies have quantified the cost of soil loss in terms of the market value of nutrients it contains. Using this approach Pimentel (1993) cites USDA research indicating that the (nutrient-loss) cost of US erosion in the 1980s was equivalent to about half the 45 million tonnes of artificial fertilizer applied annually.

R — Rainfall Erosivity Factor

This first determinant R in USLE is a measure of the energy capacity of rainfall at any site to cause erosion by the combined processes 1 - 4. This potential capacity, erosivity, of a single storm is taken as ExI_{30}, or its total kinetic energy (from Wischmeier and Smith's $E = 916 + 331 \log I$) times its maximum 30 minute intensity. There are many other ways by which storm energy can be statistically correlated with resultant erosion; this particular one EI_{30} just happens to give the highest correlation under US climatic conditions. It 'explains' in a statistical sense about 90% of the erosion due to storm energy under most storm/slope/soil conditions. EI_{30} was obtained from recordings of all storm intensity profiles over 20-25 years, totaling these and dividing by number of years of record to obtain an annual average EI_{30} for particular areas and mapping these as R values, eg R = 50 in Dakotas and 600 in south Texas; but this can go to 2 000 in parts of the wet tropics, and equally important, 75% of this can occur in only a few months of the year (as in Ivory Coast, Liberia, Nigeria). In the previous example of USLE, R is given a value of 1 000. In regions where storm energy profiles have not been analyzed and mapped by this procedure an approximation can be taken as half the mean annual rainfall in mm, eg if rainfall is 2 000 mm, R = 1 000 (FAO).

A more serious limitation is that R is based on long term averages. In the hurricane or cyclone zone what is of more concern is not the *average* rate at which erosion might occur but the great damage that might be done by a single storm, that with a return period of 10, 20, 50...years. As' an example, soil loss on wheat farms in Australia are in the order of 2-6 t/ha in an average year, but Edwards (1990) cites the example of one storm causing a 350 tonne loss, and similar severe losses to single events are common elsewhere. It's been suggested that this R rainfall intensity factory (and USLE itself) be put on a probability rather than an average basis but it's not clear whether this would be an advantage. Additional accuracy might be gained but at the cost of reduced simplicity, applicability. Expected return periods of particularly damaging storms are still only statistical abstractions: the '50' year hurricane can in fact occur in any 2, 3... consecutive years and the soil damage that rains can do, each compounding the others, can be far greater than if they actually do occur one each 50 years. Climatologists might also point out that the basis of these probability measures is what happened in the past, but with climate change this past record might be a poor guide to storm intensities, recurrence intervals, in the future. Rainfall intensities in Central America's hurricane Mitch (Nov 1998) went off the scale and were described simply as 'biblical'. Probability measures can be made to look impressive but even the best of them, no matter how well decorated with algebra, are thin defense against a biblical event.

K — Soil Erodability Factor

This factor of the USLE equation expresses the relative erodability of the subject soil in tons/tonnes of expected soil loss per acre under specified conditions of cultivated fallow with no cover or other conservation practice on a standard 9% slope of 72.6 feet length. (These conditions of slope and length only reflect the experimental conditions under which relative erodibility of the various soils were obtained: they have no other significance.) Values of K for soils in the US have been tabulated for at least a sufficient number of representative soils from which approximate K values for others may be obtained by comparison. Typical K values range 0.30 - 0.50 ton per acre (0.70 - 1.10 mt/ha). In the previous USLE example K is given a value of 0.20. The K factor is acknowledged as one of the least satisfactory in the USLE, especially when this is applied in the tropical world, mainly because K for a soil is not constant over time. It changes with land use intensity and crop management practices, especially those affecting soil organic matter content. For example Granatstein and Bezdicek (1992) present the following data (adapted from Gersmehl and others) which show how K on a similar soil can change according to land use:

virgin forest   K = 0.27;   a crop rotation K = 0.39;   continuous corn   K = 0.51

To an extent the USLE recognizes these possibly different values for K on the same soil and assigns higher K values to an eroded than to a non eroded soil condition. Young (1989) has noted that variability in K results from changes in its organic matter content; its severe depletion can lower soil resistance to erosion (measured as an increase in K) by 50%. This is particularly important on newly cleared rain forest or savanna land where organic matter levels are initially high. Roose (1996) gives the example of K value (erodibility) increasing from 0.1 to 0.3 as a soil deteriorates in organic matter, structure and permeability. In this sense, through these factors increasing K, erosion feeds on itself: any decline in organic matter increases erosion rate (via K in the USLE) which further decreases organic matter and increases K and so on.

C — Ground Cover Factor

This factor reflects soil loss under any crop or vegetative cover regime relative to that on bare land which is given a value of 1.0. Tables for C are available for most cropping regimes in the developed world. To quantify these crops and rotations in terms of C, they are broken down into their several growth/bare land stages then individual C factors are obtained for these stages according to how much erosive storm energy (EI) is likely to be applied at each seasonal growth stage, eg the crop 'corn' is broken down into five growth stages and the total C value of all stages, the entire crop, is about 0.4. This compares with 0.01 for permanent pasture under US conditions. C values are also increasingly available for tropical crops and locations.

The importance of C Where it is not possible to contain erosion to 'allowable' limits by reducing degree or length of slope S and L, or where mechanical controls are not feasible, reduction in the value of C through increased vegetative ground cover is often the only available strategy. In Table 6.18 this is illustrated by the C values for a range of crops and practices in comparison with natural systems (from Young (1989), consolidated from FAO, Roose in West Africa, Lewis in Central Africa).

Table 6.18   Values of C for a Range of Crops and Land Use Systems (Bare Land C = 1.0)

Tree crops—	C	Ground crops—	C
coffee	... 0.02	corn: poor crop	... 0.90
banana: dense	0.04	dense crop	0.40
banana + sorghum	0.14	cotton	0.50
oil palm + cover	0.10	pasture, rough	0.10
		peanuts: poor crop	0.20
Natural systems—		dense crop	0.20
dense rain forest	... 0.001	pineapple: no residue	0.50
savanna: good	0.010	residue mulch	0.01
burned	0.100	cassava: poor, sparse	0.80
overgrazed	0.100	dense	0.20
woodland + undergrowth	0.010 (80% cover)		
woodland no undergrowth	0.060 (80% cover)		

Sadly for H. sapiens the crops they are most interested in — foods and industrials — are in general the most erosive and, with a finite soil resource, the least sustainable. Such crops as corn, cotton, peanuts, cassava, potatoes, beans are always highly erosive (high C values) but this can be modified by the way these are grown, eg by the vegetative mass they can be induced to provide or by retention of residues as mulch, as with pineapple. The tree crops are markedly better but again this also depends on management, eg planting density and whether the interspaces are cultivated (as with dense banana v banana with intercultivated and erosive sorghum, corn, peanuts etc). Most efficient of all in reducing C values are the natural systems, especially the rain forest with a C value of 0.001. Considering this cover factor alone, as it would operate through the USLE, this is about 100 times less erosive than pasture and 900 times less than a poor crop of corn (the normal case). The other efficient natural system is savanna, C = 0.01, but burning or overgrazing this can quickly reduce its cover-hydrological efficiency by a factor of 10, down to that shown for rough pasture. Efficiency of the 'woodland' items is similarly reduced by a factor of 6 where the undergrowth is removed, again usually by grazing or cutting for firewood. Again one concludes that the most efficient agroecosystems in reducing erosion through manipulation of C are those which are closest analogues of natural systems. The multistorey polycultures of some forest swiddening systems are also very efficient in terms of C (Shifters).

P — Conservation Management Practices

These refer to the ways in which the land is cultivated, the spatial design used in growing crops and the addition of any supporting mechanical structures, measures. In large scale field crop farming there are relatively few of these alternatives: contour cultivation and planting (as distinct from up-and-downslope cultivation); strip cropping or arranging these contoured crops in strips of different crops (as distinct from planting the whole field to a single crop); and supporting these agronomic measures with contour terraces, diversion channels, contour ridging etc. For measuring the effectiveness of these practices soil loss from them is taken as relative to what it would be on bare land cultivated up-and-down slope. Table 6.19 summarizes the effects of these practice for land of various slopes under US conditions. Eg on a 4-7% slope contour cultivation/planting reduces soil loss by half compared to up-and-down ($P = 0.5$); if these contoured crops are grown as alternating strips this reduces soil loss by half again (to 0.25); and if this is supported by contoured terraces loss is reduced by half again to $P = 0.10$. Contouring alone ceases to be effective on slopes above 12-14%. In this example supporting terraces are not needed below 2% and not feasible above 16-18% (because of the area taken out of production by the increasing number of terraces required at higher slopes, and/or decreasing efficiency of machinery use in the narrow between-terrace lanes).

Table 6.19   Example of Practice Factors for Use in USLE

Slope	(P) Contouring	(P) Strip cropping	(P) Terracing & contouring
1.0 – 2.0%	0.6	0.30	(terraces not necessary)
4.0 – 7.0	0.5	0.25	0.10
7.0 – 12.0	0.6	0.30	0.12
12.0 – 18.0	0.8	0.45	0.16
18.0 +	0.9	0.50	(terraces not feasible)

Source: Wischmeier and Smith 1978.

Other aspects of P   These particular practices — long terraces winding sinuously across a gentle landscape, flanked by strips in matching shades of green — are an engineer's delight. In the world of small farms in the wet tropics the model has little to do with reality. The concept of P still holds and there are dozens indeed hundreds of other 'practices,' most still waiting to be described and measured. ...*This wide range of farming techniques (ie practices P) is not the outcome of chance but the long adaptation of each group to the environment in which they live* (Roose 1996). These practices often exist at micro level, down to the scale of a single plant, and their soil conservation purpose cannot be distinguished from their water or fertility conservation function. As one example, on cracking clays in the unirrigated parts of the Madiun Valley, Java, at the end of the dry season a corn crop is established by prying out some of these hard clay slabs with a crowbar and throwing 6-8 seeds into the holes spaced every two meters or so, the field is pockmarked with them. The corn germinates and grows on the moisture in the holes, but this is insufficient. The first function of the holes then is to trap running water from the earliest thunderstorms and to this extent they are micro irrigation systems. But they also trap silt and residual fertilizer from a previous crop, so to this extent are soil conservation and fertilizing mechanisms. (In this agronomic-conservation system, or whatever it is, time is of the essence as lawyers say, a race between the seeds germinating and the paddy field rats finding them. One refinement then — or perhaps it is the precursor to clod lifting — is to simply drop seeds down the narrowest cracks and trust Allah, which beats the rats but often not their little cousins the field mice.) There are hundreds of these various microsystems around the world.

Roose has described other 'minimum tillage' systems from West Africa, Haiti, Ecuador, Algeria, most more sophisticated, some not. Some of those in Africa are breaking down under pressure from increasing populations; others might be abandoned when a group migrate to a new and unfamiliar environment. But in S and SE Asia most abandonment is due to the pressures on farmers to modernize, intensify, adopt new varieties and farming methods, and in this there is often no room for the old methods of soil and water management, as inefficient as these sometimes are ... Which is unfortunate because the standardized substitutes offered by modern soil conservation engineering, which until recently seldom looked for and incorporated traditional technology, has itself in this absence made little progress. Or as Roose mildly puts it...*The failure of a number of erosion control schemes would probably have been avoided if traditional methods had been studied more closely...and (those) more in tune with the local environment (had been) proposed in place of the generalized imposition of methods developed in the 1930s for mechanized farming in the United States.*

Effect of Increasing Population Pressure and Land Slope on Erosion Rates

As populations increase in the developing world this puts enormous pressures on the remaining forests of these countries (Table 6.15) and subsequently on the lands after they've been cleared. Farming systems both intensify and shift progressively up to steeper slopes. The effect of this in terms of increasing erosion rate can be illustrated by the data of Table 6.20 for Ivory Coast (Roose 1996) for a sample of forest swiddening systems having a forest cover, a bare land and a cropping phase. (Data for the forest phase also represent the comparative situation in sedentary forest gardens of the wilder kind.)

Table 6.20   Soil Loss and Runoff in Relation to Crop Phase and Land Slope

Slope %	Soil loss, t/ha/year			% of storm runoff		
	Forest phase	Bare	Crop phase	Forest phase	Bare	Crop phase
4.5	(no data)	60	19	(no data)	35	16
7	0.03	138	75	0.14	33	24
23	0.1	570	195	0.6	24	24
65	1	(no data)	(no data)	0.7	(no data)	(no data)

Source: Roose 1996.

In the bare land phase soil loss increases about 10 times (60 to 570) by moving from a 4.5% slope to a 23% slope. Further, on the highest slope of 23% for which comparison was possible soil losses under the bare land and crop phases of the farm system are respectively 5 700 and 1 950 times higher than under the forest phase and water runoff is 40 times higher. On these steep lands engineering structures are not feasible (except possibly bench terraces at high cost). If sustainability is any consideration at all the only possible way of farming these – without losing 570 tonnes soil/ha/annually – is with some agroforestry systems which would approximate the 0.1 tonne shown for the forest phase of the table. Nonetheless, in the tropical world generally the thrust of agricultural development, modernization, is towards field crop farming, the alternating crop-bare land phases of the table, according to European-North American land use models. The comparative soil loss rates shown (0.1, 570, 195 tonnes etc) can be put in perspective by comparison with 'allowable' soil losses under good conservation conditions which are in the order of 5 t/ha/year.

Comparative Conservation Efficiencies of Tropical Forests and Agroecosystems

Table 6.21 consolidates the end results of all those subprocesses entering into the erosion process for a range of forest garden, plantation and natural forest systems in SE Asia. They're not strictly comparable, having been obtained by several workers sometimes under unspecified conditions; nonetheless the conclusion must be that the most efficient systems are natural forests, closely followed by their analogues the forest gardens.

Comparison of Conservation Methods, Slope Limits, Costs

Table 6.22 consolidates data for a range of the most common conservation methods used around the world, their land slope limitations, main uses and comparative on-farm installation costs. Except for the forest gardens they're all engineering or mechanical structures of one sort or other, ie Strategy A, although the first three (contouring) are incorporated into normal crop field operations and require no earthworks. Again with the exception of forest gardens (and some alley or corridor systems) they all aim to counter erosion at the overland flow stage by dissipating the energy of flowing surface water whereas the forest gardens dissipate both the energy of incoming rain (particle detachment) and that of overland flow. Here these gardens are representative of a wide range of those agroforestry land use systems such as forest extractivism which hydraulogically are close, analogues of natural forests. (But 'agroforestry' is now a fashionable term and many such claimed systems are erosive to a varying degree.)

Costs shown in Part A of Table 6.22 are calculated by author; those in part B are actual costs as reported around the world (De Graaff); both refer to on-farm installation costs in terms of farm labor days/ha (except for two types of terracing) and these omit subsequent annual maintenance costs.

To briefly summarize Most of the conservation methods shown in Table 6.22 are either not feasible beyond specific slope limits or incur significant costs, or both; and several of them also require considerable expertise for installation. On the other hand globally most erosion is occurring *at* steep slopes or where finance/food aid is not available to implement these works or the expertise is lacking. One concludes that the better alternative in general is through Strategy II, adopting those farming systems which don't cause much erosion in the first place and—especially in the moist tropics—these are mainly the traditional agroforestry systems.

Table 6.21   Comparative Soil Loss Rates under Agroecosystems and Forestry in SE Asia

Land use system (Wiersum 1984, Java)—	Min.	Med.	Max.
		Soil Loss – t/ha/year	
1. Multistorey tree gardens/forest gardens	0.01	0.06	0.14
2. Bush fallow phase after shifting cultivation	0.05	0.15	7.40
3. Natural rain forest	0.03	0.30	6.16
4. Forest plantations undisturbed	0.02	0.58	6.20
5. Tree crops with cover crop or mulch	0.10	0.75	5.60
6. Shifting cultivation, crop period	0.40	2.78	70.05
7. Taunga, cultivation period *	0.63	5.23	17.37
The effect of removing weeds or litter:			
8. Tree crops clean weeded	1.20	47.60	182.90
9. Forest plantations with litter removed, burned	5.92	53.40	104.80

(Krishnarajah 1982, Sri Lanka)—			
10. Forest gardens		0.05	
11. Tea, mixed crops in the same catchment as 10.		18.70	

(Wickramasinghe 1989, Sri Lanka)—	Land slope	Siltation rate
12. Home gardens, type unspecified	21-45%	9-10 t/ha
13. Row crops, tobacco	45-55	50-65
14. Grass fallow phase	35-45	10-12
15. Forest reserve, type unspecified	8-45	4-5
16. Sparse woodland	45-50	11-13

(Brunig 1977, Malaysia)—	
17. Rain forest, annual average soil loss (unusually high)	25t/ha
18. Estate tea	490

(Walsh/Douglas 1992, Malaysia)—	
19. Primary rain forest (suspended sediment in runoff)	3.12t/ha
20. Same forest after logging **	16.00
21. Secondary rain forest	0.35
22. Same forest after logging **	6.60

(Jacobs 1982) ***—	
natural rain forest	0.2-10t/ha
dense man-planted forest	20-60
slash-and-burn, crop phase	1 000

*A plantation forestry establishment system which incorporates cropping (cassava, corn etc) among the young trees for 2-3 years. ** Sediment yield would decline with forest recovery at an unknown rate. *** These total rates are very high but probably valid comparatively for these three systems.

Table 6.22 Guide to Some Conservation Methods, Practices, Limits, Uses, Costs

A. Practice/structure	Slope limit	Common use	Cost labor days/ha
Contour cultivation only	to 3%	Field crops rotation	-
Contour strips, rotational	to 6	Field crops rotation	-
Contour strips, buffers	to 8	Field crops rotation	-
Alley or corridor cropping—			
Without benching	to 15	Crop/bushes/trees	-
With benching	to 25	Crop/bushes/trees	130
Broad based terraces—			
Water disposal	to 8	Any field crop	($150)
Absorption	to 8	Any field crop	($150)
Narrow base banks, bunds—			
For soil cons. only	to 15	Crops/pasture	70
Same, with ponds	to 15	Crops/pasture/fishponds	300
Contour planting ridges	to 20	Vegetables (Andes, Asia)	-
Bench terraces, earth, irrigated	to 45	Intensive rice	320
Bench terraces, dryland	to 20	Field crops	260
Benches, level, rock walls	to 50	Any irrigated/dryland crop	1 270
Benches, sloping, rock walls	to 20	Any dryland or tree crop	1 000
Filled terraces, retaining walls	to 45	High value crops	4 000
Hillside step terraces	to 40	Rubber	90
Hillside planting platforms	none	Single trees	65
Lock-and-spill silt traps	to 40	Tea, Sri Lanka	900
Silt traps – feeding troughs	to 6	Coconut, Sri Lanka	20
Reynoso systems	to 4	Sugar, Java	45
Forest gardens	none	Text	none

B. Actual reported (De Graaff 1993)		Slope	Days/ha
Burkina Faso:	Dryland stone bunds, small	2%	80
Mali:	Dryland contour banks, type unspecified	2	50
	Permanent grass barrier strips	2	30
Mexico:	Dryland terraces + maguey plants	7–9	50
Java:	Dryland bench terraces, earth	5	210
	Terraces with waterways, ditches	30	750
Indonesia:	Bench terraces, earth, with waterways	20	570
	Same at …	30	770
	Same at …	40	850
Jamaica:	Hillside diversion ditches	10	150
	Ditches and grass barriers	20	250
	Bench terraces, earth, with drains	20	530
	Same at …	32	670
Thailand:	Bench terraces only, earth	300	500

6.9 Soil Salinization – Lot's Wife, Tree Pumps and More Solar Energy

The second main category of land degradation listed previously in Table 6.16 is nutrient loss but this is relatively easy to rectify by changing farming practices. Salinization (76 million ha) the next main cause is very difficult and expensive to rectify. Even predating that unhappy experience Lot's wife had with salt outside Sodom, from temple records in Sumaria, Mesopotamia (Jacobsen and Adams 1958) agriculture and civilization began to collapse there with rising watertables and soil salinization from about 2 400 BC. Grain yields fell from 2 500 liters/ha to 900 liters by 1 700 BC. From pottery evidence wheat and barley were of equal importance in agriculture's youth, c 3 500 BC, but by 1 700 BC wheat had been abandoned altogether, with only barley being able to tolerate the salt. In contemporary times, again following rising saline watertables from irrigation and canal seepage, Worthington (1977) has estimated losses of *only* irrigated land in recent historical times at: Iraq – 50%;   Egypt – 30%; Iran – 15%;   Syrian-Euphrates – 50%. The Australian situation is reviewed below; there are many similarities with Pakistan where modern irrigation started in late 19th century colonial times with barrages across the Indus diverting water, with this as basis for agro-modernization largely for exportable commercial crops notably cotton for UK mills then later the subcontinent's own mills. These barrages/weirs and diversion channels were engineering marvels of their time. More recently the centerpiece for modernization and expansion of these schemes in Pakistan became the World Bank financed Tarbela dam, early 1970s, at that time largest earth fill dam in the world.

Now Tarbela is 32% silted up from uncontrolled erosion in its denuded catchments. This or something like it had been anticipated and accepted by Tarbela planners. Not so anticipated was the more serious loss of farm land itself to waterlogging and salinization when this water was turned loose on the plains, with these underlain by an ancient ocean and its vast quantities of residual salt and needing only excessive irrigation to carry it to soil surface (as in Australia below). By World Bank measurement 37.6% of Pakistan irrigated areas are now waterlogged, 14% of surface soils are salinized to some degree and 6% are severely salinized. Some 4 million ha per year are now being degraded at varying rates by this process; in some places the watertable carrying these salts has risen by 25 meters to the surface since 1900.

Salinization in Australia

In the past 40 years increasing soil and water salinity has emerged as the most important agro-environmental problem in Australia and impacting also on flora/fauna of the salt affected lands and downstream river ecosystems into which the saline waters drain. Australian soils have been described as 'a film of productivity overlying a vast bed of toxic waste.' The general model is that this salt was deposited by oceans over geological time, about 10 000 tonnes under any ha of wheat-pasture land and 200-400 tonnes in the immediate vicinity of their rootzones (Ralph 1993; WA Salinity Council). Other models (Williamson 1990) accept this but emphasize salt imported in rain from the ocean and which continues at 15 kg salt/ha/year deposited on coastal lands and 5 kg in the dry inland.

Salinization occurs in all states; in the irrigated lands it's concentrated in Victoria-New South Wales and in dry crop lands in West and South Australia and starting to emerge in Queensland following large scale land clearing there for livestock. Here only a few illustrative statistics must suffice, keeping in mind that these represent the impact of European farming methods on a fragile continent in little more than 150 years. At national level up to 10% of all croplands and 25% of pasture lands or a total of 8 million ha have been damaged beyond practical reclamation (Gaertz, CSIRO, cited by Beale and Fray 1993). Total salt-affected unirrigated/dry farm lands are now 2.5 million ha, to increase to 15.5 million ha before reaching (saline) groundwater equilibrium even if no further clearing occurs (Prime Minister's Science Council 1998). In the major southeast (mainly irrigated) food bowl of the lower Murray-Darling basin, 15-25% of lands will be unproductive within 40 years (Ralph 1993). Dryland *seepage* salinity increased from 420 000 ha to 790 000 ha in 10 years 1981-92 (ABS 1992). By as early as 1975, 900 000 ha of the most productive irrigated land along the Murray River, mainly in Victoria, had been salinized to varying degree (Woods 1983). One third of Victorian irrigated lands are now salinized (Lowe 1996). Dry farmland salinization is currently expanding at 2% annually.

Taking the states separately, in NSW dryland farm salinization increased from 4 000 to 174 000 ha in the 11 years 1982-93 and is now 'hundreds of thousands of ha' (NSW - Conservation and Land Management). In the most affected state of West Australia, of 20 million ha of dry farm land (wheat-sheep), 3 million ha are now salinized to the extent that massive tree planting and/or drainage are required which will cost $A3 billion (WA – CALM). This dryland farm salinization, mainly by rising saline watertables under the state's best farm lands, is expanding at 6 000 ha annually. *All* streams and rivers and riparian strips are salt-degraded along at least 80% of their lengths (WA – CALM). As habitat, *all* river banks, wetlands and estuaries are salt-degraded threatening 260 rare or endangered plant/animal species (WA – CALM). A few examples of salt increase in streams after land clearing and the implied impact of these on stream and riparian ecosystems are summarized, salt increases by factors of 4 to 19 (Williamson 1990):

Increase in Stream Water Salt Concentration after Catchment Clearing

River:				River:			
	Dale (WA)	...	19 times		Finniss (SA)	...	8 times
	Collie (WA)		15		Bremer (SA)		6
	Axe (VC)		10		Hughes (VC)		4
	Avoca (VC)		10		Para (SA)		4

Source: Williamson 1990.

For dryland salinization, given the initial geological condition of the large subsurface salt deposits these previously were constrained by deep rooted forests which through their high evapotranspiration rates acted as evacuation pumps to keep groundwater levels low and their salt content at safe distances below the rootzone of shallower rooted vegetation. In effect incoming solar radiation (stage 1) supplied the operating energy for this pumping system, through that part of insolation directed to tree photosynthesis, growth and evapotranspiration. With deforestaton for agriculture's shallow rooted crops, typically wheat/pasture, the forest's deep rooted evacuation system no longer functions.

Many solutions are under trial—continuous low cost windmill pumping; evaporation ponds to harvest industrial/culinary salt (in place of wheat and sheep); fish ponds—but the only large scale solution is reforestation including agroforestry. This would require substantially abandoning those European notions of proper farming which caused the problem in the first place. Here there are a confluence of factors. Reforestation/agroforestry would also reduce soil erosion (both water and wind); through habitat rehabilitation protect biodiversity; and through tree carbon sequestering and storage partly offset CO_2 emissions elsewhere causing greenhouse—including those emissions from biomass burning in the state of Queensland where agro 'developers' still clear/burn 500-600 000 ha annually for their cows and woolies.

6.10 Acid Sulphate Soils (ASS)

Clearing and draining of potential acid sulphate soils has released another imp which no one knows how to coax back into Pandora's box. Total global area of these soils is unknown but they occur generally in coastal lowlands and estuaries throughout the moist sub/tropics from West Africa to Indonesia-Australia to Viet Nam-China and no doubt on to tropical America, usually associated with mangroves of the littoral. Extent of the problems which arise when these potential ASS soils are 'developed'—

primarily for agriculture but also aquaculture ponds, towns, infrastructure—is suggested by a recent bibliography containing over 2 000 references to ASS research and development problems (Havilah 2000), with most of this remedial, or how to 'fix it' after potential has advanced to actuality.

The problem arises when surface drains across wetlands or pond excavation for aquaculture expose soil iron sulphide or pyrite which combines with atmospheric oxygen and water to produce sulphuric acid; the acid then releases aluminium and the toxic combination of acid and aluminium in sufficient concentration can kill vegetation and most water and soil life forms. The pyrite was laid down thousands of years ago when iron and sulphur were deposited by rising seas. The problem is naturally contained by leaving these lands undrained and covered by water and vegetation. It becomes acute when land is drained/excavated, especially when the watertable is further lowered by seasonal drought exposing more of the pyrite to air and returning rains flush out the sulphuric acid into ecosystems (or aquaculture ponds); ie the actual problem is man-made and follows development of these wetlands.

On-going rehabilitation of the prawn/shrimp ponds of South Sulawesi, Indonesia, is perhaps representative. Here there are several thousand ha of such ponds, most developed on acid sulphate coastal soils, many from mangrove forest swamps, and some 60-70% of them are now acid affected including abandoned. Some were excavated in previous padi/rice fields (where the acid problem had been contained by crop water), others were excavated in cleared mangrove forests as part of ill-conceived *transmigrasi* settlement projects (Diversity). These and similar commercial aquaculture ponds elsewhere are very high-energy input systems, in this the equivalent of high pressure broiler farms and with management focused on the same set of *economic* efficiency parameters: maximum stocking rate per sq meter, least-cost feed formulations, maximum optimum market weights etc, often to the neglect of more fundamental bio-ecological concerns (so much so that in one Australian case many hundreds of thousands of dollars were directed at these farm 'economics' before management realized their prawns, not prospering, were swimming in a dilute solution of sulphuric acid and the enterprise abandoned). In the South Sulawesi case one solution has been to keep the ponds full (to prevent air exposure of the pyrite and acid formation), grow industrial seaweed with the prawns (Gracillaria verucosa), add milkfish (Chanos chanos) also as a cash crop and to keep the seaweed clean of parasites and to raise juvenile tiger prawns (Penaeus monodon) in this more balanced environment for on-selling to farms without acid problems. Agroecology rediscovered.

Australia has around 50 000 sq km of these soils with 30 000 sq km along the east coast actually or potentially sulphuric acid sources, typically in river valleys and estuaries where population is highest and pressures for ASS development greatest for luxury canal-side housing, roads, urban sprawl, aquaculture as well as agriculture, mainly for sugar and pasture. Water table is high (1-3 feet from surface), previously controlled by native forest-as-pump evapotranspiration.   By one estimate (Fitzpatrick et al/CSIRO 1998) these coastal ASS contain over one billion tons pyrites which when fully oxidized produce 1.6 tons sulphuric acid per tonne pyrite. Sammut et al (1995) have discussed the biological impact (the economic impact is largely unexplored).   These acid discharge rates vary widely by site, land use, crop, drain depth, management care etc.   On one representative sugar cane farm in sub tropical NSW Melville and Quirk (1999) found the potential acidity beneath the crop 'equivalent to hundreds or perhaps thousands of tons of sulphuric acid per ha'; but most of this is safely below field drain levels leaving 50 tons as the immediate problem and with shallow drains and careful management only 0.5 t/ha sulphuric acid actually drains from the farm annually.   This still exceeds neutralizing capacity of the receiving waters and there are usually hundreds of these 'per hectares' draining into a stream.   Jes Summat (Univ. NSW) has measured acid inflow to one estuary—the Richmond River where the economy is heavily dependent on fishing and tourism as well a sugar—at equivalent to 3 000 tonnes/year of pure sulphuric acid, with by far the bulk of it from sugar and cattle farm drains.

Further north in tropical Queensland a 190 ha cleared site (monitored by CSIRO Projects Office) has discharged 120 000 tons sulphuric acid since 1976 and residents of the area warned not to eat wild pigs foraging there because they're probably contaminated with soil arsenic and zinc released by the acid.   Damage is more easily measured as impact on infrastructure... *Coastal development worth over $A10 billion is threatened by ASS impact.   Costs of treating and rehabilitating ASS associated with urban development and infrastructure projects total multi millions of dollars;   many have stalled or been abandoned.   Millions of dollars of corroded infrastructure have had to be replaced.   Millions of dollars of oysters, prawns (shrimp) and fish have been destroyed, their nursery areas decimated and vast areas of land have been degraded by poor acid soil management* (Fitzpatrick et al).   Without these clean-up costs, and serious curbs on the wetland Nation Builders, it's unlikely there will be an intact river-estuarine ecosystem left on the Australian east coast in 30 years time. (Oddly, with better timber use technology the 'trash' swamp oak, mainly Laucaena glauca, which held these wetlands together for millennia is becoming a scarce and valuable resource.)

There are greenhouse and health connections. Undisturbed acid sulphate soils in wetlands usually accumulate carbon in their organic matter, but the 190 ha North Queensland site has been *emitting* 33 tons carbon/ha/year to the atmosphere. Extrapolated nationally and supported by other data it's now estimated (CSIRO) that Australia's total emissions from its disturbed acid sulphate soils have been around 10 million tons carbon annually over the past 20 years or about 1/5th of emissions from all other forms of land clearing and development (Greenhouse). Mosquitoes provide the second connection. As these move south with global warming – now in Australia bringing record numbers of mosquito borne disease infections – their larvae tolerate the acid drain water but their natural fish predators don't.

The economic and ecological ramifications continue, outnumbered only by the technological fixes proposed for them, though the only solution is to put the wetlands back to trees and reeds the way they were, or to some agroforestry system which solves the problem by evapotranspiration instead of drains. The models are in the natural palm and vine wetland forests of Kalimantan, the *aguajales* of Amazonia, and with Malay settlers who insert their coconuts and mangoes then jak trees and diverse things into the tidal nipa groves. Wetland Garden of Complete Design (Diversity).

6.11 Atmospheric Pollution: Greenhouse and Climate Change

> *Observe how system into system runs,*
> *What other planets circle other suns ...* Pope

This set of imps which escaped Pandora's box owe their liberation primarily to the Humans substituting artificial energy for natural—the fossil fuels for solar radiation and biomass—at accelerating rate since the Industrial Revolution c 1750. The greenhouse effect itself is benign and biologically necessary to maintain average global temperature at around 15C degrees in comparison with the -18C degrees which would prevail without greenhouse, a temperature regime resulting from first incoming solar radiation then redistribution of this heat by that part of insolation used to drive the winds and ocean currents. The general greenhouse model was sketched in Figure 6.2 (also see Figure 6.1 – Ramanathan 1980, 1989, Lutgens and Tarbuck 1986, Bryant 1997, Cotton and Pielke 1992, Houghton 1997).

From part A of that figure most insolation is on comparatively short wavelengths and outgoing re-radiation from the colder body earth is on longer wavelengths. Those gases which comprise earth's atmosphere through which radiation/re-radiation pass are relatively weak absorbers of incoming shorter wave radiation but stronger absorbers of outgoing longer wave re-radiation. The normal gases at their normal levels which comprise the atmosphere collectively emit out-going re-radiation in all directions, some going back into space, some directed back to earth, with this again re-radiated from earth to atmosphere about seven times before it escapes earth's systems. The net result is that the atmosphere is heated primarily by radiation from earth, from 'below', rather than by incoming radiation at the top, with the atmosphere acting as a thermal blanket (Houghton 1997). This is for normal greenhouse. The situation ceases to be normal and the effect of the thermal blanket increases when certain gases are added to the normal atmosphere (Table 6.23). Some 34 anthropogenic gases are implicated in enhanced warming, dominated by carbon dioxide (CO_2), methane (CH_4), ozone (in troposphere) O_3, nitrous oxide (N_2O), nitrogen oxides (NO_x) which are responsible for about 93%, with a range of others in the chlorine and bromine group (CFCs, HCFCs) very important on a per unit basis but this offset by comparatively very small global emissions, and all these warming gases offset to very small extent by other gases and pollutants having a cooling effect.

But for the moment, *forget six countries overhung by smoke, forget their hissing steam and piston stroke* and consider water, not included in the Table 6.23 and by far the most abundant of the greenhouse gases though its full effects are ambiguous. From one view water is just another greenhouse warming gas; however when viewed as cloud it can be both a warming and cooling agent. Global cloudiness has increased by 4-10% since 1900 (Henderson-Sellers 1992) and increases over land have been greater than over oceans (2%). Bryant (1997) attributes most global warming over the past 40 years to increased night temperatures rather than day temperatures. Night cloud more so than day cloud acts as another thermal blanket. But the ambiguities continue. Earth's main cooling mechanism has been disposal of surplus heat in the 'sweating' process or evaporation of water in the hydrological cycle and the transport of latent heat out of the global system by wind turbulence, cumulonimbus cloud towers and upper atmosphere meteorological systems (Lindzen as summarized by Walker 1992). From this viewpoint the gas-centered warming model is concerned with only about 25% of the warming problem; or stated another way, if the hydrological cooling system did not exist the greenhouse temperature increment would be several times what it is, and here there's an agricultural connection.

Most of the sweating process, the cooling towers, occur along the inter tropical convergence zone with those over land substantially fed by water pools, evaporation and plant evapotranspiration. To the extent this is terminated by their draining or deforestation for agricultural crops—in effect substituting the evapotranspiration from a crop biomass of 10, 20, 30... t/ha (at best) for one of 400-500 t/ha—such land use change must contribute to global warming via clouds and the hydrological cycle. Or maybe not. Clouds remain a mystery. As IPCC (1995) admit...*At present it is not possible to judge even the sign of the sum of all cloud process feedback as they affect global warming*; and meteorologists Cotton and Pielke (1992) concede that clouds in all their moods are the Achilles' heel of climate modeling. (But Shelley had already warned them. He was always suspicious of clouds; as ladies of the day or night they have too many moods, reflectivity and absorption capacities to be reliable ... *I am the daughter of earth and water, and the nursling under the sky; I pass through the pores of oceans and shores, I change but I cannot die.*)

Agriculture's Contribution to Global Warming

From estimates by or based on Jackson (1993) and Duxbury and Mosier (1993), global agriculture at farm level is responsible for around 15% of all warming, with a further 7% from land clearing/draining for agricultural expansion. Relative contribution by this sector varies widely among countries according to their stage of economic development: 5-10% in USA representing the developed world; 70-80% in Brazil representing industrializing-urbanizing countries and high biomass burning for agricultural expansion; 40-70% in India representing those mainly Asian countries which are both industrializing and have agricultural sectors dominated by intensive irrigated rice/cattle (both of these major methane producers). These three situations also suggest greenhouse gas abatement strategies as these concern agriculture directly. The 1st World would rely primarily on agricultural management innovation (no-till or minimum tillage, more careful use of machinery and less nitrogen fertilizer etc); the tropical deforesting world would look for alternatives to this, primarily agroforestry systems which would either avoid carbon release in forest biomass burning or replace forest carbon storage capacity with agroforest storage capacity. The rice-growing world and irrigated agriculture globally would direct priorities to reduction of methane and the nitrogen gases. All these would re-think the place of the cow as she represents global livestock. (We know that 1 cow = 0.7 automobile in greenhouse terms; but since it's about the same in sentimental terms this might be hard to do.)

The above estimates (15 + 7%) refer to only direct on-farm greenhouse gas emissions. In addition a significant but unknown part of the balance, 77%, arising from industrial-transport-urbanization sectors is also attributable to that part of industry-transport involved in up-stream manufacturing/transporting farm inputs (machinery, fuel, fertilizers, agricides...) and in the down-stream transport-processing-packaging-storage of farm outputs. (These latter include refrigeration which in using the refrigerant CFCs is also a major contributor to ozone depletion in the stratosphere.) Further, such transport-storage-packaging of farm produce is almost a necessary condition for enabling urbanization—specifically the spectacular growth of the megacities of which Mexico City, Cairo, Calcutta, Jakarta, Shanghi...are obvious examples and great consumers of fossil fuel energy and from these producers of greenhouse gases in their own right, as cities.

Here there's another agricultural connection. Much of their population increase is from rural areas, propelled to the cities by erosion of those farming systems which previously supported them, usually through excessive pressures applied to their soil and water resource base. Globally these are emerging as the eco refugees. Adding to them is another stream propelled from the countryside by economic development policies which favor growth of GDP over welfare, industrialization over agriculture, large agribusiness over family farm, city over rural village. The degree to which greenhouse increment can be attributed to agricultural breakdown, and to deliberate development policies which disfavor the rural sector, are not known. (But for some of the social impacts see World Bank Development Report 1999/2000.) In summary, if these various indirect impacts are added to on-farm impacts, one might hazard a guess that agriculture's real total impact could be 30-35% of the global warming increment.

Table 6.23 summarizes percentages of the global emissions of main greenhouse gases which are emitted *directly* by agriculture. This does not translate to actual greenhouse impact by the respective gases. On a per unit or molecule basis their respective warming effectiveness or potential varies widely relative to CO_2, eg methane = 21, nitrous oxide = 200, rising to 3 000 – 7 000 for some of the CFCs; and the long term warming effect further depends on their respective atmospheric lifetimes, eg 15 years for methane and 120 years for nitrous oxide.

Other Agricultural Emissions and Types of Pollution

Carbon monoxide (Table 6.23) is representative of those gases which themselves have only a minor direct impact on global warming but a significant indirect effect. In this case CO reacts with and destroys some (benevolent) hydroxyl OH. Hydroxyl itself plays an anti-greenhouse role in that it combines with and removes some important greenhouse gases (methane, nitrogen oxides, ozone) and to the extent that this hydroxyl is made less available by its reaction with carbon monoxide this latter contributes significantly to greenhouse in this indirect way.

Table 6.23 Comparative Warming Gas Emissions of Agriculture (direct only)

		Percent of global total emissions*	Main sources	Total contrib. of this gas to warming
Carbon dioxide	CO_2	20-25	fossil fuels, biomass burn, soil erosion	50-60%
Methane	CH_4	60	livestock, wet rice, biomass burn	18
Nitrous oxide	N_2O	60-70	soil cult, N ferts, biomass burn	6
Nitrogen oxides	NO_x	30-40	biomass burning	n.a.
Ozone**	O_3	30-40	biomass burning	14
Carbon monoxide	CO	30-40	biomass burning; farm machinery	text

*Percent of all anthropogenic emissions apart from natural emissions.
**Ozone in troposphere as a warming gas.

Other agricultural emissions have no or little direct warming impact but are important causes of other types of atmospheric pollution. One of these is the set of nitrogen oxides (NO_x, Table 6.23). These are a minor direct source of warming, a more important indirect source by also removing the atmospheric 'cleansing agent' hydroxyl, and an important cause of acid rain. In this latter role NO_x combines with oxygen to yield nitrogen dioxide then this with atmospheric moisture to result in the bio contaminant nitric acid. Most NO_x comes from biomass burning, especially of the tropical forests (see Table 6.4). Another important pollutant is sulphur dioxide and compounds. SO_2 (equivalent) emissions have increased from 3 million tonnes in 1860 to some 80-100 million by 1995, with about 10% of this now from agriculture, mainly tropical biomass burning (Table 6.4). SO_2 itself is believed to have a net global cooling effect, operating primarily through the atmospheric sulphates being conducive to formation of a cooling cloud haze and to brighter clouds able to reflect more incoming radiation (Houghton 1997).

Acid rain The negative impact is SO_2 as a source of sulphuric acid for acid rain and smog. Industrialized Europe's acid rains and their impacts are well documented, eg from Hinrichsen (1988) 80% of the lakes in southern

Norway were biologically dead or very sick and rain corrosion of buildings/infrastructure costing Europe $20 billion annually. Chris Park (1987) has described impacts on German forests, with 50-80% of trees there damaged to some extent (with 2% dead), depending on species. Similar damage levels have been common throughout eastern Europe and lesser in the industrialized parts of US and Canada. With biomass burning as the main source, acid rains are now common throughout the deforesting tropical world—Brazil to Congo. Andreae (1991) has measured rain acidity levels here as being at least as high as in the most polluted parts of North America-Europe. The full long term effect of both NO_x and SO_x are not known. To some point they act as a fertilizer in the short term and stimulate plant/tree growth; beyond that input level and over longer time their impact on microorganisms which underlie agriculture and forestry (especially those engaged in nutrient recycling, Nutrients) can only be guessed at.

Ozone as a contact pollutant   While ozone is a benign and necessary gas in the upper atmosphere (stratosphere) as radiation screen, in the lower (troposphere) it is a significant warming gas (Table 6.23) and in proximity to the biosphere excess levels are a serious agricultural pollutant. These latter levels depend much on proximity to industrial areas and local and wind patterns. Annual damage to crops in USA is several billion dollars, much of this not directly visible but occurs through poor growth or reduced resistance to infections (O'Hare and Wilby 1995, Wilby 1995). Tingey (1986) and Kruper and Jager (1996) have reviewed the impacts of ozone on crops around the world. Almost all except some small grains suffered yield decreases, in some cases (beans of up to 60-70%) at very high ozone levels.

Contact pollution by ozone on tropical crops is not well documented and as in the case of industrial-urban ozone in temperate countries this depends partly on proximity to emission source and varies seasonally. From Andreae (1991)…*very high concentrations of ozone are produced in the (biomass burning) smoke plumes, often exceeding those of industrialized regions*, and he's measured peak emission rates from tropical burning reaching 120 parts/billion or 3-4 times those of the most polluted industrialized countries. Whether ozone and the other polluting/greenhouse gases from tropical biomass burning are a serious long term problem is a matter of viewpoint. At present agro development rates there'll be nothing much left to burn within 30 years and the problem should go away.

Dust   There's only one main Pandorean imp left to confront, dust mainly from wind erosion and aerosols from industrial sources and agro biomass burning. Consensus is that this as haze is a global cooling agent, but much depends on the composition and reflectivity of the individual particles/species/aerosols which comprise the dust. If brown haze it's a warming agent, if light it's a cooling agent. In any case whether benevolent or malign it's paid for or offset by desertification—Australian agro-dust from wheat farms blanketing New Zealand, Indo-Pakistani dust blanketing that sub-continent (Origins), African dust over Italy are incidents now too common to be remarked upon.

The last imp out the door has been aerosol/soot haze. In mid 1999 a blanket of it 2km thick and extending over 10 million sq km in the northern Indian Ocean was reported by US Scripps Institute – National Science Foundation oceanographers (NSF News June 1999) and attributed to industrial activity, though it's also likely much was from the massive El Niño forest clearing fires of Indonesia-SE Asia in 1998. Preliminary consensus is that this blanket has had a regional net *cooling* effect of 10%. The specific concern is that this could impede ocean evaporation which feeds the summer SW monsoon rains in S. Asia on which about ¼ of global population depend. Prometheus would have anticipated this; so would have Georgescu-Roegen. They're examples of his 4th law of thermodynamics in full flight – of inevitable pollution from the Humans' misuse of energy.

6.12 Conclusions

This chapter has addressed criteria 23-34 of Table 1.2 of Chapter 1 as these relate to the use/misuse of energy in farming systems. Consider the forest gardeners. There's little or no exposure to the erosive energy of raindrop impact or overland flow so little soil erosion; no tree removal which might terminate their groundwater pumping role in preventing acidification; no soil compaction to impede infiltration rate; no artificial drainage which might generate sulphuric acid (ASS); no biomass burning which in other systems emits the greenhouse gases discussed above; no cultivation or artificial fertilizers to emit carbon dioxide and nitrous oxide from these sources; no NO_x or sulphur gases for acid rain; no energy-intensive farm capital or fuel and therefore no greenhouse gas emissions upstream from the farm in their manufacture. As for the downstream impact through produce processing-storage-packaging-refrigeration, a Kandy housewife's storage technology is as often by smoke-curing over the hearth fire. Her packaging dilemma is easily resolved with a folded banana leaf, though more robust commerce in pigs and chickens will require a bamboo basket; her refrigeration problems overcome with a clay pitcher buried in moist sand. For most of us how deliciously backward and impractical...Or is it?

References

ABS-Australia Bureau of Statistics (1992) *Issues and Fact.* ABS No. 4140, Canberra.

Allee, W.C. (1926) Measurement of environmental factors in the tropical rain forest in Panama. *Ecology* 7(3):273-301.

Allen, R. J. Lindsay, D. Parker (1996) *El Niño, Southern Oscillation and Climatic Variability.* CSIRO, Australia.

Andreae, M.L. (1991) Biomass burning: its history, use and distribution and its impact on environmental quality and global climate. In J.S. Levine (ed) *Global Biomass Burning: Atmospheric, Climatic and Biospheric Implications.* MIT Press, Cambridge Mass/London.

Ankeny, M.D., T.C. Kaspar, M.A. Prieksat (1995) Traffic effects on water infiltration in chisel plow and no-till systems. *Soil Science Society of America Journal* 59:200-204.

Australia Prime Minister's Science, Engineering and Innovation Council (1998) *Dryland Salinity and its Impact on Rural Industries and on the Landscape.* Occasional Paper No. 1, Canberra.

Australian Environment Council, AEC (1989) *Strategy for Ozone Protection.* Commonwealth of Australia, Canberra.

Australian National Health and Medical Research Council (1989) *Health Effects of Ozone Layer Depletion.* Canberra.

Barber, R.G. and D. Romero (1994) Effects of bulldozer and chain clearing on soil properties and crop yields. *Soil Science Society of America* Nl. 58:1768-1775.

Beale, R. and P. Fray (1990) *The Vanishing Continent.* Hodder and Stoughton, London.

Black, N.J. (1971) Energy relations in crop production: a preliminary survey. *Proceedings of the Association of Applied Biologists* 67:272-278.

Booth, N. (1994) *How Soon is Now?* Simon and Schuster, London.

Boyden, S. (1992) Biohistory: the Interplay Between Human Society and the Biosphere. *Man and Biosphere* 8. UNESCO/Parthenon, Paris.

Brown, W.H. (1919) *Vegetation of Philippines mountains: relation between environment and physical types at different altitudes.* Bureau of Science, Manila.

Brunig, E.F. (1977) The tropical rain forest: wasted asset or essential biosphere resource. *Ambio* 6:187-191.

Bryant, E. (1997) *Climate Process and Change.* Cambridge University Press.

Burkill, I.H. (1935) *Dictionary of the Economic Products of the Malay Peninsula.* Crown Agents, London.

Clapham, W.B. (1973) *Natural Ecosystems.* Macmillan, New York.

Cotton, W.R. and R.A. Pielke (1995) *Human Impacts on Weather and Climate.* Cambridge University Press.

De Graaff (1993) *Soil Conservation and Sustainable Land Use: an Economic Approach.* Royal Tropical Institute, Netherlands.

Detwiler, R.D. and C.A.S. Hall (1988) Tropical forests and the global carbon cycle. *Science* 239:42-47.

Duxbury, J.M. and A.R. Mosier (1993) Status and issues concerning agricultural emissions of greenhouse gases. In H.M. Kaiser and T.E. Drennen (eds) *Agricultural Dimensions of Global Climate Change.* St Lucie Press, Delray Beach Florida.

Edwards, K. (1991) Soil formation and erosion rates. In P.E.V. Charman and B.W. Murphy (eds) *Soils: Their Properties and Management.* Sydney University Press.

Ehrlich, P. (1985) Scale of the human enterprise and biodiversity loss. In J.H. Lawton and R.M. May (eds) *Extinction Rates*. Oxford University Press.

Ekern, P.C. (1954) Rainfall intensity as a measure of storm erosivity. *Soil Science Society of America* 18.

Ellison, W.D. (1945) Some effects of rain drops and surface flow on soil erosion and infiltration. *American Geophysical Union Transactions* 26.

El-Swaify, S.A., E.W. Dangler, C.L. Armstrong (1982) *Soil Erosion by Water in the Tropics*. University of Hawaii Research Extension Series No 24.

Evans, G.C. (1956) An area survey method of investigating the distribution of light intensity in woodlands with particular reference to sunflecks. *Journal of Ecology* 44:391-428.

FAO (1965) *Soil erosion by water: some measures for its control on cultivated lands*. FAO, Rome.

FAO (1977) Soil Conservation and Management in Developing Countries. *Soils Bulletin* 33. Rome.

Faernside, P.M. (1991) Greenhouse gas contributions from deforestation in Brazilian Amazonia. In J.S. Levine (ed) *Global Biomass Burning: Atmospheric, Climatic and Biospheric Implications*. MIT Press, Cambridge Mass.

Fishman, J. and R. Kalish (1990) *Global Alert: the Ozone Pollution Crisis*. Plenum Press NY/London.

Fishman, J.C.E., J.C. Watson, J. Larson, J.A. Logan (1990) Distribution of tropospheric ozone determined by satellite data. *Journal of Geophysical Research* 95:3599-3618.

Fitzpatrick, R., R. Merry, J. Williams, I. White, G. Bowman, G. Taylor (1998) *Acid Sulphate Soil Assessment: Coastal, Inland and Minesite Conditions*. Land and Water Div., CSIRO, Canberra.

Fraser, P.J. (1989) Stratospheric Ozone. In *Health Effects of Ozone Layer Depletion. Australia National Health and Medical Research Council*, Canberra.

Fraser, P.J. (1997) Global and Antarctic ozone depletion: what does the future hold? *Australian Institute of Refrigeration, Air Conditioning and Heating Journal*, April 23-29.

Freise, F. (1936) Microclimate of forests in subtropical Brazil. Petermanns Mitt. 82:301-307.

Georgescu-Roegen, N. (1971) *The Entropy Law and the Economic Process*. Harvard University Press, Cambridge, Mass.

Gillespie, A.R., D.M.Knudson, F. Geilfus (1993) The structure of four homegardens in Petan, Guatemala. *Agroforestry Systems* 24:157-170.

Granatstein, D. and D.F. Bezdicek (1992) Need for a soil quality index: local and regional perspectives. *American Journal of Alternative Agriculture* 1-2 1992.

Gribbin, J. (1988) *Hole in the Sky; Man's Threat to the Ozone Layer*. Corgi, London.

Hallé, F. (1974) Architecture of trees in the rain forest of Morobe district, New Guinea. *Biotropica* 6:43-50.

Hallé, F. and R.A. Oldeman (1970) *Essai sur l'architecture et la dynamique de croissance des arbres tropicaux*, Masson, Paris.

Hallé, F., R.A. Oldeman, P.B. Tomlinson (1978) *Tropical Trees and Forests – an Architectural Analysis*. Springer-Verlag, Berlin.

Havilah, E. (2000) *Bibliography of Acid Sulphate Soils*. NSW Dept Agriculture, Sydney.

Henderson-Sellers, A. (1992) Continental cloudiness changes this century. *Geo Journal* 27:255-262.

Hinrichsen, D. (1988) Acid rain and forest decline, in E. Goldsmith and N. Hildyard (eds) *Battle for the Earth*. Mitchell Beazley International UK.

Holden, S. and H. Hvoslef (1990) *Management of Tropical Forests: Towards an Integrated Perspective*. Center of Development and the Environment. University of Oslo.

Horton, R.E. (1937) Hydrological interrelations of water and soils. *Soil Science Society of America* 1, 1937.

Horton, R.E. (1938) Interpretation and application of runoff plot experiments with reference to soil erosion problems. *Soil Science Society of America* 1 3.

Horton, R.E. (1945) Erosional development of streams and their drainage basins: hydrophysical approach to quantitative morphology. *Geological Society of America Bulletin* 56:275-370.

Houghton, J. (1997) *Global Warming*. Cambridge University Press, UK.

Hudson, N.W. (1977) Research needs for soil conservation in developing countries. In FAO (1977) op. cit.

Hudson, N.W. (1987) Soil and Water Conservation in Semi-Arid Areas. *Soils Bulletin* 57. FAO Rome.

Hudson, N.W. (1991) Study of the Reasons for Success or Failure of Soil Conservation Projects. *Soils Bulletin* 64. FAO Rome.

IPCC (1990) *Climate Change, the IPCC Assessment*. J.T. Houghton, G.J. Jenkins, J.J. Ephraums (eds). Cambridge University Press, UK.

IPCC (Intergovernmental Panel on Climate Change) (1995) *The Science of Climate Change*. J.T. Haughton, L.G. Meira Filho, B.A. Callander, N. Harris, A. Katterberg, K. Maskell (eds). UNEP/Cambridge.

Jackson, R.B. (1993) Greenhouse gases and agriculture. In R.A. Geyer (ed) *Global Warming Forum: Scientific, Economic and Legal Overview*. CRC Press, Boca Raton.

Jacobs, M. (1982) *The Tropical Rain Forest: A First Encounter*. Springer-Verlag, Berlin.

Jacobson, J. and R.M. Adams (1958) Salt and silt in ancient Mesopotamian agriculture. *Science* 128:1251-8.

Jensen, M. (1993) Productivity and nutrient cycling of a Javanese homegarden. *Agroforestry Systems* 24:187-201.

Kira, T. and K. Yoda (1989) Vertical stratification in microclimate. In H. Lieth and M.J.A. Werger (eds) *Tropical Rain Forest Ecosystems: Biogeographical and Ecological Studies*. Elsevier, Amsterdam.

Klinge, H., W.A. Rodrigues, E. Brunig, E.J. Fittkau (1975) Biomass and structure in a Central Amazon rain forest. In F.B. Golley and F. Medina (eds) *Tropical Ecological Systems: Trends in Terrestrial and Aquatic Research*. Springer-Verlag, Berlin.

Konig, D. (1992) L'agriculture écologique agroforestaire: une stratégie integrée de conservation des sols au Rwanda. *Bulletin Réseau Erosion* 12:130-139.

Krishnarajah, P. (1982) Soil erosion and conservation in the Upper Mahaveli watershed. *Annals of Proceedings of the Soil Science Society*, Colombo 1982.

Krupa, S.V. and H. Jager (1996) Adverse effects of elevated levels of UV-B radiation and ozone on crop growth and productivity. In F. Bazzaz and W. Sombroek (eds) *Global Climate Change and Agricultural Production: Direct and Indirect Effects of Changing Hydrological, Pedological and Plant Physiological Processes*. FAO/Wiley.

Lal, R. (1976) Soil Erosion Problems on an alfisoil in Western Nigeria and their control. *IITA monograph No 1*. Ibadan Nigeria.

Leach, G. (1976) *Energy and Food Production*. IPC Press, Guildford, UK.

Levine, J. (1991) *Global Biomass Burning: Atmospheric, Climatic and Biospheric Implications* (ed). MIT Press, Cambridge Mass/London.

Lockwood, J.G. (1985) *World Climate Systems*. Edward Arnold, UK.

Longman, K.A. and J. Jenik (1974) *Tropical Forest and its Environment*. Longman, London.

Lowe, I./Commonwealth State of the Environment Council (1996) *State of the Environment, Australia*. CSIRO, Melbourne.

Lutgens, F.K. and E.J. Tarbuck (1986) *The Atmosphere: Introduction to Meteorology*. Prentice-Hall.

McConnell, D.J. (1986) Pepper in Sri Lanka. In D.J. McConnell (ed) *Economics of Some Tropical Crops*. Midcoast, Coffs Harbour, Australia.

McLean, R.C. (1919) Studies in the ecology of tropical rain forest with special reference to the forests of SE Brazil. *Journal of Ecology* 7:5-54; 121-172.

Melville, M. and R. Quirk (1999) *Cropping effect on sugar cane of acid sulphate soils. Workshop on Remediation and Assessment of Broadacre Acid Sulphate Soils*. Southern Cross University, Lismore, Australia.

Michon, G., J. Bompard, P. Hecketsweiler, C. Ducatillion (1983) Tropical forest architectural analysis as applied to agroforests in the humid tropics: traditional village agroforests in West Java. *Agroforestry Systems* 1:117-129.

Molina, M. and F.S. Rowland (1974) Stratospheric sink for Chloro-fluoro-methanes. *Nature* 249:810-814.

Nair, P.K.R. (1993) *Introduction to Agroforestry*. Kluwer Academie, Dordrecht, Netherlands.

NASA (1988) Present State of Knowledge of the Upper Atmosphere. *Reference Publication* 1208.

Newman, A. (1990) *The Tropical Rain Forest*. Facts on File. NY.

Nicholls, N. (1987) The El Niño-Southern Oscillation Phenomenon. In M.H. Glantz, R.W. Katz, M.E. Krenz (eds) *Climate Crisis*. UN Publications.

Noller, J.S. (1993) *Late Cenozoic Stratigraphy and Soil Geomorphology of the Peruvian Desert: a Long-term Record of Hyperaridity and El Niño*. PhD Thesis, University of Colorado.

Odum, E.P. (1989) *Ecology and Our Endangered Life Support System*. Sinauer, Sunderland, Mass.

O'Hare, G.P. and R. Wilby (1995) Review of ozone pollution in the UK and Ireland with an analysis using Lamb Weather Types. *Geographical Journal* 161(1):1-10.

Park, C. (1987) *Acid Rain: Rhetoric and Reality*. Methuen, London.

Philander, S.G. (1990) *El Niño, La Niña and the Southern Oscillation*. Academic Press, London.

Pimentel, D. (1980) *Handbook of Energy Utilization in Agriculture*. CRC Press, Boca Raton, Florida.

Pimentel, D. (1993) Soil erosion and agricultural productivity. In D. Pimentel (ed) *World Soil Erosion and Conservation*. Cambridge University Press, London.

Pimentel, D., T.W. Cullinay, I.W. Butler, D.J. Reinemann, K.B. Beckman (1989) Low-input sustainable agriculture using sustainable management practices. In M.G. Paoletti, B.R. Stinner, G.G. Lorenzoni (eds) *Agricultural Ecology and Environment*. Elsevier, Amsterdam.

Pimentel, D. and W. Dazhong (1990) Technological changes and energyuse in US agricultural production. In C. Carrol, J.H. Vandermeer, R. Rosset (eds) *Agroecology*. McGraw-Hill, NY.

Quinn, W.H. and V.T. Neal (1992) The historical record of El Niño events. In R.S. Bradley and P.D. Jones (eds) *Climate Since AD 1500*. Routledge London/New York.

Ralph, R. (1993) Restoring the balance. *Rural Research* (Spring) CSIRO, Canberra.

Ramanathan, V. (1980) Climate effects of anthropogenic trace gases. In W. Bach, J. Pankrath, J. Williams (eds) *Interactions of Energy and Climate.* D. Reidel Publishing Company, Munster, Germany.

Ramanathan, V., B.R. Barkstrom, E.F. Harrison (1989) Climate and the earth's radiation budget. *Physics Today* May 1989 pp.22-32.

Richards, P.W. (1952, 1996) *The Tropical Rain Forest: an Ecological Study.* Cambrdige University Press, UK.

Roose, E. (1996) *Land Husbandry: Components and Strategy Soil Resources Management and Conservation Service, Land and Water Development Division.* FAO Rome.

Runeckles, V.C. (1992) Uptake of ozone by vegetation. In A.S. Lefohn (ed) *Surface Level Ozone Exposures and Their Effects on Vegetation.* Lewis Publishers, London.

Rycroft, M.J. (1990) The Antarctic atmosphere: a hot topic in a cold cauldron. *Geographical Journal* 156(1):1-11.

Sammut, J., M.D. Melville, R.B. Callinan, G.C. Fraser (1995) Estuarine acidification: impacts on aquatic biota of draining acid sulphate soils. *Australian Geographical Studies* 33(1):89-100.

Sanchez, P.A. (1989) Soils. In H. Lieth and M.J.A. Werger (eds) *Ecosystems of the World.* Elsevier, Amsterdam.

Schiff, L. (1951) Surface detention, rate of runoff, land use and erosion relationships in small watersheds. *Transactions of the American Geophysical Union* 32.

Schwab, G.O., R.K. Freuert, T.W. Edminster, K.K. Barnes (1977) *Soil and Water Conservation Engineering.* Wiley, New York.

Seager, A., R.N. Green, J.J. Lawrie (1965) A technique of using earth banks for soil and water conservation in the northern region of Somalia. In FAO (1965) op. cit.

Shultz, J.P. (1960) *Ecological studies of rain forest in N. Suriname.* Nhum, Amsterdam.

Sioli, H. (1985) The effects of deforestation in Amazonia. *The Geographical Journal* 151(2):197-203.

Slesser, M. (1975) Energy requirements of agriculture. In J. Lenihan and W.W. Fletcher (eds) *Food, Agriculture and Environment.* Blackie, London.

Tingey, D.T. (1986) The impact of ozone on agriculture and its consequences. In T. Schneider (ed) *Acidification and its Policy Implications.* Elsevier, Amsterdam.

Tivy, J. and G. O'Hare (1981) *Human Impact on the Ecosystem.* Oliver and Boyd, NY.

Tomlinson, P.B. (1983) Structural elements of the rain forest. In F.B. Golley (ed) *Tropical Rain Forest Ecosystems: Structure and Function.* Elsevier, Amsterdam.

Trenberth, K.E. and A. Solomon (1994) The global heat balance: heat transports in the atmosphere and ocean. *Climate Dynamics* 10:107-134.

Walker, D. (1992) *Energy, Plants and Man.* Robert Hill Institute, University of Sheffield, UK.

Walsh, R.P.D. (1996) Microclimate and hydrology. In P.W. Richards (2nd ed) *The Tropical Rain Forest: an Ecological Study.* Cambridge University Press.

Weirsum, K.F. (1984) Surface erosion under various tropical agroforestry systems. In C.L. O'loughlin and A.J. Pearce (eds) *Symposium on the Effects of Forest Land Use on Erosion and Slope Stability.* E.W. Center, Honolulu.

Wickramasinghe, A. (1989) Environmental deterioration in the hill country of Sri Lanka. *Malaysian Journal of Tropical Geography* 19:44-51.

Williamson, D.R. (1990) *Salinity: an Old Environmental Problem.* Division of Water Resources, CSIRO, Canberra.

Wilson, E.O. (1992) *The Diversity of Life.* Belknap-Harvard University Press, Cambridge, Mass.

Wischmeier, W.H., D.D. Smith, R.E. Uhland (1958) Rainfall energy and its relationship to soil loss. *Transactions of the American Geophysical Union* 39:285-291.

Wischmeier, W.H. and D. Smith (1978) *Predicting Rainfall Erosion Losses: a Guide to Soil Conservation Planning.* USDA-ARS Handbook 537. Washington DC.

Woods, L.E. (1983) *Degradation in Australia.* Dept Home Affairs and Environment, Canberra.

World Resources Institute (WRI) (1991-92, 1994-95) Annual Reports. Oxford Univesity Press, UK.

Worthington, E.B. (1977) *Arid Land Irrigation in Developing Countries: Environmental Problems and Effects.* Permagon, UK.

Wuebbles, D.J. and J. Edmonds (1991) *Primer on Greenhouse Gases.* Lewis Pub. UK.

Young, A. (1986) The potential of agroforestry for soil conservation. *ICRAF Working Papers 42 and 43.* ICRAF Nairobi.

Young, A. (1987) Soil productivity, soil conservation and land evaluation. *Agroforestry Systems* 5:277-292.

Young, A. (1989) *Agroforestry for Soil Conservation.* C.A.B. International. Wallingford UK.

Chapter 7
Shifters ... and Forest Gardens as Alternative

By selection it's possible to prove or disprove almost anything about the shifters but five propositions seem to hold their ground: Until undone by circumstance —population, commercial temptation, scowls of displeasure from their Prince – they managed to achieve a 'balanced exploitation' of their environment, to use Hans Ruthenberg's phrase.   Through the millennia they modified it enormously, but what other macro-fauna hasn't? As resource managers they've been more successful than most – H. sapiens' survival proves it.   On almost every count they've done less damage to the natural world in 20 000 years or so than modern agriculture has inflicted in the last 200.   In economics' vision of the future, their time is running out. But there is still much to be learned from the shifters.

Swiddening (old English) or shifting or slash-and-burn agriculture is one of the two fountains of agriculture generally and forest gardens in particular. (The first in time was manipulation of the Great Forest – Genesis.) In both the tropics and temperate worlds, as technology, it was present at the birth of conventional field crop farming (Origins, the Flannery models). In the contemporary world it is considered wasteful, especially of tropical forests, obsolete and urgently in need of advancement to more productive systems. The thrust of this has been to move the shifters on to proper field crop farming; but it is suggested here a better alternative would be to pick up the ancient tree crops stream which was present in tropical shifting agriculture and develop that instead of the ground crops stream (Evolution, below).

7.1 Shifters and the Origins

Swiddening is now confined largely to the tropics but as recently as 1920 it was still common in North Europe and existed in remote places in Finland into the 1970s. It referred to the cutting/burning of some forest patch to grow a few years of food crops until fertility was depleted, then the swidden was abandoned for a longer period for tree regrowth and fertility restoration. In Europe its origins lie far back (Parain 1942). As agriculture (wheat and barley) filtered up the river valleys from its Near East hearth it required removal of the great oak forests. One of its earliest archeological records here, c 7 000 BP, is as layers of oak charcoal and various pollens, the fingerprints of shifters. There is similar archeobotanical evidence that as agriculture moved further north to colder and less fertile sandy soils wheat and barley were replaced by rye and oats and the swidden system became more deliberative, eventually stabilizing before the 17th century as a cycle of 4-6 years of crops followed by 20-30 years of restorative birch and pine forest. From then on in Europe but less so the tropics, these two land use activities became separated – the forests managed as forests and the swiddens as permanent farms. With the trees removed as fertility source this was replaced by a grass ley phase and livestock.

In the tropics, with slow contraction of the pristine forests, shifting has been extended to lesser biomes – secondary forests, bush, savanna, grassland. As biomass of these, as fertility source, declined and labor in obtaining it increased, this has represented a downward spiral on pan tropical scale. One consequence of this was emergence of constraints other than fertility depletion, notably pests and weed infestations. Another was development of many techniques other than fire to capture the dwindling nutrients and control these pests. While the trees were present another evolutionary change was to the continuous economic use of the tree fallows in place of simple periodic abandonment of the swidden (managed fallows). All shifting systems are somewhere along this evolutionary path; with increasing population pressures most are moving down the spiral at different rates. Consequently these systems now range from the most primitive and exploitive forms of land use to – in some places still – sophisticated and sustainable modifications of natural ecosystems.

7.2 Relevance, Aspects of Waste

The shifters are now an embarrassment. Governments only wish they'd go away. Stop wasting resources. Do something useful for the Prince (perhaps earn a little foreign exchange...?) Come in from the cold.

Negative Impacts of Shifting Agriculture

Space Of earth's total population of 6.0 billion some 3.0 billion continue to exist as traditional subsistence farmers and at least 300 million of these are shifters in close association with the remaining forests. Globally they occupy about 36 million sq km or 30% of the world's agriculturally exploitable soils (FAO 1978). However because this system now extends to lesser non forest biomes and the technical boundaries of the system can't be defined precisely, the real number of shifters of one sort or other could be 600, 700... million throughout most of the tropical world. Writing in 1974 Rattan Lal estimated that of all the potentially arable land in Africa between Sahara and South Africa, with this devoted mainly to dozens of different kinds of shifting, only 4-5% was actually under a productive crop in any year and 15 ha were required to feed one person. Table 7.1 offers a generalized view of carrying capacities increasing as shifting systems move, and have historically moved, through the several evolutionary stages. From this, simple shifting (with hunting-gathering) in a previously pristine forest would support only up to 4 persons/sq km, a short fallow system 16-60 etc, which seems to offer governments good reasons for pushing the shifters on to proper agriculture, stage (5), where 100-250 might be possible, at least initially.

**Table 7.1   Evolution of Shifting to Fixed Farming Systems
– Some Indicative Parameters**

Swiddening	Fallow			
Evolutionary stage	Length	Years	Type	Population /sq km
1. Simple shifting + hunt-gather	indefinite	-	forest	to – 4
2. Recurrent cultivation, long	long	40	2nd forest	4 – 16
3. Recurrent cultivation, short	short	2 – 10	bush-grass	16 – 60
4. Mixed cultivation (crops + trees)	short	2 – 5	weeds	50 – 100
5. Continuous cultivation (a) initial	season	season	weeds	100 – 250
(b) with increasing inputs	short	1 – 3	legumes	100 – 250
(c) without supporting inputs	season	season	weeds	0 – 50
OR Trees				
6. forest gardens	permanent	permanent	trees/litter	300 – 500
7. plantations	semi-perm.	semi-perm.	trees/litter	(?)

Source: Lines 1-5 based partly on Raintree and Warner 1987, Boserup 1981, Greenland 1974.

If population can be somehow stabilized and a forest cover kept and an ecologically balanced swiddening system established, Ruthenberg (1980) cites Geertz's estimate (for SE Asia) that the stable population would be 20-50 and Ruthenberg's own general estimate is 56/sq km. (This would correspond to but be greater than line 2.) Such ecologically sustainable systems do exist but they are rare.   Others have defied the spiral by developing ever more sophisticated techniques for managing their dwindling land resources. Some of these are discussed below. Most – now at increasing rate – succumb to it. The carrying capacities of the table are only snapshots at points in time. For the bulk of the world's shifters station (5), continuous cultivation, is not a feasible destination. They rapidly pass through it and without supporting inputs arrive back at station (1).   If a station (-1) were shown they'd go to that because attempts at (5C) will have so depleted their initial range that even its hunting-gathering capacity might have been destroyed.

The planning alternative would be (and one evolutionary alternative has been) to bypass station (5) and procede to stations (6) or (7) (Figure 7.6 below). The carrying capacity for (6) is based on Kandy and it is higher than this in Kerala and these farms have been there, productivities undiminished, for centuries.

Soil, weeds and productivity   A second wastage is land more-or-less permanently lost through qualitative deterioration, mainly by erosion, weed infestation, desertification.   Erosion can run to several hundred tonnes/ha/year, at least 100 times higher than 'allowable' soil loss (Energy, conservation). By one conservative estimate the grass wastelands of S-SE Asia alone extend over 35 million ha (Nature) but much more than this if smaller areas at village level are added.   Such areas cannot regenerate naturally to forest because of erosion and loss of their soil seed bank. They are now closed off to the shifters (and most forms of sedentary farming) because of infertility and difficulty of cultivating the dense grass mass with hand implements (Nature, the cogon lands).

Forests and hydrology   In recent decades the most serious charge leveled at the shifters (particularly by foresters) is that they have been the single greatest cause of tropical forest destruction. Roche (1974) estimates that 100 million hectares of closed rain forests in Africa have been destroyed by shifters in recent times. High forest has spiraled down to bush, this to savanna, to dry thorn scrub then to desert. The process has been apparent for decades...*We are witnessing slow stages in the drying up and degeneration of tropical Africa* (Roche citing Aubreville in 1947); *and the situation is similar in tropical America* (citing Watters' FAO surveys there 1971).   Globally in this regard H. sapiens have forgotten nothing and learned nothing. Parain (1942) says of 14th Century Europe...*swiddening could endure only where the forests seemed inexhaustible...it squandered natural wealth and turned forest land into moorland.* So not much has changed, except populations.

Water   A third agricultural wastage is water. Except in the humid tropics (and these now subject to climate change) water is often a more important production constraint than land. Almost by definition a shifting system cannot invest in fixed structures like ponds and distribution channels which would otherwise capture and spread storm runoff over a dry season to increase production. (But some shifters do practice a rudimentary irrigation by diverting streams.)

Impact on diversity   To this agricultural list of complaints ecologists will add wastage as attrition of plant/animal species each time the forest is iteratively cut down and burned for cropping or other biome similarly modified. This is particularly the case with managed fallows (below). Though these are later described as being 'biologically enriched' this refers to enrichment of a narrow kind – that which enhances soil fertility or quality recovery. Apart from this it has little to do with recovery, regeneration, enrichment of the species and populations which comprised the biome and whose diversity progressively declines.

Energy   Inefficiency is also present as the high amount of biomass fuel energy expended per unit of crop plant nutrients recovered (N, P, K, Ca...) from it each time it is burned or otherwise processed for this purpose. Although they are billed as 'natural' systems requiring no such external energy subsidies as artificial fertilizers, fossil fuels etc, it is likely that forest slash-and-burn is among the least energy efficient of all farming systems (Energy, below).

Shifters and Greenhouse   Climatologists with an eye to atmospheric pollution and global warming will charge the shifters on two counts: (a) the immense amounts of Greenhouse gases released globally in their clearing fires, and (b) the replacement of some initially large carbon-sequestering biomass, particularly a rain forest, with a much smaller and therefore less efficient one, their crops. In the classical model of long-fallow forest swiddening the shifters would have had no case to answer: whatever carbon is emitted in clearing is re-captured when the forest fallow grew. However this cycle of carbon emission and recapture now hardly exists. German Bundestag's best estimate is that it takes 45 years for rain forest regrowth to recover its former carbon storage capacity. Globally the fallow (regrowth) phase in forest swiddening is now down to less than 10 years. Consequently the atmospheric carbon sequestering capacity of forests subject to swiddening has much declined and this continues as the fallow/regrowth phase becomes ever shorter. Brown et al (1993) offer a general guide to the effect on biomass carbon storage of converting three types of tropical forest to shifting agriculture:

Decreases of Carbon Storage in Forest Biomass with Conversion to Agriculture

Initial carbon storage		Conversion effect by second year of cropping
primary, closed ...	283t C/ha	- 204t C/ha
secondary	194	- 106
open	115	- 36
(includes carbon in biomass and soil)		

Conversion of primary forest would, in the second year of a subsequent cropping phase, reduce initial carbon storage by 204 tonnes carbon/ha.

Summary

All these are sound reasons for somehow stabilizing the shifters. The problem is that the main technology which has been proposed for achieving this – various versions of modern North American/European field crop farming – are in most of these respects equally destructive. When actually implemented under the common tropical conditions of high population pressure on land, poverty, steep slopes, intense rainfall, attempts at fixed-in-place farming (Table 7.1 line 5) advance an iterative and partially destructive process to completeness and permanency. Much proper fixed-in-place modern agriculture is itself a sort of shifting system – at least according to that mythical American who claimed a farmer could not be considered competent until he had worn out at least three.

7.3 Trees and Shifters

In the wet tropics it is unlikely that earliest shifters ever fully subdued the trees when clearing them from swiddens. Stone age technology would have prevented that, in places nearby to the present day (1930s). The stone age 'axes' were more in the nature of hammers to beat tree trunk bark off then burn the tree without completely killing it. They would have had no more success in *complete* tree clearing from the swidden than did Kunzel's Tongan who moaned that until he bought a tractor he'd had to fight them all his life (Reconnaissance). To these initial trees, left in the swiddens willingly or unwillingly, others were and are added or replacements provided by four main mechanisms:

1. By accidental seed import in food brought to the swidden and discarded.
2. Seed import by birds/bats/animals, distributing it along their flyways, trails.
3. Natural re-supply of seed from adjacent forest reservoirs.
4. Much later and deliberately, import of selected species for biological and economic swidden enrichment (managed fallows).

In more recent times in many places the shifters have at last succeeded; the trees disappear through five main causes:

1. Loss of the fallow seed bank in soil erosion.
2. Destruction of seed viability by increased exposure to solar radiation.
3. Disappearance of adjacent forests as seed reservoirs.
4. Destruction of live stumps by the inexpert fires of 'unsettled men'.
5. Live stump removal to permit field plowing in place of hoes and dibble sticks.

Functions of Trees in Swidden Systems

The different functions by trees (or lesser biomass) in each of the three-cycle swiddening phases (fallow-clear/burn-crop) can be summarized:

Trees in fallows
1. Accumulate future crop nutrients in their above-ground phytomass (major).
2. Accumulate nitrogen/organic matter in their shallow root systems.
3. Trap incoming atmospheric nutrients (minor).
4. Restore physical soil properties, infiltration rate, reduce acidity.
5. Reduce runoff and erosion, contain nutrients on the site.
6. Produce litter for nutrient recycling to (1).
7. Provide the microclimatic conditions for microorganisms engaged in (6).
8. Shade out and kill weeds.
9. Provide a seed bank for future fallow trees, biomass.
10. Yield a continuing stream of economic produce (especially enriched fallows).

These trees-in-fallows can also indicate the end of the evolutionary progression – stations (6) and (7) of Table 7.1. The swidden becomes a sedentary agroforest. Whether this happens or not the most useful of the trees now become to some extent domesticated, with this conventionally seen as one requirement of legitimate agriculture (Origins, domestication). From these several functions the optimum length of fallow depends on which purpose is paramount. This is generally taken to be (1), nutrient accumulation for the following crop; but it might equally be restoration of soil quality (eg aggregation, permeability) or reduction of soil acidity or suppression of weeds. These several objectives can be pursued by adjusting the length of the fallow and/or changing its species composition.

Trees in fallow clearing, burning phase   In this phase, a short one, the tree fallow or other biomass and the way it is managed also can have several specific purposes, as distinct from simply burning it to clear the field:

1. obtain its crop nutrients as ash (or mulch, in some non burn systems);
2. obtain some of these more slowly through decomposition of unburned wood;
3. provide heat to mobilize other nutrients from the soil (burned earth systems);
4. provide heat to destroy weeds or their seeds;
5. provide heat to stimulate soil microbial activity;
6. recover other economic products, usually charcoal or firewood.

From this it will be apparent that considerable skill is involved in the management of fire (or other clearing techniques) to best achieve whichever of these specific objectives are relevant.

Trees in the cropping phase – managed fallows   In this cropping phase, in classical forest swiddening, although the trees are nominally cleared from the field enough (or their stumps, branches) are left to perform two sets of functions – economic and ecological. In these *managed* fallows the useful trees spared will continue to yield some stream of the 'natural product of the country' – food, fibers, large insects as food, honey, even miscellaneous products such as caterpillars-silk (Assam below), and this produce range extends if bamboos and palms have been left (or have been planted) in these economically enriched fallows. These or other tree species will also be spared for their ecological functions in providing shade, wind protection, micro climate for the current swidden crop, and their general function is the stand-by one of readiness to reemerge – from live stumps, root sprouts, bits of vegetative matter mixed among the crops. Here the shifter has the alternative of hoeing these out as weeds if they are of little future (economic or ecological) use, or protecting them. If they are entirely removed, forest regeneration for the next fallow phase will be that much more difficult (Regeneration, below). Given this it is almost inevitable that forest species composition will have changed over the centuries or millennia, even after the shifters have disappeared.

With or without this management selection, fallow regeneration after the cropping phase procedes according to some specific successional pattern (Regeneration). The short life woody weeds will emerge first and dominate the recovering forest for 15-20 years or so, and the permanent long life hardwood species will not emerge until the necessary micro environmental conditions for this have been created by the 'weeds.' If this natural succession is cut short by reduction of the fallow period for any reason, the system will degenerate. The shifters' scope for species selection will contract with it to perhaps choice among a few forest weeds. These contracting opportunities are also reflected in both reduction of total crop nutrients recoverable from the fallow and changes in their (P,K,Ca...) chemical composition (Assam fallows, below).

7.4 System Activity Spheres, Multiple Economies, Boundaries

The French term *l'agriculture nomade* should not be taken literally; it's unlikely the shifters were ever nomads. A more useful concept is that they've operated a total economic system with several integrated activity spheres, overlapping or parallel:

1. A semi permanent or infrequently shifting base or 'house' with this preferably located close to some important food source, typically a river for fishing.
2. In the tropical forest this household or clan base often evolving into a village, the houselots of this into forest gardens.
3. One or a few food crop swiddens – the 'shifting' part of the system – operated iteratively around the wider range, their locations constrained by distance from the base (for logistics and security) as well as by such agricultural factors as soil quality; and with this initially random shifting evolving into a planned circuit, eventually its scope reduced by population pressure to approximate the rotations on a sedentized farm.
4. A wider activity sphere for hunting.
5. Another sphere for forest extractivism (notably Amazonia but throughout the tropics).
6. Often a scattering of micro activity spheres within the total range,

perhaps consisting of only single trees of some particular economic importance ('bee trees' below).

7. With these spheres – although they are listed here as discrete areas – merging into each other in space and time, as when the swidden crops affect the hunting range, or the fallow phases of the swidden continue to yield economic products long after they've been nominally abandoned.

8. Finally, with this entire system, these spheres collectively, occasionally migrating to some other part of the forest (or savanna), usually compelled to this by exhaustion of the most important of these activity spheres. This is usually taken to be exhaustion of soil fertility, though equally it might be exhaustion of the hunting or extractivist spheres, or there can be a dozen other reasons for migration which have little to do with the swidden's agriculture. A forgivable comparison might be some battered German army corps wandering on the Russian steppe, its components daily realigned to face new threat, but the whole thing secure enough as long as it stays within the elastic walls of its *Wanderkassel.*

Multiple Economies – the Kantu' Iban

The shifting component of these systems, the swiddens, can also be combined with many other types of agriculture, themselves fixed or shifting. From Dove's (1993) account of the Kantu' Ibans along the Kapuas River in West Kalimantan their economy is based on a shifting food swidden emphasizing rice, with these swiddens also planted (economically enriched) to food trees and pepper vines to yield a continuing stream of produce when the swidden is in its regrowth/fallow phase, and separate rubber gardens are operated as commercial mini plantations in parallel with the swiddens. The 14 households living in the Kantu' longhouse of Tikul Batu each have an average of 24 swidden plots scattered around their territory, 2-3 in operation and the rest under economically enriched forest fallow. Each also have five rubber gardens of less than one hectare and always located close to one or more of the food swiddens. Here they set up temporary house and live away from the communal longhouse for about half the year.

Priority is given to the food swiddens but the swidden/rubber activity spheres are complementary in terms of labor, land and Kantu' culture. The swiddens require four months of work annually, theoretically leaving eight months for rubber tapping. However households in greater need can do both, tapping in the (optimum) early morning and swidden work in the rest of the day. When rain prevents tapping they can if necessary still work on the nearby swiddens. Although rubber can be tapped almost daily it is beneficial to periodically rest it to increase latex flow and for a longer period to allow the tapping scars to heal. Under the Kantu' system these short rests occur when they are necessarily busy with the peak seasonal work on their food swiddens, and the longer rests occur in that interval when they go off to open another of their 24 food swiddens (and start tapping another of their five rubber gardens). Dove also points to the cultural complementarity which exists between these two activity spheres in that when ritual sacrifices, moon phases, curing ceremonies, ill omens, prevent work on the food swiddens these proscriptions often don't apply to the rubber groves. On the more material plane of land and soil type the two spheres are also complementary: rubber gardens tolerate a lower soil quality and greater exposure to flooding than do the swidden crops, allowing the Kantu' to make fuller use of total land resources. An inverse relationship also exists between good or poor seasonal productivity of the food swiddens and the intensity with which the rubber groves are tapped to buy rice in lean periods.

Boundaries and Ownership

From almost every viewpoint – economics, ecology, soils, sociology, law – these systems can be extremely complex. A first thing to do in understanding agroecosystems is define their boundaries. But where do the boundaries of shifting systems start and finish? Weinstock and Vergara (1987) illustrate some of these complications in New Guinea and Kalimantan, socio-legal distinctions between ownership of land and plants. In New Guinea communal forest land ownership by clans allows its periodic allocation to individuals for clearing it for their food swiddens, but when its fertility declines or it becomes weed infested it is returned to the communal land pool for the fallow period and later reallocated, not necessarily to the same household. For ecological and economic reasons PNG government have deemed it desirable that tree crops be incorporated into these intensively cultivated food swiddens. But from the clan's viewpoint, tree crops planted in the swiddens would extend this traditional temporary individual land right, through individual ownership of its trees, to permanency; and this would result in *de facto* private land ownership and loss of the clan's land resources.

With the Luangan swiddeners of Central and East Kalimantan, a family traditionally obtained permanent ownership of land by clearing it of primary forest. They might clear several parcels. Regardless of how long ago this was they still own it, though they might grant others permission to again clear and farm it in its subsequent swidden cycles. These permanent rights, ownership, pass to descendents through a cascade of kinship precedence which only a Philadelphia lawyer could savor. This concerns activity sphere (3), the swidden. Also in this sphere, those useful perennial swidden plants (fruits, bamboo) which volunteer are 'owned' less rigorously and anyone can share their produce *in situ*, but not to the extent of carrying much of it away.   In those swiddens which have been permanently converted to tree crops (rubber, coconut, coffee, rattan...) these trees can be sold with or without the land, or harvest or share rights only can be sold for varying periods. Finally, in activity sphere (6) most of the plants which naturally occur within the clan's common range can be harvested but can't be privatized. These include medicinals and such wild economic species as damar gum (Hopea micrantha) and wild rattan. But also in this common range certain other prized trees can be individually privatized, by finding and tending them but the land itself cannot be claimed. These include the durian and especially 'bee trees,' claimed by clearing the bush around them (an example of the kind of 'environmental domestication' discussed elsewhere, Origins). The bee trees are privatized and 'domesticated' by driving climbing pegs into their trunks. (These bee trees also occur in crop swiddens where they and the bees are carefully protected from the clearing fires.)

Development Implications

One approach to developing the shifters – whatever that may mean – is to try to understand these complexities of their agroecosystems. The other way is to smash straight through them *mit der nachten faust* and procede directly to 'development'. Under the basic forest and agrarian laws of Indonesia (from Dutch times but not enforced with vigor until Suharto's New Order) this has meant non recognition of traditional land ownership unless such land carries a *hasil* or grows some useful crop. The village or house forest gardens have been safe since they carried permanent tree crops, and the swiddens are also safe enough in their cropping phase because their *hasil* is cassava or whatever. But in their forest fallow phase they appear to carry none, nor do the common lands or their scattered micro activity spheres, wild trees like the forest durian and bee trees. With these producing nothing of economic significance – at least no hasil that a Javanese general could understand – they are judged empty and available for Nation Building.

7.5 Cycle Configurations and Measures of Land Use Intensity

The general model is sketched in Figure 7.1. In situation (a) the length of the cropping phase is sufficiently short or fallow period sufficiently long that the system maintains an ecological/ nutrient balance and could continue on forever, but this is seen to require a wastage of land. (This would represent stages (1) and (2) of Table 7.1.) In situation (b) this balance is not maintained and the system slowly or rapidly deteriorates, on the spiral to abandonment or cows.

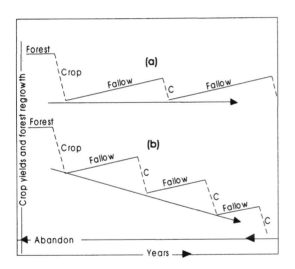

Figure 7.1 Sustainable (a) and Unsustainable (b) Shifting Systems

Measures of Cycle Intensity

Land use intensity of the shifting cycle can be measured as a simple ratio or fraction, C/F, or years of cropping C to years of fallow F, eg 1/10 for a year of cropping followed by 10 years of fallow. The more usual measure (Ruthenberg 1980) is in terms of land use intensity, R = (C x 100 / C + F) or (1 x 100 / 1 + 10) or R = 9.1 for the same example. (This can be arranged in several ways, eg as L = C + F/C, where L is a 'land use factor' which can be used to quantify degrees of 'shiftingness' and measure the degree to which these systems differ from permanent continuing agriculture. 'L' for permanent agriculture is usually taken as less than 2 and for classical shifting more than 10.) However measured and except in remote places shifting land use intensity is increasing. Beets (1990) offers the rule of thumb that the C/F ratio should not be less than 1:10 for physical sustainability. Data from Ruthenberg (1980) for 25 systems around the world in rain forest, dry forest and savanna show this was seldom achieved even 20-30 years ago: only two of the 25 systems had intensities less than 1:10; their mean 1:5. Globally swidden crop species are becoming more exhaustive as increasing household *cash* demands are added to precommercial *food* demands. These most profitable crops are often the clean cultivated and most erosive (corn, cotton, tobacco, peanuts... Conservation), and most demanded by governments for export income, in place of the traditional miscellany of ground covering/part exploitive-part restorative species combinations or polycultures.

Examples from New Guinea

The main determinant of intensity usually remains population pressure. Bourke et al (1994) obtained intensities R for 23 shifting systems in Eastern Highlands of Papua New Guinea on which the intensity/population graph of Figure 7.2 is based. (These are general systems, each supporting from 200 to 60 000 people.) The regression suggests that for a doubling of population pressure the land use intensity as defined above would increase by about 75%. (Imperfection in the correlation is due to differences among systems such as elevation, land steepness, soil types, remoteness from markets, cultural differences among these isolated valleys.)

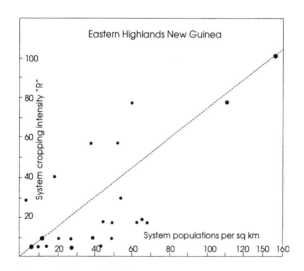

Source: Burke et al 1994.

Figure 7.2 Increasing Cropping Intensity 'R' with Population

7.6 Reasons for Shifting – Technical

Lengths of the fallow/crop phases depend on 11 main factors. Of these the technical have received far more attention than have the sociological; that shifters shift because their plots run out of fertility is almost a cliché.

Determinants of fallow period F
1. efficiency (type/amount) of the biomass used in the restorative phase (forest, grass etc);
2. population and economic pressures on land and now degree of commercialization.

Determinants of cropping period C
3. rate of soil fertility decline with cropping (depletion of N, P, K, Ca...etc);
4. declining soil quality (acidity, organic matter etc);
5. declining soil amount through erosion;
6. availability of external nutrient inputs to supplement (1), fertilizers, manures;
7. rate of weed infestation;
8. rate of pest/disease infestation;
9. increasing labor required to maintain the existing plot relative to a new plot (weeds);
10. possible gender conflict over these tasks;
11. disincentives to 'shift' (logistics, housing, fear of venturing from an established position).

Crop Nutrient Supply – Rates of Accumulation in Fallows

Nye and Greenland (1960) offer a guide to nutrients accumulated in various types of fallows in West Africa (Table 7.2):

Table 7.2 Nutrients in Vegetation of Five Types of Fallow Biomass

Forest age or fallow type		N	P	K	CA	Mg
				Kg/Ha		
Forest:	40 years	1 810	120	810	2 490	340
	18 years	550	70	400	(.......501.......)	
	5 years	390	20	340	(.......290.......)	
Savanna:	(20 years)	30	10	50	30	20
Imperata grassland		20	10	30	10	10

Source: Nye and Greenland 1960, data rounded.

From this the best situation would be a long forest fallow (line (2) of Table 7.1, but a primary forest would be even more tempting). Deterioration in potential nutrient supply as fallows become shorter and shorter, towards the now common case of 5 years, is apparent. On savannas even a fallow of 20 years would offer thin nutrient pickings, but not as poor as those contained in degraded Imperata grasslands. These data do not include nutrients accumulated in the soil during the fallow period or in the root fractions of the biomass. On the other hand they are only 'gross' potential nutrients for the following crops insofar as a good part of these theoretical supplies will probably be lost as smoke or washed away in erosion after they have been transformed to ash, or leached beyond the reach of the crop, or those thin supplies from burned grasslands blown away in the wind.

Mechanics of nutrient recovery in biomass burning   The full process of nutrient recovery in biomass burning is complex. In addition to those recovered in ash, other nutrients are obtained by accelerated mineralization as heat is applied to the soil itself, their amounts dependent on soil type and fire temperature. Several swidden systems are in fact based on 'burned earth' principles rather than slash-and-burn nutrient recovery from the biomass as in the previous tables. Finally, and except where a high degree of heat is aimed for, as in burned earth systems or to kill weeds, most forest burning is as slow and partial smoldering rather than an all consuming blaze. The result is a mess of half charred logs and branches. This continues to yield some nutrients through decomposition which in turn depends on the biological activity of the decomposing microfauna and funghi. Opening up a rain forest swidden to sunlight itself increases incoming solar radiation by a factor of about 25 relative to that on the previous forest floor. This greatly stimulates the activity of the decomposers, unless there are none left after an inexpert intense firing by those unsettled men.

Organic matter supply   In addition to their accumulation of nutrients, a main purpose in shifter fallowing is to restore plot organic matter. (This itself is an indicator of 'fertility' and general soil quality.) Charreau (1972) offers a guide to the efficiencies of several types of fallows in restoring this qualitative factor. The contributions by crops and grasses consist of roots which stayed in the ground after harvest, or after the savanna grass fallow was burned (Table 7.3).

By far the most effective of these alternatives is rain forest, contributing 10t from litter, 5t from roots. On modern farms, grasses, legumes and mixes as pasture or silage crops, with or without livestock, are also highly effective and much research is directed at using these in place of forest fallows. The problem is that they can hardly be applied to *shifting* agriculture until this is fixed in place, and then their actual adoption requires dedicated extension effort, capital, perhaps fences etc to do a job which – the shifters would argue – nature in time will do anyway.

Time rate of fertility recovery I.A. Jaiyeoba in rain forest fallows in Nigeria found that soil organic matter (and nutrients generally) increase very slowly for the first 4-5 years until the mass of supplying vegetation has developed (and to also reduce soil organic matter loss in runoff), then rapidly to about year 8, then slow down as accumulation reaches an equilibrium with extraction by the growing trees:

Fallow year:	2	4	7	8	10	13
% organic matter recovery*:	22	26	71	78	78	82

*Relative to undisturbed forest.

From this, and by rule of thumb that fallow organic matter should be allowed to recover to at least 75% of that under undisturbed forest if the swidden is to be indefinitely sustainable, the necessary fallow length is about 8 years (and this is about the length shifters do use in the area). Until all the microclimatic conditions and biological processes are in place – vegetative mass, shade, moisture, earthworms etc – attempts to shorten fallow length with use of artificial fertilizers would be ineffectual. These recovery rates of organic matter and nutrients refer only to preparation of agronomic conditions for the next crop: they have little to do with recovery of the plant species composition of the forest or dependent biosystems which take 60, 70... years, or centuries (Regeneration below).

Table 7.3   Vegetative Matter Returned to Soil by Types of Fallows and Crops

		From aerial part or litter	From roots	Total
		Additions to soil – t/ha/year		
1.	crops – from 1st phase	0.0	1.5	1.5
2.	crops – from 2nd phase	0.0	2.0	2.0
3.	savanna grasses	0.0	1.0	1.0
4.	A. albida	4.2	2.1	6.3
5.	thin forest	5.0	2.5	7.5
6.	rain forest	10.0	5.0	15.0

(4. is a large African acacia tree.)
Source: Charreau 1972.

Crop Nutrient Demands – Removal by Crops

The other side of the picture, rates of drawdown of nutrients and quality deletion in the cropping phase, are best measured in terms of declining yields. These rates depend on the nutrient requirements of specific crops (eg the three shown in Figure 7.3), but also on the soil nutrient store initially available (soil type), how much has been added by the previous fallow, how much is lost to leaching (soil type and rainfall) and to water erosion (steepness and rainfall intensities) or wind erosion (as wind-blown ash). These technical factors might determine when the cropping phase is to be terminated, or these might be overwhelmed by some of the agronomic-sociological factors previously listed (weeds, pests, conflicts), by factors having little to do with agriculture *per se.*

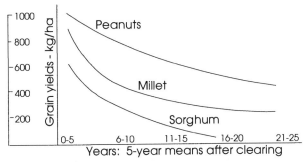

Source: Sanchez 1989. Data Nye and Greenland 1960 Nigeria

Figure 7.3 Typical Yield Declines after Clearing (Savanna)

Other Reasons, Other Views

Fertility rejected Enough of technics. Denevan (1971) among the Campa in the Gran Pajonal (foothills) of Peru observed that where meat is scarce a main reason for shifting the agricultural component of the system, the swidden, is depletion of the hunting range. Conversely the riverine Campa have greater access to fish so they don't move as frequently. Denevan also introduces a religio-spiritual basis to swiddening. The non riverine Campa like to build their villages on exposed ridges... *The reasons are not entirely clear but among them are certainly a psychological desire, with religious overtones, to be exposed to the sun and probably also a consideration of defense.*

The *chacras* (fields) are cleared on steep slopes near the houses even though these are subject to much soil erosion under the preferred crop *yuca*, cassava. The midslopes and bottom lands have richer soil but these are only occasionally planted. The poor and erosive soils of the ridges are preferred because these have more sunlight, fewer pests than do the bottom soils, and the swidden logs are closer to hand for cooking fires and warmth. With this kind of land use the Campa maintain their swiddens usually for only one crop and if they stay on in the area the plots have a fallow period of 10 years. But they usually move house to a fresh location more frequently than this, although there might still be plenty of swiddening forest left at the old site.

Shifting house Shifting house is not synonymous with shifting plots. From Ruthenberg's (1980) description... *The cultivated plots move slowly away from the previous clearing and the hut.* But this rate of departure is governed by the type and heaviness of the crops grown, especially of root crops. Beyond some point logistics suggest when a new house should be built and the old abandoned. But then, this depends on the kind of building materials available. In the rain forest these are plentiful but on the savanna a new house might have to be built of mud. A bamboo house will be so cheap in terms of labor that it might tolerate a 2-year occupancy while a mud house might permit and require a 16-18 year occupancy to justify it. This 'permanent' house might serve as a base, perhaps a well planned refuge, with a temporary shack slapped up wherever the swidden happens to be, no more substantial than a Javanese *globok* where small boys can hide from rain and rattle stones in tin cans at birds and guard their field from dragons. A main shift might occur every 5 years for a distance of 5-10 km (Philippines) or, in primary forest on the Rio Negro Amazonas each 10 years, each plot cultivated at each site for only 2 years and with an average family moving 500 km in a lifetime, with each shift facilitated by travelling by boat. The view from space? *...gradual movement of the larger community across the countryside in a manner suggestive of the progress of an amoebae across a microscope slide.*

Unsettled men In Watters' (1971) socio-economic classification the shifters, mainly in tropical America, fall into two groups. His classical type I are clan members fixed to place by custom and acknowledging need for its sustainable use, though still shifting around some familiar range. Their object is subsistence, decision making is on a group basis, fire is handled carefully, rudimentary soil conservation measures are often taken (bunds, woven fences), much traditional agronomic/botanical knowledge is applied. In Ruthenberg's phrase their system is one of balanced exploitation. Type II are outside individualists, perhaps propelled from some group elsewhere by population pressure. These arrive as colonists and must attach themselves to their new host community but this is for convenience. They obtain use of a piece of forest often on a produce sharing basis with traditional owners. They have more knowledge of the outside world and approach their new property as a future farm, with economic rather than subsistence intent...*but it is rare in the tropics to find land which lends itself to such purposes... They have no idea they are starting a cycle of shifting agriculture. They will cultivate some plot until the soil is utterly exhausted. It can still perhaps be used for grazing and this will spell its final ruin. The cultivator will then seek other bits of forest where he will once more undertake a similar operation which explains why this cultivator is, if not a nomad, an unsettled man. This type of farmer does not only waste land, he destroys it.* Which is a long way from the optimism of agronomists convinced that some technological fix is possible, some Package which will not only advance the shifters on to proper farming but also stabilize these unsettled men.

Summary Apart from soil fertility depletion there can be dozens of reasons for the shifting element in shifting agriculture: quality/quantity of phytomass available for the restorative phase (managed fallows, below); richness/depletion of the hunting or fishing range; swidden-to-house logistics; and *who* is to do the logistics, he or she; and to what extent this person is exposed to the 'primary zone of sorcery' as her plot takes her further from the safety of the house. There are a host of non material reasons: feuds, omens and probably no better reason – as when a family pack the car and go to California – than boredom and a change of scene.

7.7 Fallows and Other Methods of Nutrient Capture

Burning some forest biomass for fertility restoration has received so much attention that it obscures the dozens of other methods used around the world. These fall into three main groups – composting, use of animals and tactical use of fire.

Other Methods, and Open and Closed Systems

A useful ecological/economic distinction is between those fallows capable of supplying all the necessary nutrients for the cropping phase from within the swidden site (closed self-sufficient systems) and those systems for which a good part of the nutrients must be imported to the site (open systems). From this viewpoint closed systems occur generally in rain forests where the nutrient supply in the phytomass is large; open systems occur increasingly as the fallow phytomass becomes smaller – open forest, bush, savanna then poor grassland. Moving down this scale, animals play little role as nutrient suppliers in the forest but become increasingly important as nutrient importers to grassland swiddens (though whether this source will actually be exploited depends on cultural attitudes). At the extreme the phytomass itself is so sparse – on the site or near it and with or without animals – that few nutrients can be recovered from it, and here resort is made to the second resource, the earth itself, in burned earth systems.

Marvin Miracle (1967) has listed a range of fallowing/fertility restoration methods in Congo Basin, many of his primary sources going back to administrative reports in colonial times. A few of these and others illustrate the wide range of indigenous technologies actually used in addition to that universal cliché 'fire'. They might suggest that shifters have been as successful as any other folk in developing technologies appropriate to their place and circumstance.

Cattle In this open nutrient importing system in southern Sudan grasslands, the Dinkas drive their range cattle in and night-tether them at the future farm plot, shifting them around each night until the grass is removed and the whole plot fertilized.

Bomas In Central Africa the swiddeners often have no cattle while other tribes do; the swiddeners then erect a temporary *boma* (corral) over their future plot and might pay the cattle keepers to use it, or exchange future swidden produce for this service.

Buffaloes Although it is now only a remnant of earlier swiddening it is still possible to find in southern Sri Lanka that ancient practice of tethering 20-30 buffalo close together and driving them around a small rain wet field until it is churned up sufficient to plant rice without further cultivation. The manure they drop along the way is incidental.

Cultural attitudes Although animal manure might be plentiful, whether it is used or not depends on cultural attitudes. In Bahr-el-Ghazel in the 1970s a Chinese aid project obtained remarkable rice yields by collecting and using night soil from the nearby town of Aweil; but this alien technology collapsed immediately the aid workers left, through local revulsion at handling such material. Audrey Richards (cited by Miracle) speculates that this attitude, extended to cattle, might explain why so little manure is actually used in a continent with many cattle and goats and desperate need for fertilizer of any kind.

More praise of termites   Again among the Sudan savanna Dinkas, and because an insufficient biomass is available for slash-and-burn in the usual way, bush is cut from a larger area than the intended swidden and carried to it, as in an open system. However it is not burned but thrown down and left for termites to reduce to dust, which they do in four months and into which the crops are planted.

Ant hills   Among some Bemba people of dry SW Congo Basin the preferred method of growing corn and sorghum is to chop up ant hills and grow these crops on fertility which the ants have concentrated there.

Tethered pigs and loose pigs   In New Guinea highlands a common land use is intensive cultivation of a plot for 15 years before long fallow but this cropping phase is periodically broken by a short fallow of a year or so. In this latter phase pigs are tethered to glean what they can, to root and fatten on earth worms, and in this to cultivate the land for the next crop as well as to contribute their manure to it (Bourke et al 1994). Alternatively the protecting plot fence might be broken down in this short fallow phase to let in whatever wild pigs are roaming in the area, for the same purposes.

Falling leaves and mulch   In one system in northern Tanzania only the smaller vegetation is slashed down, the larger trees are ringed/girdled and left standing. As they die the falling leaf litter is collected and incorporated as organic manure directly into the planting beds; there is no burning. (This is definitely a hard hat area.)

Burned earth   One system used by the Bemba of Cameroon in grasslands (also in parts of Kenya and Ethiopia) is to cut and pile grass then cover it with earth for 4-6 weeks, then ignite and leave it to smolder. This process is aimed at nutrient recovery from the soil rather than the grass. The soil turns a 'light vermilion color and becomes puffy'. The other benefits are destruction of soil pests, wire worms, and decrease in acidity. In Laos the forest rice swiddeners say their clearing fires are intended to 'cook' the earth as well as provide nutrients in the ashes and kill weeds.

Burned turf   A similar practice where there is even less fallow biomass to burn is reported from western Kenya highlands and Ethiopia. In Kenya the turf is lifted off in 'large thin slabs' (in Ethiopia scratched off with an ox plow). This is stacked/heaped to dry then coals are inserted into the heaps which are left to smolder. The burned turf/earth is then scattered over the fields before crop planting. Again the second purpose is to reduce soil acidity.

Medje systems According to the description of their system left by M. Lacomblez, government agronomist there at the time (1918), the Medje of Congo forests do things in rather a backward way but still get good results. They first slash down the fallow undergrowth then the women immediately dig holes amongst this fresh debris and plant their staple plantain-bananas. Only when the field has been planted are the remaining large trees felled, except the largest and those with very hard wood. The Medje are also careful to leave all economic tree species, specially raphia and oil palms and other oil bearing species. Then when suitably dry the whole mess is burned carefully, doing no damage to the planted bananas or spared trees. After that cassava is planted in the ash, among the bananas. (In those distant times Lacomblez found the Medje using elephant tusks as digging sticks.)

Complex Bemba systems The most intricate and deliberative of the African fallow systems described by Miracle (taken from Audrey Richards' 1930-34 studies there) is that of the Cameroons Bemba in open dry forest. It illustrates how swidden technology is highly dynamic in that it is continuously adjusted to declining fertility gradient and increasing weed infestation of the cropping phase. Here the small trees are chopped down but the larger ones are pollarded, at five times the height of a man. This phytomass recovery extends over an area six times as large as the intended swidden. (In this it is an open nutrient-importing system. Or for those Cassandric souls concerned with spatial wastage, it is six times as bad as closed, nutrient self-sufficient systems would be.) The branches are carefully trimmed, carried to the swidden by women and arranged radially, and these circular piles are then edged by an extra layer of branches. When the piles are burned this outer border provides an extra thickness of ashes for specific crops planted in them, usually potash-greedy cucurbits. The second purpose of the burned piles is to destroy weed/grass seed so that hardly any weeding is required by the first crop. This is usually millet/sorghum.

When this is harvested the stalks are also burned to provide a nutrient input for the second crop and so on, with the residue of each crop providing on the one hand a declining fertility input for the next and, on the other, less burn heat to kill its weeds. By the fourth crop this procedure is almost ineffectual so the Bemba then start serious digging and scraping up mounds, on the theory that the more cultivation expended the more of the declining fertility will be captured. But they also adjust the size of the mounds to the requirements of the crop to be planted, peanuts -v- cassava etc. In this new mounding phase any crop residues are also piled between mounds, burned, the ash incorporated into a new mound, until the diminishing fertility stock is utterly depleted for even the least demanding crop, or the weeds finally triumph...or the women get sick of all this and start carping at the men to open a fresh weed-free swidden.

Old huts Where bananas-plantains are a staple food they are among the most nutrient demanding of crops, but oddly in Central Africa they're maintained for such a long period they are more of a permanent crop than a swiddening phase. With good soils in Uganda and Congo this is enabled by returning all organic matter (trimmed leaves, harvested stems) to the plants as mulch. This is also supplemented by collecting tree leaves from off the site and Miracle says it extends to breaking down old grass huts and using them too. In this way a plantation can be maintained for up to 50 years...which is a lot better than Australian banana farmers can manage because such is the rate of soil depletion there that those modern shifters have to clear a fresh patch – or sell the farm – every 18 years or so.

Improved and Managed Fallows

Contrary to early views that the fallow phase amounted to simple and periodic land abandonment, and the only 'product' of this phase was fertility restoration, closer studies the shifters have revealed that most fallow phases are integral parts of operational *systems* in that they continue to yield some stream of economic produce, or are so managed that the restoration process is accelerated rather than left to nature. Raintree and Warner (1986) distinguish between two kinds of deliberative fallow enrichment, biological and economic. At the extreme these economically enriched fallows can in fact be more profitable than the cropping phase of the system (rattan in Borneo). Kass et al (1993) offer examples of managed fallows from Latin America.

Babassu palm fallows in Brazil The *babassu* is one of South America's great palm resources (Diversity). It comes to dominate repeatedly cleared/cropped sites. For the swidden cropping phase its leaves are cut, heaped and burned, then crops are planted between the palms after which the depleted land is used for grazing cattle. The babassu recovers its full leaf mass after four years to provide another store of fire-recoverable nutrients, supplemented by those from the cattle. The cycle length varies but the babassu leaf mass available for burning is 16.7 tonnes/ha each cycle. The cycle yields food crops, cattle/meat and the range of palm products in addition to the leaves which are burned for crop phase nutrients.

Mimosa fallows in Honduras   *Mimosa tenuiflora* (carbon negro) volunteers to dominate swidden fallows which are usually maintained for 12-15 years (reduced to 7 where land is scarce). In a longer fallow it gradually gives way to longer lived tree species in the forest regenerative succession. During this 15 year period the mimosa is harvested for an age-sequence of different uses, successively fencing, small trees as fuelwood/posts, animal browse, large trees for poles. The initial fallow starts with a species composition of about 85% volunteer mimosa and this is gradually reduced by the products taken, especially by cattle browsing. Citing Landaverde (1989), Kass et al give the soil nutrients added by a 12 year old mimosa forest as greater than those added by a 32 year old secondary mixed forest, in terms of organic matter (twice), K, P, Mg, CA (three times).

Mixed ice cream tree fallows in Peru   The ice cream tree (*Inga edulis*) with its sweet bean fruit is high on the recommendations lists of many New Crops departments (Diversity). According to Page (1984) there are 30-40 Inga species of the Leguminosae family, hailing from West Indies through Central America to Peru (collectively *shimbillo*). *Edulis* fixes nitrogen like the others, is reported widely used as coffee shade in Mexico/Brazil and in Australia it is remarkably tolerant of highly acidic soils and high brackish water-tables (as shallow as two feet and 20% as saline as sea water). From Sanchez, Szott et al (1991), Denevan and Padoch (1987), Inga is one component of traditional managed fallows of the Bora Indians of Peru. Upland rice is planted as the swidden food crop then a wide range of acid tolerant food trees (Inga, peach palm/*pejibaya*, 'araza' etc) are planted in the rice. The fast growing Inga produces fruit after two years. Under commercial conditions this is sufficient to cover all land clearing and plot maintenance costs. It can later be cut down for charcoal. (The slowest growing component of these highly managed 'ice cream' fallows is mahogany-like *tornillo*, planted for its valuable timber and harvested after 20-30 years.)

Oil palm fallows in Congo   Around the Kwilu-Kasai rivers in east Congo the forest fallow is started by broadcasting oil palm nuts on the abandoned fields, and the young palms are cultivated until large enough to compete with the other regrowth. In this way oil palm cultivation spread through the Congo Basin long before the arrival of Europeans (Miracle (1967), citing Vanderyst's survey in 1923). Other people of this forest eg the Medje encourage this and a range of other naturally occurring palms by sparing them and cultivating around them when they clear their swiddens.

Highly diversified and semi permanent tree fallows in Amazonia   Padoch, Unruh, Denevan and others have described these systems around Iquitos in Peru (Reconnaissance). They are apparently one of the two origins of forest gardens and agroforestry generally (Evolution and development below).

Rattan fallows in Borneo   Perhaps the extreme in economic fallow enrichment in terms of profitability is found in the rattan cane fallows of the Luangan Dyaks of East Kalimantan (Weinstock 1983). Here unoccupied forest is slashed and burned in the usual way and after a 1-2 year cropping phase the swidden is planted to rattan, either seedlings or seed. The forest trees then reemerge and are allowed to grow up, the result being a mixed phytomass dominated and used by the climbing rattan palms as a trellis. After 7-10 years this is large enough when cut and burned to supply fertility for the next food crop phase, as well as yield a mature crop of rattan cane. While slashing down the 7-10 year fallow the farmer also harvests his rattan. The two activities are ecologically and economically synchronized. A farmer with four swiddens would have a continuous supply of food and a rattan cash crop each two years. (From Weinstock's data a 14-15 year fallow would be even better, at least for the economics of the rattan component of the system, because doubling the fallow time yields three times as much rattan and this is of larger diameter and higher unit price.) But this ecologically sound and profitable system contained the seeds of its own undoing.

Because of deforestation and destruction of wild resources in the rattan epicenters (Philippines, Malaysia, Borneo/Kalimantan) rattan supplies have historically been declining and prices increasing. Under the Dyak swiddening system yields are about one tonne/ha for the 7-8 year fallow (and three tonnes for the 14-15 year fallow). Rattan prices were $600 – 1 000/tonne, considerable wealth to the Dyaks. In fact it's one of the most profitable agroenterprises from any perspective. But it led to 'avarice and conspicuous consumption' by the swiddeners and the emergence of two new classes – professional rattan harvesters and these employed by traders. The traders introduced the *tabasan* system, purchase of the standing rattan crop before its maturity, with payment often made as rice, or outboard motors, cassettes, Swiss watches and similar glitter. Traders not farmers then decided when a rattan plot was to be harvested and the period terminated. Synchronization of the two phases of the cycle – food crops and restorative (rattan) fallow – was broken.

As rattan demand and price increased so did pressure by traders and unemployed cutters to harvest immature rattan and reduce the restorative fallow period. While their supplies lasted some Dyaks became over-dependent on rattan cash income for food purchase. These neglected the food cropping phase and became dependent on trader credit for rice, and this debt put further pressure on their remaining rattan... *Fifteen years ago the banks of the Montalat River were covered so thickly with rattan that it was impossible to land a canoe anywhere... Today the river banks are barely covered with wild grasses and small quantities of immature rattan are all that's to be seen* (Weinstock). This example illustrates how traditional swiddening can evolve through economic enrichment of its fallow phase, but how if carried to extreme it can destroy all components of the system. Sadly it's now academic. Weinstock's rattan forests, natural and fallows, have largely been consumed by El Niño's 1997-98 'wildfires', most deliberately lit by plantation companies to extend their oil palm fields. To paraphrase Marc Faber in a different context, What idiot would throw out such a perfect system?

7.8 Ecology as Agronomy

The three formal conditions for agriculture or proper farming are plant *domestication, cultivation* of these in an organized deliberative way, and these enclosed in a localized economic system as a *farm* (Origins). The shifters achieve the last of these, and although the boundaries around their systems are somewhat fluid these can comprise several activity spheres (above), where modern agriculture is often stressed in managing one. But do they qualify as farmers according to the first two criteria?

Plant Domestication and Breeding

For millennia the shifters have achieved domestication in two ways – genetic manipulation to improve the plant and stabilize this improvement, and taming it through domestication of its environment (Origins, Yen, Rindos). Katherine Warner (1981) found in Philippines the Tagbawan swiddeners on Palawan had domesticated rice to the extent of breeding 140 varieties of it, and Fujisaka in Laos reports the shifters there routinely use 16 varieties – again in contrast to a 'sophisticated' Western farm which

probably uses one – and buys that from a store. The reclusive Hanunóo of
Mindaro Island, Philippines have classified 1 600 plants in their area
(where scientific botany managed 1 200), have actually bred/improved 400
of these in their swiddens and as many as 40 species may be present in a
single field. In the Gran Pajonal of Peru, Denevan (1971) citing an earlier
investigator Weiss, says the primitive Campa routinely use 385 plants,
domesticated/cultivated and wild (192 of these for their medicinal or
magical properties). Dufour (1990) found Tukanoan Indians in NW
Amazonia having bred and using up to 48 varietals of cassava/manioc.

According to Marvin Miracle, when M'sieur Lacomblez first met the
Congo rain forest Medje they had bred and were using 27 varieties/types of
plantains/bananas, 10 of bulbous yams and 22 other indigenous root crops.
Further, the different varieties of crops like banana-plantain are developed
for a range of specific purposes – vegetable, fruit, flour in bread making,
beer, cattle fodder. In Micronesia the 150 varieties of yams on Ponape and
the 40 types of taro on Pohnpei have been the result of plant domestication
in slash-and-burn swiddens there from at least before the 8th century AD
(Reconnaissance). In brief, whatever charges might be brought against the
shifters as improper farmers these could not include lack of botanical (or
economic) sophistication.

Agronomy? ...Ecology?

How well the traditional shifters have achieved the second condition,
organized cultivation, depends on one's preconceptions.   To modern
agriculturists their swiddens are a mess, often not cultivated at all but poked
with a digging stick or hacked about with a machete, the plants apparently
thrown in at random and left to wander where they will, a botanical
happening. Beets (1990) concludes from his survey of forest swiddens
around the world that...*Extreme crop diversity is an important
characteristic of pure traditional shifting agriculture*...(though there isn't
much of this pure type left), and Unruh (1988) found much order in this
chaos.

Nutrient allocation, scheduling In polycultures with managed fallows, the
several species might be planted in succession according to their varying
nutrient demands and later weeding needs. Planting the trees/perennials
first in the ash gets them quickly started but not to the extent of excessive
competition with the fast growing food crops. By the time the food annuals
have consumed most of the ash fertility the perennials will be tapping
deeper soil nutrient reserves. This timing is also aimed at getting the

trees/perennials to a height, during the ground cropping phase and assisted by their weeding, where the perennials can survive against volunteer regrowth when the swidden is 'abandoned'. Several things are being managed simultaneously: some limited supply of nutrients is apportioned among species and over time, in such a way that later weeding labor for the economic perennial species of the future fallow is minimized.

Niche creation   Again in managed fallows, some (tree food) species will be slashed around to give them greater sunlight. Other tree species in the fallow will be harvested by felling them (Mimosa, ice cream trees, tornillo, above). This changes the structure of the growing fallow from what otherwise would have had few ecological niches to a modified one having many, in terms of sunlight, humidity, wind etc. These niches would eventually have been created by the forest itself but this would take 30, 40... years. These induced niches become habitats for seed carrying animals (birds, bats, monkeys) which act as agents for further fallow enrichment. The largest 'niche' induced is as increased sunlight now reaching the floor of the growing fallow forest and which enables germination of useful tree seeds in the soil bank, otherwise not possible in the forest darkness.

Diversity and risk management   Katherine Warner described manipulation of environmental factors by the Tagbanwa. Each household usually operates two swiddens, the smaller by the wife who grows early maturing upland rice varieties (and a wide variety of non rice foods) and a larger by the husband who might grow only late maturing rice. These swiddens are located in different places and oriented to different directions (east -v- west etc) to minimize drought risk by their different microclimatic environments. To minimize pest risk each swidden of the community is separated from the others by strips of forest (but this ideal is not pursued when there is need for proximity as mutual security). For further pest/disease avoidance and to utilize different ecological/edaphic conditions, as well as food variety, the community as a whole uses 140 different rice varieties, each family growing 7-15 on its two swiddens (most on the wife's), with this number being constrained by difficulty of storing separately so many varieties. Warner's conclusions seem increasingly relevant for much of the tropical world: *That this knowledge (of swidden management) is extremely valuable is apparent in the transformation of the east coast of Palawan within 15 years from an area of forest maintained under Tagbanwa methods to the abandoned kogon (grass) wasteland created by the profligate agricultural practices of peasant immigrants... Unfortunately the immigrants are appropriating the land and forest and may well establish in its place the exploitive system responsible for deforestation, flooding and aridity in other parts of Philippines.*

Clearings, trails and flyways   Over longer time, cutting trails to and clearing forest swiddens has two sets of effects on the plant/species composition of the forest itself and on the swidden (Unruh 1990, Fleming and Heithaus 1981). Trails and clearings alter forest architecture and this alters bird and bat flyways and ground animal trailways. These drop seeds while commuting to and from the swiddens; and new useful plants which have been added to or concentrated in a swidden are spread by these agents over a large area of the forest. Over millennia the swiddens have probably been important dispersal points for alteration of the larger forest. These architectural/flyway changes also affect swidden insect populations, and again their bird predators, though whether this is good or bad in terms of pest damage is uncertain. Research evidence has pointed in both directions.

Weeds

In advanced agrosystems the economics of weeds amount to the costs of removing them. While this also concerns the shifters (especially the deep rooted grasses) they take a wider view; weeds can have at least 11 economic outputs, uses:

1. As laid mulch around standing crops for other weed suppression, moisture conservation.
2. As biomass for fertility restoration in place of forest or savanna grass (Miracle has described such systems in Africa where weeds thrown into pits and covered provide fertility for the next crop, planted above the decomposing weeds).
3. Retained selected woody weeds which will develop into biomass for the next fallow phase.
4. As habitat for beneficial insects (in indigenous integrated pest management programs).
5. As household fuel.
6. As animal fodder. (This can be so important that farmers schedule their weeding according to livestock feed requirements, S. Asia.)
7. Standing weeds or mulches as soil erosion barriers.
8. As food, herbal tea, medicines (eg the Jola in Gambia).
9. As indicators of fertility status, decline.
10. As crops. Weeds, or 'plants out of place,' might also be described as crops before their time. (Who knows how many potential New Crops are lying dormant in the shifters' fields...?)

11. As evidence: Though this would not interest the shifters, weeds are an important archeobotanical tool for tracing the sites of past civilizations, movements of ancient peoples through the natural world (Origins).

These uses are not confined to shifters. Researchers in Bengal rice fields found that of 158 weed species present 124 had significant economic uses. (Nor are these confined to the 3rd World: a good place to watch the weed collectors at work is outside any southern European city on a Sunday afternoon, though these are mainly after item 8.)

Weed types, successions These uses acknowledged, weed infestation and labor rather than soil nutrient depletion can be the main reason for terminating the swidden cropping phase. Weeding often requires the single biggest labor input after clearing/burning, increasing in importance on swiddens in primary, secondary forest, savanna, grassland in that order (economics). The *types* of weeds also change among these biomes as well as successively through the cropping phase, with easy-to-remove broadleaf weeds predominating in forest fallows and/or in first crops but these giving way to deep rooted grasses requiring progressively heavier cultivation. The table offers a general guide to yield losses by not removing weeds:

Guide to Relative Crop Losses Caused by Not Weeding Crops

corn	...	28% loss	sweet potato	...	91
cowpeas		59	cassava		92
soybeans		60	upland rice		100
yams		73			

Source: Moody/FAO 1974.

The forest shifters counter weeds by several strategies. Especially in the humid tropics they use highly mixed polycultures of several architectural levels and with ground covering crops to shade out the weeds. A second is to arrange the sequencing of crops so that the weed-sensitive ones are planted first, on fresh ground, while the weed-tolerant ones are planted later and these are able to mature in a tangle of weeds and other vegetation. A third solution is to make the tree fallow phase permanent, convert the swidden to an agroforest (ie procede to lines (6) or (7) in Table 6.1).

Weeds and fire farming In grassland and savanna swiddens where an inordinate amount of pre or post planting cultivation for weed control is necessary (economics, swidden labor), Miracle has outlined how the southern Sudanese confront this problem with strategic use of fire. In these *hariq* systems the grass in a future swidden plot is guarded from accidental burning for up to two years to develop into a dense mass. Then, and unlike in other systems which wait to burn in the dry season, the grass fires are lit after the start of a wet season when new shoots have emerged. The mass of accumulated combustible material then provides intense heat to kill or at least retard grass growth and no weeding is needed for the first crop. (This method is possible only with certain grasses.)

Pests

Pest problems in traditional agriculture are partly a matter of perspective (Brown and Marten 1986). Globally crop losses to pests on traditional farms, fixed as well as shifting, range 10-40% depending on type of crop, about the same as in developed commercial agriculture, but with the difference that these latter losses are in spite of the vast amounts of pesticides used and the 100% losses which can occur in modern monocrop systems where pesticides are *not* used or discontinued. Traditional pest management practices are summarized; some apply on both fixed and shifting farms.

Polycultures In the humid tropics the multi-storey, multi-species crop mixes of polycultures are a most effective pest as well as weed control (on forest gardens as well as swiddens). Brown and Marten have described how they work. Polycultures offer a wider range of insect habitats than do monocultures. Increased populations of the pests are usually more than offset by the increased populations of predators. Polycultures also dilute the attractiveness for pests because they offer fewer host plants of any one species. The various plants of different densities, architectures, heights in polycultures can also provide physical impediment to pest movement. One function of the taller plants is to disrupt insect flight patterns, confusing these into landing on the wrong plant where their economic damage is minimized.

The taller varieties of similar crops (eg rice) are also used to hide or camouflage the shorter varieties. Pest concentration on some intended host crop can be repelled by the odors of nearby plants, including the weeds. Color is also a factor in pest movement; it's been suggested that some insects are more attracted by a contrasting green(plant)-brown(soil) vision than by the solid green vision of polycultures. Though increasingly much is known about these mechanisms there is still much to learn, especially in the really complex shifting and forest garden systems, where these human systems are only scratchings within the wider universe of some rain forest or savanna ecosystem and beyond experimental control. It is still possible to find villages in Java – even in the highly developed rice lands – where pests are handled by going out at certain phases of the moon and threatening them with magic incantations. But these charming methods distract attention from the fact that in traditional systems more material technologies are also present, even if the shifters and forest gardeners, like scientists, are not too sure just how they work.

Ants and rats In Thailand subterranean-dwelling ants are captured by burying coconuts in which holes have been drilled to allow the ants entry to eat the coconut meat. After five days the coconuts are dug up and the ants shaken out over a fire. After water soaking to remove the fire odor the coconuts are again buried and used repeatedly. Paddy field rats are a particular problem in continuously cropped rice lands. In Central Java one chore in land preparation between crops is to gouge out their nests in the bunds and replaster the face of the bund with mud. Elsewhere they are hunted by beaters with packs of dogs. Where the hunt is not in vogue their burrows are fumigated with sulphur-rice straw smoke, this blown into the burrows through bamboo pipes (Brown and Marten).

Birds and ducks In Java, Iskander found that most birds are insect eaters rather than grain or fruit eaters: overall their contribution in insect control is greater than the crop damage they do. However some bird species are in fact serious rice pests and these are netted to eat or sell; but this supplementary income is so attractive that it is extended to birds of all kinds and now intensifies the insect pest problem. In Congo where birds themselves can be the main threat to grain crops the shifters there outwit them by planting the most susceptible grain types to mature in that season when the birds are busy in nest building (Miracle). Especially in Philippines and Java but throughout SE Asia paddy field ducks have long been a main pest control, flocks of several thousand herded through the fields after harvest to glean and fatten and provide manure as well as eat snails and larger insects. But in the intensive rice areas they don't much thrive on the agricides of the Green Revolution package. Village duck mortality has been remarkable.

Chickens and fish Chickens play a similar range of roles, especially under forest gardens and pigs, dogs, scavenging under these systems (as livestock, Reconnaissance) must also have pest control impact though it seems not to have been measured. More deliberative poultry management for pest control in Java includes planting peanuts near the house to attract and concentrate certain insects so that chickens can more easily find them. Although fish in the standing water of paddy fields were never developed to the extent of paddy field ducks in Philippines, such fish are the aquatic pest control equivalents. Again because of pesticide residues in intensified rice systems, Otto Soemarwoto in Java found the area of fish-in-rice systems fell from 8 500 to 1 800 hectares in just four years (1974-78).

Pests as livestock, protein   From Reconnaissance one main economic output of forest swiddens is the game they attract. This is important because the larger animals of forests, like their trees, are usually dispersed. On the other hand the lesser game, as insects, are plentiful both in species and populations. Most of the 200 'animal' species used by the hunter-gatherer-shifter Mbuti of Congo forests are large edible insects (Nature).

Grassland and savanna shifters often have large livestock (cattle) but this depends on culture and incidence of tick and fly borne diseases. Forest shifters keep small stock – goats, pigs, poultry – though with exceptions like the ubiquitous pigs of Oceania and traditionally dogs in Micronesia their economic role is limited. But less orthodox sources of animal protein also are exploited, this sometimes verging on animal husbandry. Denevan found the Campa in Peru 'cultivate or at least protect' an edible grub *poshori* in piles of corn cobs; and they actively manage hills of leaf cutting ants by controlling vegetation encroaching on their nests and...*the ants are gathered and eaten in October when they swarm.* When clearing their forest swiddens the Medje of Kinshasa, Congo, spare trees whose leaves are 'liked by caterpillars' (Miracle citing the government agronomist there, 1918). Though bees don't come under the heading of pests the Luansan swiddener-rotan growers of Kalimantan are particularly partial to unhatched bees in the combs as well as the honey, and they also shape the wax into candles (Weinstock and Vergara 1987); these bee trees were noted before as micro activity spheres in swidden systems.

The dispatching of rats in parts of Thailand was noted above; but in NE Thailand they are also a protein source and this country cuisine also extends to termites and small paddy field crabs (Brown and Marten). Consequently their pest dimensions here are not taken as seriously as elsewhere. In Indonesia a grasshopper damaging to rice is trapped at night and garnished with onions-sugar-salt to make a useful meal, or is fed to pond fish (West Java) or sold as bird feed in the market as a cash income source. Farmers' yield expectations are anyway not high, the object is sufficiency not input/output optimization. Especially in Buddhist/Hindu society there is a willingness to share the table with whatever fauna drops in, though if such losses exceed 40-50% farmers would begin to resent this as abuse of their hospitality.

Silk In Assam the forest swidden fallows also contain a tree favored by caterpillars. These are spared in clearing and their leaves trimmed off and fed to adult silk worms. They are not suitable for young worms so a suitable fodder crop (Ricinus communis) is grown in the swidden for these young. In this way these systems yield 4 kg raw silk/ha in addition to their other crops (Toky and Ramakrishnan 1981).

Soil Conservation

With the shifters, erosion ranges from the minimal to the terminal and depends on two sets of factors: (a) just where the particular system is on the evolutionary progression of Table 7.1, with very little erosion in classical long fallows (line 1), but much as this period shortens (line 4); but also (b) these rates still subject to the erosion sub-processes discussed previously (Conservation) as these are expressed in the Universal Soil Loss Equation – rainfall intensity (R), length and degree of slope (LS), soil erodability (K), ground cover (C), management practices (P). These USLE factors provide a basis for considering why shifting systems should be more (or less) efficient in soil conservation than are conventional agrosystems.

Land is seldom completely cleared, the stumps, logs, large branches left provide a network of barriers to overland flow (L). In those systems which can't use fire because of almost continuous rain (Congo, Colombia) the cut-and-mulch methods of fertility recovery used in its place act as effective barriers to both rain impact erosion (R) and overland flow (L). In forests, clearing for the first crop will itself have sufficiently loosened the soil for planting. Erosion is reduced by such incidental 'minimum tillage' (a practice P); but subsequent crops might require progressively more tillage (and erosion), as with the Bemba, so that this factor is dynamic, changing/increasing with each crop of the sequence. Low intensity and careful control of the clearing fires as with the Medje results in conservation of soil organic matter which contributes to high infiltration and this to minimum runoff/soil loss (through K). Similarly runoff/soil loss is minimized by avoidance of soil compaction by draft animals, tractors, in all clearing and crop production operations (increasing infiltration).

Erosion control through polycultures Especially in the humid tropics the major advantage of shifting systems – working through all these technical USLE factors collectively – is by the common use of polycultures. Vergara

(1987) offers an example for Solomon Islands consisting of, from the bottom: sweet potato, taro, cassava, sugar cane, banana/papaya, with these ascending to a top storey of breadfruit and sago palm. Economic life is indefinite; as one species or single plant is harvested another is planted to replace it. There is no period when the land is bare (Charreau, below). Such systems are practically erosion proof. When they are eventually 'abandoned' this is because of fertility depletion, sometimes pests, not soil erosion.

Charreau (1972) offers a guide to soil loss rates under the alternative, a sequence of discrete crops separated by bare land intervals. From this, by far the greatest erosion (to 139 tonnes) occurs in these between crop/bare land phases. These approximate the between-crop phases on the fixed-in-place farms of proper agriculture.

Table 7.4   Guide to Runoff and Soil Loss for Shifting Systems in West Africa

			Annual runoff			Annual soil loss		
	Rainfall	Slope	Forest	Crop	Bare	Forest	Crop	Bare
	mm	%	%	%	%	t/ha	t/ha	t/ha
Upper Volta (DF)	850	0.5	2.5	17	50	0.1	4.5	15.0
Senegal	1 300	12.0	1.0	21	20	0.2	7.3	10.5
Ivory Coast	1 200	4.0	0.3	13	23	0.2	13.5	24.0
Ivory Coast (WF)	2 100	7.0	0.1	10	19	0.03	45.0	139.0

DF, WF dry and wet forests. Where some of these data were reported as a range their midpoints have been taken.
Source: Charreau 1972. See also Roose in West Africa (Energy, conservation).

Gone with the wind Discussion of soil erosion in these terms of tonnes or inches per ha etc is likely to overlook loss of nutrients in the fire ash. This is particularly so in grass or poor savanna fallows which yield only a few millimeters of ash nutrients anyway and it is easily washed off or blown away by the wind. This hazard is well illustrated by Ramakrishnan and Toky's (1981) studies of 30, 10 and 5 year cycle swiddens in Assam (Table 7.5). Here wind losses ranged 41-60% of the total ash nutrients made available by burning these bush and forest fallows, especially during the high winds which immediately precede the monsoons, while the farmers are waiting for those rains to plant their crops. The table also shows how the nutrient composition of fire ash changes, as tree species composition of the biomass changes with its age, evolving from woody weeds (5 years) to bamboo (10 years) to permanent trees (30 years, Regeneration below).

Table 7.5 Nutrient Losses to Wind; and Changing Chemical Composition of Ash with Age of Fallow

	30 year cycle Release – windloss		10 year cycle Release – windloss kg/ha per burn		5 year cycle Release – windloss	
P	310	150	260	160	150	40
K	1 740	820	2 070	1 230	260	160
CA	960	450	190	120	120	30
Mg	210	100	150	90	110	30
Totals	3 220	1 520	2 670	1 600	640	260
% Loss		47%		60%		41%

Source: Ramakrishnan and Toky 1981.

7.9 Economics

Farm economics has proved of limited use in evaluating the efficiencies of shifting systems (ethnobotany, human nutrition and sociology have done much better). There are too many types of biomass and fertility restoration technologies to permit comparison; too many cultural and gender constraints to tasks, activity spheres in operation simultaneously – farming, hunting, gathering. In polycultures there are so many crops being managed at once that it is impossible to assign inputs/costs to these in isolation, and too many outputs. Only the most obvious of these (typically days of labor input and kg of corn or cassava output) are measured and reported, the rest ignored. Largely on this basis consensus is that shifting systems compare poorly with sedentary systems. Pedro Sanchez has stated it as well as anyone...*The traditional system is ecologically sound but guarantees perpetual poverty...There is no such thing as a prosperous shifting cultivator in the tropics.* This view was modified somewhat by Dabasi-Schweng (1974) who from his review of the formal economics of shifting in seven countries conceded that...*The (economic results) do not show that shifting cultivators live in more abject poverty than the rest of the rural poor...*eg in Venezuela at that time shifter family incomes were $40-200 per year, but 53% of *all* rural families made less than $66. But still, this is hardly a warm endorsement.

Labor, Tasks, Gender

The only significant input in most swidden systems is labor. Typical levels are shown in Table 7.6A. From part B the labor input for land clearing/preparation increases as the site degenerates from rain forest (68 days) to grassland (137 days), mainly because in the forest systems no specific pre-planting cultivation is required. Part B, relating to a different sample of forest swiddens also shows how total production labor is typically allocated according to task and gender. Depending on culture, the most labor intensive tasks of weeding and harvesting-threshing-carrying are done by women/girls; and as noted previously the gender inequity which these imply can be a more important reason for abandoning the swidden then its fertility depletion. (In many cultures around the world the women are the operational farmers and effective decision makers: the theoretical aspects of 'fertility depletion rates' etc labored over previously are often an abstract game men play with numbers.) These data of parts A and B are relevant for system comparisons with forest gardens: from Kandy, the most onerous of these production tasks hardly exist. Governments, on the other hand, will be most (or least) impressed by the line 'yields per ha' of 1.0-1.5 tonnes/ha which seem hopelessly low in comparison with the 3, 4, 5…tonnes possible (for a little while at least) with proper farming.

Table 7.6A   Labor in Swidden Clearing and Preparation

Labor days/ha:	Borneo rain forest	Borneo regrowth	Guatemala milpa	Congo savanna	Congo grassland
slash/cut	54	30	46	37	7
clear/burn	14	14	25	57	55
cultivate (pre-plant)	-	-	-	75	75

Source: Modified from Ruthenberg 1980.

Table 7.6B   Labor Inputs for Swidden Rice, Days/Ha

			Laos [1]	S. Leone [2]	Liberia [2]	Ivory Coast [2]
(a)	Develop swidden					
	Clear/burn	(M)	66	36	68	50
(b)	Produce crop					
	plant	(M/F)	23	21	18	24
	weed	(F)	198	24	6	49
	pests*	(F/C)	na	25	4	22
	harvest/thresh	(F)	78	54	27	50
	Total crop		299	124	55	145
	Yield/ha, tonnes:		1.00	1.20	1.00	1.50
	Kg/labor day:		3.3 kg	9.7 kg	18.2 kg	10.3 kg

*mainly scaring birds. M/F/C common tasks men/women/children.
Source: [1] Fujisaka 1991, [2] Ruthenberg 1980.

Exceptions – increasing yields with decreasing fallow lengths The proposition that 'yields' (and implied value of output) decline as fallows become shorter-and-shorter is a generalization which might not hold for specific crop species. Table 7.6 (from Toky and Ramakrishnan in Assam) illustrate these exceptions. While all grain yields in their surveyed swidden farms declined markedly with shorter fallow periods, as they are supposed to do, the yields of vegetables and root crops in these crop sequences actually increase. Because of this, when prices are attached to the three systems (a), (b) and (c), their *economic* results as $ output/input ratios are from a farmer's viewpoint not much different (2.13, 1.83, 1.88). But from the public viewpoint of land use efficiency – and subject to whichever of these three sets of commodities are valued most – there would be a great difference between the 5 year cycle (and its output) and the greater grain output of the 30 year cycle which also ties up land, in fallow, for this period. But then again, and going beyond *farm* economics, the social-economic benefits of the longer cycle (as catchment protection, downstream irrigation water supply etc) might well transcend alternative productivities (a), (b), (c) on this patch of ground as farm.

Table 7.7   Efficiency Comparisons of Shifting and Sedentary Systems

	(a) 30-yr	(b) 10-yr	(c) 5-yr	(d) Terraces	(e) Rice
Yield comparisons—	Yields, kg/ha/year				
All grains (6 species)	2 590	1 340	130	1 140	3 710
All vegetables (8 species)	260	380	580	390	-
All roots (3 species)	610	1 650	870	1 440	-
Economic comparisons—					
Output/input ratios	2.13	1.83	1.88	1.43	1.14
Energy comparisons—					
MJ Output/input ratios*	34	48	47	7	18

*Megajoules; energy is discussed below. These three systems (a), (b) and (c) were cropped for only one year; their fallow periods were 29, 9 and 4 years.
Source: Toky and Ramakrishnan 1981.

Attempts at shifter stabilization to fixed farms These researchers also compared these three shifting systems with farms on dryland terraces (d) and in valley bottoms (e), Table 7.7. The terraces had been constructed and fertilizer provided by government as models in an attempt to stabilize the shifters on permanent fields but continuing to grow similar crops. (They were abandoned after 6-8 years and their occupants reverted to slash-and-burn, mainly because of necessary continuing fertilizer and labor inputs for terrace maintenance.) The bottomland farms (e) grew two crops of rice annually, with high crop labor inputs but without fertilizer, their nutrients being provided by silt wash-off from nearby hills (ie their productivity was at the cost of soil erosion elsewhere, but the cost of this upstream loss is not included in their farm economic performance, column (e)). From columns (d) and (e), both these fixed stabilized systems performed poorly in comparison with the three sets of shifters ($ output/input ratios of only 1.43 and 1.14 respectively).

This Assam example is considered because it encapsulates just about all the problems encountered in attempts to stabilize the shifters. From their viewpoint the best strategy is to get their hands on a nice fresh patch of forest, if they can find it (output/input ratio ($)2.13). Failing that, the shorter forest fallows are still better than stabilized farming even when the valley bottom farms grow two rice crops annually (and 'steal' the fertility to do this from other up-slope lands). But these $ output/input ratios can also be manipulated through adjustment of the relative crop prices mechanism, therefore economics can play a central role in leading them to that system considered most socially desirable. Other factors are erosion rates, hydrological efficiency and water yield of the larger catchment, energy output/input ratios, and the social prices which must be applied to these if the 'optimum' system, (a) – (e), is to be determined and implemented (Energy, below).

Economics Expanded

When economics are approached more broadly and things like sustainability, absence of debt, conservation of human culture, freedom from landlordism, group cohesion, risk minimization are admitted as dimensions of 'prosperity' it is doubtful if Sanchez' generalization would stand up. One might start with the Kantu' Ibans whose system collapsed because of *excessive* prosperity in a society not attuned to prosperity of this financial nature. Or if their swidden rattan is ignored, with Dove's (1993) note that these shifters were getting 7.9 kg of unmilled rice per labor day compared to an average of 4.2 kg in the irrigated and meticulously tended rice fields of Java.

The Tiburay

Schlegel's (1981) comparison of two groups of the Tiburay in Mindanao, Philippines, is especially relevant. He tested the assumption that settled plow agriculture is 'better and more advanced' than shifting, and since its superior technology (carabao plows -v- dibble sticks) results in greater productivity this must result in greater prosperity.

For historical reasons one group of Tiburay, 10 000 or so folk, had remained in their forests as swiddeners. The second had been acculturated and advanced on to become typical Philippino buffalo plow peasants. Part of their acculturation process (under US occupation c 1910) had been indoctrination in the concept and sanctity of private land 'ownership' and some of this group, becoming more acculturated than others, advanced on still further to become landlords of their fellows. When all this had settled down and Schlegel arrived to measure comparative economics he found the primitive forest swiddeners producing 52 *cavans* of rice per hectare while the proper farmers were managing 38 *cavans*. With corn the swiddeners were getting 20 cavans while the plow peasants were getting 19 cavans. Since it is difficult to compare plowed field grain yields with the yields on irregularly planted and mixed swiddens Schlegel repeated the comparison on a yield output-per-seed input basis and found that the swiddeners' advantage was even greater, 30% and 23% for rice and corn respectively. Such comparisons also have to take into account respective labor inputs. To get these higher grain yields the swiddeners were expending 2 500 labor hours/hectare over their one year cycle while the peasant tenants were expending twice as much, 4 970 hours – though their total annual output was presumably higher because they had to spend most of it on buying supplementary food.

This comparison is of measurable agricultural output. In addition the swiddeners still had their forest for hunting and collecting...*while the Kabakaba men (the farmers) are kept busy planting, plowing, caring for the carabao and dragging sacks of grain to their landlord's house...Their exploitation of wild resources is virtually non existent.* They had to buy their fish and meat in the market and with their forests gone there was nothing left to gather.

The Less Formal Economics of Collecting

So far discussion has concerned the *farm* component of these swiddens; but this is only one of their several activity spheres. Fleeing from formal economics to get closer to these realities, Clare Madge (1994) moved in with the Jola rice swiddeners of Gambia, West Africa. She found that the 'collecting' extends to food processing, preservation and storage. At each of these stages a large body of indigenous knowledge is employed and these stages also have dimensions of gender, place, season, purpose. As elsewhere a wide range of items is collected (specifically by men, women or both) for a variety of uses and purposes (food, materials, preservatives, medicines, trade), here from five ecozones (forest, mangroves, swamp...) according to their seasonal and the competing agricultural calendar of operations. Some collected food plants are emphasized at the height of the farm work season because they require less cooking and therefore minimize time expended in fire wood collecting. Although these are rice farmers, more than 60% of meals taken, in addition to snacks, have a significant 'collected' component. In contrast to the conventional view of economic systems being a dichotomy of farm -v- domestic activities, in Madge's opinion here and in Africa generally they are a *bricolage of interacting activities,* farming hardly distinguishable from the others. This mosaic includes such things as pounding the fruit of boabab, Adansonia digitata (later met with as *tebeldi* in Kordofan) to a flour then lightly smoking this to preserve it for up to a year, or adding its leaves to yet other foods as a preservative; and preparing the fruit of Borassus (African fan) palm as a breakfast porridge. (The situation is similar on Kandy farms: post harvest crop processing and storage can't be distinguished from routine kitchen work.)

Following Madge, we once again encounter the limits of agriculture, shifting or fixed. Aid projects directed at these societies are almost invariably concerned with the *farming* component of their economic systems – increasing crop yields, introducing more nutritious vegetables etc. But citing previous workers (Rogers 1980; Barrett and Browne 1989) Madge points out that these attempts by outside experts are often based on ignorance of the role which collected produce plays in the total domestic economy and of its nutritional quality, eg...*While NGO workers recommended villagers to eat imported oranges for vitamin C they failed to recommend local fruits which had higher vit. C levels*; and again...*In Tanzania increased consumption of prestige European cabbage lowered vitamin A intake because consumption of (superior) wild spinach was reduced* (Rogers 1980). This is partly due to the gulf between formal scientific knowledge obtained through nutrient analysis of a food under laboratory conditions, and informal knowledge of the actual and varying values of that food when it is processed, stored, combined with other ingredients in a hundred and one different ways at village level.

This devaluation of knowledge extends to collecting from the indigenous trees. Madge quotes a Jola boy...*People with knowledge of trees and how to use them are said to be backward*, which is nearly what Okafor says of trees in Nigeria (Reconnaissance) and pretty well sums it up. There is only one outstanding matter. The next chapter which goes off in search of the Origins in Africa now seems to have got it wrong. The real origins of agriculture must surely be not in those BP dates and lists of domesticated grains the archeologists have provided, but buried somewhere in this...*bricolage of women's invisible activities* (Genesis).

7.10 Energy

An alternative to $ economics for evaluating and comparing shifting with other systems is in terms of their energy efficiencies, energy produced per unit of energy expended or E ratio (Energy). For these systems some relevant parameters are: 1 kcal = 1 000 calories = 4 186 joules; 1 MJ = million joules = 239 kcals, which can be put in context as eg the food energy content of a kilogram of rice being about 3 630 kcals; the daily food needs on a person dependent on this rice as 2 400 kcals or 10 MJ; and this person expending 250-550 kcals per hour of work in growing it (depending on task and bodyweight). Some E values for different shifting systems are listed and comparative efficiencies for a range of other agrosystems are discussed elsewhere (Energy).

Table 7.8 Energy Efficiencies of Some Shifting Systems

Sarawak	upland rice	...	E	=	34	Philippines	upland rice ... E =	18
Ghana	corn/maize				25	Gambia	swamp rice	12
Ghana	sorghum				20	Gambia	upland rice	3
Shifters	cassava				60-70			

Source: Black 1971.

Example from New Guinea

When Rappaport (1971, Bayliss-Smith 1982) dropped in on the Tsembaga Maring to measure their efficiency and pester them about their joules and kcals, he found these 200 isolated clan folk just emerged from the Stone Age, 30 years previously, now brandishing their new steel axes. They admitted the energy embodied in this farm Kapital was considerable, about 15 000 kcals per kg of steel, but on the other hand it had increased their overall system efficiency by 200%. (Rappaport didn't bother them about the monetary dimension of their economy, it didn't have one. They used pigs instead.) Their community energy expenditures are summarized by task and activity in Table 7.9, most absorbed by farming polycultures of 19 plant species on 38 ha (of their total range of 820 ha), with this food output also supporting the pigs. Their GDP in energy terms is summarized in the bottom lines: a whole-system energy efficiency of 14.2, the women doing okay with an E of 20.9, the men wasting a lot of time out hunting and coming home with an E of only 5.4, but still doing better than the pigs, 3.2 (ie pig meat output/labor and food input).

Table 7.9 Energy Expenditures and Efficiency Ratios of a Maring Village, New Guinea

Activity and task	Energy input rate MJ per hour	Community* work input Hours per year	Community* energy input MJ per year
Agriculture (38 ha)			
clear-and-burn	1.27	7 340	9 330
fence plot, soil con.	0.95	4 460	4 240
plant and weed	0.62	18 960	11 740
harvest	0.49	7 340	3 580
carry	0.88	6 440	5 640
Herd, tend pigs	0.40	17 000	6 750
Hunt and gather	0.83	1 340	1 110
Totals....		62 880	42 390
Efficiency ratios			
Whole system	... E = 14.2	pigs ...	E = 3.2
crops	20.9	hunt-gather	5.4

*for whole community of 200 persons of all ages.
Source: Rappaport 1971.

Returning to Assam with Toky and Ramakrishnan (1981), their output/input data of Table 7.7 – now put in terms of energy efficiency in Table 7.10 – illustrate how the best situation from a shifter's viewpoint

Table 7.10 Energy Efficiencies of Assam Swiddens and Fixed Systems

Swiddens – fallow length	(a) 30 yr	(b) 10 yr	(c) 5 yr	(d) Terraces	(e) Wet rice
		A. Energy MJ/ha/cropped			
Inputs:	1 670	1 180	510	6 510	2 840
Outputs:	56 770	56 600	23 860	43 600	50 600
Efficiency ratio:	34	48	47	7	18
		B. Average annual outputs over whole cycle (30, 10, 5 years)			
Outputs MJ:	1 890	5 660	4 770	43 600	50 600

Source: Toky and Ramakrishnan 1981.

would be to continue swiddening (E = 34, 48, 47). It will be recalled that the terraces of system (d) were installed to advance them on to proper farming but these were abandoned. The low energy efficiency of E = 7 possible with them (partly due to their required maintenance labor as well as smaller energy output in two cases) explains why. But as society sees it, since these higher shifter energy efficiencies are achieved respectively only once each 30, 10 and 5 years, this is at great cost in terms of land wastage (the bottom line).

Amendments and apologies So far the calorific energy expended in burning (or mulching) the swidden biomass for cropping has been ignored. When this is included the picture changes drastically. Jordan (1987) offers an example from Amazonia. Based on kcals of food produced per kcal of (cultural labor) energy expended, this system had an energy efficiency of 13.9 and was close to those of shifting systems elsewhere; but it required burning 324.6 tonnes of forest to capture the nutrients necessary to obtain this crop yield and ratio. When the calorific value of this biomass input is added (at 5 300 kcals/kg of forest) the output/input energy efficiency ratio falls to only 0.005:1 or a return of 5 units of crop food energy for every 1 000 units of human cultural plus forest biomass energy expended...*The energetic value of the trees dwarfs all other inputs necessary for shifting cultivation. When their services in mobilizing the nutrients are considered the energy efficiency makes a dramatic shift for a net gain to a net loss. From this viewpoint shifting cultivation is a very inefficient system* (Jordan).

Efficiency engineers would agree. The shifters wouldn't worry as long as they have plenty of biomass to burn; to them its energy is a free good. They'd still say that the ratio of 13.9 is the more valid...and besides, their forest offers renewable energy whereas the E ratios of only 2, 3, 4... obtained by modern agriculture, especially the Green Revolution, are based on non renewable fossil fuel...But then again, what about those other social energy costs implied in the Greenhouse gases released when they burn that 324.6 tonnes of forest? Should these be charged against their system as a cost? Energy analysis, like economics, suffers from too many points of view. Perhaps a Jesuit could sort out what the shifters' efficiency ratios really are.

7.11 Forest Regeneration ... *So like a wheel around they run from old to new* (Spencer)

When the forest has been cut-and-burned (or mulched) it's necessary that it regenerates to enable a future similar cycle. If regeneration doesn't occur or is inadequate the fallow biomass contracts over time in size and quality, and the cropping phase based on this spirals towards abandonment. There are tens of millions of ha in the tropical world which are now in this process.

Forest Swidden Regeneration and its Necessary Conditions

Regeneration is possible through four main mechanisms:

1. from seed previously deposited on or in shallow soil storage;
2. sprouts from live roots left in the clearing process;
3. seed produced by on-site live standing trees;
4. new seed imported to the site by animals (birds, bats, rodents).

Clearing, seed viability and soil temperatures The first of these depends on the viability of seed dropped and held in soil storage. But the large temperature differential which exists—24-28C degrees under forests v 38-40C under cleared land—often destroys such viability. Temperatures in this soil seed storage can go to even 50C degrees after clearing. Myers (1979) notes of tropical forests that they are *little able to tolerate the abrupt and broad-scale (micro-environmental) disruption that modern man can inflict.* Specifically...*its seeds have little or no capacity to remain dormant.* His examples are seed from tropical American forests which can survive for only about 25 days in comparison with 10 years or so in temperate forests. (This comparison does not apply to many volunteer non-permanent forest species, below.) Seed viability is very sensitive to even small temperature variation from their normal level of 24-28C degrees in forest soil storage, eg the Dipterocarps a major tree group in SE Asia require the narrow range of 24-to-26C degrees (Myers). If this is almost doubled, as when the forest is cleared and left exposed, the seed will not germinate or the seedlings will die. Forest regeneration without these species, unless they are supplied by one of the other regenerative mechanisms, will result in a different botanical composition and consequently all that this means for zoosystems which might have depended on the now excluded plant species.

Effects of clearing fires This first effect of higher soil temperatures on viability of the future fallow seeds is compounded by the much higher soil temperatures caused and sometimes maintained for several days under the clearing fires themselves. Unruh (1988) on heat of swidden fires in Peru describes it as *deadly* on seed at 2cm, *extremely serious* at 5cm, and *selective* at 15-20cm. Sanchez (1989) notes that air temperature just above ground surface in a clearing fire ranges 450-650C degrees and decreases by about 100C degrees for each centimeter increase in depth below the surface. Temperature at any level below the surface also depends on the kind/intensity of the fire. Three types are shown in Figure 7.4; the moderate, heavy and very heavy/concentrated re-burn fires would respectively result in soil temperatures of about 75, 220 and 350C degrees at 2cm depth.

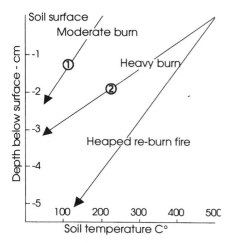

Source: Sanchez 1989, Zink, Thailand.

Figure 7.4 Effect of Fire Intensity on Soil Temperature

Seed import and clearing methods   In the presence of these two adverse temperature effects regeneration must depend on the other regeneration mechanisms. The efficiency of these is governed by the method and extent of forest clearing:

1. that these faunal seed importing agents exist (ie that these also have not been removed in the clearing process);
2. that the swidden's biosystems, as food sources for these seed importers, also continue to exist. (The importers will require some *quid pro quo* of the site for this service);
3. once the seedlings begin to grow, or the residual twigs or stumps to sprout, they must have a supply of fast-acting nutrients. A substantial part of these are provided by soil mycorrhizal fungi; but these are as much subject to the adverse effects of high soil temperatures as are the

seeds. If this nutrient source is not present only the most tolerant of the regenerating plant species (weeds) will thrive, perhaps preventing the regeneration of the others and again leading to compositional degradation of the forest;

4. there must be a *source* of fresh seed for the potential importers. This requires some area of more or less undisturbed forest in the vicinity of the site.

Deterioration over time   All four regenerating mechanisms and their enabling conditions have been present in traditional small swidden clearings in the deep forest; elsewhere they're generally breaking down. Because of population and now commercial pressures the swiddens are opened too frequently and cropped too long, their soil seed banks depleted. Seeds, ground biosystems and the funghi are cooked under the inexpert clearing fires of those 'unsettled men,' as are the live stumps and sprouts from which regeneration might have been possible. There are too many adjacent swiddens, the larger forest seed reservoir disappears. The seed importers are harvested in acts of attrition (Diversity, forests).

At regional scale, historical example of these process is suggested in Newman's (1990) review of the agroecological causes of collapse of the Maya empire, Guatemala-Mexico c 800 AD: population increase; unsustainable pressure on the swidden fields;   massive soil erosion; liquidation of the soil micro organisms exposed too long to strong sunlight/radiation and too frequently to the clearing fires *(...At these temperatures and with exposure to ultraviolet light the soil chemistry degrades and mycorrhizal and other funghi and decomposer organisms perish).* Collapse of the forest's regenerating mechanisms. Of the forest as fertility supply. Of the *milpa* food fields. Of the Maya.

Phytomass recovery rates   Where natural forest regeneration is still possible its time rate is important parameter. Saldarriaga and Uhl (1991) offer a guide to forest recovery rates after swidden abandonment (Figure 7.5) from a mosaic of swiddens at various stages of regeneration on the upper Rio Negro in Colombia-Venezuela.   Rates of (above ground) recovery decrease with time: 34 tonnes/ha by year 5; 58 tonnes by years 9-14; 150 tonnes by years 60-80. Full recovery to the condition of primary forest, containing 255 tonnes of above-ground vegetation, was projected to take between 140 and 200 years (middle curve). Recovery rates vary widely among forests, soil types. Saldarriaga and Uhl note that their Rio Negro swiddens were 5-7 times slower than some other tropical American forests; and Sanchez cites the case of a forest in Congo recovering to 175 tonnes/ha or 90% of a fully recovered secondary forest in only 18 years. Rapid recovery rates are also much preferred by power line companies, miners and developers who cite them ceaselessly as proof that whatever liberties they are obliged to take with their bits of this biome it will soon recover, in 15 or 20 years or so…or by and by.

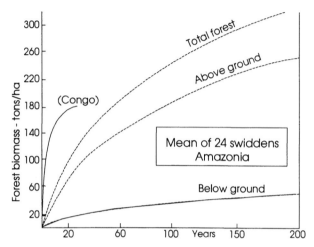

Source: Sandarriaga and Uhl 1991.

Figure 7.5   Recovery of Forest Biomass in Swiddens Over Time

Biological recovery and species composition   Phytomass recovery rates have little to do with recovery of the species composition of a forest and less with recovery of its faunal components, ie the forest as comprehensive biosystem.   Saldarriaga and Uhl found that recovery of plant species composition falls into four phases. First to appear are the grasses and forbs. Also from the first year a wave of short lived volunteer trees begins to appear but by about year 5 mortality of these exceeds their establishment rate. This wave is replaced by a second wave of volunteers which become dominant by about year 50. Finally, the permanent primary forest species which also have been slowly accumulating from the early years of regeneration begin to dominate the forest by years 40-80. The total number of tree species present keeps increasing through time. In brief, it took 50-80 years for the forest to recover *most* of its tree species. (The non tree synusiae weren't considered.) In this regeneration succession the role of the first wave of woody weeds is as forest 'healer', providing temporary ground cover.   They do this through rapid germination of their large amounts of previously fallen seed which, in contrast to the short viability of the seed of larger and permanent trees, can lie dormant and viable for many years. For this function their growth rate is very high, 3-4 meters annually. However the 'price' they pay for rapid growth is often their production of unusually soft or light wood, eg in Amazonia these species include the famous *balsa* (Ochroma). The other functions of this first wave then the following waves are to provide the necessary micro environmental conditions of shade/light, humidity, temperature, wind protection for the next wave. Whether these successions will ever result in full recovery of *all* the forest's components again depends on how many of the four regenerative mechanisms are left to perform this task, ie the length of time, intensity and geographical extent of the previous swiddening. Newman cites the case of Maya forests and others around Ankor Wat, Cambodia, farmed long and intensively to support great Princes before their abandonment 560 and 1 200 years ago, and still not recovered from that experience.

Recovery successions and nutrient yields   Recovery successions, and the time at which they are terminated by once more clearing them for crops, have implications for both the total amounts of crop nutrients obtained from them and their P-K-Ca-Mg compositions. In the Assam fallows (Table 7.5), Ramakrishnan and Toky explain these differences by the 5 year fallow consisting largely of woody weeds, with these yielding a fairly even balance of nutrients in their ash. The greater percentage of potash of the 10 year fallow was due to bamboo which by then was dominating the succession. By 30 years the large broadleaf trees had taken over and these greatly increased the supply of calcium. It is not known if the shifters actually allow for these changing *qualitative* aspects of nutrient supply in deciding when to cut and burn the fallow trees but since they seem to have though of just about everything else, in their agronomy or ecology or whatever it is, they probably do.

7.12 Evolution and Future Development

Figure 7.6 sketches how forest shifting systems have evolved in two directions in the past, towards field crop farming to tree crops, agroforestry, including forest gardens. These paths also define the alternatives for future development. (The sketch is an elaboration of Table 7.1.)

Starting with some intact forest this can be (and historically has been) modified in two directions, track (F) or (A). Track (F) results in forest gardens directly by leaving the economically useful plants and integrating others into the forest over time. Other activity spheres are also preserved (hunt, collect); this system can continue on indefinitely. Track (F) then is one of the three fountains of forest gardens. If track (A) is followed, forest conversion to swidden, this can subsequently branch in many directions. (The numbers on these branches refer to the lines of Table 7.1.) As long as a long fallow is possible (enabled by low population pressure) this also preserves the other activity spheres and these too can continue indefinitely. But this is now seldom possible (and historically became impossible) because of increasing pressure, fertility depletion, soil erosion, grass infestation, transforming the progression to ever shorter fallows and eventual abandonment.

The alternative is (and was) track (4) to a cropping phase now more strongly supported by managed, improved, enriched tree fallows; the use of trees now becomes deliberative, selective. From here, either of the system's two components – its trees or its ground crops – can be emphasized to become dominant. If the trees are followed, track (6) again

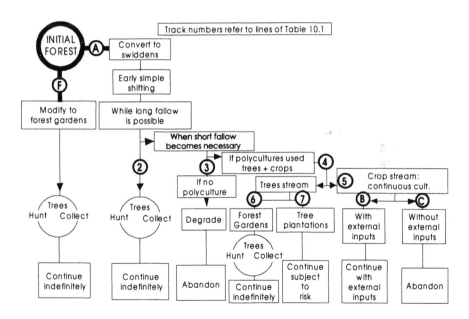

Figure 7.6   Evolution and Development of Forests to Forest Gardens –Swiddens–Tree Plantations–Field Crop Farms

leads to forest gardens (their trees now economically supported by a little ground cropping, and perhaps some localized collecting/hunting is still possible. This track (6), the swidden trees, is the second fountain of forest gardens. The other 'trees' alternative track (7), in very recent times, is towards specialist agroforestry, tree monocropping, industrial plantations, orchards, which also can continue indefinitely but always subject to risk (of disease, pests, prices).

If the ground crops have been emphasized, leading to proper farming on track (5), this will be possible only with high and continuing external inputs of fertilizer, agricides, soil conservation terraces, fences, irrigation, HYV seeds, extension support, credit ie The Package. Station (5B) is the destination to which agro developers and governments sell tickets when they promise to move the shifters on to proper farming; but globally there aren't enough resources to go round so they end up at Station (5C)...or, in terms of Table 7.1, back at Station –1.

Development Track (5) to Ground Crops as Proper Farming

'Managed' shifting One most extensive attempt at shifter stabilization was introduction of the *couloir* system in former Belgian Congo (1930s). Appalled at the shifters' impact on the forests, officials there first attempted to apply modern European agronomy, centered around use of green manure crops in place of forest fallow...*The results were disastrous, the yield of every crop fell rapidly and the experimental area had to be abandoned* (Roche 1974). The traditional shifting system was then re-studied, the scientific validity of its periodic shifting accepted, the new agronomy was incorporated into it and the new version formalized as 'managed' shifting. In this *couloir* or corridor reorganization, narrow strips about 100 meters wide and largely following land contours (to minimize erosion) were surveyed through the forest, but these also as close to an east-west alignment as possible to maximize sunlight. Farmers were allocated plots along a strip, cleared and cropped it to a planned coordinated sequence for three years then moved 'sideways' to clear and crop the next strip, the shifters shifting in unison, like a large family turning over in bed. There were enough strips in any village/planning unit to permit a forest regrowth fallow of about 12 years on any strip before the shifters returned to it. By early 1960s 200 000 households occupying two million hectares had been organized.

Unless imposed by authoritarian regimes such attempts at development have seldom succeeded; after Congo's independence this *couloir* system was abandoned. Ruthenberg (1974) suggests they are a waste of time...*Research and development efforts in this direction seem to be a waste of resources. Much more relevant is the gradual changing over to other systems...*ie to nail the shifters down to proper farming. Where

either are possible it then becomes a question of which model to offer them
– ground crops or trees, tracks (5) -v- (6) or (7), Figure 7.6.

City lights   Shifters like other folk are attracted by social infrastructure,
and in moving to be close to these clinics, wells, stores, churches, roads and
especially markets they concentrate their population in a smaller area. No
one would deny them the benefits of this development. It can be formally
planned, as with 'integrated rural development programs,' or occur
spontaneously when someone sets up a store and on this base the rest just
happens.   In either case this stabilization can have far reaching
consequences.

Fosbrooke (1974) in Botswana found the chiefs attracted by these
facilities to the extent they built permanent houses near them. As their
subject populations grew their increased food demands were met at first by
changing from swiddening hoe to ox plow and farming larger nearby areas.
But as these wore out and livestock accumulated and the local nobility were
reluctant to abandon their fine houses, poorer families were obliged to take
up their old ways and send some family members out to colonize bush
swiddens. In consequence the families were split and had to maintain two
houses. The 'town house' was occupied by old people caring for young
and school age children, and the bush swidden by work-age family
members, usually including the mothers who are the actual farmers. In
Fosbrooke's view this fragmentization and absence of mothers was
responsible for serious social consequences. The need to somehow get
water and supplies out to the distant colonies and tote farm produce back
were also detrimental to the logistics and economics of the community.
These situations face governments with a dilemma. Such development
infrastructure must be provided (or let happen); but this requires or leads to
place stabilization of the agrosystem and this might make a poor situation
worse...*Who is really more developed?   The community which can
maintain itself in perpetuity in a harsh environment, or the man who by
undisciplined land use ruins his farm in less time than he wears out his
tractor?* (Fosbrooke).

Development Track (6) to Trees

Swiddens as agroforestry   Following this alternative it is often unclear just
when the swidden trees – spared in clearing or added to managed fallows –
become so economically dominant that the system is best described as
agroforestry. One such system from Tamshiyuca near Iquitos, Peru, was
noted previously (Reconnaissance, Padoch and De Jong, 1987) and further

described by Padoch and Chota Inuma et al (1985) who call it 'cyclical agroforestry'. Here, after forest clearing a polyculture of ground food crops are planted and from the second year perennials/tree crops are added; the short term crops are phased out by 2-5 years then the trees continue producing a wide range of foods and cash crops (Reconnaissance) for the next 25-50 years. Of the seven produce categories taken from these swidden-agroforests five are completely or partly from the trees/plants of the fallow phase, ie the managed and economically enriched fallow.

The Tamshiyaquinos are *ribereños* of mixed European-Amazonian ancestry and now well attuned to the market economy. But from Padoch's description of their system it has ancient origins as...*cyclical agroforestry which bears close resemblance to and was obviously developed from the swidden-fallow agroforestry techniques developed by tribal groups throughout the Amazonian lowlands.* In the same area Unruh counted 77 economic species in these managed fallows, almost all were food trees and palms indigenous to Amazonia. The whole thing is cut down at the end of the fallow phase and often converted to charcoal before starting the cropping-tree planting cycle again, with each household operating several plots. But where land is not so plentiful it is easy to visualize these long tree fallows becoming permanent as one of the fountains of forest gardens (Track (6) of Figure 7.6).

Swiddens to tree crops and the settling of unsettled men   The evolution of agrosystems can be slow or rapid, the destination indeterminate, the pathway like a random walk. When the rain forests of Luzon's southern Sierra Madre were logged out (1957-80) their remnants attracted migrants from all over Philippines – agro adventurers, unsettled men whom Watters would easily recognize from Venezuela. By the time two of their communities were studied (Fujisaka and Wollenberg 1991) they had gone through it all. Obtaining occupancy of their plots by a variety of means (including 'grants' of these government lands by local leaders) the newcomers first logged whatever timber was left, some of them becoming plank sawyers and carpenters. They then opened food swiddens; but in the meantime for cash income they also became competitive forest scavengers of whatever they could get their hands on – fuelwood, rattan, ferns, trees for charcoal making – the planning principle being to grab it first. The next stage was addition of get-rich-quick crops to the rice swiddens, mainly market tomatoes.

At this point farm economics and ecology caught up with them. The market crops required high nutrient inputs (urea and chicken manure), paid for by the forest scavenging. Their prices also fluctuated, boom or bust.

The upland rice was hopeless – average yield ½ tonne/ha in spite of high fertilizer inputs, with high crop losses to weeds, rats, birds, insects, disease (in spite of their second largest input being agricides, returns negative ($-113/ha). The rain kept falling (5 400 mm/220 inches/year) and the soil acidity kept rising. These factors compelled progress to the next stage, adjustment of the swidden mix to crops which would tolerate these conditions – cassava, sweet potato, pineapple, banana. In this case the 'greater knowledge of the outside world' of these now settled men stood them in good stead. They then added tree crops to this polyculture until it evolved into...*diverse, multistorey, intercropped plantations* containing up to 15 tree crops (coconut, coffee, jak...as in Kandy) and 28 understorey crops.

The Cuyunon of Palawan Eder (1981) offers an example of two Philippino groups following these alternative paths (tracks 5 and 6 of the sketch) in evolution of their farm systems. These Cuyunon were shifter immigrants from Cuyo Island in the Sulu Sea where they had long existed as forest swiddeners, fishermen, scavengers of the littoral, until forced out by overpopulation and attracted to fresh land on Palawan. Eder found they had developed two sedentary agricultural streams. The first, track 5, continued as slash-and-burn farmers, but these swiddens now bush rotations on fixed-in-place small farms and growing mainly upland rice. The swidden cycle time had declined to 3 years fallow/one year crop, and with this the *cogon* grass infestation of their swiddens had become a major problem. By others' estimates its eradication requires about 10 years of fallow under forest; but the swiddeners' poverty and attempted land use intensity rendered this control measure no longer possible. Eder found cases where these farmers had gone off to hire fresh land elsewhere for their rice rather than stay home and fight the cogon grass.

The *idea* of polycultures and tree intercultures is also an ancient one among the Cuyunon who brought this with them to Palawan too. The second group of settlers followed this stream, track 6, early incorporating trees into their rice swiddens, planting them as 'orchards' *(huertos familiares)* then over time extending them to all or a good part of the swidden, replacing the ground crops. (Some also had enough land to run cattle.) Alternatively, Eder noted, the main pillars of the tree crop system were erected by planting coconuts in a grid pattern (as areca palms are used in Kerala) then gradually filling the interspaces with other trees, a process which might take 20 years. This enrichment is more or less continuous, with new species added according to opportunity and commercial prospects. On one 0.5 ha farm where the conversion had been completed

Eder found a total population of 302 trees/palms:

Mix of Economic Tree Species on a Cuyunon Farm

coconut	40	cirguela	4	soursop	4
banana	100	mango	2	alligator pear	8
papaya	3	canistel	4	tamarind	5
cashew	6	kapok	11	star apple	50
caimito	10	jakfruit	3	guava	50
mandarin	2				

The economics of the two groups are compared in Table 7.11 below. On a per ha basis the net value of produce is about 1.6 times higher, household net returns 5.5 times higher and labor inputs 3.6 times less on tree farms than on the swidden rice farms.

One reason why the rice swiddeners had smaller areas is that these were all they could manage in their struggle against the cogon grass. But factors of personal preference were also present. Some preferred to find work off their farms, including laboring on the more prosperous tree farms of the second groups (*...That their own farm lies in relative neglect may distress us but should not surprise us* – Eder). Another way of controlling cogon grass is by cattle; but remaining poor they could not afford cattle. Also, one avenue to prosperity on the tree farms is by feeding surplus produce to pigs; but the rice swiddeners, being poor, could not afford these either, so they didn't have this incentive to plant trees...

Table 7.11   Comparative Returns to Land and Labor in Two Cuyunon Farming Systems, Palawan

			Rice swiddens	Trees, mixed
Size of farm (average)		ha	0.648	5.500
Net returns:	per farm	pesos	259	3 427
	per ha of land	pesos	400	623
Labor input:	per farm, family	days	84	199
	per ha of land	days	130	36
Family returns:	per labor day	pesos	3.10	17.20

Source: Adjusted from Eder 1981; data are means of two groups of farmers.

In the end the reasons they offered Eder for not planting trees, as the others had done, are largely circular: No money – no cattle to control the cogon – must work elsewhere – no time to plant and look after trees – no food for pigs – no money. And it is so much more difficult to do these things now than before. Now there are too many people and too many cogon fires which would kill the young trees if we planted them. And wisdom comes with hindsight. Who could have known that those old plants we had on Cuyo Island, collected from around the Sulu Sea, would also interest merchants and turn out to be so valuable...?

References

Ahn, P.M. (1974) Observations on basic and applied research in shifting cultivation. In *Shifting Cultivation and Soil Conservation in Africa, Soils Bulletin* 24. FAO Rome.

Barrett, H. and A. Browne (1989) Time for development. *Scott Geographical Magazine* 105:4-11.

Bayless-Smith, T. (1982) *The Ecology of Agricultural Systems.* Cambridge University Press, UK.

Beets, W.C. (1990) *Productivity of Smallholder Farming Systems in the Tropics: a Handbook of Sustainable Agricultural Development.* Ag Bé Publishing, Netherlands.

Black, J.N. (1971) Energy relations in crop production: a preliminary survey. *Proceedings of the Association of Applied Biologists* 67:272-278.

Boserup, E. (1981) *Population and Technology.* Basil Blackwell, Oxford.

Bourke, R.M., B.J. Allen, R.L. Hide, D. Fritsch, R. Grau, P. Hobsbawn, E. Lowes, D. Stanndard (1994) *Agricultural Systems of Papua New Guinea – Eastern Highlands.* Australian National University, Canberra.

Brown, B.J. and G.G. Marten (1986) Ecology of traditional pest management in Southeast Asia. In G.G. Marten (ed) *Traditional Agriculture in Southeast Asia: a Human Ecology Perspective.* Westview Press, Boulder, USA.

Brown, K., D. Pearce, C. Perrings, T. Swanson (1993) *Economics and the conservation of global biological diversity.* Working Paper 2, UNDP-UNER-World Bank.

Charreau (1972) Problems Posed by the Agricultural Use of Tropical Soils for Annual Crops. *General Agronomy Series Technical Studies No. 9.* IRAT, Paris.

Dabasi-Schweng, L. (1974) Economic aspects of shifting cultivation. In *Shifting Cultivation and Soil Conservation in Africa. Soils Bulletin* 24. FAO, Rome.

Denevan, W. (1971) Campa subsistence in the Gran Pajonal, Eastern Peru. *Geographical Review* 61:496-518.

Denevan, W. and C. Padoch (eds) (1987) *Swidden-fallow agroforestry in Peruvian Amazon. Advances in Economic Botany.* New York Botanical Garden, New York.

Dove, M. (1993) Smallholder rubber and swidden agriculture in Borneo: a sustainable adaptation to the ecology and economy of the tropical forest. *Economic Botany* 47(2)P:136-147.

Dufour, D.L. (1990) Use of tropical rain forests by native Amazonians. *Bioscience* 40:652-659.

Eder, J.F. (1981) From grain crops to tree crops in the Cuyunon swiddening system. In H. Olofson (ed) *Adaptive Strategies and Change in Philippine Swidden-based Societies.* Forest Research Institute College, Laguna, Philippines.

FAO (1973) *Shifting Cultivation and Soil Conservation in Africa. Soils Bulletin* 24. FAO Rome.

Fleming, T.H. and E.R. Heithaus (1981) Frugivorous bats, seed shadows and the structure of tropical forests. *Biotropica: Reproductive Botany* 13:45-53.

Fosbrooke, H.A. (1974) Socio economic aspects of shifting cultivation. In *Soils Bulletin*, FAO, Rome.

Fournier, F. (1967) Research on soil erosion in Africa. *African Soils* 12:53-96.

Freedman, J.D. (1955) Iban Agriculture. *Colonial Research Studies No. 18.* HM Stationery Office, London.

Fujisaka, S. (1991) Diagnostic survey of shifting cultivation in northern Laos: targeting research to improve sustainability and productivity. *Agroforestry Systems* 13:95-109.

Fujisaka, S. and E. Wollenberg (1991) From forest to agroforest and logger to agroforester. *Agroforestry Systems* 14:113-129.

Greenland, D.J. (1974) Evolution and development of different kinds of shifting cultivation. In FAO (1974) *Shifting Cultivation and Soil Conservation in Africa.* FAO, Rome.

Jordan, C.F. (1987) *Amazonian Rain Forests: Ecosystem Disturbance and Recovery* (ed). Springer-Verlag.

Kass, D.C.L., C. Foletti, L.T. Szott, R. Landaverde, R. Nolasco (1993) Traditional fallow systems of the Americas. *Agroforestry Systems* 23:207-218.

Lal, R. (1974) Soil erosion and shifting agriculture. In *Shifting Cultivation and Soil Conservation in Africa. Soils Bulletin* 24. FAO, Rome.

Landaverde, R. (1989) *Ecology and Use of Mimosa teniaflora in the Dry Zone of Honduras.* Secretariat of Natural Resources, Honduras.

Madge, C. (1994) Collected food and domestic knowledge in The Gambia, West Africa. *Geographical Journal* 160:280-294.

Miracle, M. (1967) *Agriculture in the Congo Basin: Tradition and Change in African Rural economies.* University of Wisconsin Press Madison.

Moody, K. (1974) Weeds and shifting cultivation. In *Shifting Cultivation and Soil Conservation in Africa. Soils Bulletin* 24. FAO, Rome.

Moutappa, F. (1974) Soils aspects in the practice of shifting cultivation in Africa and the need for a common approach to soil and land resource classification. In *Shifting Cultivation and Soil Conservation in Africa. Soils Bulletin* 24. FAO, Rome.

Myers, N. (1979) *The Sinking Ark: a New Look at the Problems of Disappearing Species.* Permagon, London.

Newman, A. (1990) *Tropical Rain Forests.* Facts on File, New York.

Nye, P.H. and D.J. Greenland (1960) The Soil Under Shifting Cultivation. *Technical Communication* 51. Commonwealth Bureau of Soils UK.

Padoch, C., J. Chota Inuma, W. De Jong, J. Unruh (1985) Amazonian agroforestry: a market-oriented system in Peru. *Agroforestry Systems* 3:47-58.

Padoch, C. and W. De Jong (1987) Traditional agroforestry practices of native and ribereño farmers in the lowland Amazon. In H. Gholz (ed) *Agroforestry: Realities, Possibilities, Potentials.* Martinus Nijhoff/ICRAF.

Page, P.E. (1984) *Tropical Tree Fruits for Australia.* Queensland Department of Primary Industries, Brisbane.

Parain, C. (1942) Evolution of agricultural technique. In *Cambridge Economic History of Europe.* Cambridge University Press.

Pimentel, D. and W. Dazhong (1990) Technological changes and energy use in US agricultural production. In C. Carroll, J.H. Vandermeen, P. Rosset (eds) *Agroecology.* McGraw-Hill, New York.

Raintree, J.B. and K. Warner (1986) Agroforestry pathways for the intensification of shifting agriculture. *Agroforestry Systems* 4:39-54.

Ramakrishnan, P.S. and O.P. Toky (1981) Soil nutrient status of hill agroecosystems and recovery pattern after slash and burn agriculture (Jhum) in NE India. *Plant and Soil* 60:41-64.

Rappaport, R.A. (1971) The flow of energy in an agricultural society. In *Scientific American, Energy and Power*. W.H. Freeman, San Francisco.

Roche, L. (1974) Practice of agri-silviculture in the tropics with special reference to Nigeria. In *Shifting Cultivation and Soil Conservation in Africa. Soils Bulletin* 24. FAO, Rome.

Rogers, B. (1980) *The Domestication of Women: Discrimination in Developing Societies*. Kogan Page, London.

Ruthenberg, H. (1974) Agricultural aspects of shifting cultivation. In *Shifting Cultivation and Soil Conservation in Africa. Soils Bulletin* 24. FAO, Rome.

Ruthenberg, H. (1980) *Farming Systems in the Tropics*. Clarendon Press, Oxford.

Sanchez, P.A. (1989a) *Properties and Management of Soils in the Tropics*. Wiley-Interscience.

Sanchez, P.A. (1989b) Soils. In H. Leith and M.J.A. Werger (eds) *Ecosystems of the World*. Elsevier, Amsterdam.

Sandarriaga, J.G. and C. Uhl (1991) Recovery of forest vegetation following slash-and-burn agriculture on the Rio Negro. In A. Gomez-Pompa, T.C. Whitmore, M. Hadley (eds), *Rain Forest Regeneration and Management, Man and Biosphere* No. 6. Parthenon/UNESCO.

Schlegel, S.A. (1981) Tiruray traditional and peasant subsistence: a comparison. In H. Olofson (ed) *Adaptive Strategies and Change in Philippine Swidden-based Societies*. Forest Research Institute College, Laguna, Philippines.

Szott, L.T., C.A. Palm, P.A. Sanchez (1991) Agroforestry in acid soils of the humid tropics. *Advances in Agronomy* 45:275-301.

Toky, O.P. and P.S. Ramakrishnan (1981) Cropping and yields in agricultural systems in the northeast hill region of India. *Agroecosystems* 7:11-25.

Unruh, D. (1988) Ecological aspects of site recovery under swidden-fallow management in Peruvian Amazon. *Agroforestry Systems* 7:161-188.

Unruh, J.D. (1990) Iterative increase of economic tree species in managed swidden fallows of the Amazon. *Agroforestry Systems* 11:175-197.

Vergara, N.T. (1987) Agroforestry: sustainable land use for fragile ecosystems in the humid tropics. In H. Gholz (ed) *Agroforestry: Realities, Possibilities, Potentials*. Martinus Nijhoff/ICRAF.

Warner, K. (1981) Swidden strategies for stability in a fluctuating environment: the Tagbanwa of Palawan. In H. Olofson (ed) *Adaptive Strategies and Change in Philippine Swidden-based Societies*. Forest Research Institute College, Laguna, Philippines.

Watters, R.F. (1971) Shifting Cultivation in Latin America. *FAO Forestry Development Paper* 17. FAO, Rome.

Weinstock, J.A. (1983) Rattan: ecological balance in a Borneo rain forest. *Economic Botany* 37(1)58-68.

Weinstock, J.A. and N.T. Vergara (1987) Land or plants: Agricultural tenure in agroforestry systems. *Economic Botany* 41(2):312-322.

Chapter 8
Origins ... of Mainstream B

This chapter lights a fevillea lamp and wanders off into the long previous night looking for the origins of agriculture, but always heedful of George Orwell's reservation ... 'If one were obliged to write a history of the world, would it be better to record the true facts as far as one could discover them, or better simply to make the whole thing up?' (Essays iv). To be convincing it must find Witnesses. 'But,' protests Jean Racine, 'Witnesses are expensive and not everyone can afford them.' So pour me a drink and I'll tell you some lies.

Regardless of its apparent production success few will deny that conventional agriculture historically has had a devastating impact on the natural world, cannibalizing its own resource base of earth, water, plant, animal (and now sky, through atmospheric pollution), a process which under the pressures of increasing populations, poverty or affluence, a mindless agrotechnology, trade globalization, continues at accelerating rate. Many will agree with Chandran and Gadgil and the conservation biologists that more thoughtful and benign forms of agro land use are urgently needed and some can be found in the traditional tropical world (Nature, Kandy, Shifters, Conservation). But governments and development agencies show little interest, indeed insisting that the ancient agroecosystems and their folk technologies be swept aside as hindrance to imposition of their own 'monolithic visions of resource management.' This must be partly due to that received Western agroideology which assumes that for an agrosystem to be in some sense legitimate and worthy of furtherance it must be in line of descent from the kind of agriculture which supported the classical civilizations of the past or can be traced to the biblical world. Wheat, cows, sheep; all else must be chaos and barbarism. This chapter reviews these notions of agriculture's origins; they're largely Judeo-Christian cultural mythology. The following chapter, Genesis, ventures into the Great Forest to get closer to reality.

8.1 The Iceman Cometh

Any quest for agriculture's origins must probably be futile. Darwin warned somewhere...*it would be impossible to fix on any point where the term 'man' ought to be used,* so it should be equally difficult to say just when a recognizable agriculture emerged from previous subsistence-economic systems. Whether one is attracted to the search by curiosity or potential for low farce there is much time to work with, great leaps of it. Using it liberally Lee and De Vore (1968) in an oft quoted passage remind us that *cultural man has been on earth for some two million years (and) for 99% of that time has lived as a hunter gatherer...Of the 80 billion people who ever lived out a span on earth over 90% have lived as hunter gatherers.* So whenever it was that ManWoman tired of this shiftless way of life and took up farming they would have had more than ample time and experience in the ways of plants to prepare themselves for Agronomy. Tentatively guided by the box the facts seem to be about as follows:

Dramatis Personae

Evolutionary stage	Braincase capacity	Time frame
Smaller Australopithecus (man-ape)	440 cc	4.0 – 1.5 million BP
Robust Australopithecus (man-ape)	520	4.0 – 1.5
Homo habilis	640	2.3 – 1.3
Homo erectus (Java)	880	1.6 – 0.6
Homo erectus (Peking)	1 040	0.50 – 0.30
Homo sapiens	1 450	0.35 – 0.00

Sometime before two million years ago ManWoman's ancestors walked out on the other members of the hominid family but like them continued to subsist by chewing plants (human dentition is one proof). The general scenario is that they came out of central and southern Africa as H. erectus and dispersed to others of the six old breeding grounds of modern man – West Africa/Europe, Africa north of the Congo basin, east China and SE Asia. By c 350 000 BP they had emerged as archaic man, H. sapiens, and well before 100 000 BP they had occupied all the African-Eurasian landmass, including those parts of it now drowned, like SE Asia's Sunda plains (between Malaysia–Sumatra–Java–Borneo).

In one version of the evolutionary model which might appeal to economists and climatologists (extended from Pfeiffer 1973), man's ancestors the gorillas and apes were forest dwellers because that's where food was most plentiful; but some were forced from that earliest Eden by the push of population pressure and pull of the expanding East African savannas due to climate change. With this savanna pressure valve in place, those that stayed in the 'green caves' of the forest had no need to adjust their economic system and remained apes. Those that left had to develop a new economy – initially wild grain collecting, then their descendents

(Australopithecus 4 million BP) hunting – and with the greater mental skills demanded and slowly acquired these became H. erectus then sapiens. As Singer says of free will: they chose it this way because there was no choice.

There are many versions, reservations and extensions of this basic model. One of these would point out that H. erectus was certainly wandering around the rain forests of the Solo Valley at least as early as 1 600 000 BP as Java man. It would query that long interval between this date and the supposed date of arrival (65 000 BP) of modern man in Sahul, the once joined New Guinea–Australia landmass. Surely it would not have taken 'man' of some sort – even that dull Java fellow with a cranial capacity of only 880 cc – 1.5 million years to find his way 2 000 km along the easy stages of the eastern islands to Sahul? (Genesis).

Similar trouble with timetables underlies the reluctance of many to accept modern man's arrival in North America as the conventional 15 000 BP. Peking man was more modern and apparently a much brighter fellow than his Java colleague (box); and after coming to America H. sapiens, an even brighter fellow, was able to gallop through it north to south at the mean rate of 16 km per year. Why then, with his ancestral mental advantage and the rich technology bequeathed by Peking man, would it have taken Archaic H. sapiens 350 000–minus–15 000 or 335 000 years to figure out a way of crossing the shallow and not formidable Bering Strait...? In this immense period of time even the accident of a raft in a storm could have done it, or they could have followed bears across the ice.

In the course of time the diet of H. erectus broadened to include meat. But there is no evidence that they became primarily hunters except for certain periods. That they did is based largely on crafted axes and skinning tools and their mastery of hearth fire, more necessary to singe/cook meat than to boil plants. But this does not rule out the near certainty that their main diet continued to be plants, for which the evidence is mainly organic and most does not last. One supporting theory is that the animals they ate were only or mainly those found dead or sick (in the same way but for a different reason that contemporary African farmers commonly find it more economical to eat their expired cattle rather than kill live ones). But in time H. erectus did become omnivorous and their successors H. sapiens even more so, eating anything and everything, and their increasing mental capacity they were in time able to mount that 'hunting blitzkrieg' which Richard Leakey calls the Sixth Event.

But again this doesn't prove or disprove anything regarding dietary preference. These were economic opportunists, dispatching the megafauna where these existed and if they had to but always scavenging and always –

constrained by supply and risk avoidance – remaining most partial to the plants. Nonetheless the stereotype of *hunter*-gatherer has almost universal acceptance, for no better apparent reason that it is a convenient brief description of that economic system and in English it is easier to say that way. Carter (1977) also insists on some gender clarification. It was Ms. Early H. Sapiens who did the core activity of gathering. And later when agriculture was thought of it was she who fiddled with the plants which became vegeculture and with their seeds which became granoculture; and it was she who nurtured the young animals spared in the hunt which became livestock. He did his macho thing of bearding mammoths, but only when he had to. She disappeared into her bricolage of invisible activities. One wonders what really happened, what previous agroecosystems must have existed over this great leap of missing time, humanity's *temps perdu*.

8.2 Streams

Forest gardens differ from conventional farms in so many ways and have been intimately associated with human development for so long that it seems justifiable to distinguish them as a part of one of the two received agricultural traditions, stream A of Figure 8.1. Some of the salient technical and economic differences between these streams are listed in Table 8.1. Crops/species *per se* can't be assigned to either stream, this depends on the way they're approached and handled. Fundamental difference between the streams is in their agro-ideological approach to the natural world. Stream A grafted itself onto that world; Mainstream B replaced it, after its 'gene pool had been sounded' and its space was needed. Most of Mainstream B's species (and ideas) have been borrowed from A. The New Crops development industry is a contemporary example, a sort of vacuum cleaner that ingests good prospects and spits out the bits it can't digest, claiming new advances. But there are very few new crops of which Mainstream B can boast which have not passed, centuries or millennia ago, through the hands of *obscure and unknown men*, as Candolle and Oakes Ames have said. Offsetting this, Mainstream B can often act as a transfer mechanism enriching A, when its plantation crops imported from another region escape the fence and become incorporated as a 'people's crop' in traditional systems of local *campesinos.*

Stream A

Stream A is the oldest but its origins lie so far back in time and were protracted over so great a period that, unlike the confident attribution of

Mainstream B's origins to discovery of a few grains of emmer in some Anatolian midden, they can never be known. The essence of Stream A when it emerged was and remains vegeculture, that of Mainstream B is

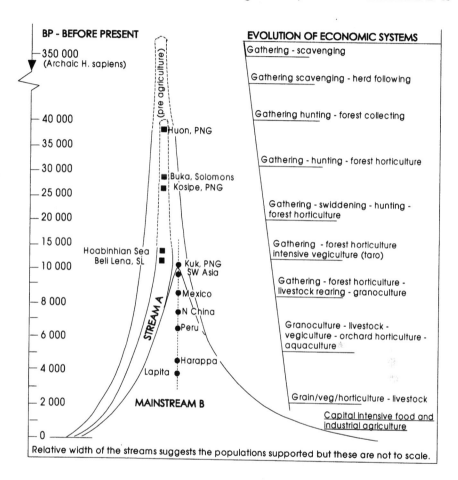

Figure 8.1 Streams A and B

granoculture. Vegecultural produce – leaves, roots, soft tissue – bequeaths little archeological record. Unless the durable artifacts which might have been associated with it can also be found – tools, vessels etc – its previous existence must be largely cultural speculation; and this itself, this stealing of the evidence by time and climate, greatly prejudices the wet tropics as repository of the Grail, and misdirects that quest towards Mainstream B and the biblical *wadis* of Jericho. In the course of time the practitioners of A did manage a few simple skills – yam planting, protective cultivation, plants to pound for fish stunning, edible insect husbandry, a pharmacopoeia – but their problem was they were always confusing agriculture with their wider economic system, the natural world. And it is well known (Zohary 1986, 1994) that the intricacies of horticulture are such that whatever they did with plants could not be considered proper farming. That had to wait for domesticated granoculture. Nonetheless the more pompous of the elders began to refer to their system as Stream A. Of course it was never very far from the folk ways of Archaic H. sapiens...nor all that far from H. erectus either. Ethno curiosity. Obsolete. Irrelevant in the booming Elmer Gantry world of Mainstream B.

Table 8.1   Some Distinguishing Characteristics of the Agroecosystems of Streams A and B

Characteristic	A	B
Dominant culture	vegeculture, trees, roots	granoculture, cereals, oil
Crop production	polyculture, mixed, random	monoculture, specialized intensive
Enterprise diversity	very high	very low
Climatic risk	very low	very high
Soil impact	erosion low	erosion high
Ownership	family, clan, not recognized	individual, corporate
Orientation	subsistence	commercial
Operating objective	survival, sufficiency	profit maximization
Planning horizon	long, indefinite	short, medium
Farm inputs source	internal, generated	external, purchased
Farm outputs disposal	internal, used, cycled	external, sold
Energy source	internal, labor, solar radiation	external, fuel, purchased
Energy consumption	low, very low	high, very high
Livestock (if any)	mixed, multipurpose	specialized, commercial
Pre-existing ecosystem	modified	replaced

That's a view from the research station and commodity exchange. One does not have much of an idea how large stream A might now be. There are thought to be about 3 billion subsistence/traditional farming people in the world, farming 60% of its cropped area (Shand 1997, FAO data). One of the largest such groups would be the forest/savanna shifters usually counted at 300 million. The hunter gatherers proper (though these are usually also part time farmers) account for only around 0.001% of total population. Others would be the littoral and river fisher-scavengers, the sea gypsy-sometimes-farmers of Moluku and similar folk of no statistical significance, like the ever-moving camel people of Somalia and the yak herding–buckwheat growers of Haa. Available estimates offer little guide.

World Resources Institute puts the shifters and hunter gatherers, still largely dependent on the forest in India/Nepal/SE Asia alone, at 600 million.

Mainstream B

Since its emergence at Jarmo or Ali Kosh or wherever Mainstream B has become so dominant it now practically defines agriculture, though curiously it has also reduced the scope of this agriculture to a minute fraction of what it was and is in stream A. Historians with an eye to the geometry of the sketch will interpret Mainstream B's expanding legs as a series of inflexion points, each enabling greater impact by the next. One of the earliest of these would be the Will and Testament of that departed Egyptian agro-entrepreneur which brags that by the time he was ferried across the River he had accumulated 834 oxen, 2 234 goats, 974 sheep... According to the time scale of the sketch that would have been about 7 000 BP, the Middle Kingdom, with the necessary Mainstream B technology if not the acquisitory habit inherited, most say, from Sumeria. Moving down the time scale they'd identify Hesiod's *Works and Days* as another, particularly its references to the impoverished valleys of Greece, as indicator of stream B's progress in the ancient world by 2 800 BP. This agro-devastation prompted Hesiod to issue his guide book calling for a more restrained and careful husbandry, reinforcing this with his plea for a return to social righteousness. By then the two had diverged, and in spite of Hesiod's book and his pleas and final prayers to Zeus they were seldom to unite again.

Search for these great moments in Mainstream B's procession would also turn up Mago of Carthage c 2 300 BP. His economics inclined treatise is the essence of Mainstream B, foundation Principles of Agribusiness – large estates, professional management, Berber sharecroppers (slaves if Berbers were not available), specialized export crops, vertical integration with Phoenician trading houses and shipping lines...the whole nine yards, and all this claimed by Mago as 'scientific farming according to the most modern Greek ideas'. (Apparently Mago had not sat down with Hesiod.) This claim of a 'scientific' basis is possibly the first made on stream B's behalf; it has been repeated ever since. Rome, until then a land of small farms, approved so highly of Margo's Carthaginian methods that the senate had his work translated into Latin and urged it as a model to solve Rome's ills, but it only hastened Rome's decline. No one now cares too much. About all that's remembered is that Mago's Carthage was destroyed as Cato had kept insisting that it should be, and that Mago's Mainstream B eventually destroyed much of North Africa and the Mediterranean basin, and Mago is long dead, but his Principles of Agribusiness elsewhere flourish greatly.

Some crops and agrosystems are dragged from camp A to B quickly and without much ceremony. Weinstock (Shifters) has described how the Dyaks and their rattan were enticed by greed into camp B and what happened to them there in only 15 years or so. Other rattans (Draecaena propinquus and micracanthus) would have been captured too had the Greeks known about them, their seed yielding that ancient and persistent marvel Dragons Blood, though it is now used for oriental medicines and not the coloring agent (D. draco from Canary Islands) which turned Greek housewives on. These also come from the Borneo rain forests – or did, before the Pharaoh burned them down. Only fools like Hesiod might mourn them. Hesiod would mourn the buffalo too. When whites arrived in North America there were 60 million bison, a complete economic and agropastoral system in itself. By 1910 there were 300 left, most slaughtered by Nation Builders in just 10 years 1865-1875 on the Great Plains as pests and for their hides and bones as fertilizer, replaced immediately by cows then Donald Lathrap's monstrous corn and hogs, perhaps the most rapid and complete 'adjustment' of an economic system in history (and of its people).

Other agrosystems made the transfer to B millennia ago then changed their minds. One of these is taro, operated with such intensity in Kuk Swamp, Papua New Guinea, 9 000 BP that it there formed the main pillar of Mainstream B (Genesis), advancing Kuk on to a veritable frenzy of land clearing, leveling, draining, irrigating, weeding, pest chasing, all this to counter those universal banes of monocropping, pest infestation and soil impoverishment. Disciplined this way the taro needed an astonishing 3 000/ha hours of labor; and the purpose of this mad activity seems not to have been food production but accumulation of wealth, in the new currency of pigs, and such wealth or lack of it then created social class divisions...All of which makes Kuk Swamp 9 000 years ago undeniably a fountain of Mainstream B. How Karl Marx would have delighted in Kuk Swamp and its taro and its pigs and its energetic bourgeoisie had he but stumbled on them in his search for the roots of Kapitalism.

Plantation crops   More recent inflexion points would have to include Renaissance Europe's discovery of Walter Prescott Webb's Great Frontier of the new worlds, with these providing the wherewithal for her later Industrial Revolution c 1750, and this leading to almost every subsequent expansion event from the acquisition of black labor out of Africa to work the new plantations, to the US Morrill Act of 1862 (to bring back a little sanity as Hesiod would have wanted it), to refrigerated ocean transport (1880s) and Vavilov's urgent journeys of just yesterday. The scale of the graph is too small to illustrate the great expansion in Mainstream B which began occurring with its acquisition of a plethora of beverage/medicinal /condiment/industrial crops, particularly the tree crops. Intensifying through the 18-19th centuries, hundreds of new expansion points began appearing and made possible by Western crop science of a sort, and the legal fiction of *terra nullis* – that the Great Frontier was empty and unowned, and often the scientific fiction that these New Crops, these unknown corners of the Garden were, like their ignorant custodians, sorely in need of domestication. Opening this era, among the first of these were the spices of Ternate...*When the Dutch established their influence in these seas and relieved the native princes of their Portuguese oppressors, they saw the easiest way to repay themselves would be to get this spice trade into their own hands. They adopted the wise principle of concentrating the culture of these valuable products in those spots only of which they could have complete control...Banda was chosen for nutmegs...Amboyna was fixed for cloves* (Wallace, 1869, though he does not inquire too deeply how this was done and steers clear of the God-fearing homicidal Governor Coen).

There are too many other moments in Stream B's progress to contemplate. Representative would be Kew's abduction and brief but intense affair with Brazil's rubber (1876) before passing her on, well pregnant, to Peradeniya and Singapore; and Clements Markham's pursuit of cinchona. (...*Towards the end of 1859 I was entrusted by Her Britannic Majesty's Secretary of State for India with a commission to secure seeds and plants of the Red Bark Tree, and I proceeded to take the necessary steps for entering on this performance*...CRM, India Office files, 1862.) Some, discounting malaria, might now think lightly of Markham and his C. succirubria and Charles Ledger's C. ledgeriana as serious inflexion points in Mainstream B's march. They might sympathize more with the perils of Lena Batu...*The nomenclature of Ceylon and Java types (of citronella) is misleading. Both originated in ancient Sri Lanka as descendents of the wild mana grass. One, C. Nardus mahaparingi, is thought to have been the first type cultivated in historical times. But in Sri Lanka after 1880 it had caused such soil impoverishment that it was progressively replaced by the second type C. lenabatu...However the Dutch had previously taken both types to Java, discarded lenabatu, improved mahaparingi, and this subsequently spread throughout the East and to tropical America until in 1977 it returned to its Sri Lanka homeland as 'Java' citronella, and possibly now because of its superior commercial attributes it will displace lenabatu* (Peradeniya crop records, McConnell 1986).

Moving down the time scale historians might find it was the recent raising up (1920-1950) of that galaxy of monocrop and monodimensional Research Institutes – for tea, rubber, cacao, sugar, coconut, abaca, pineapple, rice, oil palm, cows...which in the tropics finally lit the table for Mainstream B's cannibalization of Stream A, and with this out of the way, for its consumption also of the natural systems of the East.

8.3   Perceptions, Definitions ... Preparations at the Caravanserai

Seekers of the origins can attach themselves to any of the three caravans posted Time, Place and Circumstance. Interline transfer is easily arranged and most who start off asking When and Where end up asking Why. They should bring along their own handbook. The thicket of definitions and conceptual models left by previous travelers is now so dense it threatens to obscure the Grail. The field is rich in disputation, much of it of a Scholastic angel counting kind.

The Formal Conditions for Agriculture

The orthodox position is that three conditions are necessary for proper agriculture: the development and use of *domesticated* plants (or livestock), their *planting-cultivation* or tending in a systematic way as crops, and this occurring in a limited area spatially differentiated by system boundaries from the rest of the natural world as a *farm*. The problem is that these conditions have largely been laid down by modern Western (agro)science – botany, genetics, zoology, agronomy etc and farm economics. When applied beyond their specialist range, and especially when used for uncovering and adjudicating on human subsistence systems of more than ten thousand years ago, these conditions are often irrelevant. Indeed the possibility of earliest systems is pretty well defined away. The seekers ask for universality as guide. Science gives them carbon dating and lectures on polyploidy. But the Quest must always be a journey of the mind.

Crops and domestication   Harlan (1975) defines a *crop* as plants which might range along the agroevolutionary spectrum from wild (undomesticated) landraces at one extreme to those that have been fully domesticated at the other. He defines *domestication* as the evolutionary process operating under human control. It is achieved when plants have been altered genetically to be of greater service to man. The final result are plants which when fully domesticated become completely dependent on man for their survival. Lathrap (1977) calls such plants genetic misfits. He also says the real situation is that plants have domesticated man; and in the example he offers of the US Midwest and its farmers shackled to their '*monstrous corn*' (and regional economies shackled to corn growing farmers) he is probably right.

David Rindos (1984) distinguishes three stages in general domestication. *Incidental* domestication occurs in nature, in a pre-agricultural environment in which the plant takes advantage of man's presence to change morphologically and become 'more fit'. Animal and human behavior towards the plant will be unspecialized but to some extent consistent: predators will come to value some morphs as food source relative to others, perhaps even protecting them from other predators (as many ants do). The ultimate development is *an obligate relationship between man and plants.* In Rindos' view acknowledgment of the processes and relationships of this first kind of incidental domestication is extremely important for an understanding of the origins...*they are the very basis, the ultimate cause of agriculture's origin, elaboration and spread.*

Specialized domestication then involves intensification of certain tendencies already present from incidental domestication; more important, it involves change in the behavior of the plant's *agent*, of man, and in the basis of his subsistence strategy. The third stage, *agricultural* domestication, is the culmination of the previous processes and in its highest form is practically equivalent to modern plant breeding in which the breeders are...*aware of at least some of the processes that control the evolution of plants and attempt to take advantage of them.* Rindos' scheme is important from the viewpoint of forest gardens (and Stream A generally) because it bestows on them a scientific legitimacy they otherwise are thought to lack. In many of these systems modern accelerated *agricultural* domestication hardly exists. That which does is closer to *incidental* domestication and operates largely through alterations to the plant's micro environment, accidental or deliberate, though this is a few steps (or many) in advance of what happens through man's casual impact on some natural ecosystem such as a forest. Piperno and Pearsall (1998) offer the example of West African forest yams quickly changing their shapes from long thin tubers to the familiar globular ones simply by being repeatedly harvested and replanted.

In spite of attempts to liberalize thinking about just what domestication is, there's a growing impatience with the whole thing – with domestication however defined, with 'cultivation', even with the concept of 'farm'. Matthew Spriggs (1993) has called it *fetishism*, probably because these concepts – especially 'domestication' – are unhelpful to an understanding of the economic systems which existed in the Melanesian and other rain forests 20-, 30-, 40 000 years ago. Spriggs also points out that the processes leading eventually to full domestication (ie when the plant is entirely dependent on man, Harlan above) are cumulative. At what *point* in time can a plant be considered domesticated? In this he would get support from Yen who earlier had noted that many of the important Pacific species had still not been genetically fixed, fully domesticated, and anticipated great difficulty in using technical criteria to categorize degrees of domestication.

Consider the sago palm Metroxylon in Seram (Ellen 1988) which provides most of the human food energy there. About half this food is from wild thorny palms, half from planted suckers of non-thorny types and to this extent they might be called domesticated. Both types are treated the same and have about the same yield. Is this difference sufficient to describe one part of these plantations as 'farms' and the other as wild groves? – some palms as a crop but others not? What would be the practical significance of such differentiation? *The differences between cultivation, incipient cultivation and non-cultivation are quite indistinct* (Ellen).

Or the *tebeldi*, boabab, Adansonia digitata which in Sudan is not only a tree but a micro system in itself (McConnell 1986). In the wet season the seasonally nomadic clans move north through Kordofan and Darfur, going as far as they can with their cattle and goats then with the end of the wet season and its pasturage they turn and retreat as the country again dries up. In its march north the clan will find a tebeldi (or several) which they have previously developed. If it still contains water they will pause there and refresh their stock from it. If it is empty they excavate a shallow earth reservoir around its base, wait until a storm has filled this and after watering the stock they use the rest of the water to fill up the tebeldi. Sometimes this takes several storms; and while they are waiting the clan might also grow a crop of fast maturing melons at the margins of the small reservoir.

If they find an unclaimed tebeldi on the march they convert this to a water tank too, by cutting a man-hole in the top of the trunk, scooping out the decaying wood to enlarge the hollow trunk, filling this with storm water, then replacing the lid after they have cut the clan's mark on it to indicate ownership. In this way if the season to the north turns adverse the several tebeldi offer a safe line of retreat. If it is good the march south can be leisurely, the tebeldi tanks offering stock and domestic water points and even opportunity for a little more melon farming along the way. It is true that the tebeldi offers little conventional agronomic product, only edible seed, camp fuel and tough inner bark for making livestock ropes and cordage (to the disadvantage of Nile hippos), and in that *tebeldi* is not selected nor planted nor cultivated nor in any way maintained it would not be considered from a scientific viewpoint to be domesticated. But in a more realistic sense, since its water storage (and micro-irrigation) functions provide the basis of semi-nomadic life in these places, and have done through time immemorial, the gaunt tebeldi must be counted among the most domesticated of all the plants. (Inconceivable a few years ago, it's now been counted by IUCN in its Red List of endangered plants.)

Insistence on formal domestication as a condition for the official arrival of livestock husbandry (cattle, goats, Jarmo and Agissa c 9 000 BP) seems particularly absurd. Specific-and multi-purpose animals as pets, food, clothing, scavengers, hunting partners, exchange commodity, guards, friends, have been part of the human *menage* as long as this has existed, with Ms Early H. Sapiens commonly suckling their young, as Ms Modern H. Sapiens still does with small pigs in New Guinea. And you can't get much more domestic than that. These relationships extend to reptiles. In Sri Lanka the cobra is often brought into the household (the literal meaning of domestication), or more precisely into shops and grain stores and allocated a jug or box as daytime sleeping quarters and occasionally fed tidbits, in exchange for her night services in rat control and anti burglar device. This symbiotic relationship between merchant and serpent has existed for centuries; but it's always a careful and tentative one, and not too clear whether it's contributed much to development of either species.

Planting and cultivation Similar problems undermine this second formal requirement, 'cultivation', all the agronomic steps in establishing and caring for a plant, from ground preparation through planting, cultivation, weeding, watering, fertilizing to harvesting. The concept is clear when it refers to weeding a patch of corn or fertilizing a stand of coconuts. These operations are directed at the subject crop. However there are other indirect kinds of cultivation. One has been discussed above on forest gardens (Kandy) where because of the heterogeneous and unorganized botanical mix it is impossible to cultivate one species without also positively or negatively affecting the others. It is the near environment which is cultivated – if at all – not its separate crops. A second kind of indirect cultivation in a forest environment, in prehistoric and present times, is to intentionally or unintentionally modify the micro-environment of some subject plant. Gather-hunters do more than simply collect: they also encourage (or slash down) the bush around the plant to maintain a favorable micro-environment of shade/sunlight, wind protection, warmth, ground litter/humus etc. Even in cutting or maintaining an access path to a tree the gatherers will incidentally alter these physical conditions by opening up the canopy. Once at the plant the first thing they will do is clear a collecting area, a safe working area if some processing of produce is intended (eg atap weaving). All these activities have agronomic effect, often deliberate. A contemporary example is the harvesting bush tea tree leaf in Australian forests. One important task is to slash down the competing vegetation which might otherwise shade out and kill the defoliated tea tree. (Official forestry policy frowns on this old practice because it also causes compositional changes in the forest.) For examples of more advanced non-or pre-agronomy Harlan (1975) offers Steward's (1941) discussion of gatherer-hunters in Nevada scattering grass seed after rudimentary land preparation, and other North American groups constructing extensive irrigation systems to water natural vegetation. Tindale (1974) has described the non-farming Aboriginal people in Australia constructing temporary dams to flood irrigate their (grassland) grainfields, and doubling the carrying capacity of the country by these 'non-agricultural' practices. These also have never been considered farmers, although in Tindale's view...*many of the activities of northern Australian people were akin to those associated with the earliest garden cultures, lacking principally the idea of deliberate preservation and sowing of seed.* Yen (1988) has applied the phrase *agronomy of hunter-gatherers* to this type of system in prehistoric Australia (in places continuing into the 20th century). Other examples are the use of localized fire on cycads to 'even out' their flowering and fruiting patterns, and to induce extended stands of selected grass species for grain; and the continuous 'cultivation' of natural yam fields as a simple result of digging up the mature tubers. Yen says of these indigenous economies...*The element of plant domestication is absent from the Australian hunting-gathering systems but the domestication of the environment in which these plants grow is not.*

Farms and farming Definitional problems also beset this third formal condition; Bennet Bronson's (1977) is probably representative...*true farming (is) committing one's resources to the establishment of an artificial ecosystem to yield a staple food supply (and) filled with subtle risks and calculations, but small-scale non-staple cultivation is elementary.* While this is a reasonable description of how Mainstream B sees itself in comparison with Stream A many will question its general validity. 'Non-staple cultivation' is far from elementary (Nature, Shifters, Diversity). Many traditional farms, particularly agroforestry systems, structure themselves to most closely resemble natural ecosystems which are by definition extremely complex as are the farms which mimic them (economically as well as biologically – Chandran and Gadgil; McConnell and Dillon 1997).

Summary These three concepts of domestication, cultivation and farm are of vital importance; they will determine what the seekers are looking for. They will lead to Ali Kosh and Jarmo, and the Three Rivers and the high barrancas. But they won't lead to the Grail.

8.4 The Caravan of Circumstance

Based on abundant direct archeological evidence (plant remains, artifacts, settlements), increasingly supplemented by proxy 'evidence of varied character' (Candolle, below), consensus is that the first signs of formal agriculture begin to appear almost simultaneously (Near East, Africa,

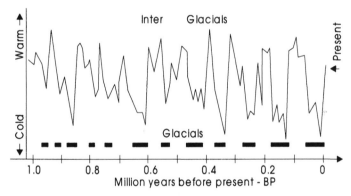

Figure 8.2 Glacials (Ice Ages) and Interglacials of Past Million Years

China, Americas) about 10-8 000 years ago (BP). Explanations of the socio-technical circumstances which led to the sudden emergence then of food production (farming) replacing procurement (scavenging, collecting, fishing, hunting) are innumerable but most are music orchestrated around two themes, continuing human evolution and climate change. By the first (not universally agreed), although modern H. sapiens had emerged by 350-300 000 BP they became thoroughly modern only about 120 000 years ago with final reorganization of the human brain. By the second the greater cultural and technical abilities could not be fully exploited until the Last Ice Age had passed its maximum (c 22 000 BP) to terminate completely around 11 500 BP. The (geological) Pleistocene Epoch had lasted approximately

1.0 million years and corresponded to the period when H. sapiens were emerging. It had consisted of some 12 cold or glacial periods alternating with warms or interglacials (graph), though with many shorter peaks, troughs and false recoveries to the warms. After the usual flickerings or rapid temperature oscillations and one of these false starts (followed by a sudden plunge back into ice age conditions, the Younger Dryas) the present or Holocene Epoch warm began suddenly, in only 30 years or so (Bryant 1997), one circumstance suggested for the 'suddenness' of agriculture's emergence in this early Holocene warm.

The Last Ice Age (the Wisconsin in North America, Würm in Europe) lasted approximately   50 000 years from 80-30 000 BP and the interglacial warm which preceded it from 130 000 to 80 000 BP. The ice ages differed (graph) but typically lasted 40 000 years and had far greater impact on the northern hemisphere than on the southern. Taking the last as representative its continental ice sheets covered much or all of North America, Greenland, Europe, Siberia, Tibet-North China and elsewhere isolated mountain chains, while in the tropics and south only a few regions (apart from Antarctica) were glaciated, the mountain chains of New Zealand and Andes-Patagonia. The ice sheets also locked up immense volumes of water which lowered global sea levels by 100-130 meters which globally exposed 20-25 million sq km of land; and they also set up strong high pressure wind systems which altered global rainfall distributions.   In brief the glacials are associated with cold temperatures, droughts, animal and human retreat, high winds and dust deposits from these winds (which has provided one major way of measuring the glacials in North China wind-blown loess/silt deposits).

Conversely the interglacials, specifically the present Holocene, reversed those conditions, raising temperatures, lifting sea levels by 100-130 meters and drowning 20-25 million sq km of continental shelves, coastal plains, estuaries, river valleys. Among these drowned areas are Sunda (the once plain between Sumatra-Malaysia-Java-Borneo) and Sahul (now the Arafura Sea, New Guinea-Australia), the now shallow Palk Strait (between Tamil Nadu-Sri Lanka), Peru-Ecuador coast, Mexico-Guatemala Gulf plain, Amazon estuary…all places where H. sapiens must have been and now largely beyond archeology's reach.   Human consequences followed warning and altered rainfall regimes: movement of the large herd animals north following their 50 000 years of accustomed cooler climate, expansion of other rich game resources on the northern grasslands which the ice sheet had left behind, and increasing mobility and skills of the human bands which followed them. Subsequent interpretations are social and economic: increasing human populations applying ever-improving technologies to the changing natural world until familiar *procured* resources were depleted and had to be replaced by production, ie early Holocene farming, varying only in local circumstances (eg the Flannery models below).

What if...? This is much a Northern Hemisphere interpretation of agriculture's incubation and it sits well with archeological discoveries there. But a southern model is also possible. This sees the Last Ice Age, high drama in the north, as almost unremarkable event. Much depends on glacial-interglacial temperature differences, their geographical extent and the kind of economic system which was actually followed and here there's no clear consensus. In one view the tropics also experienced climate change but these effects diminished away from the land mass ice sheets. Glacial cooling impacted on the continental Great Forests of Amazonia and Central Africa, changing their botanical structure and opening their canopy and in Africa diminishing an area (now savanna) which never recovered; but forests were nonetheless able to survive in small corners and re-emerge with the return of warmer temperatures and higher rainfall (Wilson 1992). Bryant (1997) has described this relative immunity in the true tropics...*Temperature, evaporation and rainfall remained high in the tropics but within a restricted band around the equator;* and in support of this Anna Roosevelt (1999) cites recent archeobotanical evidence that the forests of east Brazil before 11 000 BP were denser than they are today. Bryant further calls attention to the paradox of some tropical oceans actually *warming* by as much as 2.0 C degrees during the last ice age. In one of these ocean corners, the Coral Sea to 15 degrees south, temperatures remained within 1.0 C degree of present levels. In another nearby corner (Philippines Sea), Hope and Golson (1995) cite estimates that this El Niño 'warm pool' was only 2.0 C degrees cooler than now even at the depth of the last ice age. From this it can be assumed that the *maritime* forests of the S-SE Asia-Melanesia littoral and probably elsewhere, and the human economic systems associated with them, were not much affected by the drama to the north.

The northern model of agriculture's early Holocene incubation also depends much on H. sapiens following large game resources then reacting to their scarcity, the economy progressing from hunting to wild grass collecting to grass domestication as cereals. This is not the southern or tropical forest tradition. The model here is H. sapiens as opportunist: littoral scavenger, fisherfolk, forest collector, later vegeculturist. This does not lead them suddenly to early Holocene proper farming: it gives them tens of thousands of years of so-called Last Ice Age to work towards it.

That some did so is clear, for example from taro starch on stone tools at Buka Island (Solomons) from 28 000 BP (Spriggs 1993) ie trough of the ice age. So why not much earlier, 40-50-60 000...BP? What if one day someone had thought to kick a yam back in its hole and seen it flourish, instead of casting it aside...then planted two yams? Would they have been farmers this long ago? What if the 400 000 year old tools in India are seen as leading somewhere, to an agriculture in the south, safe from glacials? And where better to have done this than on the now drowned plains of Sunda or Sahul...or Tamil Nadu and Lanka? (Genesis).

The Origins as Granoculture

Kent Flannery's (1973) model concerns emergence of grain farming, core of Mainstream B. The actual circumstances of its origins were different in each of the main hearths(Near East, North China, SE Asia, Mexico-Meso America, Peru-Andes) but the basic principles apply to all. Central of these is Binford's (1968) earlier thesis that agriculture arose as response to disruption of some preexisting social-physical-economic equilibrium. 'Hunter-gatherers' would have stayed that way until the equilibrium of their population (food demand) with their physical environment (food supply) was disturbed by either an increase in population or its geographical distribution or by a change in physical environment leading to decrease in food supply or change in the kinds of foods available.

Near East granoculture   In the Near East such environmental change occurred as a result of and recovery from Last Ice Age last glaciation after c 20 000 BP and there is also archaeological evidence there of changes in population distribution. Climatic-environmental crises attributed to previous glacials had been met by migration; but by the end of the last H. sapiens had occupied all the Afro-Eurasian land mass and emigration was no longer a general option though it remained a local one. (Redistribution of population pressures from this last glacial are one explanation of H. sapiens finally packing up in the Gobi desert and moving to America). The two remaining options were the old one of population control through infanticide and senilicide and the novel one – the New Crops option as it were – of exploiting a range of previously neglected foods, particularly the wild grasses as grains/legumes, emmer and einkorn wheats, barley, lentils, field peas, flax, vetch and the weeds that accompanied them such as oats and other crop weeds which later became valued pasture species (ryegrass) after livestock had been concentrated on farms.

The technical circumstances (and some gender corrections) In the Near East the technical circumstances responsible for the emergence of granoculture have been canvassed by Halbeck, Zohary, Hopf, Harlan, Flannery, Bender and many others (references) and need only be summarized. In this nuclear zone (approximately Tauros mountains of S. Turkey through the Zagros mountains into Iran) some of the natural stands of wild wheat and barley and many other grasses are as dense as if they had been cultivated and yield 500-800 kg/ha (Zohary). Harlan (1967) found that a family of four could have harvested about 1 000 kg in three weeks, enough for their year's supply. But these wild grains are available only within a short seasonal time window between maturity and their grain heads shattering and falling to the ground. In the worst seasons this was only a few days. This problem could be overcome by two strategies. The first would have been to select among the many grass/grain strains those which had a tougher rachis from which the mature grains did not quickly fall (domestication); the second was to translocate these selected strains to damper areas in valley bottoms, along stream banks, where the adverse effects of rainfall variability would have been minimized, this initiating cultivation. The third leg of farming was put in place by transportation in wicker baskets (of which a few fragments remain from 10 500 BP) and central grain storage, usually in underground holes plastered with lime (from c 11 000 BP, especially in Israel). The actual sequence was probably different: H. sapiens would have needed baskets anyway to carry their things around and grinding/storage was used on wild grains millennia before these were domesticated. In any case, with all this Kapital to guard as farms H. sapiens' roving days were over.

These technical accounts are always gender-neutral which encourages the assumption that 'early man' means exactly that. There is now increasing reasoned speculation that in initial preagricultural economic systems human bands as primarily scavengers-collectors had moved across the landscape in unified extended family groups or clans, without gender distinction among activities. When and where hunting was added to the economy such specialization emerged, the women and girls continuing their grass grain collecting which became sedentized on 'farms', the men and boys continuing their hunting but from an ever contracting circuit of base camps around the (women's) farms as game became depleted and/or the produced farm supplies increased in importance/reliability relative to the procured game, culminating in the more amenable game species being added to the (women's) crop fields as livestock. But as fields became depleted in soil fertility or weed-infested it would have been necessary to shift them around (Shifters), and necessary to re-integrate men into the system to do the heavy work of land clearing and later drainage and guarding against animals (and other predatory bands). Such agriculture was initially women's invention.

This is for the classical hearths of granoculture. In wet tropics arboriculture where there never was a hunting activity of comparable importance, or need for male specialization in heavy farm/development labor there was probably never such male/female activity specialization. On contemporary forest gardens, 'field' is essentially the same thing as dooryard garden plot, 'crop' the same as family food supply, 'packing shed' the same as kitchen, woman economically the same as man. As Stearman later says of technology, it is not destiny. Neither is gender. It became destiny, artificially, through imagined or forged religious sanction – purdah restricting inferior females to one set of economic activities on one part of the farm; or through male ego which reduced females to objects of wealth display, or mutilated them – the tittering bound-feet songbirds of Chinese nobility – as evidence of farming prowess and household conspicuous consumption, waste affordability.

The origins, shifters and the marginal lands   In Flannery's model grain *collecting* would have continued on the better lands simply because there was so much to collect. But increasing population, itself made possible by the wild grass granoculture, would have pushed the excess population out onto the marginal more arid, stony, less fertile lands. Pressure on the wild grains there would have led soonest to selection of those strains most amenable to domestication and to their planting. Production began to replace procurement. Although Flannery's model does not include them here or in the Mexican *barrancas* (below), at this point also we would probably meet the Shifters. Whether they were marginal or not it would have been necessary to clear these first fields of competitive vegetation for the selected grass-grain crop; and if they were 'marginal' with respect to fertility or susceptibility to weeds, bush regrowth, it would have been necessary to move these fields around. These first grain farmers were almost certainly shifters, as were the first vegeculturists (manioc – South America below).

The Origins as Tropical Vegeculture

To paraphrase Tim Flannery (1994) on the origins of New Guinea agriculture as vegeculture – *To propagate these plants one simply needs to grub them up and stick some part back in the ground, and this simple act has been part of the human repertoire for 100 000 years or more. Clearly it does not constitute a farm or farmer. But what if this person were to return to the plant occasionally and weed it? – Or grow ten taro plants instead of one? – And fence them in? At what point would farming begin?*

To construct his model of the vegeculture origins David Harris (1969, 1971, 1972, 1973, 1988) goes back to the circumstances of the *kinds* of initial ecosystems on which agriculture was superimposed. These form an array according to their degree of specialization or generalization. Specialized natural ecosystems are those with a small number of plant/animal species but these have large populations, ie they have a low diversity index or ratio of number of species to number of individuals. At this extreme are deserts and sub-polar habitats supporting only mosses and reindeer, but many of them. Less specialized are grasslands of many species and the animals which they support. Further along the array are grass-woodlands with much more diversity, and most generalized of all are the tropical rain forests with highest species diversity (and also stability) but with small populations. Societies in highly specialized ecosystems developed specialized economic systems; those in generalized ecosystems developed diversified systems. With a wider spectrum of resources these latter would have moved less frequently than did the gatherer-hunters of specialized ecosystems, they would have had more opportunity to gain deeper knowledge of plants and are more likely to have been first to use this, aided by their sedentism, to modify and domesticate the most promising.

The ecological basis for granoculture or vegeculture In asking where the most favorable ecozones existed Harris came to almost the same conclusions as did Vavilov in his search for centers of maximum genetic diversity. These places would be transition zones, ecotones, between uplands and lowlands, grasslands and forests, the littoral between aquatic and dryland resources. Any of these transition zones would have provided the maximum yields of small game, fish and plant foods. These biologically rich zones exist throughout the world and fall into two groups: those in drier temperate-subtropical environments with abundant and varied grass resources (the future granoculture), and those of the tropics with a more generalized plant *ensemble* including the root starch plants (the future vegeculture). In the first group of these favorable environments agriculture as granoculture emerged in SW Asia (wheat, barley, emmer...), in Mexican valleys as corn and its companion beans and squash, in Peru-Andes as a root crops and corn complex, and in E. Asia as millet and rice. The second group gave rise to agriculture as *vegeculture* in three major tropical forest/forest margin zones: S and SE Asia-Melanesia, South American lowlands with extensions into Central America and the African wet tropics. Some representative vegeculture crops are manioc, sweet potato, yams, taro, arrowroot, arracacha (South America, below).

Ecosystem-to-agroecosystem   transitions   From Harris' ecological viewpoint one of three things can happen when an ecosystem is replaced by an agroecosystem: on path one regardless of its initial diversity it can be replaced by a farming system which reduces the degree of generalization (as eg when a woodland is replaced by cotton monocropping); or on path two the ecosystem can be further enriched, generalized (as when irrigated mixed farming is introduced into a desert environment); or on path three the agroecosystem can be constructed within a highly generalized/diverse ecosystem without replacing it by substituting economic species which occupy the same niche as do the natural species. The classical example of this is provided by forest gardens and the forest shifters (*...Swidden cultivation and fixed plot horticulture manipulate the generalized ecosystems of tropical forests in this way and come closer to simulating the structure, functional dynamics and equilibrium of the natural ecosystem than do any other agricultural systems man has devised* – Harris).

Timing   Harris' ecological concept of specialized and generalized ecosystems suggests to him two main reasons why highly diversified tropical farming systems should have preceded specialized systems such as granoculture. First, farming systems which specialized in one or a few species would have required a long time period to bring about the *far-reaching genetic and morphological changes* which these specialized species required (as with the domestication of wild grasses, palms, fruit trees etc). Further, to then maintain these improved specialist systems would also have required a long time to bring about the necessary advanced and complex human socio-political systems. On the other hand the process of substituting more efficient species into their appropriate ecological niches within an already generalized or diversified ecosystem (path three) would not have required these conditions, at least to the same extent. This approach to agriculture...*is likely to have occurred long before specialized agricultural systems emerged.* Harris does not pursue this point but this type of agriculture would have depended on little more than selective *tending* of existing wild but productive plants, the same as is done today in Borneo (Shifters). If this is admitted – which amounts to the kind of indirect 'domestication' discussed by Rindos and Yen – it would push the origins back to 20, 30, 40 000...BP, or nearly as far back as Darwin's definition of 'man' would anyway permit.

8.5 The Caravan of Place

Travelers in this company have been drawn by five main aspects of place:
1. the origins of plants as object in itself (plant geography);
2. these as indicators of the origins of agriculture, the Grail;
3. place origins as pointer to new crops;
4. as pointer to new genetic material to upgrade existing crops;
5. to genetic material for defense of old crops against pests, disease, deteriorating environment.

So the patrons are a mixed bag. *Something lost beyond the ranges. Lost and waiting for you. Go!*

Travelers

Alphonse de Candolle was one of the earliest of recent travelers. Along the way he took two sets of notes. The first, Geographie Botanique Raisonee, is said to have greatly influenced Darwin's Variation of Animals and Plants under Domestication. In the opening sentence of the second, Origin of Cultivated Plants (1882), Candolle confesses his wider humanistic motives...*Knowledge of the origin of cultivated plants is interesting to botanists and even to historians and philosophers concerned with the dawnings of civilization.*

Candolle is one of those travelers equally at home in the caravan of circumstance (he probably preferred it to the others). In preamble to the place origins of the separate domesticated plants he sets down five main circumstances:

1. Agriculture arose in those places where potentially useful plants were most numerous; conversely it did not occur in places of indigenous floristic poverty (eg Australia);
2. moderate climate;
3. not excessive seasonal drought;
4. some degree of security and prior settlement; and
5. necessity for food production over and above what could be achieved by hunting and fishing (ie disturbance of the equilibrium in the Binford/Flannery model).

These plant resources and the other necessary conditions existed with rice and legumes in South Asia, grains in Mesopotamia and Egypt, Panicum in Africa, corn-potatoes-cassava in tropical America. Then anticipating

Vavilov's preoccupation with centers…*Centers (of agriculture) were thus formed whence the most useful species were diffused. The greatest of these were China, Southwest Asia and intertropical America* (the Mexico-Peru axis).

It's at this point, some later critics found, that Candolle's classical European scholarship had led him onto sticky ground. The kind of agriculture he had in mind was that capable of supporting the 'great civilizations' at each of these centers, only conceding that…*in Africa and elsewhere savage tribes may have cultivated a few species locally at an early epoch, in addition to hunting and fishing, but the great civilizations based on agriculture began in the three regions.* Even accepting the first part of this – the view from Geneva that outside the classical civilizations all was barbarism – the second part is more an explanation of these civilizations than of the origins of the crops which supported them. It has little to do with the natural centers of origin of the plant species or where they were actually domesticated, which could have been within these civilizations or far from them and millennia earlier. But this risks misinterpreting Candolle. He was aware of the hazard of birds carrying seeds long distances and messing up plant geography; and within the constraints of archaeological knowledge of his time he also says…*The history of the Dravidian (south Indian) and Malay peoples does not extend very far back and is sufficiently obscure, but there is no reason to believe that cultivation has not been known among them for a very long time, particularly along the banks of rivers* (Genesis, Carl Sauer).

Candolle's quest is most fascinating because while its basis was botany this had to be supplemented by 'evidence of varied character' and he looked for this in five directions: general reflection or lateral cultural thinking, archeology, paleontology, human history, philology (plant names in living and dead languages). Like Thoreau in Concorde, Candolle traveled widely in Geneva, through 'much research in books and hibaria,' correspondence with learned societies, letters from friends, the eyes of travelers pressed into his service. His accounts of the origins of the then 249 most important economic plants are a *tour de force* through classical and nineteenth century scholarship…correcting Linnaeus as he goes *(three out of four of Linnaeus' indications of the original home of cultivated plants are incomplete or incorrect)*, chiding the Greeks and Romans for their sloppy botany. This lateral thinking – or side excursions from the quest, depending on one's patience – was too much for some (Vavilov complained it was *saturated with a multitude of facts*).

Consider his pursuit of the olive, Olea europea: He first excavates 23 names or linguistic traces of it in the classical languages both living and dead, then wades through a swamp of mythology to track it down to Genesis and Noah and the olive branch and it turns up again as one of the promised trees in the land of Canaan. Consulting with one of the 35 authorities he enlists for the search he rules out Mt Ararat of the bible as the olive's home, for that mountain had recently been relocated by scholarship from Armenia (the ancient land of Masis) eastward to the Hindu Kush and India. But appeal to Sanskrit linguistics reveals this could not be the olive's true home either; nor could it be in the north otherwise the Aryans would have known of it, nor Babylonia for that country is too wet and Herodotus would have gossiped about it, and though it was present on the black ships of the Iliad and Odyssey it had but newly come to Greece (for Minerva in Attica had just planted an improved cultivar in the New Crops section of her garden)... To arrive with Candolle at the final destination (Old Syria) is anticlimax. Far more informative is his sifting through evidence of varied character – expeditions with him on Phoenician ships to the Canary Islands, quick visits to Egyptian tombs for visual proof, to the Berbers for linguistic instruction, time wasted with Tartars, Arabs and Pliny who by then could not remember the olive as anything but a long domesticated plant.

Nicolay Ivanovitch Vavilov (from 1924) was another distinguished traveler in this caravan, another who had 'so long suffered in the Quest, Heard failure prophesied so oft...' With Vavilov both references are more than allegorical. Of the first he describes how...*Caravans of Soviet expeditions have traveled along the Cordilleras from California to Chile* (and of the second, after Nikolay Ivanovitch fell from official grace he was starved to death in a Stalinist prison in 1943). With Vavilov search for the origins ceased to be an academic exercise. As founding director of the USSR All-Union Institute of Agriculture (the VIR) Vavilov directed and often accompanied expeditions which scoured the agricultural world for genetic resources (and New Crops) on which to base what Harlan later described as...*one of the most dazzling and ambitious plant breeding programs ever attempted.*

Vavilov concluded that agriculture's origins lay in eight geographical areas, centers (later amended to seven) with some sub-centers (Figure 8.3). He estimated that in 1930-1940 a total of 1 500 plants were under cultivation, with 1 000 accounting for 99% of the world's agriculture. He also observed (1931) that all the centers were in or adjacent to mountainous areas of the tropical and subtropical world (map) so that...*in essence, only a narrow belt of the world's continents plays a major role in the history of global agriculture,* and was aware of predominance of the true tropics as contributor to plant species. As one of Vavilov's colleagues puts it...*In the inexhaustible womb of tropical flora, treasures are hidden that are in many cases still unknown...*

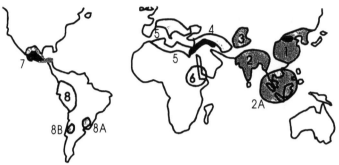

Vavilov centers and % world crops

1.   E Asia (Chi-Kor-Jap-Tai) 20%
2.&2A. S and SE Asia 33%
3.&4. Inner Asia-Caucasus-
        Asia Minor 14%
5.   Mediterranean 11%
6.   Ethiopia-Som-Yemen 4%
7.   Cent America Highlands 17%
8.   Andes-Peru-Ec-Col (17%)

Harlan's centers

The three centers shown; and associated non-centers in all of SE Asia (as Vavilov's A2), across northern savanna Africa, and from Argentina to N. Brazil coast in South America. Note: Vavilov's centers were frequently revised, Ethiopia later dropped.

Figure 8.3   Vavilov and Harlan Centers

However Vavilov's main aim was to find centers of maximum diversity of economic species and their wild relatives; finding the geographical origins of crops as agriculture was incidental, although...*the basic centers of origin of cultivated plants appear, as a rule, to be found where a striking diversity of types is accumulated.* His later critics (Harlan, Zohary and Hopf), apparently discounting the qualifying phrase, were to point out the many exceptions where place of greatest diversity does not correspond to place of geographical origin of a crop (exceptions which Vavilov himself was aware), and claim they'd demolished the 'elegant simplicity' of his theory. It might also have been due to selection because he also said...*Diversity alone does not determine what is the initial center or origin of a particular plant.*

This criticism has been summarized by Harlan: *Basically what Vavilov did was to draw lines around areas in which agriculture had been practiced for a very long time and in which indigenous civilizations arose.* What Vavilov had actually said was...*It is of comparatively little interest to us that wheat and barley have been found in the tombs of pharaohs...It is much more essential...to learn how the Egyptian wheat originated and where to find the basic elements, the building blocks, from which the present cultivated species and strains were built.* One of Vavilov's VIR associates, P.M. Zhukovskiy (1969) later found it necessary to increase the number of 'centers' to 12 (including all of tropical Africa, most of North America) and expand others into megacenters until they took in just about the entire habitable globe south of Siberia. In effect this tolled the bell on 'centers', and a good thing too because Candolle's idea was becoming a game of semantics – centers, non-centers, focii, hearth areas, affected areas, megacenters...applied with no great clarity in whether these meant geographical origins of useful plants or where these had first been brought into use or where as crops they first came to archeological attention as supporting civilizations.

Jack Harlan (1971, 1975, 1986) like Vavilov was concerned with the location of genetic gold for (US) plant breeding, approaching the idea of centers skeptically but nonetheless finding enough evidence – technical and of varied character – to designate three (Near East-North China-Mexico), each with associated regions but these so large he called them non-centers (Figure 8.3):

Harlan Centers

Center	Associated non-center
(A1) Asia Minor – Zagros – Euphrates	. . . (A2) African savanna – Ethiopia
(B1) NE China	. . . (B2) SE Asia – India – to New Guinea
(C1) Middle America	. . . (C2) S. America (3 regions)

with center/non center relationship consisting of mutual stimulation and feedback of ideas, farming methods and plants between the two, eg between Near East and savanna Africa, North China and SE Asia.

Nonetheless in spite of Harlan's skepticism the perceived Candollean link between 'first crops' and civilizations persists. Of the Near East...*It appears that all the characteristics of a center can be established on archeological evidence alone* (Harlan 1971). But this doesn't apply universally...*A center must be established one crop at a time*...which seems to be saying that there are as many 'centers' as there are crops, although in some ecozones (eg Near East grasslands, the root crops of the Andes) these were often naturally concentrated. From a human cultural rather than a technical botanical perspective the main weakness of the 'center' concept is that it implies that nothing much in agricultural development was happening outside them. The real situation is probably that humans had manipulated plants from the very earliest times, everywhere, wherever they happened to be. Further, there's no need that these places had to be particularly rich in potential crops (as eg the Near East was rich in the future grains). They could equally have been places of botanical poverty where this fact would have exerted greater pressure for earlier and/or more intensive manipulation of those few plants at hand. (A later model for this would be found on Polynesian atolls which at the time of human occupation had hardly anything at all except weeds and pandanus but which, with much imagination applied to them, became comprehensive economic systems – Genesis, Lapita.)

8.6 Limits of Archeological Evidence

*We are continually enjoined to accumulate more facts (preserved plant remains from secure archeological context) as the road to God's truth...*Donald Lathrap

To avoid more disputation it might be prudent to pause here by the roadside and check the *bona fides* of our guides the archeologists, though they're indispensable.   Archeology's reach is limited by evidence then its interpretation. (Two grains of sago underneath a rock that shouldn't be there are evidence of *what*?)   Paleoenvironment's is longer, finding disturbed habitats. (But again disturbed by what? El Niño wildfires? Shifters...?) Climatology's is longer still but blind without the others, finding only might or could or possible, and even with them the darkest corners, the country further out, can be reached only by imagination. But how much imagination or departure from the literal facts should be allowed? Using essentially none the date of H. sapiens' arrival in Americas has been put at around 13 000 BP. Using slightly more and applying it to the same facts the ETA becomes 20 000 BP. Using more again it becomes 290 000 BP. Who to believe? In SE Asia strict literalism puts the ETA of modern people at around 40 000 BP and can see no link between this and the presence of previous 'people' from at least 600 000 BP in the same area (Genesis – Java, Flores).   These wide bands of possibility don't inspire much confidence in archeological literalism as pointer to prehistoric economic systems, one of which was agriculture, the Grail.

Early archeology through the 19th century was in the tradition of attempts to discover linkage with the classical Greek and Roman world. Continuing into the early 20th century its pioneer discoveries of agriculture's traces were accidental. Smith (1986) observes that workers in this tradition remained art and antiquity oriented until well into the 20th century. Initially plant remains ...*were not sought for but brought to public attention by some unplanned event –* plant remains recovered from drought-exposed Swiss lake beds, others accidentally discovered in Egyptian and later in Peruvian tombs. Even then their true significance from the viewpoint of agriculture's evolution for a long time went unrecognized. By the late 19th century a second post-Darwin school had emerged which was concerned primarily with man's evolution regardless of proximity to Europe's own cultural roots. But it is likely that Smith has mainly the first or Indiana Jones school in mind when he complains...*Unfortunately many excavations are being made today (1986) without any effort to recover plant remains and information they convey.* As more serious archeology (partly) superceded tomb robbing and treasure hunting it became clear that the great and lesser civilizations of which knowledge was sought could be understood only through their agricultural means of subsistence. To achieve this, and as the size of the task began to be appreciated, it later (1940s) became necessary to co-opt the expertise of botanists, zoologists, climatologists, ethnographers in developing what Smith calls *integrated theories on the ecological parameters of prehistoric people.*

Some constraints are imposed by what Carter calls regional patriotism, others by cultural inheritance. Until quite recently it was assumed that since indigenous Australians lived as 'hunter-gathers' in a difficult arid environment this must have been so over their long occupation of that continent. But modern paleoenvironmental opinion now supports the likelihood that they also had lived by a range of various economic systems, including aboriculture of many kinds (eg the cycads above) in a once rain forest environment, initially or intermittently, over the past 60 000 years or more (Edwards and O'Connell 1995). Other limitations are self-inflicted: professional preoccupation with one kind of evidence (bones) at the expense of ignoring others (plants) or *vice versa*; or with evidence selected to substantiate a previously declared position. There's also intrusive evidence – some bit of bone or a seed that has slipped down an earth crack due to some past earth tremor or drought and this, with the archeo-agricultural era it represents, being attributed to a date several thousand years in error...*I wish I could explain to every botanist that while an archeologist looks the other way for one minute, a packrat can bury an intrusive bean 50 centimeters deeper in his favorite cave* (Flannery 1973). (Un)healthy competition is a problem too. This same authority castigates those...*botanically naïve, sensation-seeking opportunists more concerned with finding the oldest domestic plant than with clarifying the processes by which agriculture began.* He was anticipated by Higgs and Jarman (1972) who called for a redirection of archeology's resources away from...*search for first instances and rival hearths of domestication* to an understanding of the *economic mechanisms* of human development (Genesis).

Preconceptions

These start with extension of such terms as stone age, neolithic etc out to the tropical world from Europe.   In large parts of the tropics, notably savanna Africa and SE Asia, there's no stone and therefore no artifacts fabricated from it and no durable lithic evidence.   More accurately one would speak of the Old Wood Age (which continues in Africa), or in SE Asia-Melanesia the Recent Bamboo Age which continues – or did till yesterday – with bamboo used for about everything in daily life and agriculture;   but succumbing to termites, and the vegeculture it once handled disappearing too, their absence suggests nothing was ever there at all.

Archeology's initial and desultory search for the Grail (c 1860-1950) was in a field early prescribed: initially the Nile Valley, Greece (not overly productive), the Holy Land, the Tauros and Zagros foothills and Fertile Crescent generally; the same classical world to which educated Victorians would journey (in winter) in search of the origins of their culture, religions, ethical systems, much language, and to which literalists of the jarring sects still journey to in search of God.   Carter (1977) and others have remarked on its religio-cultural significance as determining direction of the Quest...*Western orientation of research has given us a near-fixation on the Near East as the Garden of Eden from which all ideas flowed.*   And since God's people now receive His bounty as wheat and vetch and cows and woolly sheep, this is where the one True Grail must be.   It's nature, agriculture-as-granoculture, and direction of its dissemination out into barbarism had early been pronounced by Halbeck (1959)...*defined by circumstance that all the initial cultural developments and diffusions within it* (the Old World, Europe to Indus) – *from the appearance of the food producing stage onwards* – *depended on the cultivation of wheat and barley for subsistence.*   And from Hutchinson (1965, 1972, 1977)...*The oldest Indian agriculture of which we have evidence was based on the...crops and farming patterns of West Asia,* but he also points out the limitations of this evidence...*It seems unlikely that Harappan agriculture could have arisen, equipped with a repertoire of plants and skilled textile craft, without a long gestation period.*

Darlington (1969) was a notable exponent of the classical Near East-granoculture model and its subsequent gifts.   Starting from the usual premise that agriculture arose in the nuclear zone of Anatolia-Iran-Syria before 9 000 BP he sees archeology's task as explaining why it did not 'reach' India and China until 4 000 years later. Since archeological evidence of its migration to China is not otherwise forthcoming this is likely due to destruction of the logistics evidence, drowning of ancient caravan routes from Near East to China by shifting sands of the Tarim basin. Another proposal also based on archeological evidence was that pottery too originated in this nuclear zone and similarly diffused. The need as cooking vessels arose with granoculture, and the skills involved in pottery making prepared the way for metal smelting. Not only agriculture but civilization as pottery and later metal fabrication were due to the granoculture of...*a single district of origin on the northern side of the Nuclear Zone in Anatolia.* Expanding this Nuclear Zone-granoculture model globally to explain failure of the origins elsewhere, vegeculture could not have provided an alternative. In contrast to the...*skill and forethought* required of the temperate grain farmer the few simple tasks facing the tropical vegeculturist provide him with no challenge...*for Nature takes care of everything for him.* Further, tropical agriculture itself stemmed from a prior granoculture stream, at least in Asia. Having migrated from the Near East to Harappa in Baluchstan-Punjab-Sind as wheat it continued on to SE Asia as rice. While tropical vegeculture as root crops and bananas/plantains had supported large populations in the tropical world (South India, South America, SE Asia, Polynesia) it arrived relatively late, c 2 000 BP, and it was derivative from the earlier imported *idea* of grain farming which provided the model for other kinds of farming. In summary, all the great civilizations of the past were based on the *industrious and prudent* grain farmer. On the other hand – here Darlington enlists the thoughts of Burkill on manioc farmers in colonial Malaya as his authority – *vegeculture advances that part of the population which contributes least to the common good.* With this support he arrives inevitably at his conclusion...*Easy cultivation does not select for either foresight or industry. That seems to be why SE Asia was not a center of origin of agriculture.* So that's all cleared up. (For an updated version of Darlington's model in full diffusive flight see Jared Diamond (1998) who finds in these biblical lands the fountain of just about everything from proper agriculture to writing, cities, empires and civilization.)

Fitting out the Expedition

Daniel Zohary and Maria Hopf (1994) have prepared a comprehensive list of the 24 types of archeological evidence the modern traveler should look for, to handle every situation from bogs to tombs, and this is accompanied by about 20 other types of evidence of varied character, again Candolle updated. Their book, with location maps of all the significant archeological sites, is a most comprehensive account of agriculture's origins in the Near East and its relationships with that of surrounding territories. Delaying only long enough to despatch and bury Nikolay Ivanovitch *(...Other suppositions such as Vavilov's equation of centers of diversity of cultivated plants with places of origin proved to be incorrect and can no longer be used)* they load this improved apparatus on the camel and set out. It's worth following a while to see what conclusions they arrive at. For present purposes these are three, relating respectively to place, time and circumstance.

First...*it is evident that the crops domesticated in the Near East nuclear area were also the initiators of food production in Europe, central Asia, the Indus Basin and the Nile Valley...All over those vast areas the start of food production depended on the same Near Eastern crops.* The second conclusion concerning time is that indigenous agriculture outside the Near East did not start until the Near East package arrived at these places, and then any local agriculture became only an appendage to the imported package...*Signs of additional domesticants start to appear soon after the introduction of Near East agriculture to Europe, central Asia and the Nile Valley. The addition of some of these crops obviously took place outside the Near East but within the already established cultivation of the Near East crop assemblage.* Both these propositions must be partly right. Of course there was a tremendously important early Near East agriculture and it did spread out in all directions; but this seems to be a view that Near East enlightenment spread into a vacuum. Even Candolle from the academic insulation of Geneva had conceded that 'savage tribes' in Africa and elsewhere must have been capable of growing *something.* (Faced with these conclusions he might have reissued his warning about Greeks...*who were disposed, like certain modern writers, to attribute the origin of all progress to their own nation.*)

Near East Trees and Wider Economic Systems

The third set of conclusions (Zohary and Hopf, condensed) concern the emergence of tree foods:

In spite of these uncertainties the following statements can be made -

1.   The earliest definite signs of fruit tree cultivation appear in the Near East.

2. *Horticulture developed only after the firm establishment of grain agriculture.*
3. *Like grain, several local wild fruits were taken into cultivation about the same time.*
4. *Domestication of fruit crops relied heavily on vegetative propagation.*
5. *Planting fruit trees is a long term investment, promoting a fully sedentary way of life.*

Viewed globally and as continuing from the previous two conclusions these propositions are untenable: there are 'signs' of tree/palm cultivation before 10 000 BP in South America, to 15-12 000 BP in S and SE Asia, to 35 000 BP in Melanesia (Genesis). Domestication, if that's important in early and prehistory at all, can occur in half a dozen ways apart from vegetative propagation (Rendos, Yen, Shifters). Sedentism, though useful to a modern apple farmer, is not necessarily promoted by planting trees (Shifters, enriched fallows); sometimes they promote the opposite, mobility (Posey's Indians and their resource islands).

In Near East itself these propositions can at least be questioned. From archeological evidence Spiegel-Roy has given the apparent dates and sequence of domestication of all the main Near East (and many Chinese, Indian, American) fruits; and Zohary and Hopf (1994), Zohary (1986), Zohary and Spiegel-Roy (1975, 1986) have listed the apparent domestication sequence (c 5 700 – 5 500 BP) of the grape, olive, fig, date, from Palestine sites and these must be accepted. However in a wider ecological and economic context Flannery has described the then Near East landscape as a mosaic of grasslands (the future granoculture as wheat, barley, rye, oats, safflower, chickpea...) and of woodlands providing a rich variety of tree foods: pistachio, walnut, almond, hazelnut, acorn (with their dependent fauna). This landscape, or close to it in Central Asia, is also the native place of fig, quince, apricot, cherry, pomegranate, apple, pear and many berries. In addition there were the olive and the adjacent Mediterranean trees and vines (plum, grape etc). Piperno and Pearsall (1998) also offer a reminder that this general landscape was not always a goat-chewed wilderness of camel thorn, from records in recent historical times of Roman legions marching through most of North Africa in the shade of trees and lions still a rural hazard in Greece. Given these tree food resources and their attributes of volume, variety, storability, transportability, staggered seasonal supply pattern, and that intelligent H. sapiens had been wandering through them for at least 120 000 years, it's almost inconceivable that the trees did not form a major economic basis millennia before the first proper (neolithic) farmer stripped off her first crop of HYV emmer c 10 000 BP, or that their use had to await domesticated certification by archeology from just 5 000 years ago.

8.7 The Great Diffusion

So agriculture's roots were discovered largely by accident and within a frame of time and place comfortably close to the Biblical and in recognizable form. The good news came in from Tell Murybit and Tell Abu (wheat's ancestors, 8 100 BC), from Ali Kosh (barley, einkorn, emmer, 7 500–6 700 BC), Jericho (einkorn, meat gazelles and emmer, 7 000 BC), from Jarmo (pigs, two row barley and einkorn, 6 750 BC), Argissa (more emmer, lentils and cattle, 6 000 BC), the farms of Ur (barley, flax and woolly sheep, 3 000 BC) and Tepe Sabz (vetch and flax 5 500 – 5 000 BC) – from all these dusty places with sonorous names.

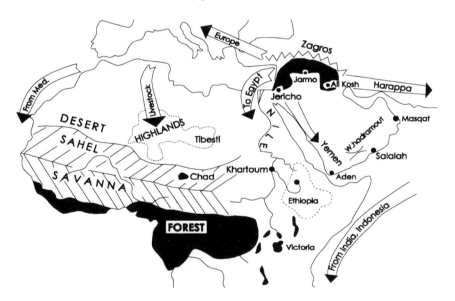

Figure 8.4   The Great Diffusion

Archeology's heralds, historians in diffusive mode (Figure 8.4), trumpeted these wonders up the Danube into Europe (clocked at an average speed of one kilometer per year) – Then south into Lower Egypt with wheat and barley, sheep and cattle before 7 000 BP – On up past the cataracts with wheat and goats and barley for Nubia (the last not really welcome for they had their own) – Past Malakal on the Nile where the hippos start with these things to feed the Cushites of Ethiopia also (but here they had to contend with local barley, small grained teff (Eragrostis tef) and Ensete ventricosa which were the staples of that place) – Checked, retreating, then wheeling west through Cyrenaica about 6 800 BP still with bulls and sheep and wheat and barley, to Carthage then it came, burning, burning... Then south into the once-moist Sahara uplands with these grains still and livestock but south from Dhar Tichitt the grains withered and the woolly sheep expired and many of the kine also but enough survived the terrible crossing to bring Pastoralism to West Africa.

Then once more to the south to build the farms of Yemen by the wells of Shebwah (later dowry of a certain queen) and crossing the Red Sea to offer yet more grain and vetch to the Cushites but here nutrition prevailith not over the peace brought by a local gift which is *khat*, Catha edulis. And so we come to the holy city of Dire Dawa. *(In Dire Dawa we graze on khat and if our mother dies on Saturday while we commune with God through khat we would not bury her.)* Reversing east into the Wadi Hadramout. Wheat for the people and wheat straw for the bricks to raise up the mud skyscrapers of Say'un and this straw also for the conical hats of the black-clad ladies of the fields. Right comely they are too, these ladies of the Hadramout, white teeth smiling in copper faces through dust under the copper sky. But they smile too much. They are inclined to levity which the orthodox know passes so quick to ribaldry then unbelief. They laugh at everything and everyone as relief from tedium of their Agronomy. *She had a heart, how shall I say, too soon made glad.* Back out again to Al Ghaydah, on to Salalah. Salalah where they build the dhows and the old Jews denied land once beat out their souls in silver in the souk, but that was later and it's still there. In Salalah in the souk. Hurrying on past ramparts of black rocks which hurl the great heat at Masqat and Matrah (but fertile ground for the new dates carried down from Basrah), to leap Ormuz where Europe's black ships later came, riding the monsoon east to Lanka. Or Taprobane or Serendib. To Char Bahai and then to steal, yapped at by dogs, through the verminous villages of Baluchistan. At last the banks of Mother Indus, the empty waiting fields of Mohenjo-daro. And all these gifts from the Sumarians unto the Harappanites unloaded there by 4 500 BP. Give or take a bit.

There's only one thing wrong with this MGM epic, it didn't happen. Not the way the script has read so far. While the great diffusion is based on archeological evidence this like gold is where you look for it and by now archeology has reduced the classical nuclear zone to a rabbit warren while all else is comparative ignorance. Enough evidence has been found to allow patriots by thoughtful interpretation to prove anything (including that bizarre complaint of Chinese water management and irrigation knowledge stolen from Babylon). While the great diffusion rests here by Mother Indus, girding itself for the final push on to Cathay there is just time to poke around, to double back and try to fill in some blanks, to consider Harlan's reservation... *Were the Europeans, the Indians and the Ethiopians entirely without cultivated plants at the time?* To see if Candolle's savage tribes might have had something of their own. To haggle with Witnesses.

8.8 India ... *a story of fragrant exotica, and as real as a snake charmer disappearing up a rope*

The absence of archeological information regarding early agriculture in geographical India (including Pakistan, Nepal, Bangladesh) is even more remarkable than in Africa. According to Vavilov the southern part of Asia, with SE Asia, has contributed 33% of all the world's now relevant plants. Also in the tropical south (Reconnaissance) the range of crops still grown is enormous, but there's little archeological evidence of this diversity in the past or of the ancient farming systems built on it.

Indus – Harappa

Most archeological information concerns the Harappan civilization of the Indus Valley with its main cities of Harappa and Mohenjo-daro (Figure 8.5). Romila Thapar (1966) sees Harappan civilization as culmination of a long slow cultural evolution involving successively at least four racial groups (Negrito, proto-Australoid, Mongoloid, Mediterranean). On this foundation the time scale is for an indigenous pre Harappan civilization which arose before 5 000 BP and spread in a remarkably uniform and organized way from the Arabian Sea to Delhi (over 70 sites have been uncovered) and was absorbed by waves of Sanskrit speaking pastoral-warrior types from the trans Caucacus plains via Iranian plateau beginning 3 700 BP and continuing to 3 000 BP. Lannoy (1971) notes that with a total area of half a million square miles Harappa was the largest unified empire until Rome. The geographic similarities with Near East and Egypt include semi arid climate, settlements along the Indus river system as along the Euphrates and Nile and some dependence on annual inundation for maintenance of field fertility, with the similar flood hazard. The crops assemblage is also similar, built around the cereals. If archeologically recovered species from other north Indian areas (Rajastan and upper Ganges Plain) are added to what has actually been discovered at Harappan sites the range of crops grown there was wide but limited to the cultivated field crops, except for the date palm, though other vegetative crops like sugar cane and probably some root crops which leave no trace would also have to be assumed, Table 8.2.

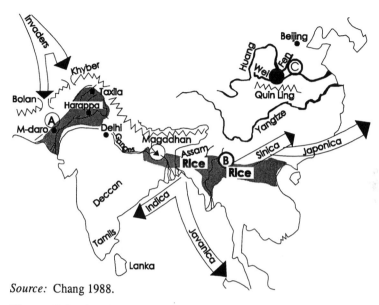

Source: Chang 1988.

Figure 8.5 A – Harappa, B – Rice Origins, C – North China Center

Indus (and Indian) agriculture before c 5 000 BP is practically a blank page. Few Witnesses have come forward. About all that's generally agreed is that the Harappans were indigenous to the Indus not previous immigrants, dark skinned Dravidians (*probably resembling modern Dravidians in South India* – Lannoy), and as Hutchinson (1977) has noted of their agro-civilization...*this must have been preceded by a long period of farming development.* Harappan cities were already models of uniform town planning, and two wheel carts, ox plows, domesticated elephants and camels, ceramics, writing (undeciphered), standardized bricks, all indicate a well advanced agro-technology and even more a tightly organized socio-political system. Further indications of comparative modernity of this Indus farming (ie of that so far brought within archeological reach) is in the significant number of crops which by this period had already found their way to Harappa from Africa (table), and in an established cow cult, and in a general field crop agrotechnology which is said not to have changed essentially to this day. By Harappan times proper (2 500 – 1 500 BC) there was also a well established cotton textile industry, and most intriguing of all is Hutchinson's suggestion that the crops of African origin indicate

Harappa had trade links with civilizations in Africa which are still unknown (Carl Sauer's 'lost corridor of Mankind' – Genesis).

Table 8.2   Main Crops of Harappa/Indus/North India

	Crop	Archeological period	Origin
wheat	Triticum sphaerococum	4 400 – 3 800 BP	SWA
barley	Hordeum vulgare	4 400 – 3 800	SWA
rice	Oryza sativa	4 400 – 3 800	IND
sorghum	Sorghum bicolor	4 400 – 3 800	AFR
pearl millet	Pennisetum typhoideum	4 400 – 3 800	AFR
finger millet	Eleusine coracana	3 600	AFR
peas	Pisum arvense	4 400 – 3 800	SWA
chick peas	Cicer arietinum	4 000	SWA
lentils	Lens culinaris	4 000	SWA
grass peas	Lathyrus sativus	4 000	SWA
gram	Phaseolus spp	3 600	IND
horse gram	Dolichos biflorus	3 600	IND
flax	Linum usitatissimum	3 600	SWA
mustard	Brassica spp	4 400 – 3 800	IND
cotton	Gossypium	4 400 – 3 800	IND
sesame	Sesamum indicum	4 400 – 3 800	AFR
dates	)		
melons	) also present but no archeological dating		
madder*	)		

*For vegetable dye, similar to European madder Rubia tinctorum. The origins of these species are in India (IND), Southwest Asia (SWA), Africa (AFR).
Source: Compiled from Hutchinson 1977, Allchin 1969.

The Indian circumstances   The archeological record of Harappa's crops will change and is of passing interest, but the various accounts of the Harappan agrosocial system are important.   One starts with the geographical theory (of Lannoy who bases it substantially on the Allchins' interpretation of their archeological evidence) that agriculture emerged at Harappa rather than elsewhere because land cover in the Indus Valley was then predominantly open moist woodland easy to clear in comparison with the peninsula which was then covered with 'jungle,' and with the Ganges plain then dense jungle swamps. (Given this favorable condition and lack of evidence to the contrary the other Indian subsistence systems in those times have been dismissed as gathering and hunting.)   In the Lannoy/Allchin sociological model of Harappa, the seeds of India's future system of castes and occupational specialization had already germinated in Harappan agro-society well before the arrival of the Aryan invaders...*We can infer that the working class population deferred to a trader class who had a monopoly of profits through the enforcement of religious sanctions...Thus it was in the interests of this (superior non farming) class to maintain a static tradition (with) a degree of control undreamt of elsewhere.*

Plows, cows, goats and dust   At this point the archeological shutters come down on what happened to Harappa, although Mohenjo-daro ended as a pile of skulls which throws some light on the *modus operandi* of Mother India's new guests.   Gurdip Singh (1971) and Bryson (1975, 1977) pried them partly open with the crowbar of climatology and found Harappa had subscribed to Mainstream B and ended as a dustbowl. It was on the skids centuries before the Aryans arrived.   Singh reconstructed the Indus rainfall record from 10 000 BP from pollen deposits, finding a general decline in summer monsoon rainfall from that time to about 4 000 BP, though with a strong 1 000-year dry-wet cycle.   These declining long term conditions corresponded to the inception of field crop farming, the greater population this made possible, development of the Indus civilization and an always increasing pressure on land resources met by lateral expansion of Harappa into marginal lands then farm intensification.   Both led to wind erosion and dust cloud formation on massive scale and this, superimposed on one of the long cyclical droughts, spelled the end for Harappa.   The Aryans took over its remnants and learned their agriculture from them.

The mechanics of Harappa's decline are of modern land use and Greenhouse relevance (Energy).   When it comes to India from mid Indian Ocean, curving in across the Arabian Sea, the summer SW monsoon air carries much moisture, 80% of that over a rain forest (Bryson and Murray, 1977).   But little of this potential rain falls over what used to be Harappa. The permanent and high dust clouds set up local conditions the opposite of global warming: a daytime earth *cooling* blanket sufficient to prevent formation of strong convectional updrafts and their thunderstorms.   At night the dust acts as a warming blanket preventing dew formation which might otherwise hold down the dust and lightly water desert vegetation. Although Harappa's farms and towns were abandoned (approximately the dry steppes and desert of Rajputana), over following centuries other great Princes returned whenever the climate seemed to grow more moist or when the vegetation had recovered, promising fresh visions. But always nature's healing process, revegetation, was quickly undone by new dust clouds from the plows and cows and camels they brought back with them – and always by Ms H. Indicus chopping firewood – in a downward spiral which continues to the present day and expands the drylands perimeter by one km per year. (These terrestrial dust clouds are now compounded by a 10 million sq km cloud of aerosols over the Indian Ocean from industrial activity and agro biomass burning – Energy, greenhouse, dust.)

Glimpses from the Vedas The archeological shutters don't open again until the Magadhan empire of lower Ganges (Figure 8.3) 6th century BC and Mouryan empire of Chandragupta c 300 BC. Others have forced their way towards the missing post Harappa/pre Magadhan chapter of the origins using the crowbar of Great Lit. This was the period (1 000 to 600 BC) in which the great folk literature of the four Vedas, the Upanishads and the Epics were written down. It is said that the Vedas required 1 000 years to write (ie after the Aryans had mastered that skill (700 BC?), for when their war chariots rolled into Harappa they were innocent of it). One of the Vedas alone eventually ran to 4 000 pages. These began as Aryan bardic folklore, some fact much myth, eventually becoming in the hands of the Brahmin priestly caste, who alone had time to memorize and judiciously edit this mountain of theological supposition and garbled folklore, the fountain of all knowledge and priestly power. The Rig Veda runs to 1 028 hymns, some songs of praise to the old Iranian-Indo-Aryan tribal gods, many in celebration of the righteousness of the various Aryan tribes themselves. One such hymn divides Primordial Man into four parts: the head became the Brahmin or learned priest, the arms became the warrior, the trunk the merchant and land owner, the feet became the agricultural laborer or *sudra*...which left no part of the anatomy for the even lower species like those who tended and skinned animals or fished or hunted, ie the indigenous hunter-gatherers not yet enmeshed in this form of divine revelation. These became out-castes. Such vertical division of society was to be extremely convenient for future agricultural development when large amounts of subservient labor were needed to clear the forests and grow the food for the empires which arose.

Other glimpses from the epics The epics also are illuminating. The *Mah~bb~rata* is said to have been, of obvious necessity, the longest poem ever constructed, the story of bloody battles over farm land rights between two sets of country cousins near present Delhi, extending over about 20 years and ending with a final 18 days slaughter of one set by the other. The *Ramayana* of lesser length but still requiring several days and nights to tell (though longer when listened to) is ostensibly about the banishment of Prince Rama and his wife Sita from a North Indian principality to the forests of South India where she is captured by the demon king of (Sri) Lanka. But Thapar says all this, with flourishes like Rama's enlistment of the forest monkey king Hanuman to rescue Sita, is probably the result of later editing. The story really represents an earlier conflict between the new Indo-Aryans pressing east into the lower Ganges forest lands and there meeting resistance by the original dark skinned forest dwellers. Rama and his Indo-Aryan entourage represent light and godliness. The demon king of Lanka represents the blackness which the light skinned Aryans found, and heroically overcame, in the *indigenes* of India.

Mainstream B and the two-footed beasts Meanwhile back on the farm, this period of the great literature was also one of increasing agricultural success. Once they had settled down these Indianized Aryans (or these guided by Aryanized Indians) showed remarkable aptitude for a farming way of life. Greater food production enabled a greater population which spilled out of Indus/Harappa to the east, down the Ganges Valley, chased by Singh-Bryson dust devils and Georgescu-Roegen's 4th Law (of inevitable pollution/landscape degradation accompanying the use of energy for human purposes). Apart from their improving agrotechnology the other requirement for this expansion was an organized and subservient farm laboring class. From the Lannoy-Allchin politico-social model this class had already emerged in pre Aryan Harappa and their lowest position became solidified by priestly authority as political Brahmanism developed. Lannoy notes, on the authority of the Rig-Veda, that the Aryan-Sanskrit word for the dark skinned indigenes was *dasa*, 'two-footed beasts,' something less than chattel slavery. With this in place agricultural expansion to the east could procede...*This servile class was incorporated into the new Aryan social order under the name Sudra, laborer or cultivator, while Aryan or Aryanized tribal chiefs constituted an oligarchy with virtually absolute power over the labor force employed in the clearing and cultivation of virgin lands.*

One geographical coincidence is that these 'virgin lands', mainly the lower Ganges swamp forests, were part of the original home of Asian rice (China, below), although from the previous table there are archeological records for domesticated rice from c 4 400 – 3 800 BP, ie a thousand years before the main Indo-Aryan expansion into this crop's natural homeland. With rice added to the other crops listed (local and those brought by the Great Diffusion and from Africa), and the cow (starting as an economic asset and unit of exchange and progressing quite logically to an object of veneration then abuse), and a Prince (Chandragupta then Asoka then their thousand successors, petty or magnificent), and this Prince blessed by ten thousand priests (chanting 1 028 edited hymns of sanction), and these Brahmins obsessed by the potential for using skin color as a weapon for class oppression, and these also not rejecting reformers like Gautama Buddha (b 566 BC) but silently absorbing them, and with all this supported by the two-footed beasts who cleared the forests and grew their food – Mainstream B had triumphed.   It ushered India through her turbulent centuries, offering a prize which few wandering adventurers or more thoughtful Nation Builders (Kusharas, White Huns, Gurjaras, Scythians, Achaemenids, Greeks, Macedonians, Mongols, Arabs, Turks, Afghans, Europeans...) could resist.   Mother India absorbed them all.   Her farms fed them all.

Central and south   Evidence from other sites in the (dry) Deccan and more recently deforested wet south reveals even less antiquity than does Harappa.   These sites have been mapped and their crops and livestock discussed by Allchin (1968, 1969, 1977) and Hutchinson (1977).   The crops assemblage are similar to the field crops listed for Harappa, grains and lentils with the addition of date and a few fruits.   They generally occur c 3 800 to 2 500 BP.   Allchin has described some Deccan settlements as early as c 4 300 BP based on cattle and millet and their hillside house sites and small fields already terraced with dry stone walls, again proving little more than that these were far from the origins.

Dissent – the View from Daksina

Srinivar Iyengar (1983) and other southerners have vigorously protested these northern interpretations, especially those by writers who...*with the self-confidence of a Cooks Tour guide conduct the ancient Dravidians through the mountain passes and drop them with a ready made culture on the banks of the Vaigai River.*   The ancient Tamils were indigenous to India, not immigrants.   Their earth consisted of five natural regions, and in passing through each of these successively all Tamils – and all humanity – have passed through five corresponding economic-cultural stages: (1) Kurinji forest hills and rain forests (gatherer-hunter);   (2) Palai desert (nomad);   (3) Mullai midslope forests (pastoralist);   (4) Marudam lower river valleys (farmer);   (5) Neydal the ocean littoral.   According to this Tamil mythology, universalized, paleolithic people emerged in the Kurinji rain forests as collector-scavengers, only later becoming hunters and then for defense.   Their first great invention of fire was inspired by watching the bamboo violently rubbed together by monsoon winds until it burst into flame.   A second was observation of and then participation in 'natural' agriculture by women, as collectors of bamboo-seed rice, wild rice and panicum seed – again the visible parts of their *bricolage* of often invisible activities.

Tamil mythology is not a good place to shop for facts. Although bargains can sometimes be picked up ('Pearls are made by transformation as rain drops impact upon the sea') these shed little light on the origins, or on H. sapiens' earlier adventures among the palms and wild buffaloes of their home in Tamil Nadu, that universal homeland of Kurinji. But then, even less light seeps out from Archeology's shop across the street.

An ecological view Those approaching the Indian origins from ecological considerations would probably agree with Srinivar Iyengar, though they might want to come in *via* Marudam, the river valleys; and they might not confine themselves to the Tamil country. The Indus rivers – but the eastern and southern Indian rivers even more so – would *par excellence* have met Harris' conditions of ecological diversity for agriculture's emergence. The necessary 'highly generalized' initial ecosystem would have been provided by the five great rivers of Punjab (Indus, Sutlej, Chenab, Jhelum, Ravi), each seasonally bigger than most in Europe. In the dry season they are not so much rivers as braided networks of channels, reedbeds, residual swamps. On ecological grounds a more likely place could hardly be found...and to satisfy Vavilov all this is close to the other transitional ecotone between the plain and high mountains, with their valleys and still more diversity.

The people would have been there too, immune to Glacials in an always Interglacial afternoon, somewhere in those five provinces of the Tamil-universal earth. After all, H. erectus had wandered through them on their way to Peking and Java (and Flores Island 800 000 BP) about a million years before Harappa, and there's no reason why some did not stay in India. There are stone tools, Soan pointed stones, scrapers, choppers, from 400 000 BP (from which is inferred among other things a uniformity of human culture Europe-Africa-India-East), though encyclopedias still .insist agriculture when it was invented did not 'enter' India until 5 000 BP. More recently, regionally, there's archeobotanical evidence from the Hoabinhan economic-technical complex from 14-10 000 BP, centerd on Thailand-Vietnam but extending into Myanmar/Burma and probably into Indian Assam as well (Genesis). From this the origins of Indian agriculture have always been in India. To use Romila Thapar's phrase, each civilization is its own miracle. One does not need successive waves of intruders – of Aryans, White Huns, Scythians and similar footloose rabble from the Russian steppes – to explain it, nor Centers to pretend paternity, nor Cooks Tour guides waiting at the Khyber Pass to conduct the Great Diffusion down to the forest farms of Kerala. Problem is, there's little hard archeological evidence, no Witnesses. At Taxila near Mother Indus there's an excellent museum. One can wander through it all day sniffing in the dust of Greek-Buddhist memorabilia without achieving the slightest insight of how these ancient folk and their ancestors subsisted, except by that vague (and modern) Near East cliché to wheat and cows.

8.9 China ... Man Farming Woman Weaving

Chinese farming has been described as essentially gardening. (One of the best Western descriptions of its general structure remains F.H. King, U. Wisconsin/USDA 1911.) The term becomes increasingly appropriate towards the south where more favorable climatic conditions and wider range of crop species permit intensive year-round relay planting of often spatially mixed crops, integrated with animals, poultry, fish, and for added interest turtles, eels and silk, and their endless cycling of fertility, soil conditioning and energy. *Man farming Woman weaving.* In Harlan's view Chinese agriculture as a system has historically been expansive; and it might be claimed that some of the best 'Chinese' farms are to be found in such places as the mountains of Java.   Although not based on trees they are another form of Soemawoto's gardens of complete design.

Vavilov considered practically all of non-tropical China with Korea and Japan as one of his seven centers and attributed 20% of the world's crop species to it (*...as far as the composition of wild and cultivated fruits is concerned China occupies first place in the world*). This area, much reduced to a smaller area of North China is also one of Harlan's three centers. It occupies a small rectangle along the northern slopes of the Chingling Mountains which roughly divide China into the cold millet-growing north and the warm rice-growing south. This part of Kansu-Shensi provinces is drained by the three rivers (Figure 8.5), the Huang Ho (Yellow), Fenho and Weishu or Wei and most of it is deep loess (fertile wind deposited silt).

From archeological evidence, supported by presence of these fertile soils, this 'area of the Three Rivers' is taken as the birthplace of Chinese agriculture (An Zhimin1988), and the Chinese cultures with which this earliest agriculture is associated are:

To	7 000 BP	...	Dadiwan – Peiligang – Cishan	5-4 000	...	Longshan
	7-5 000		Yangshao	4-3 000		Shan

The Yangshao in the Three Rivers are generally acknowledged as the first farmers. But this area has been subject to the most intensive archeological investigations (over 1 000 sites) and these have uncovered a picture of the Yangshao as an already sophisticated society producing such things as 'elegant pottery' and large houses by 6 000 BP. As at Harappa this kind of evidence only proves that these were not the origins. Very little direct evidence of crops is available from earlier Dadiwan times; An Zhimin (1988) mentions only foxtail and broomcorn millets, with kaoliang (Andropogon sorghum), Chinese cabbage (Brassica), mulberry trees/silk, hazelnut (Corylus) and waterchestnut (Trapa natans) being added in Yangshao times. This authority refers to...*abundant remains of farming cultures* from 8-7 000 BP but does not cite his evidence when he says that the real origins...*can be traced back to a still earlier period.*

Agriculture as Rice

On firmer ground, at least archeologically, the view is that the Langshan culture (and agriculture) which apparently developed out of the Yangshao expanded both eastward and to the south, Bender (1977) says as far afield from the Three Rivers as to Honan, Shantung, Hupei, Kiangsu and Taiwan. Remains of millet and rice became abundant by 4 500 BP. Chang (1988) attributes this rice addition to northern Langshan millet cultivators – or their technology – moving out in these directions and then applying millet technology to rice. Pursuing agriculture as rice, Chang goes back to the original home of this genus Oryza in Gondwanaland. There are some 20 wild species throughout the tropical world and two cultigens, the Asian O. sativa and the West African O. glaberrima. (Most of the world's wild rices occur from the lower Indus across to New Guinea-Northern Australia.) In the Asian area O. sativa emerged as three ecogeographical races (indica, sinica/japonica, javanica) in a long zone from the Ganges across into South China (Figure 8.5). Indica and sinica eventually entered the rest of China to 53 degrees N latitude. The earliest archeological date Chang cites for both indica and sinica in China is 7 000 BP in Zhejiang. (By comparison the other earliest archeological evidence of domesticated rice in Asia is 6 570 BP in U. Pradesh India, 6 000 at Xom Trai cave in Vietnam and 4 000 BP in Taiwan.) In Chang's model both types were domesticated in temperate China rather than in the southern homeland out of seasonal necessity: there would have been more reason to select, retain seed and improve the species here than in the tropical south where wild rice was easily available simply by collecting it from its naturally occurring stands.

But as usual, evidence invites interpretation: that rice was domesticated by 7 000 BP in Zhejiang is not proof that it was not domesticated somewhere else at some previous time; it is so widely cultivated that it was probably domesticated independently across S and SE Asia. Chang's reasons why rice was first domesticated in temperate China are plausible if this means deliberative genetic improvement. He also advances the possibility that rice could have been first managed or encouraged as a useful plant to have around by some housewife collecting wild rice seeds and noticing that this wild grass thrived in wet spots. She then would have thrown a few seeds into such a place near her house on the way home. The next step would have been to weed and protect the emergent seedlings, also proposed by Chang. But one would think that these first accidental steps would have been more productive if taken in the warm humid environment where rice occurred naturally in the South, in contrast to the alien (and seasonal) environment further north.

The Southern Hypothesis

Based on archeological evidence alone Chinese agriculture is of no great antiquity. If this indeed presents a true picture Reed (1977) attributes its late emergence possibly to the fact that the Three Rivers are further north than is the Fertile Crescent of SW Asia and emergence of agriculture would have been correspondingly delayed by the retreating last glacial ice fields. Putting archeology aside on grounds that its activities have anyway concentrated in the north and center, an alternative scenario would be that much older forms of agriculture developed in the south and that the *idea* of farming (and many plants) later spread to the north. The basis for this would amount (as with the origins in South America) to Harris' distinction between prior specialized and generalized ecosystems. For this hypothesis a 'greater south' would be defined as tropical South China through Indo China to Java. The ecozones here – again forest and the littoral, mountains and lowlands including now drowned Sunda – were and are much richer in potential crops than in the north. For example, about 33% of Vavilov's 1 000 modern economic species are from S and SE Asia; and if one half of plants from his 'East Asiatic' center are attributed to the southern part of East Asia it would mean 43% of the world's economic species were available for domestication in the greater south, only later stimulating agriculture's development in the north...unless Darlington is right and these lazy fellows were dissuaded by providence of nature 'who takes care of everything'.

8.10 Africa ... *Any consideration of the beginnings must be a survey of our ignorance* (Thurston Shaw)

If the Grail exists it must still be in Africa. Between the two rivers Nile and Congo and by the chain of lakes next to the Great Forest are likely places.

While H. erectus wandered off to the five other breeding grounds of modern people they also stayed continuously in Africa. Speculation about their economic systems is made difficult by the lack of archeological activity away from the Cairo-Khartoum beltway, confusion of indigenous crops with imports and the diversity of ecological zones ranging from desert to rain forest. Considering these factors and comparing them with the unfavorable conditions in the other regions (Near East, cold North China, the high valleys of the Andes) one might guess that agrosystems *should* have developed in Africa at least as early as they it did elsewhere (and probably 100 000 years earlier, but that depends on one's vision of the Grail and confidence in archeology).

Something Borrowed, Something New

Based on archeological evidence and its interpretation the African origins were until recently approached as borrowings from the Near East, part of the great diffusion (Figure 8.4)...*It was for long assumed that cultivated wheat, barley and flax were all introduced into Africa from the Near East. But it is now known that wild barley and perhaps einkorn wheat occur in several areas of Egypt where their food value was recognized from very early times...Most of the crops which are or have been cultivated in Africa are species indigenous to that continent and which must presumably have been first cultivated there* (Phillipson 1993). Jack Harlan (1971) also, with his intimate knowledge of African field crops, seems to subscribe to Near East diffusionism to Africa. There is the problem of reconciling Nile Valley agrotechnology (grain grinding implements) of c 20-15 000 BP with the 'first' official Near East agriculture of 10-8 000 BP (Jericho, Jarmo etc, previous list). Harlan canvasses the possibility that this Nile technology didn't represent proper agriculture because it was applied to wild not domesticated grain. It was much later taken to the Near East, applied to domesticated crops there, then these returned to the Nile so that proper agriculture could then officially commence there c 9-8 000 BP, slightly later than Near East. These tortuous migrations seem to be demanded by that fetish of domestication. A casual observer might think it simpler to admit that an African agrotechnology of 15 000 BP in fact represented an African agriculture of that date and most probably long before.

Import of livestock to Africa could be on firmer ground, especially cattle (Figure 8.4) which moved from Babylon into North Africa via that chain of *plazas de toro* through Crete to the modern one in Madrid, which Harlan calls the Mediterranean bull belt, with an eastward extension of the cow cult to Harappa. Received wisdom has it that the only *indigenous* animals domesticated in Africa were the guinea fowl and cat (in addition to those believed to originate in SW Asia). This seems to have overlooked oryx in upper Egypt and meat gazelle...*delicate wanderer, Drinker of horizon's fluid line.* However H.S. Smith (1969) has listed a total of 39 that were being used by c 3 000 BC in classical Egypt for either fattening/food, sacrifice, household pets, park decoration or cult object; and methods of handling these species for their various non economic uses would be practically equivalent to their domestication for food. Since these uses often overlapped (sacrifice/pet/food), and by 3 000 BC Egypt was already a highly developed civilization, and many of the animals were being imported from Sahara and those mysterious regions upriver to the south, it might be inferred that animal production/domestication of some sort, of many species, had been occurring there among Candolle's savage tribes millennia before evidence of it turned up on Egyptian temple paintings – hyenas being bottle-fed, and others run with more conventional stock in integrated grazing systems; stork-based 'broiler' farms; ibis farms to supply Horus' temples after his partiality to mummified ibis had caused a market shortage; giraffe ranches in Sahara to supply other cults (or Memphis butcher shops?). By that time also there's evidence of modern high pressure livestock systems in lower Egypt – a sheep engineered to grow a tail so large and fat it had to be supported on a small trailer which the sheep pulled, and similar now increasingly familiar grotesqueries (Diversity, genetic engineering).

Smith's uses don't include animals-as-spectacle (and it would be stretching the imagination to see the pussycat domesticated as a substitute for lions after the Sunday morning lion hunt in royal parks faded from vogue, or these larger felines were exhausted); or animals-as-weapons systems. And Duncan Noble's (1969) account of the onagar for this last use might easily be taken for low comedy (though this is from Mesopotamia before c 4 000 BC). For battle these 'half-asses' of little more than a meter high were hitched four abreast to a sort of miniature narrow and unmaneuverable Polish-German farm wagon, with a pole. A groom pointed this contraption towards the foe then jumped clear and it cut a swath 2.5m wide through the opposing team; and since the onagar is fleet though lacking stamina the one man crew (plus onagar prodder) had time to jab only one or two of the opposition in his passage: the main armament of this platform were the 16 flailing hooves of the half assed onagar. All this suggests how plastic one's concepts of early 'farming' must be, in Africa or anywhere (Genesis).

Pride and prejudice As Portères and Barrau (1981) and many others now see it, for a long time the real origins of agriculture in Africa were obscured by two perceptions. The first was that recurring theme of agriculture, as handmaiden of civilization, having to originate in the Near East; and consequently – with a few inescapable exceptions – whatever legitimate agriculture Africa possessed must have been imported from that garden of Eden...*Ideas about the origins of agriculture have for a long time been pervaded by some measure of ethnocentricity.* The second was colonialist prejudice which...*allied to ignorance, led to a long period when the participation of Africa in developing agricultural techniques...was either played down or disregarded altogether.* Here Portères–Barrau might have had rice in mind, for although Strabo the Greek wrote about this indigenous West African crop (Oryza glaberrima) about 20 BC, and Ibn Battuta penned a reminder (postmarked Niger Bend, 14th century) it was believed in the West until the 20th century that African rice must have come from East Asia, Indonesia. These perceptions have been reinforced by two further factors: lack of archeological evidence of anything much at all, as Shaw has complained, and again that devil domestication which sees the Origins through the eyes of an Iowa corn farmer.

NE Africa – Nile Valley From archeological evidence uncovered so far (eg Phillipson 1993) the pre agricultural system of NE Africa (Sahara-Nile) was a broad based one which emphasized fishing and/or hunting, depending on proximity to rivers, lakes. However this apparent dominance is based on the durable evidence of bones and stone/bone harpoon heads. Plant foods leave far less evidence, and as Phillipson points out a further cultural bias might be added in that although meat/fish are important today they might not have so dominated food supply in the past. By at least 20 000 years ago populations had concentrated in the Nile Valley and were also exploiting whatever wild plant foods were available. Generally this meant wild grass grain as precursor to granoculture; but from grindstones and charred remains (Wadi Kubbaniya, Aswan) it also included the tubers of nut-grass *(...a salutary corrective to our tendency to emphasize ancient use of cereals at the expense of other plant foods).* Of these cereals, by at least 12 000 years ago wild barley was being collected in Upper Egypt, and there are enough animal bones lying around in concentrations to suggest that animal management of herds, if not their 'domestication,' could by then also have been part of the economic (or theological) system. From other sources there is also evidence of grain processing/grinding from Abu Simbel and Luxor c 15 000–14 000 BP and further indications of 'domestication' of meat gazelles and oryx in this same period.

Earliest domesticated farming   From the archeological record so far these economic systems dating from 20 000 BP consisted of *collected* grains and roots. It is impossible to proclaim when proper agriculture arrived, partly – in addition to incompleteness of archeological evidence – because of the perceptions introduced by the formal agricultural requirements of domestication and cultivation. If these formalities are observed the first proper farming seems to have occurred in the then more moist Egypt-Sudan Western Desert at Nabta and at Kharga oasis (Egypt) some time before 9 000 BP. (*The Nabta barley is the earliest confirmed occurrence of a domestic plant in the continent of Africa* – Phillipson), and sheep and/or goats were also then present at Kharga, upper desert Egypt. Also at Nabta from this time the remains of dom and date palms have been found, along with certain weeds which flourish on cultivated ground. The significance of these is that they are species indigenous to Africa, not Near East. Further out in the Sahara improved shorthorn cattle were kept at Uan Muhuggiag in SW Libya around 7 000 BP. From the archeological record this (domesticated) agriculture occurred first in the Sahara before the Nile, explained by favorable climate change after the Last Ice Age passed (Bender 1975).

Alternative interpretations   Considering only archeological evidence of field crop farming in this part of Africa there's a very long period over which the origins can be contested. Should they be dated from 20 000 BP when the technology of harvesting and processing the wild grasses and roots was present? Or from 9–8 000 BP with grass domestication? Or from 7–6 000 BP with the archeological evidence of agricultural towns? To consider this more fully it's necessary to list the general evolutionary stages, as technical operations, leading to fully developed formal grain agriculture (Harris and Hillman (1988) offer a more comprehensive treatment). The model of Table 8.3 suggests how the technology of the operations in crop production accumulated over these 13 000 years or so, and how unreasonable it would seem to wait until all the bits are formally in place (the stage 'bred/domesticated') until proclaiming the safe arrival of farming in NE Africa c 7 000 BP. As Phillipson sees it...*It now appears highly probable that the emergence of fully food-producing economies was the result of a far longer period of intensive exploitation and experimentation than was previously realized.*

Ethiopia   Vavilov was so impressed by one small corner of Africa, the Ethiopian highlands (with Yemen and Somalia) he made it one of his initial eight centers, though he later concluded it did not warrant this status and dropped it. Of the world's then (1935) 1 000 main agricultural plants only 40 were indigenous to this smallest center (Table 8.4) but the diversity of their forms was remarkable...*Abyssinia must be put in first place with respect to the number of hard wheats. It is also the center of formation as far as barley is concerned...nowhere else is there such a variation of its forms and genes as in Abyssinia. This is also the native land of the peculiar cereal called teff and that of the oil bearing noog (Guizotia abyssinica). Flax is represented by a special form (Linum usitatissimum) used not for its fibers or oil but exclusively as a cereal.* There was also...*a special kind of banana* (Musa ensete, used for bread flour) and arabica coffee and...*in part this is also the native land of grain sorghum.* Ethiopia was until recent times largely covered by mountain and deep valley forests. Only wisps of these remain. Such has been the concentration on grains and oil seeds of modern relevance that there is little information on what these economic forest plants might have been in prehistory (or for that matter what they are now).

Table 8.3   Technologies Emphasized in Successive Agricultural Evolutionary Stages

Operations	Evolutionary stages, technology advance		
	(1)	(2)	(3)
Harvest:	hand strip	knife	sickle
Process:	rocks	stick/trough	grindstone
Store:	none	baskets	house/mud bins
Plant/sow:	none	infill gaps	whole field
Cultivate:	none	scratch gaps	whole field
Irrigate:	none	wild flood	channels
Clear land:		burn for removal	burn for nutrients*
Seed type:	indiscriminate	selected	bred, domesticated
	COLLECTING 20 000 BP		FARMING 7 000 BP

*Shifters

West Africa   Harlan (1971, 1972, 1975, 1986, 1988) has listed the crops arising in West Africa (west coast to Ethiopia) but refused to designate it as a 'center' because it is so vast and...*the archeobotanical evidence is essentially nil.* Shaw (1977) offers a generalized evolutionary model based on 34 archeological sites from Atlas Mountains south to the humid forest, but half these are concentrated in the south west corner (Guinea-Ghana) leaving only 14 or so to speak for the other 5-6 million sq km of savanna-sahel-rain forest. The general picture is of farmer-pastoralists moving south from the Mediterranean to the central Sahara uplands where they left evidence of temperate grains and livestock. Here they met Negroid people going north. The woolly sheep were dropped of necessity and the temperate grains replaced by savanna grasses, and with cows and tropical grasses as its basis agriculture continued on south to humid West Africa. This is the great Near East diffusion, adjusting itself as it went along. Another version is that these Saharans, however they got there, were driven out as the Sahara moisture peak passed (there were three of these in the period 10–2 000 BP) and the country again dried up.   This climate-landscape change also forced retreat of the tse-tse fly, allowing cattle to survive in larger areas of West Africa.

Table 8.4   Some Indigenous African Food Cereal Grasses

Nile – Ethiopia—

finger millet	Eleusine coracana	Ethiopia–East Af. Highlands
oats, Ethiopian	Avena abyssinica	Ethiopia (intro. Near East Weed?)
teff (small grain)	Eragrostis tef	Ethiopia
**wild barley		Libya–Egypt–Ethopia
	**Echinochloa	Nile Valley
	**Aristida	Nile Valley
	**Cenchrus	Nile Valley
	**Panicum	Nile Valley
	**Setaria	Nile Valley

West Africa—

sorghum (Guinea corn)	Sorghum bicolor	Chad–Sudan savanna
rice, African	Oryza glaberrima	west savanna waterholes
fonio ('hungry rice')	Digitaria exilis	west savanna
black fonio	Pennisetum typhoides	Sahara–Sahel (desert margins)
guinea millet	Brachiaria deflexa	Guinea highlands (restricted)
*sudan grass	Pennisetum purpureum	savanna generally

*representative of grasses which might have been used as a cereal food but this use is not known;   **long-grain grasses probably used prehistorically as cereal food in Upper Nile–Ethiopia.
Source: Compiled from Bender 1975, Shaw 1973, Harlan 1972, 1975, 1988.

Diffusion by Idea - of Watermelons, Xylophones and Roosters

Carter (1977) offers a model of the African origins based on the diffusion of *idea* or stimulus to domesticate, this being a half-way house between the migration of things and their independent local invention, ideas travelling much faster than things. (This is not new, eg Hutchinson 1965, 1972.) To a large extent it bypasses archeology and relies on cultural history (Candolle). In Carter's scheme the three main contact points or zones of infusion of both animals-plants and ideas would have been the West Africa bulge, the great northern loop of the Niger River, and Ethiopia (Figure 8.4). The ideas would have been brought to these diffusion points by Phoenicians who had long visited the West African coast, by caravans crossing to the Niger bend, and by ships passing down the Red Sea. By the first route it would not have been necessary to bring melons to Africa, only the idea of melon domestication from Near East to be applied to cultivation of the indigenous West African watermelon; granoculture could have arrived as the idea of applying Near East wheat and barley technology to the savanna grasses (Table 8.4), and this also could have accounted for the first cultivation on the Niger of the indigenous West African rice.

Rice also presented Carter with more intriguing possibilities. It has long been known that the Malay people had ancient contact with the East African coast. By at least 2 000 years ago they were settling Madagascar. The proto Malay language groups had already extended from their hearth (which some hold was the Ganges Basin) west to Madagascar, north to Taiwan and east well into Oceania, ie over 55% of the globe's circumference. And although Western obsession with Troy and Jericho has largely obscured the fact it will not be forgotten that well before the Christian era the seas from China through Indonesia to India and Lanka and beyond were already a network of maritime trade channels, and all this implies for movement of plants and ideas. It would require no great leap of imagination (Carter's model) to see these Malay seafarers pressing on beyond Madagascar and bringing the idea of rice cultivation from, say, Indonesia, to be applied to the domestication of wild African rice on the West African coast or Niger River – as later they did with the domesticated Java chicken when it turned up on the East African coast bearing an Indian name and by then – as one might anticipate from such experience – its use well circumscribed by Hindu ritual.

The Levant. Carthage. More Phoenician galleys. Desert crossings of the great heat in Berber caravans. Proto Malay linguistics. Comparative musicology of Indonesian v West African xylophone styles. Hazardous voyages with drunken mariners to Punt. The adventures of Hindi-squawking Javanese roosters. Some of this might have happened. Perhaps it all did. One follows Douglas Carter not for his facts, though these are considerable, but for the Quest as it always must be. A wondrous journey through the possible.

Concentric Zones around the Great Forest

Roland Portères and Jacques Barrau (1981) construct their model of the African origins on ecological principles, based largely on concentric ecozones around the Great Forest (Figure 8.6). These are successively forest, savanna, desert, Mediterranean, the northern sequence repeated in the south. Applying David Harris' terminology to this, the central core of the Congo-Nigerian forest is the richest and most generalized natural biome and the forest dwellers did not need to become agriculturists because of these resources. (This is the same excuse the more thoughtful of the great apes offered to Pfeiffer for turning down evolution's invitation to come out of their 'green caves' and be the ancestors.) The outer Sahara and Kalahari desert zone is one of natural plant/animal specialization and human economic systems here also became specialized in gathering-hunting (eg Kalahari Kung). The middle savanna zone, especially that part adjacent to the forest and half way between the biome extremes of generalization and specialization had both resources and need and gave rise to agriculture, and here also but in recent times, citing Randall...*it is at the margins of the two savannas...that are to be found the most prestigious of the Bantu (black African) civilizations.*

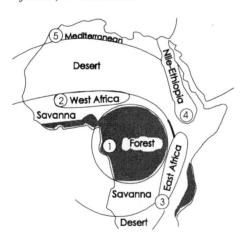

Figure 8.6 African Biomes and Centers

These are only broad biomes. Within them Portères-Barrau reintroduce the concept of Vavilovian Centers, now the broader cultural term 'agricultural cradles.' Indeed, far from dismissing Vavilov as had become fashionable they acknowledge they were completing in Africa what Vavilov had started. Where he had found two centers (Ethiopia and Mediterranean coast) they accept these two (adding Nile Valley to Ethiopia) and add four more: East Africa highlands of Kenya-Tanzania-Malawi but extending west around the Congo forest to Angola; West Africa savanna – Niger River; West Africa and central Congo forests. The Mediterranean cradle is both a center of origin of some plants and an import zone of Near East plants (some grains and bulls from Anatolia); and in later historical times this outer peripheral circle could be extended all around Africa to reflect imports/exports between Africa generally and Asia (Carter) – sorghums out via Ethiopia-Somalia (Punt?) to Harappa, bananas and taro in via Madagascar from Far East. The initial simplicity of this zonal-cradle model rapidly became obscured by exchanges among them – forest vegeculture plants to savanna and oasis farmers, and always the savanna people with their panicums and oil seeds and cows encroaching on the Great Forest.

It's long been recognized that Africa has two agricultures, the hoe of the savanna (granoculture) and the forest digging stick (vegeculture, yams). The digging stick is no doubt oldest; but Shaw's (1977) discussion of the archeological evidence also includes stone 'hoes' from West Africa forests c 12 000 BP; and Sangoan stone 'picks' have been recovered from around Lake Victoria thought to be from c 40 000 BP. For vegeculture? Or dispatching fierce beasts? Or – since similar blunt stone implements were used almost universally to beat the thin bark off forest trees and burn not cut them down – are these more finger prints of Shifters? Whether stick or hoe, vegeculture belongs to the forest margins, clearings. This still sheds no light on *aboriculture* within the forest. As in Amazonia this strand has not been followed assiduously; eg of Harlan's list of 60 African crops of modern relevance only about five are from the forests and none specifically from the Congo forest. Such is the lack of information about these ancient forest agroecosystems that one can only acknowledge their existence (Reconnaissance). But leaving Roland Portères and Jacques Barrau the Quest doesn't seem to matter much. There *was* no sharp divide between agriculture and the previous darkness...*Even in pre agricultural times man carried with him in his migrations, implements, techniques, modes of understanding and interpreting the environment, a whole range of attitudes and behavior that had grown out of his relationship with nature*...which seems a more sensible view than that enlightenment can't be proclaimed until two grains of barley are discovered in a tomb.

8.11 Mexico

H. sapiens' (or H. erectus?) ETA in the New World is still a matter of vigorous dispute;   the archeological evidence is open to various interpretations and a good part of the game has been casting suspicion on other claimants' Witnesses.   By the most conservative interpretation (Lynch 1990, Piperno and Pearsall 1998) the immigrants arrived from Asia not much before c 13 000 BP; the first signs of recognizable agriculture began emerging in upland Mexico around 8 000 BP (Table 8.3). By less cautious interpretation H. sapiens as 'hunters' had occupied all the New World to Patagonia apparently before 20 000 BP (Texas 19 600 BP, Mexico 24 000, Monte Verde in southern Chili 14 000 BP) and were hunting alpacas and llamas at Ayacucho in upland Peru by 20 000 BP, with respectable claims extending to human presence (H. erectus?) at Esperança Cave in eastern Brazil by 295 000 BP, one commentator taking this further: *It is not surprising then that H. erectus who occupied the Chinese mainland from at least 700 000 years ago and had domesticated fire 400 000 years ago would have crossed the Bering Strait a number of times.* Much then depends on the rhythms of glacials and interglacials (graphed previously, the Caravan of Circumstance) in creating the land or ice bridge assumed necessary for crossing that barrier. In Betty Megger's (1979) scheme a second barrier to the south would have been the 'monstrous glacier' that cut across middle North America, west to east. While the land bridge was open at many times in the past, eg between 50 and 40 000 BP and 28 and 10 000 BP, and allowed entry of Asian caribou and mammoths, the great glacier might have barred H. sapiens moving south from Canada. Or maybe not (Genesis, Carl Sauer).

Highlands – the Origins as Corn – Beans – Squash

Nowhere more so than Mexico have the agricultural origins been as keenly pursued as origins of crops of contemporary importance, in this case corn/zea mays. Interest in the Mexican origins increased following the first discovery of ancient domesticated maize from 4 500 BP at Bat Cave, New Mexico, and leading one of the excavators (MacNeish 1958, 1964, 1965) to extend the search further south into Tamaulipas State, Mexico, then further south again into the Tehuacán valley of Puebla State. Another early worker in Tehuacan was Kent Flannery who moved further southeast into the valleys of Oaxaca State. Most of what is known about earliest Mexican-Central American agriculture is from these sites. The several volumes of information from the archeological sites have been summarized by

Bushnell (1977) and in an American continental perspective by Hammond (1977). The more important of the early Mexican crops and their approximate archeological records are listed, Table 8.5, though some of these are common to most of the American tropics.

Table 8.5 Some Early Domesticated Crops in Tamaulipas State and Tehuacán (Peubla State)

gourd (bottle)	Legenaria siceraria	8 800 BP
chilli peppers	Capsicum annum	8 700
summer squash	Cucurbita pepo	8 700
walnut squash	C. mixta	8 700
avocado	Persea americana	8 700
amarinth (cereal)	Amaranthus leucocarpus	8 700
maize	Zea mays	7 000
cushaw squash	Cucurbita moschata	7 000
black zapote	Diospyros ebenaster	7 000
white zapote	Casimiroa edulis	7 000
common bean	Phaseolus vulgarus	6 000
runner bean	P. coccineus	6 000
tepary bean	P. acutifolius	5 500
cotton	G. hirsutum	4 000
lima beans	P. lunatus	3 500

Source: Mainly from Bender 1975.

The circumstances Barbara Bender (1975) has described the economic system in these semi arid Mexican uplands before 9 000 BP as primarily hunting of larger herd animals supplemented by wild plant food collecting. Then with further retreat of the last glacial the larger animals also moved north following the cooler and wetter climate, and some became extinct leading to concentration on the smaller animals (rabbits, rodents) and on plant foods. By 8 000 BP as the large herd animals disappeared the plants were accounting for about 40% of all food supply. These were mesquite seed (Prosopis julieflora), setaria grass seed (Setaria macrostachya), teocinte grain, prickly pear in the dry season (Opuntia spp), acorns, pinyon nuts, hackberry, avocado (Persea americana). Flannery (1968, 1973) has called attention to the many skills needed by these pre farmers and the close knowledge they must have had of plants before they formally domesticated them – how to leach tannic acid from the acorn, make tongs to harvest fruit from cactus, extract syrup from mesquite pods... Paleo-nutritionists who know about these things reckon it then took three days to de-prickle, skin, broil and otherwise bully a cactus pad into edible condition. Taking a cultural rather than an agronomic view more recent commentators will argue that the origins should properly be dated from the emergence of these plant-applied skills rather than that troublesome thing domestication (which to some extent happened anyway by human presence – Rindos).

Also about this time (Table 8.5) the first signs of plant cultivation appear – of setaria, avocado, chili peppers, amaranth (grain), squash and some authorities say this deliberate cultivation extended to maguey, prickly pear, setaria and hog-plum (Spondius mombin). Because the large animals had disappeared or been hunted to extinction (the early horse and antelope) livestock never became a part of Mexican farming until after the Conquest. (Apparently the only meat animal domesticated was the dog from at least 5 200 BP at Tamaulipas). This gradual evolution from gathering-huntering to sedentary civilization based on formal farming in one highland valley is sketched in Table 8.6.

Table 8.6   Population and Economic Changes in the Tehuacán Valley: Pre 8 700 BP to 450 BP

Phase date BP	Valley population	Subsistence from (%):			Socio-economic development stage
		Hunting	Plants	Crops	
8 700 BP	20	%	%	%	wild plants, game drives
7 000	70	55	40	5	incipient agriculture
5 400	180	35	50	15	agriculture proper
4 300	720	30	50	20	some full-time farming
2 800	1 020	30	30	40	villages, temples
2 100	2 200	30	20	50	irrigation starts
1 200	18 000	20	20	60	full irrigated farming
500	90 000	20	10	70	cities, commerce

Data are rounded, last column simplified.
Source: From Reed 1977 based on MacNeish.

The Mexican Origins and Farm Economics

As he did for Near East agriculture as wheat/barley granoculture, Kent Flannery (1973) offers a model of formal Mexican agriculture developing as largely corn and its companions beans and squash.   In this the grasses/grains were at first relatively unimportant in comparison with the other wild foods.   Two of these grasses are foxtail/setaria and teocinte. Teocinte is the ancestor of modern maize. (In later Nahuatl-Aztec times it was called *teocentli* or 'support of the Gods' – Hammond.)   The natural home of these grasses is in the slightly more humid tributary valleys, *barrancas*. In normal years teocinte was apparently the less important and most difficult to harvest but when they both began to be cultivated in the barrancas instead of being simply harvested it, not setaria, responded better and was eventually developed into recognizable maize. Another favorable factor Flannery notes is the habit of teocinte to rapidly and thickly colonize any clearing. It could be that initially it planted itself on disturbed land as a 'crop', a habit which might have induced Indian observers to later mimic with their cultivation.

At this point farm economics intervened. Natural teocinte yields are only around 80 kg/hectare of useful food. From the archeological record it could have taken up to 3 000 years for this to increase to the 'threshold level' of 200 – 250 kg/hectare. According to Flannery this threshold would probably have been set by the 160-180 kg of edible pods recoverable from the other main food of wild mesquite. Only when teocinte-maize exceeded this natural mesquite yield would it have been economic for these earliest farmers to bring this primitive maize out of the small barranca fields, clear away the dominant mesquite of the main valleys and expand maize cultivation there. Farm adjustment had been invented and Mainstream B was knocking at the door.

Crops as systems The most interesting aspect of this ancient Mexican agriculture is that for the first time it admits the concept of wild plants emerging as integrated crop *systems*. Flannery calls attention to the natural habit of wild runner beans (Phaseolus sp) growing in association with teocinte to use it as a live trellis, with wild squashes providing a ground carpet. When teocinte came into the village so did these others, induced or as weeds. With their later cultivation...*the Zea-bean-squash triumvirate is thus not an invention of the Indians; nature provided the model.* This integrated system became near-universal as part of Stream A's repertoire wherever corn is grown and this continues. But generally the old association has been abandoned, the triumvirate developed separately. Specialist technology and agro-economic agonizing over 'optimum' input levels of fertilizer, agricides, soil moisture, replaced the plant symbiotic relationships which the Indians had found in the *barrancas*. Mainstream B as monoculture (and Donald Lathrap's future Midwest 'monstrous corn') had triumphed. Mutuality rejected. No more *e pluribus unum*. It's a good story. Whether it has much to do with the real Mexican origins is doubtful (Genesis).

The Mexican Lowlands and Drowned Evidence

Because the Mexican highlands are main repository of the agricultural evidence here (Table 8.5) this has obscured near certainty that farming systems of greater antiquity must also have developed in the coastal lowlands (partly for reasons of their greater species diversity – Vavilov, Harris). Hammond (1977) has called attention to the remains of the corn-beans-squash complex being found where conditions for preservation in dry, cool upland caves are optimum. He also notes maps prepared by Bartlett (1969) show that sea level rise (with melting of the last glacial and particularly in the Gulf of Mexico) has drowned a coastal plain which varied from 20 to 120 km wide, along with its evidence of all prehistoric economic systems. In support of a lowland Mexican origins, while most of the recognized Mexican civilizations were from and in the highlands (Teotihuacan, Zapotapecs, Toltecs, Aztecs...) one earliest group were the Olmecs who flourished on the coastal lowlands of Vera Cruz state; their abandoned centers exist at San Lorenzo and La Venta. One might easily imagine the ancestors of these people slowly retreating in face of a post glacial rising sea level, rising water tables and waterlogged fields. There's at least an even chance that the same post glacial climatic changes which stimulated the origins in the Near East (Circumstances) also drowned evidence of these origins in Mexico. More likely the so-called pioneers of the *barrancas* were refugees from something better. Surely no one would really have gone to live in those dry inhospitable uplands until pressed by circumstance. What sane Mexicans, Neolithic or modern, would willingly turn their backs on *la dolce vita* of the Vera Cruz coast to go up and shiver and live on grilled cactus in a cave? At least two eminent archeologists, Piperno and Pearsall (1998), have now rejected the barranca origins...*We conclude that the humid tropical lowlands were the major settings for the origin and development of agricultural systems...The time has come for the Tehuacán (and Oaxaca) plant remains to relinquish their role as the core data base...They probably represent encampments of small mobile groups, essentially hunters and gatherers, who received plants from their lowland neighbors that were already domesticated...*

The Central American Origins as Huertos Selvas, Forest Gardens

Moving further down this other side of paradise, Dolores Piperno (1988) offers a vision of the origins as they might have been in Panama, based on archeobotanical evidence from 500 sites in more burnable semi-deciduous forest on the Pacific side, ie forest more easily coverted to farming by the later shifters:

1. Forest foraging-collecting to 9 000 BP: Some 80% of the archeo-botanical evidence for this and the following period is of tree species. These earliest subsistence systems terminated because of limitations of the forest as food supplier: the wide spatial dispersion of food-bearing plants; small size of the then undomesticated fruit; marked seasonality of supply; and this latter resulting also in seasonal scarcity of the dependent fauna as the other human food resource.
2. Gradual conversion of forest gathering to some form of tree management, perhaps horticulture.
3. Further conversion to slash-and-burn field crop farming, probably first manioc/yuca (from Venezuela?) then later c 7 000 BP teocinte/maize from Mexico (on its way to South America, clocked by archeology as travelling at 2 km/year, or twice the speed of Near East granoculture spreading into Europe). This slash-and-burn period lasted 4 000 years, allowing a greatly increased population but devastating the landscape.
4. Retreat of the shifters to more fertile and less susceptible valleys, from 2 900 BP; consolidation of maize-dominant sedentary field crop farming. Piperno doesn't explore the continuing role of the agroforests as *huertos selvas* or *huertos familiares* in stages (3) and (4). Perhaps in Panama they're not conspicuous; but further north in Guatemala-Belize-Honduras, towards the Corn Belt proper, Turner and Miksicek's (1983) review found that agroforestry had continued in parallel with field crop farming to classical Maya times (before 1 000 AD). By around 2 000 BC a highly developed civilization as pottery, town plazas, highly developed canal systems already existed. The main field crops were maize-squash-chili-cotton; but 12 of the 19 recovered plants (by 1983) were tree foods and 3 others were tree dyes and resins. This seems to take us to the Central American origins as forest gardens or hueros selvas, not cornfields, in Piperno's stage (2), as Wiseman (1978) had previously described them...*artificial rain forests, selective clearing of the forest so that useful species are retained for food, fiber and other support. The result is not a cleared landscape but a 'used forest'.*

8.12 South America

Thurston Shaw's 'survey of our ignorance' in Africa continues in South America: in this vast area only the NW corner of the Farm has been reasonably well turned over, Peru-Ecuador-Colombia and smaller bits in

Venezuela, replacing the Nile-Nubia beltway. Elsewhere few Witnesses have come forward. As late as 1979 in Betty Megger's map of the main South American archeological sites, all of the Amazon Basin carries the warning 'no evidence'; and in Piperno and Pearsall's (1998) more recent survey this situation has not appreciably changed. What evidence there is has been subject to more than the usual warmth of disputation.

Vavilov attributed 17% of the world's agricultural plants to South America, with another 17% to Mexico-Central America; Table 8.7A and B is a reminder of their global importance and some indicative dates from Andean archeological sites are noted. With a few exceptions, notably the Andean root crops of part C, the great majority of these crops are from the tropical lowlands and must have been developed there as crops millennia before they were carried up to the classical Peru/Andes 'center'. Also, in Harlan's (1975) short list of 46 most important South American food plants only 8 of these are not from the lowlands. From the perspective of prehistory these and similar lists-of-earliest crops are probably grossly misleading, relating to plants which are of obvious contemporary importance. Contemporary ethnobotanical surveys through Amazonia – much more reliable guides in these matters than archeology – have identified hundreds of the probably thousands of plants which must have been important in ancient times – foods, beverages, fibers, medicinals, magicals (Genesis).

Contending hypotheses   Nearly all accounts of the origins in South America are concerned with getting the Asian immigrants up to the Andean valleys as quickly as possible so that they can commence proper farming and leave archeological evidence. The Seekers can follow and be confused by several hypotheses, pointing in various directions but sharing the virtue of being untestable. By the first it would have been impossible for the immigrants to pass down through Central America as forest hunters and foragers without becoming skilled in the uses of the new plants, adding these to their expanding botanical repertoire and finding this greatest forest no barrier. By a related hypothesis, as hunters they'd have soon encountered the effects of climate change: temperatures and rainfall, recovering from the last Ice Age maximum (c 20 000 BP) were still lower than they would become when post-glacial recovery was complete (c 11 500 BP). During that glacial the neotropical forest had been more open and accommodated many more large game but these now starting to disappear soon after the hunters arrived as the forest canopy again started closing in (Piperno and Pearsall 1998).

A few of the elders would have started muttering about going home to

Table 8.7 Some Representative South American Lowland/Forest Crops and Upland/Andes Root Crops

(Dates according to those recovered from Andes archeological sites)

A. Some tropical lowland ground crops BP

manioc	(R)	Manihot (to 100 spp)	(3 800)
achira [1)]	(R)	Canna edulis (10 more)	(4 400)
arrowroot	(R)	Maranta arundinacaea	
leren	(R)	Calathea allouia	
cocoyam	(R)	Xanthosoma spp (to 40 spp)	
yam	(R)	Dioscorea (many spp)	
sw.potato	(R)	Impomoea batatas (15 spp)	(3 900)
taro	(R)	Colocasia esculenta (etc)	
tobacco		Nicotiana spp	
corn		Zea mays	(5 000)
peanut		Arachis (40 spp)	(3 800)
jack beans		Canavalia (20 spp)	(7 600)
lima beans		Phaseolis lunatus	(5 600)
common beans		P. vulgaris	(5 600)
cotton [2)]		G. barbadense	(5 000)
bottlegourd [3)]		Lagenaria siceraria	(7 000)
squashes		Cucurbita (5 spp)	(4 800)
peppers		Capsicum (29 spp)	(4 500)
pineapple		Ananas comosus	

(R) root crops. [1)] also subtropics; [2)] also dry tropics; [3)] ancient in South America but ex-Africa.

B. Some representative lowland tree/bush crops

rubber	Havea brazilensis	cherimoya	Annona spp
coca	Erythroxylon coch	inga	Inga (40 spp)
cacao	Theobroma cacao	papaya	Carica spp
brazil nut	Bertholletia excelsa	mombin	Spondias spp
cashew	Anacardium occidentale	sapote	Diospyros spp
avocado	Persea americana	Others	(at least 100 for food)
guava	Psidium guajava	Palms	(up to 250 – Diversity)

C. Some upland/Andes root crops

potato	Solanum tuberosum	olluco	Ullucus tuberosus
oca	Oxalis tuberosa	anu	Tropaeolum tuberosum
maca	Lepidium meyenii	ajipa	Pachyrrhizus ahipa

Source: Compiled from Bender 1975, Patterson 1971, Hammond 1977, Harlan 1975, Piperno and Pearsall 1998.

Africa, their discontent based on memory that back home on the savannas the large animal biomass had been 20 000 kg/sq km whereas this enclosing forest could now manage only 1 800 kg, and small elusive ones at that. Those who didn't want to go back would have protested they were not really hunters, never had been. That idea was closer to Near East-Eurasian folklore than tropical fact (and those who'd been inclined that way had lagged behind to chase mastodons in Texas). By now their 'hunting' was along rivers, directed at aquatic animals – cayman, manatee, rays, turtles and turtle eggs – more like foraging, not much affected by the glacial/interglacial drama to the north. But they all agreed – what with the increasing population and the supply of wild *curagua* fibers becoming exhausted and the *curandero* having to go further out to find his medicinals and the women wandering off in search of spices and magicals and getting snatched by rival bands – they all agreed it was time to concentrate all these resources around the village and invent agriculture. It came about like this:

By the disturbed habitat hypothesis economically useful plants would have emerged and concentrated more or less spontaneously as weeds in disturbed primary forest, with this created accidentally by human presence, and leading to deliberate forest disturbance, with the byproduct of attracted game. By a second (they're all complementary) the most disturbed of these habitats was the village itself and the future crops emerged there from plant remnants and discarded food brought back by forest collectors (Genesis, Edgar Anderson). By a third (Sauer, Stearman) nature herself provided the idea of the greatest plant productivity and diversity in disturbed habitats through forest tree falls, riverbank erosion, land slips (Flannery and teocinte in Mexico). By a fourth (unclaimed) disturbed habitats are not necessary: while the species and resources of a tropical forest are typically thinly dispersed the great palm groves are models of their concentration, suggesting the possibility of farms as places of similar resource concentration (Diversity; the aguajales). By a fifth (also unclaimed but based on Stearman) women thought of it first during their duties as 'beaters and retrievers of game,' filing away their knowledge of the plant resources they encountered, to be later collected, then thinking how nicer it would be to take these home and grow them closer to hand so they wouldn't have to go out retrieving game for the men (Genesis).

From whichever of these routes the plants were moved from forest to village garbage dump, then to village household garden (and with the plants unknown but probably including manihot/cassava), some proved so successful that more space was needed, obtained by forest slash-and-burn clearing. From about 7 000 BP the new fire-and-digging stick economy spread through the tropical American forest 'like a radiating organism'...*At this point slash-and-burn took off and never looked back* (Piperno and Pearsall). Eventually it reached the high cool valleys, of necessity dropping some plants and replacing them with others (the root crops, Table 8.7C) as it went, to be intensified over time and transformed under population pressure into an agronomy more recognizable and biblical, its remnants more easily dug for. Mainstream B, the economy of peasants and great Princes.

To be convincing this roughest of outlines would need more Witnesses. Some would have remembered what use was really made of the palms, whether as farm models or simply resource islands to navigate among millennia before the first New Crops as. corn arrived in the Andes classical 'hearth' from Mexico or the idea of root crops arrived there from Venezuela as manioc. Other Witnesses would be wise in the ways of radiating organisms and recall exactly how they worked, though Piperno and Pearsall drop tantalizing hints of brides delivered long distances from their forest bands and carrying their secrets – and no doubt baskets of their favorite medicinals and magicals – who could have spread the new agronomy as fast as a bush telegraph. Other Witnesses would have called their memoirs 'The True Origins as Agroforestry' and been amused by agriculture's pretence of hurrying up to the high valleys to establish itself as proper field crop vegeculture when, in the trees-palms-vines of places like Santa Rosa, it had already got there, a climax agro destination in itself (Genesis).

References

Allchin, B. and F.R. Allchin (1968) *Birth of Indian Civilization: India and Pakistan before 500 BC.* Penguin, London.

Allchin, F.R. (1969) Early cultivated plants in India and Pakistan. In P.J. Ucko and G.W. Dimbleby (eds) *The Domestication and Exploitation of Plants and Animals.* Duckworth, London.

Allchin, F.R. (1977) Hunters, pastoralists and early agriculturists in South Asia. In J.V.S. Megaw (ed) *Hunters, Gatherers and First Farmers Beyond Europe.* Leicester University Press, UK.

Anderson, E. (1952) *Plants, Man and Life.* University of California Press, Berkeley.

An Zhimin (1988) Prehistoric agriculture in China. In D.R. Harris and G.C. Hillman (eds) *Foraging and Farming: The Evolution of Plant Exploitation.* Unwin Hyman, London.

Bender, B. (1975) *Farming in Prehistory: from Hunter-Gatherer to Food Producer.* John Baker, London.

Binford, L.R. (1968) Post-Pleistocene adaptations. In S.R. Binford and L.R. Binford (eds) *New Perspectives in Archeology.* Aldine, Chicago.

Bronson, B. (1977) The earliest farming: demography as cause and consequence. In C.A. Reed (ed) *Origins of Agriculture.* Mouton, Hague.

Bryant, E. (1997) *Climate Process and Change.* Cambridge University Press, UK.

Bryson, R.A. (1975) The lessons of climatic history. *Environmental Conservation* 2(3):163-179.

Bryson, R.A. and T.J. Murray (1977) *Climates of Hunger: Mankind and the World's Changing Weather.* University of Wisconsin Press, Madison.

Bushnell (1977) The beginning and growth of agriculture in Mexico. In J. Hutchinson and others (eds) *The Early History of Agriculture.* British Academy/Oxford University Press.

Carter, G.F. (1977) Hypothesis suggesting a single origin of agriculture. In C.A. Reed (ed) *Origins of Agriculture.* Mouton The Hague/Aldine Chicago.

Chang, T.T. (1988) Domestication and spread of cultivated rices. In D.R. Harris and G.C. Hillman (eds) *Foraging and Farming* op. cit.

Darlington, C.D. (1969) The silent millennia in the origin of agriculture. In P.J. Ucko and G.W. Dimbleby (eds) *The Domestication and Exploitation of Plants and Animals.* Duckworth, London.

De Candolle, A. (1882) *Origin of Cultivated Plants.* Hafner Pub. Co., New York (1959).

Diamond, J. (1998) *Guns, Germs and Steel.* Vintage, London.

Edwards, D.A. and J.F. O'Connell (1995) Broad spectrum diets in arid Australia. *Antiquity* 69(265):769-780.

Ellen, R. (1988) Foraging, starch extraction and the sedentary lifestyle in the lowland rain forests of Central Seram. In T. Ingold et al (eds) *Hunters and Gatherers 1: History, Evolution and Social Change.* Berg NY/Oxford.

Flannery, K.V. (1968) Archeological systems theory and early Mesoamerica. In B.J. Meggers (ed) *Anthropological Archeology in the Americas.* Anthropology Society, Washington DC.

Flannery, K.V. (1973) The origins of agriculture. *Annual Review of Anthropology* 2:271-310.

Flannery, T.F. (1994) *The Future Eaters.* Reed Books, Melbourne.

Halbeck, H. (1959) Domestication of food plants in the Old World. *Science* 130:3372, 365-372.

Hammond, N. (1977) The early history of American agriculture: recent research and current controversy. In J. Hutchinson (ed) *The Early History of Agriculture.* op cit.

Harlan, J.R. (1967) A wild wheat harvest in Turkey. *Archeology* 20(3):197-201.

Harlan, J.R. (1971) Agricultural origins: centers and non centers. *Science* 174:468-474.

Harlan, J.R. (1975) *Crops and Man.* American Society of Agronomy – Crop Science Society of America. Madison Wisconsin.

Harlan, J.R. (1986) Plant domestication: diffuse origins and diffusions. In C. Barigozzi (ed) *Origin and Domestication of Cultivated Plants.* Elsevier, Netherlands.

Harlan, J.R. (1988) The tropical African cereals. In D.R. Harris and G.C. Hillman (eds) *Foraging and Farming: the Evolution of Plant Exploitation.* Unwin/Hyman, London.

Harlan, J.R. and J.M. Dewet (1972) A simplified classification of cultivated sorghum. *Crop Science* 12:172-176.

Harris, D.R. (1969) Agricultural systems, ecosystems and the origins of agriculture. In P.J. Ucko and G.W. Dimbleby (eds) *The Domestication and Exploitation of Plants and Animals.* Duckworth, London.

Harris, D.R. (1971) The ecology of swidden cultivation in the Upper Orinoco rain forests. *The Geographical Review* 61(4):475-495.

Harris, D.R. (1972) The origins of agriculture in the tropics. *American Scientist* 60:180-193.

Harris, D.R. (1973) The prehistory of tropical agriculture: an ethnoecological model. In C. Renfrew (ed) *An Explanation of Culture Change: Models in Prehistory.* Duckworth, London.

Harris, D.R. (1988) An evolutionary continuum of people-plant interaction. In D.R. Harris and G.C. Hillman (eds) *Foraging and Farming: the Evolution of Plant Exploitation.* Unwin Hyman, London.

Harris, D.R. and G.C. Hillman (1988) *Foraging and Farming: the Evolution of Plant Exploitation.* Unwin/Hyman.

Higgs, E.S. and M.R. Jarman (1972) The origins of animal and plant husbandry. In E.S. Higgs (ed) *Papers in Economic Prehistory.* Cambridge University Press, UK.

Hutchinson, J. (1965) Essays in Crop Plant Evolution. Cambridge U. Press.

Hutchinson, J. (1972) Conclusion: the biology of domestication. In E.S. Higgs (ed) *Papers in Economics Prehistory.* Cambridge University Press, UK.

Hutchinson, J. (1977) India: local and introduced crops. In *The Early History of Agriculture,* A Joint Symposium of the Royal Society and British Academy, Oxford University Press.

Iyengar, P.T.S. (1983) *History of the Tamils: from earliest times to 600 AD.* Asian Educational Services, New Delhi.

King, F.H. (1911) *Farmers of Forty Centuries.* Rodale Press, Emmaus, Penn. USA.

Lannoy, R. (1971) *The Speaking Tree: a Study of Indian Culture and Society.* Oxford University Press, UK.

Lathrap, D. (1977) Our father the cayman, our mother the gourd: a unitary model for the emergence of agriculture in the New World. In C. Reed (ed) *Origins of Agriculture.* Mouton, The Hague, Netherlands.

Lee, R. and I. De Vore (1968) *Man the Hunter.* Aldine, Chicago.

Lynch, T.F. (1990) Glacial-age man in America: a critical review. *American Antiquity* 55(1):12-36.

McConnell, D.J. (1986) Boabab in Kordofan and Darfur, Sudan. In *Economics of Tropical Crops: the Trees and Vines.* Northcoast Printers, Coffs Harbour, Australia.

McConnell, D.J. and J.L. Dillon (1998) *Farm Management for Asia: a Systems Approach.* FAO, Rome.

MacNeish, R.S. (1958) Preliminary archeological investigations in the Sierra de Tamaulipas, Mexico. *Transactions of the American Philosophical Society* 48(6).

MacNeish, R.S. (1964) Ancient Mesoamerican civilization. *Science* 143:531-537.

MacNeish, R.S. (1965) The origins of American agriculture. *Antiquity* 39:87-94.

Meggers, B.J. (1979) *Prehistoric America: an Ecological Perspective.* Aldine, New York.

Noble, D. (1969) The Mesopotamian onager as draught animal. In P.J. Ucko and G.W. Dimbleby (eds) *Domestication and Exploitation of Plants and Animals.* Duckworth, London.

Patterson, T.C. (1971) Central Peru: its population and economy. *Archaeology* 24(4)316-321.

Pfeiffer, J.E. (1973) *The Emergence of Man.* Cardinal, London.

Phillipson, D.W. (1993) *African Archeology.* Cambridge University Press, UK.

Piperno, D.R. (1988) Non-affluent foragers: resource availability, seasonal shortages and the emergence of agriculture in the Panamanian tropical forests. In D.R. Harris and G.C. Hillman (eds) *Foraging and Farming* op. cit.

Piperno, D.R. and D.M. Pearsall (1998) *The Origins of Agriculture in the Lowland Neotropics.* Academic Press, London.

Portéres, R. and J. Barrau (1981) Origins, development and expansion of agricultural techniques. In J. Ki-zerbo (ed) *General History of Africa; Methodology and African Prehistory.* Heineman, UNESCO.

Reed, C.A. (1977) Origins of agriculture: discussion and some conclusions. In C.A. Reed (ed) *Origins of Agriculture.* Mouton, The Hague.

Rindos, D. (1984) *The Origins of Agriculture: an Evolutionary Perspective*. Academic, New York.

Roosevelt, A. (1999) *Peopling of the Americas*. 1999 Yearbook, Encyclopedia Britannica.

Shaw, T. (1977) Hunters, gatherers and first farmers in West Africa. In J.V.S. Megaw (ed) *Hunters, Gatherers and First Farmers Beyond Europe*. Leicester University Press, UK.

Singh, G. (1971) The Indus Valley culture. *Archeology and Physical Anthropology in Oceania* 6(2):177-189.

Smith, E.C. (1986) Import of paleoethnobotanical facts. *Economic Botany* 40(3)267-278.

Smith, H.S. (1969) Animal domestication and animal cult in dynastic Egypt. In P.J. Ucko and G.W. Dimbleby (eds) *Domestication and Exploitation of Plants and Animals*. Duckworth, London.

Spriggs, M. (1993) Pleistocene agriculture in the Pacific: why not? In M.A. Smith, M. Spriggs, B. Fankhauser (eds) *Sahul in Review: Pleistocene Archeology in Australia, New Guinea and Melanesia*. Dept. Prehistory, Australian School Pacific Studies, ANU Canberra.

Steward, J.H. (1941) Culture element distributions: XIII Nevada Shoshoni. Univ. Calif. *Anthropology Record* 4.4(2):209-359, University of California, Berkeley.

Thapar, R. (1966) *A History of India* (I). Penguin, London.

Tindale, N.B. (1974) *Aboriginal Tribes of Australia*. University of California Press, Berkeley.

Turner, B.L. and C.H. Miksicek (1983) Economic plant species associated with prehistoric agriculture in the Maya lowlands. *Economic Botany* 38(2):179-193.

Vavilov, N.I. (1991 translation) *Origin and Geography of Cultivated Plants*. (Ed. V.F. Dorofeyev, VI Lenin All-Union Acad. Ag. Sci.; an edited collection of Vavilov's most important papers from 1924.) Cambridge University Press, UK.

Wallace, A.R. (1869) *The Malay Archipelago: the Land of the Orang-Hutan and the Bird of Paradise*. Macmillan London. Reprinted Oxford U. Press 1986.

Wilson, E.O. (1992) *The Diversity of Life*. Belknap/Harvard University Press, Cambridge, Mass.

Wiseman, F.M. (1978) Agricultural and historical ecology of the Maya lowlands. In P.D. Harrison and B.L. Turner (eds) *Pre Hispanic Maya Agriculture*. University of New Mexico Press, Albuquerque.

Yen, D.E. (1988) Domestication of the environment. In D.R. Harris and G.C. Hillman (eds) *Foraging and Farming* op. cit.

Zhukovskiy, P.M. (1969) Worldwide gene bank of plants for plant breeding – megacenters and microcenters. In *Problems of the Geography of Cultivated Plants and N.I. Vavilov*. VIR, Leningrad.

Zohary, D. (1986) The origin and early spread of agriculture in the Old World. In C. Barigozzi (ed) *The Origin and Domestication of Cultivated Plants*. Elsevier, Netherlands.

Zohary, D. and M. Hopf (1994) *Domestication of Plants in the Old World*. Clarendon, Oxford.

Zohary, D. and P. Spiegel-Roy (1975) Beginnings of fruit growing in the Old World. *Science* 187:319-327.

Chapter 9
Genesis ... of Stream A.
The Caravan of Time

*The Quest continues—abandoning the concept of Classical
Hearths as a false trail and following instead the Tropical
Tradition.*

Oh what blessed relief it is to scramble out of those biblical *wadis,* to flee
their fratricidal certainties, divorce Domestication from our bed and meet
up with Higgs and Jarman, Carl Sauer, Donald Lathrup, Edgar Anderson
and other more reasonable travelers in the Caravan of Time. Archeology
tries to help with hints, suggestions, interpretations of a few charred seeds
but the Grail is too far back for her to reach, there's little she can offer.
Higgs and Jarman (1969, 1972) rearrange the loads, throwing out three bits
of unwanted baggage from the previous trip (Origins).

9.1 White Horses and Black Horses

With Higgs-Jarman the concept of there being a prior gatherer-hunter phase
in human development leading suddenly at end of the last ice age to
agriculture is simplistic and inadequate. While gatherer-hunters did exist
this is only a convenience summary term for all those many human
economic systems which led eventually to Ali Kosh, the Three Rivers and
the high barrancas. These systems would have consisted of a range of
mutualistic relationships between man, plant and animal. They had always
changed over time according to local population pressures, climate,
environmental conditions, discoveries, accumulating skills, expanding
imagination...*We do not feel it useful to break this development into a
hunter-gatherer stage, from the origins of humanity until 10 000 years ago,
and an agricultural 'food producer' stage from then on* (Higgs/Jarman).

The People with Fight

Higgs-Jarman's rejection of this classical idea that agriculture evolved
through a series of clear economic advances (scavenger → hunter
gatherer → plant exploiter → farmer) is supported by Phillip Guddemi's
(1992) sojourn among the Sawiyanö people of Sepik River district, PNG.

This system represents many which incorporate all these 'stages', but finding some optimum combination of them (relative to place and circumstance) then stabilize short of becoming proper farming as the classical model (Near East, Mexico, North China) supposed they inevitably would. The Sawiyanö meet all the conditions for being hunters and their value and social system reflects this, especially the personal attribute of possessing 'fight' or hunting maleness (which women also can have and apply to smaller game). But the Sawiyanö are also horticulturists (taro, yams, sweet potato, sago palm, plantains, coconut, pandanus, betel vine), each man having several polyculture swiddens scattered around their forest range, though without much tending except to collect produce and deliberately left unfenced where the 'gardener'...*will stand for hours at night, bow and arrow ready to wait for a pig.* Sago processing places (by women) are also good places to hunt pigs (including the family's 'domesticated' ones).

Having 'fight' also applies to horticulture. Lazy men who lack it are only mildly rebuked, though they find difficulty in getting wives and might have to resort to hunt-and-capture of these assets too. Both hunting and horticulture require magic (which incidentally serves to introduce Carl Sauer's later argument that earliest agriculture had more to do with magicals, fibers, medicinals, than with food). Hunting magic comes by carrying magic plants (or ancestors' bones) and those plants are revealed in dreams. Aimed more directly at pigs, a sago palm spath suitably painted and hung up in the bush spells doom for any pig foolish enough to pass beneath it. As Guddemi summarizes this Sawiyanö world (first contacted in 1964) and its implications for Seekers of the Grail: *Knowledge of planting in and of itself does not mandate a social transformation of so-called hunter gatherers to farmers.* Equally, farmers can transform themselves to mobile hunter gatherers without losing their horticultural knowledge. *Technology is not destiny* (Stearman on the Yuquí, below).

Domestication With Higgs/Jarman this also is an inadequate concept when dealing with the economic systems of prehistory, even a self-serving one when technically interpreted and wielded by Candolle's 'Greeks' and other regional patriots. Domestication when and if it occurred was, in Higgs-Jarman's view...*by a series of imperceptible steps* (anticipating Douglas Yen's question of Oceanic agriculture. (At what *point* can a plant be considered domesticated?) Jack Harlan's answer was when it can no longer survive without human support, which is good enough for things like modern corn and soybeans but hardly relevant to the tens of thousands of other plants which have been used, protected or somehow managed over the past 500 000 years or so, most in a wild condition. Here the ideas of ecologists about relationships among organisms, including humans, are more useful than the definitional boundaries erected by biologists and agriculturists. Higgs-Jarman suggest that to divide plants/animals into domesticated/non domesticated is like segregating all horses of the world regardless of their degrees of color into white horses and black horses then using these statistics to draw a line across a map to explain that white horses will be found in the East, black ones in the West. This parable has serious purpose. It would be no more reasonable to classify horses this way than to classify plants/animals in prehistory as domesticated/non domesticated or to classify people who use these plants as proper farmers or non farmers. Such division is...*an artificial taxonomic line which obscures the reality that there is...a steady gradation from one class to another.* There's also the problem of 'horses', or more literally elephants, elk, gazelle, onager (once a draft and war animal), pigs, dogs and thousands of plants dropping into and out of the human *menage* as needs and opportunities changed or climate did, probably several times before they caught archeology's recent attention. The large herds of 'wild' horses, buffalo, donkeys, camels, goats, pigs along with discarded dogs and cats in places like inland Australia suggest the size of this detritus fall-out from previous domestication in only a century or so; and to count the weeds along a railroad track is to survey many crops of yesterday, Candolle's 'plants now fallen into disrepute'.

Agriculture -v- husbandry With this most important adjustment, Higgs-Jarman's term 'husbandry' offers a single enlarged category for all economic systems in which...*some forms of intentional conservation were practised,* regardless of whether these conserved species were domesticated or not (though from Rindos, Piperno, the mere fact of human presence would have led to some kind and degree of domestication). If accepted it bestows legitimacy on all 'pre agricultural' systems which have this resource conservational quality – and re-directs the Quest from Ali Kosh and Jarmo to others of the old human breeding grounds. Carl Sauer will be the best of guides.

9.2 In Pious Times

In Sauer's (1952) vision of the origins, dramatic ice age economic pressure is pushed into the background, the stimulus of population pressure and agriculture-as-necessity (Origins) is rejected and the idea of 'hunting-gathering' leading to field crop agriculture is replaced by fishing and littoral scavenging leading to sedentary village gardening, to what would now be recognized as diversified *huertos familiares* or mixed village gardens, though Sauer didn't use these terms.

Fishermen and Keepers of the Fire

The necessary conditions for this new economic system converged in the littoral of the monsoon tropics, especially along freshwater lakes and rivers close to but not in the Great Forest.

Leisure instead of economic pressure ... People living in the shadow of famine do not have the time to undertake the slow and leisurely experimental steps out of which a better food supply is to develop in a somewhat distant future...That necessity is the mother of invention largely is not true. Needy and miserable societies are not inventive (Sauer).

Fishing and sedentism replacing hunting and nomadism   Here Sauer appeals to observation of what economists would call the disutility of moving. Once they had found some rich fishing place and staked out an extractivist range around it, why hazard this security by moving? As in Africa 'fishing' is often aquatic hunting – of turtle, dugong/manatee, crocodile, rays – replacing the northern herd animals and with the great advantage of the tide bringing them to the hunter. He also introduces a feminist dimension, elsewhere among the Seekers so obviously lacking. *Women were the keepers of the fire. They were the ones most loath to move, and home makers and accumulators of goods...*and this in turn introduced H. sapiens to Energy. On sea coasts and river banks this would have been provided by driftwood, but eventually and elsewhere...*Through the ages man* (ie woman) *has moved more tonne-miles of fuel than any other commodity.*

Through fibers and poisons to food production   This search for energy as fuel and food leads to agriculture, though by a long and curious route. Fire was the earliest and universal tool (mastered even by 'obtuse Tasmanians'), fed by woman but teaching man the possibilities of woodworking and leading in time to building boats; and when the fallen supplies around the campsite were depleted the getting of this wood with stone axe suggested uses of its bark and bast as containers, and incidentally how to kill trees by ringing and bark stripping without cutting them down, knowledge of great importance to the later Shifters. (Such use of trees, of the *tebeldi*, Adansonia digitata, with slabs of bark pried from them for pretty much these same purposes of cordage, building panels, canoes, is still a common sight around the African lakes.) Since these were fisherfolk and scavengers the most important products were probably fibers for baskets, nets, cordage, and in retting these in streams the uses of some (the *barbasco* alkaloid substances) for stunning fish was learned, then the desirability of protecting then growing these plants.

The Old Planter culture   In this Sauer model, with plentiful littoral produce available it's unlikely the first plants to be added to this economy around the campsites were for food. More likely they were fibers, poisons, coloring agents; and still more likely preference was given to multi purpose species incorporating several of these uses, including food, as the breadfruit is still a source of cordage, bark cloth and yellow dye as well as food, and as the coconut has always been a veritable supermarket in itself. This first agriculture was the pan tropical Old Planter culture not granoculture (often divided into root crop vegeculture and aboriculture).

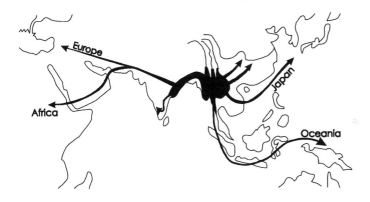

Figure 9.1 SE Asia Hearth and Dispersals (after Sauer)

Dogs, pigs, chickens and other invisible activities   As well as being keepers of the fire, haulers of Energy and custodians of the GDP, women (and children) were the domesticators of young animals initially as much for interest, entertainment and company as for food which led to livestock. In temperate Eurasia these were the larger herd mammals but in the humid and monsoon tropics there are few of these (except the bad tempered buffalo) which are not solitary or aquatic, leading instead in SE Asia to domestication of the dog, pig, duck, goose and above all the jungle fowl. This suite still describes just about every village street from Timor to the Yangtse, with the dog starting as friend and later source of fibers in place of Ur's woolly sheep then elevated to object of ancestor worship before ending up as food – or as pariah, with the pig, depending on how far Pleistocene economics gave way to theistic bigotry. *In pious times ere priestcraft did begin, Before polygamy (and pigs) were made a sin…* Then add the mulberry and the silkworm, the willow and the turtle. Man farming Woman weaving.

With Nokolay Ivanovitch … The hearths of domestication are to be sought in areas of marked diversity of plants and animals…where there was a large reservoir of genes to be sorted out and recombined (Sauer) … which needed leisure and sedentism made possible by 'fishing' and the expanding *menage* and above all some nearby generalized ecosystem which could (by Harris' later model, *Origins*) be transformed into generalized agroecosystem to provide plant foods as starch and sugar (from the palms, with protein from the fish), fibers, dyes, cloth, poisons, pharmaceuticals, magicals, oils, scents, spices, fuel, timber…accumulating to a Garden of Complete Design. The most likely place, Sauer thought, would be the river valleys, estuaries, lake outlets around the Bay of Bengal, Orissa to Burma into southern China (Figure 9.1), also partly on grounds that the reversing SW-NE monsoons here would have added more wet/dry seasonal diversity needed for those plants which needed such a dry season break (yams and other root crops) rather than perpetual moistness. Vavilov had attributed 33% of the world's agricultural plants to this area and others before Sauer (Haudricourt and Hedin 1943) had believed the origins had to be here. This was also the world 'center' of planting techniques and vegetative propagation…*Planting methods were elaborated to the most sophisticated garden culture in the world. South Chinese and Japanese agriculture are advanced developments from the original hearth to the South. Rice, bamboos, bananas, taro, persimmons and yams brought from India or Indochina were largely remade and diversified in East Asia before we can speak of Chinese or Japanese people* (Sauer).

What, are you Stepping Westward?

Sauer's parting blow to classical ideas about the origins was to wheel the Great Diffusion around and send it home, compounding this insult by making it carry the Old Planter Culture and its horticultural ideas west instead of granoculture east, and rations for he journey in baskets and gourds instead of the clay pots which Darlington later thought necessary as precursor to metallurgy then civilization. There wasn't much of the Old Planter culture suitable for those westward chimes and most going by land expired west of Mother Indus, though taro in profusion and sugar cane made it to the Arabian south coast. But at least the basic idea got through, with the sago and palmyrah and the vines transforming themselves to date palms, olives, grapes along the way (guarded by the dog). The bananas and the pigs and chickens later had to go by sea.

A few called in at Al Mukalla and Salala on their way to the Sennar country and Ethiopia, heading for the West African forest and the Niger Bend. They didn't disturb the ladies of the Wadi not yet awakened, and they bypassed Sheba's wells which were not dug. Just who the westward porters were remains a mystery...*a lost corridor of mankind;* but there are 'elf-like roundheads in the Yemen', Negritos in the Hadramaut. And Veddhas so far west from Serendib.

Since Sauer's time there's been increased awareness of the many crops indigenous to Africa (Origins), with these eroding his concept of the Old Planter culture with its plants and the 'general configuration of life' migrating here from SE Asia. His evidence found only taro, most African bananas and plantains, some yams, the chicken and pig were from SE Asia and in fairly recent times. The proper pig made it to the ancient land of Sennar between the White Nile and Atbara Rivers, only to be cast out by Hamitic nation builders and eventually reduced to Set who murdered and dismembered good brother Osiris and who—with other members of this extended family pantheon variously into resurrection, law reform, incest, hiding babes in bulrushes, virgin births—laid so many biblical foundations. What a family. Poor Set. Pig-headed god of evil. But the culinary part of pigness got through to West Africa as the indigenous bush pig better able to withstand the tsetse fly and swine fevers (a transformation much in the spirit of the Great Diffusion replacing woolly sheep with goats before attempting the terrible crossing south from Dhar Tichitt).

9.3 To America – *Not on Evidence But on Authority*

This authority of arrival in the New World not much before 13 000 BP was noted previously (Mexico) but Sauer believed it far more likely that H. sapiens had slipped in through the glacial/interglacial revolving door much earlier, when archeology wasn't looking (Figure 8.2, glacials and interglacials, Origins). The best time, he believed, would have been partly into the previous glacial (Illinois/Riss, 200-150 000 BP) when climate was still warm enough to travel around the North Pacific coasts from SE Asia. But it could have been at many times, the itinerary determined more by accumulation of skills than glacials; and dipping into their previous Higgs-Jarman stages over 350 000 years or so as scavenger-collector-tool maker-hunter-planter they would have found these skills aplenty. Survival and travel would have presented few problems, the 'planting' temporarily dropped and the sea hunting redirected from crocodiles and turtles to seals and walrus, though Ms Incipient American – coping with their ever changing GDP and toting their driftwood Energy – might not have liked it much. The crossing could have been rapid too and purposeless, suggested by the later Vikings leapfrogging quickly down the American coast from Greenland for no better reason than curiosity in what was around the corner. Perhaps Sauer's strongest argument for an early crossing were the animals which made it in both directions, the mammoths and ancestors of bison and deer going from Old world to New and the early horses and camels from New to Old...*The same invitation to pass through an easy gateway was available to animals, plants and man...No reason has been offered as to why he should not have gone where went the bear, bison and deer.*

Factual and Historical, the True Parts Allegorical (and More Diversity)

Once the distractions of Tijuana had been left behind or they'd drifted down the Mississippi with one of the post-glacial meltwater surges – and those inclined to hunting had gone off to wait at Clovis until 10 000 BP to become PaleoIndians – the New World immigrants pushed on, still preferably littoral scavengers and fisherfolk and the threads of the Old Planter Culture of SE Asia were picked up again to provide the fibers and poisons, though the plants were new. As in SE Asia they found few animals to domesticate; these were always coming and going across the ice bridge – only a tree-dwelling duck that's since been untaught how to swim and fly, a local version of the pig and later in Caribbean a 'low slung dog' (Sauer).

The palms rediscovered The Sonora desert of Mexico is not remarkable for its fishing or river bank scavenging but some who took that shortcut to the south would have started meeting the palms and palm-like plants again, 14 or so (Elaine Joyal 1996), especially Sabal uresana, looking like Borassus, remembered so well from Sunda (Diversity). Joyal has described these palm resources in Sonora – present and historical in order categorical – from 1692 but they must date from the very beginnings of human occupation: baskets, ropes, brooms, sleeping mats, blankets, whole large leaves *(pencas)* as house thatch, trunks for house beams and corrals, fresh fruit (not universally admired), tender young leaves pit-baked and now fine double woven sombreros and always palm heart or *corazon* raw or cooked and 'feeding four people'. They're harvested where they occur, not planted although there are signs they have been in the past, with the women weavers now working in excavated semi-subterranean earthen rooms (a *juqui* or húuki) which maintain a humidity in this desiccated environment necessary for easy fiber handling. As with the pandanus and bamboo weavers of middle Java, these plants now often provide the only source of household income ($5-6/day from fine hat weaving) for some of the poorest people in Mexico. Not much today, but in Pious Times it was quite a lot and the product quality was (and is) much better than the hot tin roofs which increasingly replace the pencas and the cabbages which replace the corazon and the plastics which replace the sleeping mats and the other store-bought junk you buy today – although it's now wanted by the campesinos and displayed as evidence of progress.

Agaves and yuccas Those who didn't like weaving sombreros in a húuki would have gone off and found the agaves (A. lecheguilla) and the yuccas (Y. carnerosana) and collected their hard durable fibers, *ixtle* (now for industrial brushes) and become *ixtleros*. These could hardly have been missed; according to Sam Sheldon (1980) they occur over about 11 million ha in 7 states of arid north Mexico centering on Coahuila, the Zona Ixtlera; and according to archeology their economic antiquity extends to at least 8 000 BP, from ixtle sandals discovered in that state. Much later...*when the Spanish arrived in northern Mexico in 1500s they encountered nomadic hunters and gatherers making arrow shafts from agave flower stalks and bow strings from ixtle. Rural folk who currently gather uncultivated plants in the Zona Ixtlera are therefore perpetuating a centuries old way of life that has, in part, been reinforced by commercialization* (Sheldon). There have usually been 400 000 to 650 000 of these ixtleros (1970s), many moving to farming in the good rainfall years and returning to the more dependable old desert plants in the bad.

Revenge of the cactus Another great resource the immigrants were led to in the 'wasteland' was cactus (Opuntia, 1 500 species), like the others not thought of as having much to do with the origins of formal agriculture except that these, with mesquite, were replaced by corn (the Flannery model), and came back later to exact a sweet revenge. Russel and Felker (1987) found a small valley called Milpa Alta or Upland Corn Field outside Mexico City where...*the entire cropping area of the valley was in nopalito production with a cactus known as nopal de Castilla.* The tender young pads, nopalitos, are grown as a commercial vegetable and had replaced corn because of their greater productivity and profitability, higher water use efficiency and lower climate risk. Densely planted at 40 000 plants/ha the nopalitos come into production within three months and yield 80-90 tonnes/ha. A symbiotic relationship also exists between cactus and dairy cows: dairy farmers bring up their cow manure for the nopalitos and back-haul unsold pads for their cows...who get a premium price for their milk and a better color to their butter. The best known cactus food now almost world wide, California to Chile and Italy, is the fruit (in Mexico 8 tonnes/ha by the 4th year after planting then increasing and lasting 30 years), and no doubt a major item when the immigrants passed through.

An even sweeter revenge was taken earlier in Canary Islands where...*the cochineal dye industry became so important that between 1831 and 1874 whole grain fields and vineyeards were converted to cactus plantations to grow the cochineal insects* (Russell and Felker). The different species, some spineless, have now become integrated into alley cropping systems, Mexico to Chile (cactus + vegetables, cotton, sorghum...), agroforestry orchard systems (cactus + almonds, grapes, apricots...), agropastoral systems (cactus + mesquite, Leucaena grass...), wildlife restoration or enrichment systems (cactus + brush + grass with game and cattle). From Russell and Felker's global survey 300 000 ha of these various cactus 'agroforestry' systems had been established in NE Brazil alone by 1980s and in Texas 'mixed cactus as wildlife habitat' had translated, through hunting fees, into per ha ranch incomes 2-3 times more than from range cattle. None of these modern developments undermines Kent Flannery's evolutionary model of the origins of corn in Mexico; but all suggest with hindsight in the arid lands what an ecological and economic mistake that probably was.

Pulque drinkers and mescaleros Having crossed the great ice sheet and come so far the future campesinos would be thirsty; and now, with assured incomes rolling in to the semi-subterranean húuki weaving ladies and to the cactus fruit sellers and the ixtleros and arrow makers and the nopalito farmers, they were threatened by prosperity. Their economist said they were in a dangerous inflationary spiral; they'd have to pause while he sopped up some of this excess liquidity. He'd open a cantina. It's not known exactly what he sold or how he came by it but according to Bahre and Bradbury (1980) in the central plateau it could have been *pulque*, the sweet fermented undistilled juice of A. salmiana, another gift from the agave, based on murals of ritual intoxication at the Great Pyramid at Cholula, 200 AD. By then Borassus had been forgotten and it was well known that the agaves were the first plants God created, perhaps for pulque, perhaps for food and fiber...*The baked hearts of certain agaves were the most important aboriginal food in arid and semiarid Mexico and*

American southwest before the development of agriculture (Bahre and Bradbury, citing Bruman 1940).

Or the cantinero could have sold mescal, distilled from the cooked and mashed stems of some agaves; but according to Bruman that didn't happen until some later boat people – Philippinos travelling on Spanish ships to the Pacific coast – brought the distillation technology from Manila...These Carl Sauer links with SE Asia keep recurring. Whenever, applied to two agaves in Sonora (A. pacifica, maguey, and A. lechuguilla) it turned out to be a good thing and still mops up much liquidity...*The state of Sonora is particularly renowned for its bacanoro, a bootleg mescal made from wild Agave pacifica...in some villages nearly a quarter of the households make and sell mescal...Its flavour and proof are superior to some of Mexico's finest (legal) tequilas*, made from cultivated A. tequilana. The economics are also more enticing (Bahre and Bradbury): one burro carrying 7 loads of maguey stems – and assisted by one mescalero to find and chop the plants off at ground level and trim off all the leaves except one for tying to the burro and to cook the maguey in pits then mash and distill it – can make 40 liters of high quality mescal in 14 days.

But even so, back at the Treasury in those times the economist's plan had come unstuck. The mescaleros' rising incomes more than offset the declining liquidity of the ixtleros, and their superior product had potential to undermine Europe's 'finest wines and brandies' which the Prince himself was planning some day to bootleg in from Spain. Before that happened it was time to move on. Going out the door about the last thing on everyone's mind was to go up to the high barrancas and coax corn and hogs out of teocinte. They were not that desperate yet. They'd think about it later.

The cantina is still there, still sopping up excess liquidity. Sometimes these days it's adobe wearing a cerveza sign. The more respectable places are built as a *jacal*, of almost any shape with walls of upright yucca stems and yucca roof beams supporting agave flower stalk laths then sod. And in the backroom of this joint the cantinero's daughter is still flirting with the boys – chopping *corazon* and frying nopalitos.

Ramón The north wind was blowing from the glaciers. They hurried down to Vera Cruz and Yucatan. Again the plants were new and the women wanted to rest from coping with the always changing GDP. There they sat down and spent the winter studying the new agroforestry under Ramón. Described by Peters and Pardo-Tejeda (1981), *ramón* or *capomo* or *ox*, Brosimum alicastrum, is a large buttressed rain forest tree and from their particular abundance around Maya ruins it was a main food resource of the

Maya before – like the Maya – falling into disrepute...*Until recently the significance of this (archeobotanical) association was not realized...It may also have been deliberately cultivated. Traditionally the fruits were eaten raw or made into juice or marmalades. The seeds were dried, ground into a* masa *and mixed with corn to make tortillas, or the seeds boiled then mashed and eaten as a substitute for root crops* (Peters/Pardo-Tejeda). Its classical association with corn comes from the Maya name 'ox' meaning a stock of shelled maize grain; but whether 'domesticated' or planted or simply collected it's almost certainly much older food than teocinte-corn (for the general reasons given previously by Piperno-Pearsall). For one thing the seeds have crude protein content higher than corn and some amino acids are four times higher.

Ramón is still a common component of forest gardens especially in Yucatan, but generally throughout nearly all lowland humid Mexico Peters/Pardo-Tejeda found it no longer a staple, having been replaced by government subsidized maize and now carrying 'a sociological stigma associated with consumption of its seeds'. Its other current or recent uses include timber, several parts as medicinals (some now commercialized) and its latex mixed with water as a substitute for milk. Estimates of seedfall in natural forests are 80 000 tonnes/year in Vera Cruz state alone or 10 000 tonnes of potential human food, protein, most wasted because of lack of organized collecting-processing system. If the New World immigrants had brought Agissa's cattle with them, or Ur's woolly sheep, they would have found ramón also yielding much good grazing from this seed fall and volunteer seedlings and fallen leaves (its present main use), with better feed quality than many conventional pastures; or they could have lopped it three times a year or planted this large tree as intensive hedgerows or in alley cropping systems for fodder as now done by Mexican researchers. But they didn't. Some stayed themselves to browse on *ox*, some moved on.

Pacaya In Guatemala and also Mexico to Colombia the immigrants found their first corn growing on the pacaya palm Chamaedorea tepejilote, easy to harvest because of its small size, 2-7 meters. Mont, Gallado, Johnson (1994) have sketched its modern economics in Guatemala. In its natural habitat it grows as an understorey plant and thrives on poor limestone soil in lowland rain forests to cool upland forests. Its main subsistance and commercial products are the unopened inflorescences emerging from the top of the trunk, about 30-40 cm long and resembling an unhusked corn cob. This resemblance is expressed in the Nahuatl Indian name tepejilote = mountain maize; but again, as with *ox*, it's likely to much pre date teocinte-maize as a human food (one of Wallace's 'natural products of the country'). The pacayes (cobs) are pulled off and shucked to obtain the internal edible portion and chopped as salad or fried in egg batter. Palm heart is cut from the unproductive female palms which kills them. Other products are the leaves as dry season cattle feed and the palm itself is valued as a graceful shade tolerant decorative. Pacaya has long been incorporated into Central American household forest gardens, and in field systems commercially intercropped with coffee, cacao, bananas etc for the domestic fresh vegetable market and preserved/bottled for export. Pacaya is clearly an important food since earliest times...*the status of pacaya as a traditional food and its sources of supply elsewhere in Central America and Mexico are scarcely known...What is the antiquity and nature of the agroforestry system under which the pacaya is grown in Guatemala?* (Mont et al). It could also offer example of incidental or plant domestication through changes in the near environment (Origins). The wild pacaya is usually not now collected in the forest, partly because of the small size of the inflorescences/cobs (3-30 cm); but...*Because of selection and to a degree less competitive environment, pacaya palms under cultivation have much larger male inflorescences (and) the trees are taller than in the forest understorey* (Mont et al). One likes to speculate that this less competitive environment, resulting in accidental domestication (Rindos), was first found in the disturbed habitats of Sauer-Lathrap riverside forest gardens, or on Edgar Anderson's village garbage dump, or on the Shifters' chacras.

Loroco Morton, Alvarez and Quiñonez (1990) have suggested the loroco vine, *Fernaldia pandurata*, as one last representative of the many minor foods the immigrants would have found when passing down through El Salvador and Honduras (and southern Mexico and most of Central America), though it's more in the nature of a Carl Sauer multipurpose plant than serious food. It produces 'particularly odoriferous buds and unopened flowers' and these are traditional vegetable additives to foods ranging from tamales through 'hen soup' to more recently pizzas, or the growing vine tips serve as greens. Like ramón and pacaya it's long been incorporated into forest gardens and is now also a minor plantation export crop for the 'ethno nostalgic' market. From village gossip, repeated by Julia Morton et al, the Lorocos are related to the Poisonous-Dogbane family and their roots are poisonous too. So the immigrants would have added them to their pharmacopoeia, though only for rats around the camp, not fishing. In Brazil they're recognized as a snakebite treatment and since the migrants were going that way this could be a good thing. By now, absorbing these things through Guatemala and Nicaragua, the pharmacopoeia had expanded wondrously (Diversity). A good number of the older people settled in as curanderos; the obeah men with their magicals. The others hurried on, anxious not to miss Dolores Piperno's opening of their Stage 1 in Panama.

New Venice

Once past Panama, Carl Sauer had proposed, the best place for this SE Asian fishing-planting to re-emerge as the Neotropical Tradition would have been Colombia-Venezuela, again not quite in the rain forest but at its margins, partly because of seasonal diversity *(The seasonal rhythm about much of the Caribbean is not very different from the monsoon lands of Asia)*, also because of the great diversity of landscapes from near-deserts in Venezuela to rain forests and mountains in Colombia (a bow to Vavilov), of soils and natural floristic resources. Geography also made it a human crossroads to West Indes; and the skills he observed in boatbuilding, use of fish stunning and arrow poisons also suggested past links with SE Asia, especially Indonesia.

In Venezuela   Ted Gragson (1997) has offered a glimpse of how Sauer's timeless SE Asian fishing-collecting-planting village might have worked when it arrived in Venezuela, describing the contemporary uses of only underground plant organs (tubers/corns/rhizomes/roots) by Pumé Indians of the Venezuela llanos, the seasonally flooded savannas, these also collecting from forests and operating swiddens and house forest gardens (though poor ones – cashew, mango, beans, tubers, magicals, medicinals). In terms of time spent their economy consists of: fishing (46%), swidden/house gardening (16%), collecting wild roots/tubers (11%), hunting (8%), foraging (5%). Eighteen of these root plants have 20 uses by the Pumé themselves but at least 35 uses of these same plants have been reported from the region. Since fishing also provides 64% of dietary protein and 17% of energy the most important plant (apart from swidden manihot which supplies 60% of Pumé energy) is the fish stunning poison *barbasco* (Tephrosia sinapou, toxic agent rotenone) grown specifically in the house gardens for the reasons Sauer anticipated. Another is a Discoria/yam which is both a human food and poison depending on whether it's collected in the dry or wet seasons. But importance is a matter of viewpoint: this same tuber is macerated and rubbed over the body to 'prevent measles'. The *hayo* (Banisteriopsis sp) is also sufficiently important to be grown close by in the house gardens and indispensable – when its roots and stem are chewed to induce hallucinations – for enduring all-night sessions of shamanic 'curing and dancing'. The humble manioc/cassava tubers when chewed, spat out into a trough of water and fermented over-night offer a more plebian route to conventional intoxication. While the corms and tubers of Dracontium margaretae can be boiled or roasted, other groups value them as a decoction against diarrhea, a paste against snakebite and as an amulet against murder. Gragson notes suggestions by others that the easily accessible roots and tubers might represent an economic link – an early table shared – between ManWoman and their Hominid ancestors. Why not? (Perhaps the Grail is underground...?)

But this is still a long way from Santa Rosa. While the Old Planter culture is usually followed as root crop vegeculture the tree crops stream isn't, possibly leading to Sauer's conclusion that...*All the important food plants are grown for starch and sugar. Vegetable proteins and fats were as much neglected as across the Pacific.* There's now evidence that while this is true of root crops, through the trees and palms this tropical American/SE Asian system was not as ill balanced towards starch and sugar and lacking in protein, calcium, phosphorus and fats and oils as Sauer suggests. Smith (1986) lists fruits of the Sapotaceae and Ebenaceae families as 'concentrators of these invaluable elements' generally lacking in lowland tropics...*The archeological record tells us that the peoples of the American tropics, knowing nothing about the niceties of chemical analysis, moved the sapotaceous fruits throughout the Americas.* Only a few of them from Mexico-Honduras to Ecuador-Amazonia are becoming known (sapodilla, caimito, mamey, green sapote, canistel, lucmo, abiu etc) and one of the four genera of Ebenaceae alone contains some 200 species (Page 1984). Some of the many South American palms yielding high quality oils and fats were discussed above (Diversity), and on the other side of paradise Afek's People in New Guinea highland forests still get their's from the pandan.

There's also the main root crop manioc itself, presumably from Venezuela-Colombia and now a major food pantropically. It's only an assumption that tubers not leaves were its first food use. From their survey of leaf uses around the world Lancaster and Brooks (1982) found these leaves now a greatly underutilized food resource and recommended them as a 'source of protein and vitamins for supplementing starchy diets', this based partly on the comparative protein levels: manioc leaf 7% (of edible fresh weight), Chinese cabbage 2%, spinach 2%, unhulled rice 6%, wheat 12%, soybean 18%. They could find references to only 7 Amazon Indian groups then who ate manioc leaves but suggest their much wider actual use was overlooked by surveyors interested in use of the tubers. (The similar contemporary food value placed on pumpkin leaves as 'relish' (greens) in Africa suggests how other usually ignored parts of plants, and unknown plants, might once have been most used, in pious times.)

9.4 To Santa Rosa

Donald Lathrap (1977) extended this general model of tropical American vegeculture/aboriculture as a story set partly in myth with a core of agroecology and interglacial climate change. It starts in Central Africa instead of SE Asia with some groups there, normally content with the game foods (and perhaps grasses) of the savanna following its incursion into the Great Forests during one of its glacial contractions to be there cut off, or choosing to remain when 'normalcy' returned, and the forest again closed in and the savanna animals disappeared. Their survival there required gravitation to the river banks and lakes to apply their hunting skills to aquatic animals and fish, and mastery of the necessary plants (fibers, poisons) which Sauer has outlined. In addition they needed the bottle gourd Legenaria siceraria as net float (as well as food, cooking pot, vessel, narcotics container of coca in America, areca nut and lime in Asia, penis sheath in New Guinea). More central to the story, since the domesticated gourd can't exist in forest clearings without human support and is an ancient plant from archeological sites in many places in the tropical world (Mexico, Spirit Cave Thailand, Ayacucho Peru from 13-9 000 BP) it must have been distributed to a good part of the world by people and provides the unifying element in Lathrap's unitary model of the origins. It's most likely circumstances of domestication was as part of the 'broad Sangoan culture' ie around the Central African lakes before 40 000 BP (Origins, Sangoan picks) from where it spread down the Congo rivers to West Africa and eventually crossed the narrow Atlantic to NE Brazil with a boat load of drifting fishermen (along with cotton and fish poison plants). It follows Sauer's basic model, but with fishing communities and their still few plants migrating west along the Brazil coast and up the Amazon (instead of, with Sauer, arriving from the north).

Lathrap calculated that in order to keep its rendezvous with archeology at Ayacucho in upland Peru c 13 000 BP the gourd and the riverine economic system it represented would have had to begin their Amazon trip about 16 000 BP. In the course of it four things happened. First, moving upriver they continuously encountered more useful plants and brought these into the household to expand the plant suite. But while the species of a rain forest are numerous they're widely scattered. These selected plants then would have been both the most useful ones and less easily available by simple collecting. Second, they mastered agroecology. These selected plants like the gourd would have needed some degree of weeding and tending, ie farming, and some specific microclimate of light or shade provided by clearing unwanted trees from a new camp site, leaving useful ones and planting others (notably Ceiba, the kapok tree) for these environmental purposes as well as economic products. Third, they were pursued and hastened upriver by increasing flood inundations as the Last Ice Age retreated (after c 20 000 BP) – at least fast enough to get the bottle gourd to Ayacucho in good time, but far more important, to bring the tropical forest tradition as Gardens of Complete Design to places like Santa Rose...*On a time scale calibrated in millennia there should have been little lapse between practices necessary to maintain plants crucial to efficient fishing and achievement of that remarkable artifact, the house garden* (Lathrap). A few of its South American plant components were noted previously (Nature, Reconnaissance); the full inventory typically runs to 50-100 species per household and includes also perfumes, medicinals, spices, snuffs, drugs, ornamentals, magicals, foods, as well as the all-important fishing fibers eg curagua in Amazonia.

From House Garden to Chacra

Finally, in the Amazon forests they invented proper field crop farming. The vegecultural/horticultural house gardens were also experimental plots (Anderson below). Lathrap's thesis is that most of the important food crops of tropical America were introduced and developed in this way – manioc, sweetpotato, arrowroot, yams, peanuts, aroids. As these developed and became more efficient as food source larger populations were planted until these expanded to become crops and it became necessary to go beyond the riverside gardens and open up special purpose fields, the *chacras*, characterized by the new idea of large populations of only a few plant species. But these quickly exhausted soil fertility or were invaded by weeds. The Shifters had arrived. Two types of farming now existed, or as Lathrap calls them 'two distinct categories of artificial floral arrangement.' The forest gardens continued; but later globally when under pressure these gave way to the larger and more efficient (in terms of calories per unit of land) chacras. The complexity of agroecosystem gave way to simplicity of agrosystem, at first as field crop vegeculture then this, in most of the world, to granoculture.

Curagua, a Flax more Flexible

By now if they'd kept to Donald Lathrap's timetable the New World immigrants would be closing in on Santa Rosa, the garden almost complete. As fisherfolk and forest collectors their indispensable acquisition had been the pineapple-like *curagua, Ananas lucidus*, its modern economics in

Venezuela described by Leal and Amaya (1990), fruit barely edible but narrow meter long basal stem leaves yielding fine fibers still used for fishing lines and nets and also hammocks, ropes, 'carpets', string bags. It's thought to originate along the Amazon and have been carried to Colombia, Peru, Caribbean, Venezuela, Guianas in prehistoric times. The curagua is also remarkable for the wide range of soils and environments where it's now found, from sea level to cool cloud forests, all suggesting that this must be an anciently domesticated crop dispersed widely by people. ('There's now no such thing as a pristine rain forest.')

Leal and Amaya found it interplanted with manioc, yams and other subsistence food crops in swidden chacras, no more than 100 ha in all Venezuela. Slips (as pineapple) are planted about 2 meters apart with the other crops. As the plants develop they expand in size and start producing their own suckers. From plant maturity the leaves are snapped off year-round as they reach full length and are needed and after 3 years the chacra is abandoned and some slips are taken off to open another. This applies to chacra production. Beyond the fact that...*a handicrafts industry of great importance is based on these products* (Leal and Amaya) the actual situation is vague. The curagua must also be part of household gardens. As long ago as 1782 a certain Fr. Gilij – apparently keen on New Crops or maybe clothing noble savages – was chastising Venezuelans for growing only two or three curagua when they should be planting it abundantly...*I have never seen a flax more flexible and white than the caraguate.*

Solanaceae

With the curagua added it was time to take stock; but no sooner had they started counting than someone brought in the Solanaceae family who said that here on the outskirts of Santa Rosa – the eastern foothills of the Andes, Colombia to Bolivia – they'd reached the richest place of all, even better than the 2nd Epicenter back home (and certainly better than the 3rd if they'd come from Africa). The agronomist wanted some for his *chacra* (tomato, eggplant, chili pepper, tobacco) to develop as monocrops. Others thought they'd do better as polycultures in forest gardens. While they were arguing about this and threshing around in the bush looking for more they created disturbed habitats, and still more of these tree foods/fruits (naranjillo, tomatillo, tree tomato etc) started popping up just about everywhere...*most inhabit moist forest at elevations 500-2 000 m; a few are found along the coast or in the Amazon basin at 100 m or less; still others prefer cloud forests up to 3 000 m. The plants are generally fast growing trees of the forest understorey...they exploit light gaps in the primary forest and also occur in secondary sites such as the margins of trails* (Lynn Bohs 1989). Of the 45 or so species of Cyphomandra alone, Bohs lists about 18 thought to be eaten by indigenous groups but only one or two (notably the tree tomato) have been cleaned up and allowed into the Big Store. Most of the C. species and Solanaceae more broadly are also known for abundance and diversity of their alkaloids and widespread traditional medicinal use – leaves/fruit/roots for tonsil poultice, mouth sores, skin disease, intestinal worms, analgesic, sedative, vermifugal tea. A few have been scientifically screened as precursors for synthesis of corticosteroids and contraceptives (Bohs), and in Colombia some are used in hide tanning and traditionally in Ecuador as dyes and pottery paints, most barely known – either the modern scientific potential or the paleoeconomics. It was a long time ago.

Vacuuming the Forest

There's been some skepticism that prehistoric people could actually survive on forest resources alone without agriculture. Allyn Stearman (1991) addressed this question to the Yuquí Indians of the Bolivian Amazon moist forests. The answer is yes, but survival requires acute awareness of what these plant and animal resources are, their seasonality, continuous forager mobility (as in the Garden of the Rainbow Serpent) and some adjustments to the concepts of what a food 'supply' is…*Although I have documented the amounts of fish, game, honey and plant products brought back following foraging episodes…a great many items never reach camp and are virtually impossible to include in any inventory…Very little escaped attention and would be snatched and eaten without losing stride…*as with Campa boys snacking on the countryside (Shifters), and one of Stearman's colleagues called it 'vacuuming the forest'.

Women, accompanying the hunting parties for 'tracking, calling and retrieving game' are especially aware of what plant resources they encounter and file them in a mental inventory for future collecting. Perhaps the Yuquí had a head start in this plant resource recognition. Stearman is convinced they were once forest horticulturists then turned to mobile foraging for defence against Spanish invaders…*apparently as a result of extreme isolation, population decline and the need to move almost constantly to procure food as well as avoid detection, the Yuquí lost their knowledge of cultivation as well as many other traditions…*which among other things supports the possibility that Australians were not always hunter-gatherers but shaped their economic system in response to changing circumstances (the Garden of the Rainbow Serpent).

While 'retrieving game' for the band (as one of Clare Madge's invisible activities) these Yuquí women had much resource information to file away. Stearman cites other research in Amazonia from which 36% of 94 tree species and 25% of 135, respectively, on 1-ha sampled forest plots bear food consumed by indigenous populations.   There's also acute awareness of such mysteries as gap dynamics...*Canopy trees at the edges of gaps tend to produce larger fruit crops than when in closed forests* (Stearman). This female sensitivity in resource definition, and the ways of plants and their animal consumers (the band's quarry) must be the link between hunting-foraging and horticulture.   Surely – when extended through all the Great Forests – this must be the One True Grail...?

That's about it. After a few millennia when the women had got all this sorted out they went off to the Paraguay Chaco (with Arenas and Giberti 1987) to pull some Jwiye'lax vines (Odontocarya asarifolia) down from trees for a Last Incipient Agriculture Supper. Most of the 30 species of this genus occur across Amazonia so some of these might have been used there similarly as food although some discretion would have been needed. This edible vine is of the Menispermaceae family, also famous for their alkaloid medicinals and notorious for their poisons including curare. If ever there was a prehistoric staple and a first crop in the Garden this vine must be it. New growth stems of about 20 mm thickness are pulled down, chopped into 15-30 cm sections, thrown on a fire and singed until black, this scraped off then they're boiled for half an hour, consumed, wasting only 3% of the vine stem diameter and a few internal strings. Perhaps in deference to Carl Sauer's fears of a deficiency in dietary fat the Chaco Indians dip these vine-sticks into fat while eating them. Fish fat is preferred or failing that the iguana makes a splendid dip, or the rhea or peccari.

That night they switched on the fevillea lamps and there was a big knees-up at the shaman's place, then they went to sleep. Tomorrow they would carry all these ideas they'd stolen from the Great Forest up to Ayacucho to begin proper farming. And establish a Classical Center...to be later dug up and sanctified as according closer to the biblical tradition. It would be a hard day.

9.5 Of Garbage Dumps and Open Habitats

Few if any ideas about the origins are new;   most are progressive elaborations inviting new interpretations as they go. Edgar Anderson (1954), now remembered for his garbage dump theory and early surveys of forest gardens in Guatemala (Reconnaissance), acknowledged this debt to Harvard biologist Oakes Ames (1939) for three of his: uneconomic botany, multiple and changing plant uses and an early Pleistocene beginning. By the first, pursuit of the origins of wheat or corn or beans or some other crop of current relevance misses the point: many of the really ancient economic plants have long been abandoned, or their uses were not as we would now imagine them. They now might have no economic use (uneconomic botany) or are passed by as weeds. By contrast...*Ames saw all the plant products of the world as important natural resources,* both at some time in the past and in the future as New Crops though probably these will be for different uses. (Incidentally this was an early call for large scale plant conservation on pragmatic grounds...*In modern technology an ancient love potion or an obscure fungus are as likely to present us with as great commercial success as are the plants suggested by narrow-minded practicality.*) These earliest plants are as likely to have been fibers, poisons, dyes, magicals, medicinals, containers...as for food (Sauer, Lathrap).

By the second, modern mass production of single plants as crops for a single purpose has obscured the likelihood that...*with our oldest plants, the ones which have traveled with man the longest, multiple uses are probably the rule rather than the exception,* and their current uses might offer little insight to why they were first brought into the household. Anderson offers as examples the early Mexican pumpkin, its flesh 'repulsively bitter' and barely edible but its seeds still valued in Mexico nearly as much as its conventional qualities; universally the gourd for its many uses Lathrap listed; the fish poison derris root since rediscovered as a modern insecticide; the para rubber tree probably first 'domesticated' in Sauer/Lathrap Amazonian gardens for its seeds rather than its latex. (Sometimes these first uses reemerge as modern by-products: in 1970-80s there was extensive research on rubber estates in Malaysia-Sri Lanka-Indonesia into its seed oil, from an annual seed fall 250 kg/ha containing 15-16% oil; and into oil residue as poultry and pig protein meal, with 30% meal recovered per tonne of seeds crushed and with 25-29% protein.)

Third, in the Ames-Anderson view and given the vast number of plants that people have worked with, and that when these came to archeology's attention they were often already 'domesticated', it would have been impossible to compress this (Higgs-Jarman) stage of agriculture into the conventional period of 10 000 years or so, since Ali Kosh (Origins). Experiments with them must have begun well back in the Pleistocene and...*In the 5 000 years of recorded history man has not added a single major crop to his list of domesticates.* One would now go further and say that since modern ManWoman set about developing proper agriculture they have lost almost infinitely more crops than they've discovered (Diversity).

Anderson's household-village garbage dump has two dimensions. In the literal sense it was a place of concentration for seeds and discarded bits of collected produce to emerge as weeds then crops, fertilized by household refuse. Extended it represents an open habitat, open to all-comers where some plants can find a niche and thrive, often in the company of others which they now find congenial so that the garbage dump generates not only new crops but sometimes packages of new cropping systems eg corn-beans-squash in the new mutuality. *(Species which had never intermingled might do so there.)* Further extended, similar but natural open habitats are created by forest tree falls, land slips (which teocinte/corn found pleasing, according to the later Kent Flannery model, Origins), and by river bank erosion (*Rivers are weed breeders; so is man, and many of the plants which follow us around have the look of belonging originally to gravel bars and mudbanks* – Anderson). He offers the amaranths as the dump-heap plant par excellence...*common in barnyards, middens and refuse heaps around the world*, for grain, vegetable, food coloring, scaring devils away from companion crops and when mixed with human blood good for Aztec sacrificial alters (though the Inquisition later disagreed). Habitats favoring some plants more than others are also opened up by mere fact of human intrusion (Piperno and Pearsall 1998), or more drastically and deliberately by swiddening (Shifters), culminating through *chacras* in modern farm fields which from this viewpoint are only habitats so arranged that they are opened to one plant but closed to all others. Or extending this to farms *in toto* and the few favored crops and livestock they now contain, selection gone mad, mutualism fatally abandoned. No more e pluribus unum.

9.6 SE Asia and Hoabinhian

So far the main archeological evidence which could support Carl Sauer's theories of a SE Asian origins has come from excavations of the Hoabinhian 'technocomplex' (Gorman 1969, 1973; Glover 1977; Higham 1977; Pyramarn 1989; Reed 1977; Pookajorn 1984; Solheim 1972, 1973; Yen 1977; Bender 1975) though this is still far from 'proving' them (nothing could). From Glover's review of Hoabinhian sites the term was first used by French archeologists in 1932 in reference to their discoveries of flaked pebble tools and plant/animal remains in limestone caves near Hanoi (Figure 9.2) and applied since to ancient occupied inland limestone/karst caves and coastal shell middens, some dating back to 12 000 BP, about 200 sites in all scattered throughout Viet Nam, Thailand, Burma, Laos, into China as far as the Yangtse River and east to Philippines.

These sites probably exist also in the similar karst and riverine landscape into Indian Assam. There's no reason why Hoabinhian should not have existed throughout all of SE Asia and into the now drowned plains of Sunda ie approximately where Sauer believed the origins to be although he favored the river valleys to the north in Assam-Burma-Thailand-South China (Figure 9.1).

A few of the sites are located in Figure 9.2, together with the shorelines of Sunda (and Sahul) during the last glacial and also approximately during the others previous to that (graphed previously, Origins). During these glacials, the last maximum c 20 000 BP, the then single land mass of Sunda was about four times the area of lands within it presently, the inundation having terminated about 7 000 BP after putting the human record in this largest lost bit of SE Asia beyond archeology's reach. Glover (1977) finds two explanations for the transition towards a Hobinhian economy as in the inundation itself and in the SE Asian landscape also then changing from dry open forest and savanna to a more moist forest as the glacial receded. Together these would have...*provided a massive dislocation of existing hunting communities*, creating resource stresses similar to those proposed by others (Binford, Flannery, Harris) as explanations of agriculture's emergence in Near East and Mexico (Origins). Some plant remains from Hoabinhian sites in Thailand are summarized in Table 9.1. They've been subject to more than the usual bickering over technicalities, dates and interpretations, more so by archeological literalists than by cultural prehistorians. Much ego has been at stake. To some (Gorman 1969, Solheim 1972) the Hoabinhian discoveries have represented early plant management, even forest horticulture as the first agriculture going back to before 14 000 BP ie about 4 000 years prior to agriculture in Near East. To others, especially those entrenched with barley and goats in the biblical tradition, they represented a threat to established position; eg Charles.Reed (1977) firmly dismissed the Hoabinhian evidence from places like Spirit Cave, Thailand, as representing anything more than hunter-gathering...*To date (1977) not a bit of evidence emerges which one can use to claim incipient horticulture or incipient cultivation or tending any kind of plant, or slash-and-burn or domestication of any animal...* However since this same authority also finds it impossible – again on formal archeological grounds – to see an agriculture in India before 5 000 BP (and this introduced from Near East as wheat, barley and cows, although 'more primitive cultures' in the Indian south might have been capable of 'local crops'), perhaps it's best to put aside this white horses -v- black horses view of prehistory and move on.

Most of the plants of Table 9.1 are forest trees-palms-vines still having

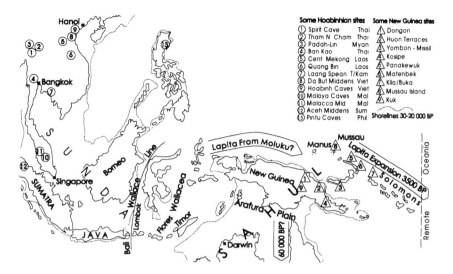

Figure 9.2   Sunda (SE Asia) and Sahul (Melanesia-Australia) Archeological Sites

multiple uses in this area. They lend themselves to three main interpretations: (a) simple gathering only (Reed) continuing today in the economies of the remaining gatherer-hunters of the region; or (b) some sort of plant tending or early horticulture, with this continuing in unbroken stream to develop into the contemporary mixed forest gardens of SE Asia (ie Stream A, Origins); or (c) since rice and beans were present at some sites, this horticulture leading and eventually giving way to proper Mainstream B agriculture as rice-beans cropping in valley bottoms where soils and irrigation water supply proved more suited to these crops than in the karst hills. These streams are not mutually exclusive, they're parallel in time; eg Yen (1977) notes the likelihood that wild rice continued to be collected as it is today, long after improved cultivars had (in principle, from archeology) become available. With general utility plants like bamboos and palms the concept of deliberative domestication as indicator of antiquity in use and management is even less relevant. One can't imagine the Hobinhian ancients sitting around in council and deciding to 'domesticate' them, or things like the candlenuts for their camp lighting when they were so freely available in wild condition. If this is relevant at all it would have occurred over a great period of time in Higgs-Jarman increments and incidentally.

Pyramarn (1989) offers the example of fan palm Licuala spinosa dated 8 400 BP from Ban Kao caves, its fruit still used medicinally by village healers and its leaves as sleeping mats by North Thailand Mlabri people who are said (by Pookajorn 1984) to be cultural descendants of Hobinhian. Glover (1977) lists the Andaman Islanders, Semang in Thailand, Senoi of West Malaysia, Punan and Penan of Borneo, Kubus in Sumatra, Toala in Sulawesi, the few bands of Tasaday in Philippines as little bits of Hoabinhian who are with us yet. One can't believe that over this great leap of time (before 12 000 BP to 7 500 BP in the case of Spirit Cave) they did not 'make experiments' with plants and with at least *some* of the 43 animals listed by Gorman (1977) as also present in these sites. Candolle's obscure and unknown men.

Table 9.1   Some Plant Remains from Thailand Hoabinhian Cave Sites

Trees – palms – vines—

Genus	Present uses	Date BP
Aleurites	candlenut – light, oil	9-8 000
Areca palm	chewing, drug, fibers	9-8 000
Licuala palm	fan palm – mats, medicine	8 400
Canarium	nut – food, oil	11 000
Madhuca	poison	11 000
Piper vine	spice, medicine	9-8 000
Calamus vine	fiber, fruit	11 000
Prunus	food	11-9 000
Terminalia	nut – food	11-9 000
Gramineae	bamboo, utensils	11 000

Other plants—

Croton	oilseeds	4 000
Cucumis	melons	11-9 000
Lagenaria	gourd	11-9 000
Trapa	water chestnut	11-9 000
Luffa	food, cloth	11 000
Glycine	beans	11 000
Oryza	rice	8-7 000

Source: From Gorman 1969; Yen 1977; Pyramarn 1989; Glover 1977.

9.7 Sahul – More Likely and Unlikely Tales from the South Pacific

Saturated with Candollean facts, confounded by substitution of carbon dating for common sense and too weary to fend off more vigorous scholarship, the Seekers in Sahul have desperately needed a Sauer, a Lathrap, an Edgar Anderson and fortunately found Les Groube and Alan Thorne.

The Dreamtime – Alan and Eve

Though this question concerns the emergence of modern people from the hominid ancestors universally, their arrival in Sunda (Figure 9.2) and Sahul (the periodically joined single land mass of New Guinea-Australia-parts of island Melanesia) is conveniently approached under the two competitive claims of the Eve or out-of-Africa theory and the multiregional continuous evolution theory. This dispute rages; the out-of-Africa theory is more or less the majority opinion based on interpretation of genetic evidence and the more impressive archeological support which Eve's champions have mustered. This sees all modern people descending from a common female ancestor 'Eve' in Africa about 200 000 years ago then dispersing over the Afro-Eurasian land mass to displace H. erectus in Africa, and/or the not quite modern hominids, descended from or in parallel with erectus in Africa and elsewhere, notably the Neanderthals of Europe.

In Sunda and its eastern margins (Java to Flores Island) both theories accept the presence here of H. erectus or something like them since Eugene Dubois found the first Java Man 1890-96 at Trinil in the Solo Valley. Others have followed in Java from Ngandong (dated c 300 000 BP), Sangiran 750 000 BP, Mojokerto claimed 1.8 million BP or about the same dating as similar evidence from Africa. Erectus must have ranged through all of Java and Madura at least as far as Bali. Mike Morwood (1998) and with others (1998), one of the current excavators on Flores, finds that in their long residence in Java from at least 1.2 million to 300 000 BP...*they clearly show evolutionary changes in a modern direction, becoming less robust and having larger brains.* Since Verhoeven's (1958) discoveries on Flores there's accumulating evidence that this H. erectus not only crossed the Bali-Lombok Strait (25 km) but three lesser channels to Sumbawa-Komodo-Flores as well.   Morwood reports stone artifacts 8-700 000 BP from several sites in Soa Basin, Flores, together with the remains of early elephants (Stegodon), then of these becoming miniaturized before disappearing, and of lizards growing into Komodo dragons which – on small isolated islands and especially with the expanding dragons preying on the contracting elephants – Darwin suspected they both would have to do. Perhaps this also partly answers Carl Sauer's question about the crossing to America *(Why should ManWoman not have gone where went the bear and bison?)*, though elephants are advantaged as large flotation bags with snorkels to assist their crossings in rough weather.

According to Eve's champions, east of Flores Island H. erectus when faced with the wider passage on to Sahul and not yet wise in the ways of monsoon winds and bamboo rafts gave up and went extinct, pushed to oblivion by Eve's more modern and aggressive Africans, H. sapiens, on the model of the elephants and dragons. History had to wait another million years or so for H. sapiens, by declaration far brighter fellows, to come from Africa and solve this raft problem and continue the great adventure. Archeologist Sandra Bowdler (1993) has explained the first failure as due to H. erectus' lack of material culture...*This H. erectus also appers to have been, compared with the populations of Africa, Europe and China, virtually without culture.* Just why it was that this particular family of H. erectus, having crossed half the world to get to Java-Flores, were still without 'culture' while all the others had it, is not explained. This long delay applied to SE Asia as well...*There is no evidence that H. sapiens was present in SE Asia before 40 000 BP...Thus we may see SE Asia, Wallacea* (Figure 9.2) *and Australia colonized virtually simultaneously by modern human beings some 40-50 000 years ago,* and for support Bowdler has mapped all the important SE Asian archeological sites bearing occupation dates 40-20 000 BP, with authority.

Continuity The second theory puts archeological literalism aside in favor of physical anthropology (Thorne 1980; Thorne and Wolpoff 1981, 1992), finding continuity, arguing that once H. erectus were out of Africa they evolved separately but always as part of the human family wherever they happened to be...*some of the features that distinguish major human groups, such as Asians, Australian Aborigines and Europeans, evolved over a long period roughly where these peoples are found today* (Alan Thorne). Specifically in China and greater Indonesia, represented by Java-Flores, the first modern people there evolved there, though this also allows very early 'pulses' of people south from China to Indonesia and through Indonesia to Sahul. As Thorne extends a deathless phrase...*The mark of ancient Java is on the early Australians. So too is the mark of ancient China.* More recently (reported Science, May 2000) skeletal remains of a young man and woman designated H. ergaster have been discovered near Tbilisi in then tropical Georgia, Russia, with abundance of simple tools reliably from 1.7 million BP. This pair were already then about one third taller (1.52m) and heavier (50 kg) than their more primitive African hominid predecessors. If in Georgia – with the elephants and rhinoceroses – why not in India, Central Asia, China, continuing to evolve to meet new challenges as they went? (the Thorne/Wolpoff model). One speculation by the Georgia excavators is that these larger chaps with their larger energy demands were propelled out of Africa by hunger. As likely it could have been idle curiosity.

If this multiregional evolution model is accepted there was no need for 'porters' to carry agriculture anywhere: it evolved wherever people did, though later they did ply their trade in 'domesticated' plants along such routes as Carl Sauer's lost corridor of mankind to the Yemen; and as Austronesians they toted New Crops down from Sulawesi-Moluku into Melanesia (with back-loading to South China); and as Lapita their conoes carried what had long been proven, possibly by erectus, out through Ocean's gates at Santa Cruz.

Logistics   As with the (presumed) first crossing of modern people to America via the Bering Strait, their arrival as Austroloids in Sahul ('carrying the marks of Java and China') is generally seen to depend much on northern glacials which in locking up ocean water in the ice made these crossings possible at several times in the past – 20 000 BP (too late), 50 000 BP, 160 000 BP... Theories regarding the most likely routes have been summarized by Flood (1995). One sees them island-hopping on from Flores and Timor to New Guinea, completing the journey H. erectus had started then walking south across the Arafura Plain. Another, favored by Alan Thorne (who built a bamboo raft to test it), proposes that any such craft caught offshore Timor in the monsoon season would almost certainly reach Australia within a week, and the necessary Indonesian bamboos don't grow there to enable a return. These logistics are with present high sea levels, so glacials and their low water would not have governed the Sahul itinerary. It could have been at any time, perhaps best determined from the other direction. Australia's oldest citizen so far is Mungo Man firmly dated 68-58 000 BP. Since the Mungo site is in inland SE Australia and 2 500 km from the northern coast (and Mungo himself would long since have lost interest in boating), Mungo's folks must have made the Arafura crossing much earlier, and their's defied that paper tiger of a barrier the Wallace Line much sooner. This pushes H. sapiens' Sahul-Australia ETA back to probably before 100 000 BP...toward where Ms H. Javanicus thought it had been that day when she was gathering mussels by the Solo River, though she was never much good with dates and couldn't distinguish young H. sapiens from the others. She heard later when they got there they'd organized three kinds of gardens: one for the pig, one of complete design and another on instructions from the Rainbow Serpent.

New Guinea – The Taming of the Rain Forests, an Introductory Model

Les Groube's (1988) model of the origins in New Guinea is based materially on discovery of waisted stone axes in raised (and still rising) coastal terraces on the Huon Peninsula on the northern New Guinea mainland (Figure 9.2), a fortunate occurrence in that shoreline emergence here has preserved such evidence while elsewhere it's been lost to rising sea levels after the Last Ice Age, in this region 90 meters. These are real hafted axes, as distinct from hand-held choppers, waisted with double blades and weighing to 2.6 kg and 20 cm long, dating firmly to 40 000 BP. Similar axes have been recovered at Kosipe in the New Guinea Highlands from 20 000 BP, with pollen and charcoal evidence of nearby forest disturbance from around 30 000 BP. These axes quickly lose their edge but their sheer weight would have made them useful to split sago palm trunks, thin out pandanus groves to promote fruiting, cut down the smaller trees for a clearing and ring the larger ones for fire (Shifters) and much later, if used as hoe or mattock, to drain swamps for the Garden of the Pig. Groube's 5-stage model has much in common with Dolores Piperno's for South America (Origins) and both offer archeological evidence for and owe much to Carl Sauer's wider vision.

1. *Arrival – foraging and hunting* From Paijmans' (1976) botanical surveys the coastal flora of northern New Guinea belongs with that of SE Asia while the upland forest plants belong with Gondwanaland. The newcomers would have found nearly the same suite of forest resources which they had passed through and mastered in Sunda. Only later when they pressed on into the mountain forests would the plants be new but the same technology, represented by the waisted axes, would have been used to tame these Gondwanaland forests too.

2. *Food-plant promotion* This stage consisted of food procurement through modification of the environment, not aimed at plants directly (Yen, Rindos), using the axes to cut back vegetation encroaching on the economic wild plants – forest yams, the giant swamp taro Cyrtosperma spp, the bush taro Alocasia spp, cycads, sago palm, bananas, many nut trees, sugar cane...and applying the detoxification methods necessary for some of these which they'd learned in SE Asia. As in Africa (Origins, Jacques Barrau) it used to be though that the indigenous useful plants had been few and most had been carried in by the immigrants, but...*what is abundantly clear is that the assumption of a decade ago, that man brought the bulk of his food plants with him from Asia, can no longer be sustained* (Groube). Indeed some plants

eg pandanus were later carried *to* Asia; and Douglas Yen found such a richness of economic plants in New Guinea-Melanesia that it could warrant the status of a Vavilovian center, if such things were still fashionable. (This preoccupation with food plants, a notable weakness in discussions of the origins here, does not rule out the likelihood that since they'd crossed so many rivers and had to walk along so many mud banks these ancestors had been Sauer's fisherfolk and aquatic hunters and that the first plants they'd valued were fibers, poisons, medicinals.)

3. *Forest management* At this stage effort was directed to selection and tending-management of stands of the more important but still wild economic plants, sago and pandanus groves. It's not elaborated by Groube but no doubt this extended to all useful individual trees wherever these occurred within the forest range, as the Kalimantan Luangan still find and manage the wild rattan and damar, durian and 'bee trees' (Shifters). Perhaps this stage still lingers: Groube offers the anecdote of laborers attracted down to Ok Tedi mine sites in 1980s who, although they maintain proper food gardens in their high altitude home villages and have company rations, still wander off into the surrounding forests...*and exposed all the natural stands of food plants...connected by faint trails running in all directions through the dense dripping rain* forest.

4. *Gardens* These are places where the most useful of the plants were eventually concentrated and formally domesticated. It's unknown whether these were of natural plant concentrations (wild taro swamps, yam patches) to which people came and settled, or if the plants were transported elsewhere to form artificial concentrations of plants, gardens, close to concentrations of people (initially Sauer's fisher villages). It's also possible that stage (4) should belong to the Shifters (reversing Edgar Anderson's concept of plants domesticated in 'house gardens' and then expanding out into the greater space of cleared *chacras*). It's just as likely that people first went to concentrations of wild economic plants, domesticating them *in situ* and only later transporting them off to form more formal gardens elsewhere. In any case plant concentrations, whether of domesticates or wild, led in two directions: in the lowlands to aboriculture and in the highlands to the intensive garden of the pig.

5. *Pigs and garden fences* Initially – in pious times – there were no pigs in New Guinea to threaten the gardens which went unfenced, unditched (though this ignores the wallaby which in biased opinion is as bad in this regard as the pig). From failure of the pig to cross the

Arafura Plain to Australia, before this was made impossible by rising sea levels, her New Guinea ETA from Asia has been put at c 10 000 BP, at first no serious hazard because the pig isn't partial to the natural toxins of some taros, presumed to have been one of the earliest crops in gardens. But as the garden's crop diversity increased so did the pig's prospects and the need to fence her out (especially after the 17th century(?) when the pig's favorite food arrived from South America as sweet potato). Then the gardeners had to invent fences and ditches and settle down to proper farming... But all this is much a modern farmer's view of things. Somehow it's more appealing to see the pig in pious times (and wallaby, cassowary, bird of paradise) valued almost as much as the garden's produce, and this first agriculture little more than innovative device for inviting pigs to garden hunting, by the 'people with fight'.

Corroborating Evidence – Axes and Some Stone Age Economics

Toth, Clark and Ligabue (1992) have offered fascinating contemporary support for Groube's stage (2), the stone axes and how they were made and used to tame the rain forest and prepare it for the garden of the pig and the garden of complete design. They found the Kim-Yal people, small (4½ ft) tough mountain horticulturists (sweet potato, taro, pigs, chickens) still making them in the cloud forests of the central cordillera of West Papua. Starting with cracking a river boulder by tapping it with another or fracturing it with fire, the axe flaking operations take about an hour per axe and another hour for grinding the sharp edge, followed by binding on a handle. With these a man can cut down a tree the size of a telephone pole in about 10 minutes and they're easily sharpened each few hours. They're also used to build fences against pigs (Groube's stage 5) and fashion wooden digging sticks, tools, containers, and are much used for trade and social harmonizing. Today 3 stone axes = 1 metal axe; one bride costs several, the exchange rate somewhat variable, as no doubt it should be. More firmly in pious times, not long ago, one axe bought one pig. (Currently in far less pious times the mesolithic axe trade flourishes in PNG with tourism.) At one stage in the flaking of these long fire axes the artisans work them from blanks about 24 cm long, about the size described by Groube for his heavy blunt Huon axes, and of the heavy Sangoan picks from c 40 000 BP in Central Africa. It's possible that these latter heavy products are intermediate stage of the fire axe technology described by Toth. But putting *stone* age technology aside, more significant as clue to the real Origins in the tropics the Kim-Yal use bamboo knives for slaughtering their pigs (as razor-sharp bamboo machetes are still used in Kalimantan for despatching intruding transmigrants), which archeologists of the classical tradition don't like because this organic non lithic technology decays and leaves no evidence, except perhaps forensic.

9.8 Kuk Swamp and the Garden of the Pig

The picture emerging for field crop agriculture in New Guinea is of Groube's 'forest managers' (stage 3) finding and domesticating lowland crops (yams, taro, banana) then as shifters carrying these up to the high inland valleys; and in the process so depleting the resource on which they depended – the forest and its soil fertility – that they ended up with proper agriculture at places like Kuk Swamp (Hope and Golson 1995; Bayliss-Smith and Golson 1992; Yen 1990; Hope, Golson, Allen 1983). Kuk at 1 600 m elevation in the upper Wahgi Valley (Figure 9.2) is one of numerous similar highly populated and intensively cultivated basins in PNG highlands extending also to the Baliem Valley in West Papua/Irian Jaya – close by Jo Mun Nerek's grave at Freeport's Copper Mountain and occupied continuously for at least 34 000 years (Hope and Golson 1995; Swadling and Hope 1992). The dates at several PNG inland sites range from 30 000 to 16 000 BP; the evidence is pollen, charcoal, grassland expansion, erosion, waisted axes, more fingerprints of Shifters.

By one interpretation people first went up into the highlands as hunters and foragers, though in the Gondwanaland high cloud forests there wasn't much to forage except pandanus nuts. By then, towards the end of the Last Ice Age, two push-pull mechanisms were in operation. People were pushed up by increasing coastal populations and declining lowland game resources, and pulled by the larger game resources of the alpine grasslands, above these mist or cloud forests. Also the upper margin of the forest was itself being pulled further up by 600 m or so as the Northern glacial retreated and temperatures started to rise. This upper forest advance also reduced the alpine grasslands and dependent animals; and post glacial temperature recovery (by 3-4 C degrees) brought changes to plant species composition of the forest generally, and more useful tropical plants into the high valleys. But hunting fires simultaneously pushed the forest margins down again, a process which expanded and accelerated after the shifters arrived with their waisted axes and clearing fires. From the fact that the plants at Kuk were truly tropical and at about their low-temperature limit it's been proposed that the earliest farmers (shifters then sedentary) literally followed increasing post glacial temperatures up the mountain. Populations continued to increase made possible by the expanding swiddens and the declining game resources were eventually offset by arrival of the pig, at first semi wild and foraging around their range with the clan. The main significance of places like Kuk (itself 9 000 BP) is that they require a much earlier parent agriculture in the lowlands, which parallels Piperno and Pearsall's conclusions in Mexico and Peru. Who indeed would abandon the biological richness of the lowlands to go up to these difficult places, Kuk or the high *barrancas*, until compelled by circumstance.

With populations always expanding and needing more swidden area, and this generating more land degradation which itself contracted the basis for swiddening, and the pig now starting to grow in stature and economic importance, a new system was needed which met the needs both of pigs and their people. This was found in iterative taro cultivation around the valley swamps. From evidence of drainage ditches (described in detail by Hope and Golson) this started at Kuk before 9 000 BP. The essence of forest swiddening continued (plot shifting because of fertility depletion, weed and pest infestation); but here by 6 000 BP it came to be geographically curtailed to rotating taro around these fairly small 1-200 ha inter montane swamps, with one or two crops on any drained field followed by 20 years of restorative abandonment to the swamp.

Bush swiddening in the surrounding higher country was not terminated but continued in parallel with wetland farming. For one thing, there wasn't enough swamp to go around...which turned out to be a good thing, at least for the swamp owners, and offers a curious reminder of Ramakrishnan's findings in Assam (Shifters) – the upland shifters, through their erosion fed continuously as nutrients to the swamp in its fallow phase, subsidized the more advanced farmers down there at the big end of the system. Apart from weeds and taro beetles attracted to this monocrop in increasing numbers the swamp was now Class A land. But to exploit it needed 3-4 times more labor (3 300 person hours/ha) than did the upland swiddens, although yields were doubled. Bayliss-Smith and Golson offer the detailed reconstructed economic budget. These swamp economics depend heavily on assumptions made for crop loss to taro beetles which was probably the main reason for drowning them by so frequently shifting the taro patch around the swamp. (Currently Kuk yields of 18 tonnes/ha with use of beetle insecticide can fall to 3 tonnes without it.) On the one hand in early Kuk times taro yields would have improved progressively after 9 000 BP with better cultivar development; but on the other actual recovered yields could have declined with increasing beetle populations and damage as the surrounding forest was cut back by shifters and replaced by grass clumps and open habitat which are much more to the beetle's liking. To this labor load of 3 300 hours/ha in ditching, draining, pest chasing, weeding, harvesting…must be added the work of carrying the crop – net of the beetle's share – back to the village to share it with the pigs. By now these had been disciplined (through Les Groube's stage 5) and brought more firmly into the household. As their reward the people got 5.1 MJ of energy per kg of taro, although the pigs got slightly more because they didn't waste the peels or have to work so hard.

Kuk Swamp when seen through these depressing economics doesn't have to represent paradise lost. Without so many people to do the work, and cooperative shifters scratching away in the catchment and topping up the swamp's fertility, and pigs suggesting to the swamp aristocracy their ownership as a new and more plausible basis for social status and political power and store of surplus value, this agronomic great leap forward might never have been possible. (Karl Marx, dear, are you listening?)

9.9 Toward the Garden of Complete Design – Aboriculture in Eastern Melanesia

Speculation regarding the true origins in lowland New Guinea and island Melanesia (Bismarks and Solomons) as far east as Santa Cruz can now go much beyond Carl Sauer's general model as this has been regionalized by Les Groube. There's now much archeological evidence that modern people had occupied the north New Guinea coast and passed on to island Melanesia before 35 000 BP (Table 9.2).

Table 9.2   Occupation Dates at Some Lowland New Guinea and Island Melanesian Archeological Sites

Mainland:	Huon Peninsula	40 000	(+)
	Lachitu	35 000	
New Britain:	Misisli	11 000	
	Yombon	35 000	
	Mopir	*	
Manus:	Pamwak	12 000	(-)
New Ireland:	Matenkupkum	33 000	BP
	Buang	31 000	
	Balof	14 000	
	Panakiwuk	15 000	
Solomons:	Buka/Kilu	28 000	

*Source of obsidian found at New Ireland sites. See Figure 9.2.
Source: Gosden 1993, 1995.

Remains of those occupations are mainly stone artifacts associated with 'gathering' and littoral foraging, not agriculture. Pavlides and Gosden 1994 have described these simple tools from Yombon 35 000 BP. These, and the proper stone axes which had already existed for 5 000 years in Huon Peninsula where these voyagers came from, and the boat building needed for these longer ocean crossings (especially to Buka) pre date the first firm evidence of plant manipulation in these islands, taro starch grains on stone tools from 28 000 BP at Buka (Loy, Spriggs, Wickler 1992). This still doesn't prove settled agriculture or aboriculture. Gosden's (1993) interpretation is that most of these early 35-30 000 BP sites were temporary camps of people moving around and going to resources; but from about 20-15 000 BP the system changed so that resources were brought to people, manifest among other things as domestication of one of the most important Melanesian food-nut trees, Canarium indicum in the Sepik area around 14 000 BP (Yen 1990).

There's a long time gap between these lowland New Guinea/island sites of 40 000-35 000 BP and the two streams of agriculture which emerged from them, ground crop vegeculture and aboriculture and both later carried on further into Polynesia and Micronesia. Until Yen's (1974) studies in Santa Cruz this aboriculture stream had been noted then dismissed as 'coconuts and bananas' and neglected. The economic trees Yen found in Santa Cruz, a group of high and low islands, Melanesian and Polynesian, 200 miles east of Solomons proper, are summarized in Table 9.3 because they illustrate the capacity of trees in ancient times to supply everything from food and fibers, clothing, medicinals to poisons and money. (The uses noted were those current in 1974 or then remembered and with variation they still occur throughout Oceania/SE Asia.)

Most striking are the large number of foods (Table 9.3) which lend themselves to processing/storage and the high food energy concentrates represented by the many nut trees. Not listed are the modern imports of orange, papaya, kapok, mango, soursop etc. Yen found this aboriculture incorporated into all agricultural systems as...*at first glance a haphazard enterprise conducted by men, in contrast to the field gardening by women*, the trees planted in upland root crop swiddens; in taro swamp fields; around houses ('the village gardens are virtually tree-gardens') and the larger and shade trees along roads and on public lands. Yen later (1993) called attention to these and other tree and plant foods of wider Melanesia (and Australia) as potential New Crops, listing 33 of these but mainly the nut trees – 5 canarium species, 4 pandanus, 2 terminalia...though it's difficult to call them new when some have been in cultivation for 14 000 years. There are also the 'bush' leaf vegetables which include some small trees (and he found it odd, as Clare Madge later did with European cabbages in Africa, that New Zealand lettuce were being flown into Pacifica when Melanesia had since ancient times better things of its own).

Table 9.3   Main Economic Trees and Uses Modern and Prehistorical in Santa Cruz (summarized from Yen 1974)

Gnetum gnemon, medium tree: Bark for fiber cords, ropes; fresh fruit and seeds as nuts; leaves and flowers as vegetable; bark strips covered with feathers to make belts of traditional money.

Pandanus: several species, mainly P. dubius, bush: Large leaves for umbrellas; fruit and edible seed 'almonds;' aesthetic plant house gardens.

Areca catechu palm: Betel nut narcotic. India to Fiji, then in Polynesia replaced by Kava.

Cocos nucifera, coconut: Used to be main pig feed in west Melanesia; universal human use Polynesia.

Metroxylon solomonensis sago: In Santa Cruz main use leaf atap roof panels; some fruit as food; trunks occasionally split for starch but this main food New Guinea north coast and Moluku (Diversity).

Cordyline fruticosa Ti plant, scrubby tree: Variegated red-green leaves as body ornaments in 'pagan ceremonies' India to Polynesia; now ornamental but root baked in earth ovens as emergency food.

Musa banana/plantain: In Santa Cruz 20 varieties differing by color, texture, fruit, vegetable.

Antiaris toxicaria medium multi storey tree: India, China, West Pacific; main use live ladder next to and for harvesting breadfruit; traditional source of tapa cloth Melanesia; latex as adhesive in feather dance costumes; poisonous fruit worn as body ornaments.

Artocarpus altilis breadfruit large tree: About 18 varieties seeded/seedless Santa Cruz; most important energy source Polynesia/Micronesia except where root crops can be grown; fire-dried for long term storage, also ensilaged in ground pits; some varieties grown for their seeds; formerly main source of tapa cloth as substitute for Antiaris; cordage and dye. (Probably a main food carried on the Lapita/Polynesian voyages.)

Inocarpus fagiferus Tahiti chestnut, small tree: Edible fruit roasted, stored. Especially important on poor atolls, with coconut.

Canarium spp large tree: One of many C. species China to Polynesia; very important; fruits dried over fires for storage; inner bark resin for house lighting.

Spondias dulcis Otaheite apple, large tree: Large yellow fruit from Indonesia to Micronesia/Polynesia; in places the second most important 'fruit' after breadfruit.

Corynocarpus cribbeanus large forest tree: Large fruit cooked; bark as medicine.

Pometia pinnata Oceanic lychee, large tree: Important fresh fruit but can't be stored. In Solomons a successional volunteer tree in swiddens.

Hibiscus manihot, bush greens: Extremely fast growing leaf vegetable. Representative of many others, including edible ferns as vegetable in place of taro and sweetpotato leaves (or as pumpkins are grown for their leaves not fruit elsewhere).

Polyscias fruticosa and scutillaria (small tree): Leaf vegetables or to add aroma to oven baked pig and fish.   Variegated colors and leaf shapes garden ornamentals and in China medicinals.

Sterculia wild forest and cultivated trees: Fresh fruit; important food tree New Guinea prehistory.

Barringtonia procera and novae-hiberniae, trees: Important nut trees, nearly year-round production and only 3 years from planting to fruiting. Another species B. asiatica was a traditional fish poison.

Terminalia catappa sea almond, tree: Important nut tree, fire dried, stored.

Eugenia malaccensis Malay apple, tree: Fresh fruit; one of many food species throughout tropics.

Burkella obovata large forest tree: Large fresh fruit eaten raw but elsewhere also preserved by pit ensilage as is breadfruit.

Plants in Low Places

Jacques Barrau (1959, 1965) early called attention to the continuing importance of swamp trees and aroids in Oceania as indicator of the role they must have played in pious times. As wild stands the several sago palms Metroxylon species cover several million ha of coastal New Guinea swamps where they probably originated (thought to center on Moluku) and remain one of tropical world's great food resources (Diversity). Their domestication where this occurs consists of transplanting thornless types from natural stands to around villages. Although also now grown on commercial plantations, under wild conditions their modern and neolithic economics (from Barrau) are: 60 good palms/ha/year 'worth felling', taking 8-15 years to maturity then yielding 4 000 – 6 000 kg starch/ha (dry basis), with one adult person needing about one kg of this energy food per day. Or alternatively, an average family needs to work 10 days per month in the sago swamps to obtain their basic starch food requirements, supplemented with fish or 'vegetables grown on platforms above water' (reminiscent of caboclos growing their spices and medicinals in baskets above flood waters in Amazonia). The sago can be stored and prepared in many ways, including fire-drying into briquettes or baked as cakes on hot stones (if Pottery has not yet arrived) or boiled as a paste in a section of bamboo.

Less conventional are the long seed pods (to 20 cm) of several coastal mangroves, eg Bruguiera eriopetala, as a vegetable. These fruits germinate while hanging on the tree, are peeled, sliced, soaked for several hours to remove high tannin amounts then steamed or boiled with coconut cream or keep several months if sun-dried. This mangrove, representative of many similar, occurs from India to Melanesia-Micronesia and must have been well known in Carl Sauer's timeless fishing village.

Still following Barrau we meet some ancient root crops, also most happy in low places. The giant taro-like Cyrtosperma chamissonis seems to have come from Malaysia-Indonesia and migrated through Philippines to Micronesia then much of Oceania (though it's unlikely Lapita would have had much patience for it). It grows 4 meters tall and produces tubers up to 40 kg but on atoll coral rubble can take 10 years to do it (14 kg dry starch). Here, taken down the declining resource gradient out to Micronesia /Polynesia, the amount of labor growing it is as impressive as its size, with this agronomy aimed at duplicating the swamp conditions it likes but left behind in Sunda's wet forests. A pit is dug down to the water table, bottomless baskets are placed in the pit, any organic rubbish is thrown into the baskets and the *biah* are planted there...*These pits almost perfectly recreate the natural habitat of Cyrtosperma, a striking adaptation to an environment where the essential fresh water* (the sea water emerging as fresh water lens) *and the organic matter of the tropical forest are lacking* (Barrau).

The taro itself (Colocasia esculenta) is more reasonable, growing in most moist but not perpetually wet places and economically a rough equivalent of manihot/cassava in tropical America, its origins vaguely India-Malaysia (Sauer's fishing village), its antiquity greater and its travels (to Japan, China, all Oceania, Arabian coast, Mediterranean) nearly as remarkable. More interesting to the Seekers will be Barrau's suspicions, like those of many others, that taro cultivation preceded rice cultivation in wet sawah padi fields, with the rice – Harris' always aggressive granoculture – later taking over as a weed. These root crops and the yams (also from SE Asia mainland monsoon forests) and their Oceanic division into 'male' and 'female' crops, with the male plants also used to sop up excess moisture in female gardens, and sugar cane as a magical along field bunds to ward off evil, and hymns to earthworms to leave the earthworks intact, and village magician-meteorologists arranging the agronomy – and long male yams always threatening the shy triangular female taros – still hint at how things were in pious times.

9.10 To Remote Oceania—Lapita and the Country Further Out

Relying on linguistic evidence as well as Chinese archeology (Peter

Bellwood 1980), sometime before 6 000 BP proto Austronesian speakers (of Thai-Kedai) began to leave the South China-Indochina coast to settle Taiwan and from there, evolving culturally as Austronesians proper as they went, took the Old Planter Culture to more than half the tropical oceanic world from Easter Island in the east to Madagascar in the west and last New Zealand in the south, although by then it was not really old on Carl Sauer's scale of time. When it appeared in the islands north of New Guinea-Solomons to pass on through Santa Cruz into Remote Oceania the central strand of this movement, designated Lapita technical complex, was characterized by distinctive corded pottery, horticulture, long-distance voyaging in ocean-going canoes and large settlements on the littoral of small islands. Although Lapita took ManWoman on their first and greatest maritime adventure there's some hazard that the romance of the Polynesian sagas and Lapita's 'high archeological visibility and state of explosive expansion' (Irwin 1989) might obscure the fact that the Indonesian islands which Lapita skirted had already been occupied by H. erectus/sapiens for about 1.8 million years, and that the stone axes it carried past the Melanesian coasts had already been used there for 40 000 years to 'tame the rain forests' and prepare them for the Garden of the Pig (and Complete Design).

In Bellwood's Lapita model the impetus to move on from Taiwan had possibly been ecological destruction and soil erosion induced by forest swiddening, compounded by the usual pressures from increasing population. This earliest Taiwan swiddening had consisted of both vegeculture rootcrops and rice-millet granoculture (Origins, China). Both were carried south by the Austronesians but in the true wet tropics, and especially later on small oceanic islands, the granoculture was dropped of necessity, vegeculture came to dominate supplemented by aboriculture as the immigrants met new useful trees while passing down through Philippines then Sulawesi and Moluku to New Guinea. Bellwood concedes the likelihood that the Austronesians absorbed both plants and agricultural ideas from people already long resident in eastern Indonesia...*It is possible the resident Australoid hunter-gatherer groups were already exploiting in a fairly intensive fashion the aroids, yams and the major tree fruits*...and that inter-island trade networks already existed long before the Mongoloid migrants arrived.

There are several versions of Lapita origins, none claiming to be adequate. One while accepting an ultimately South China-Taiwan origin sees them coming out of Moluku and sailing briskly east, north of the New Guinea coast in 500 years or so to the oceanic settlements (40 of these have been mapped by Kirch 1989, Bismarks to Samoa, and they continue to be discovered – Allen and White 1989). From this they were essentially seafarers, fisherfolk, traders though also horticulturists. One interpretation of their agricultural role is that they were primarily porters (equivalent to the unknown porters along Sauer's other lost corridor), absorbing the eastern Indonesian and New Guinea plants which had long been domesticated by the resident Australoids, taking some of these on with them to east Melanesia-Oceania (yams, areca palm, taro) and sending others from New Guinea (listed by Yen 1990) back through Moluku and Sulawesi to SE Asia. Another version sees Lapita proper formulating itself in and near the Bismark Islands before the voyaging horticulturists moved on. Geoffrey Irwin 1989 has called these waters around New Britain, New Ireland and eastern Solomons the Lapita 'voyaging nursery' where they learned the open water sailing skills necessary for colonizing most of Oceania and the navigating skills to later maintain contact with these colonies in the vast Pacific (and continue the inter-island exchange of plants into European times).

As in Taiwan, Lapita vegeculture required forest clearing, accelerated as previously by increasing populations and often having devastating consequences in soil erosion and bird/animal extinctions, with this compounded by uncontrolled pig, dog, rat which accompanied the voyagers. Spriggs (1996) cites the archeological evidence of erosion at several Lapita settlements in the Arawe Islands off New Britain starting around 3 000 BP, and similar evidence of 40% of the birds, a primitive crocodile, horned turtle, lizards, extinguished by the Lapita colonizers in New Caledonia and others in Vanuatu and Eastern Solomons...*Such major impacts on the landscape by human groups were unprecedented in the region and suggest a radically different attitude to the environment and the place of humans in it.* Sometimes the voyagers moved on to repeat the process elsewhere, abandoning an island for centuries. Those who stayed took up the tree crops stream, on high islands eventually mixing these with root crops in the conservational polycultures which prevail today (Reconnaissance, Shifters). On low islands, atolls, where most root crops were seldom feasible, aboriculture always dominated (eg Yen's list for Santa Cruz). Proper root crops farming, Lapita's Mainstream B, would have been – and often was when tried – suicidal.

Lapita Settlements – Glimpses of Venice

Patrick Kirch (1989) offers a weekend in one of these Lapita villages on the lagoon of small Eloaua Island in the Mussau/St Matthias group between Manus and New Ireland. (It would have pleased Carl Sauer who was always partial to fisherfolk in Venetian villages on stilts, whether in SE Asia or Venezuela.) This village occupied 7.2 ha, the largest so far found in Melanesia and was settled around 3 600 BP, the house posts and plant food remains preserved in anaerobic conditions of shoreline muck, 20 tree species (most listed previously for Yen's surveys in Santa Cruz), some 'showing signs of human modification' (Canarium nuts), the coconuts definitely improved into the large modern drinking type, and others – cycad nuts and the fruit of the *su* tree (Pangium edule) – though still today important foods from Malaysia to Vanuatu requiring some skill in processing to make them safely edible. This wealth of tree foods attests to the antiquity of aboriculture in island Melanesia east to Vanuatu. *(It) appears to have its origins in Lapita agriculture of the second millennium BC* (Kirch), though the extent to which this was melded into far older pre Lapita Melanesian developments from 40 000 BP is still unsettled. The most common food found under these Lapita houses has been canarium nut (Yen found five varieties of it still important in Santa Cruz, Table 9.3). Matt Spriggs (1993) has described their food importance in Francis Ona's village on Bougainville as late as 1950s which suggests the role they played in the Lapita voyages...*The shiny black fruit are gathered and kernels cracked, smoked and tamped into bamboo tubes and sealed with leaves and a cement of wet ashes and they keep for three or four years.*

Another find was Terminalia catappa (also on Yen's Santa Cruz list and now candidate New Crop), described by Julia Morton (1985) as one of 250 species with Old and New World tropical distribution, this one commonly called Indian or Bengal or sea almond in Indo-Pacifica in deference to Indian's greatest modern use of it and the seed's ability to travel long distances as flotsam from its presumed home in Malaysia. Though billed as fruit and nut Morton lists other current/traditional uses from all its parts as mainly medicinal (from leprosy to vermicide), its kernel oil for soap and the residue as pig feed, the leaves black dye and ink, the roots tannin, with the tree itself able to withstand salt winds, poor soils and drought – all reasons why the voyagers might have thought it useful, and with its wood for drums, its flowers as bee range and its leaves a ready made poke for carrying small amounts of sugar it would have met Carl Sauer's ideas of fitting out the Expedition perfectly.

With a canoe load of coconuts for water, baskets of dry fish and turtle meat, tubes of canarium nuts, breadfruit chunks to chew, a sea almond pharmacy, two pigs, two rats hiding behind the pandanus pot plants, ten fathoms of bark money for incidental expenses, the Voyagers could have sailed on from Santa Cruz forever. *Tis not too late to seek a newer world. It may be that the gulfs will wash us down; It may be we shall touch the*

Happy Isles...

The Happy Isles – and Declining Resource Gradients into Remote Oceania

There seems to be a principle that while forest gardens are richest in crop species where natural plant diversity is greatest (Amazonia and Kalimantan-Malaysia-Java), as this diversity decreases the economic uses of it increase, so that this resource cline is rectified by applying progressively more imagination to it. At the extreme societies have managed very well with only things like date palm, borassus and the coconut. Yen and others have pointed to this declining resource gradient out from SE Asia through Melanesia to Polynesia and Remote Oceania (Reconnaissance). Albert Robillard (1992) sketched this resource poverty which voyagers found when they came to Kiribati 2-3 000 years ago. These 33 atolls and small reef islands were themselves newly risen form the sea only 6 000 BP, formed out of coral rubble and are still without any meaningful soil or surface water; the only plant/animal life they supported until the colonists arrived had been that carried in by winds and ocean currents. The only significant food plants Robillard found, presumably carried or later imported, were the giant taro-like Cyrtosperma described above, a few breadfruit also mulched against the frequent droughts, coconuts (origin unknown), a small fibrous fig (F. tinctoria – flotsam?) and one pandanus (P. tectorius – more flotsam?). Nonetheless they worked this pandan up by vegetative propaganda into 160 varieties.

Arthur Whistler (1987) surveyed the contemporary plant resources and uses in another of these impoverished Happy Isles, the three atolls and sand islets (*motu*) of Tokelau 300 miles north of Samoa, then supporting 1 600 people on 750 ha. He found only about 80 plants but collectively these had 123 uses, 25 as medicine/poison, 2 for hair shampoo, 5 for clothing etc. A large proportion had been brought from Samoa and this inventory includes the grasses and 'weeds,' though economically there are very few of these because multiple and specific uses had been found for almost everything: some crushed weeds for scenting coconut oil; the inner bark of *fau* tree (Pipturus argenteus) a general cordage but used specifically for hook leads in deep sea fishing and lashing up babies' umbilical cords; the *gahu* bush yielding lei flowers, pigeon food, medicine, the leaves for baitfish lures, and the soft branches when hollowed out yield pop guns for the baby when it gets big. Whistler found only one pandan, again P. tectorius, but they're almost as important as the coconut and 21 varieties had been bred with about this many uses, to pigs as well as people. Not much else happens on Tokelau.

Food preservation and storage Food preservation and storage methods were as necessary and sophisticated as the production agronomy. Jennifer Atchley and Paul Cox (1985) and Cox (1980) have described these fermentation and pit storage methods for breadfruit (and banana) throughout Oceania. One common method is to skin and quarter the fruit, bury it wrapped in leaves for a month until the breadfruits have melted into a dough-like mass, then bake this (as cakes or sheets), wrap or sew in leaves and again bury for storage. Such storage of the dough or *masi* can extend for 100 years. In Society Islands and Tonga these strategic food reserves against hurricanes, drought and war were large enough to supply the whole community. (When hidden underground from enemies they were often lost and only accidentally rediscovered by later generations.) With this technology on even poorest islands like Kiribati, and the pandan nuts and 'figs' preserved also, and the coconuts tapped for toddy and sugar in droughts when they couldn't set fruit, and dry shark fins to chew and guano birds to net and Carl Sauer's handbook on lagoon fish farming – and dynastic wars to fight in the sailing season – *no resource went unutilized* (Robillard).

9.11 The Garden of the Rainbow Serpent

*It is, after all, merely an assumption that Australia has always been a continent of foragers...*Matt Spriggs.

When archeologists and prehistorians speak of the first agriculture they usually mean the first agriculture which resembles our own; and being guided in these concepts by modern agronomy go on to list its signs or necessary formal conditions as induced morphological change in plants/animals (domestication); cultivation which is shorthand for a sequence of deliberative production steps from land clearing through dig-plant-fertilize-weed-irrigate-harvest-store as need may be; with these occurring in a bounded space or sedentary farm according to a regular calendar of operations. Conveniently, most of this agronomy leaves archeological evidence. Since Candolle's time there is also a lingering assumption that this farm must produce a storable staple crop which yields a surplus product which can be taxed or otherwise diverted to the common good, with this defined by elders or their Prince who with the surplus will sponsor learning and philosophy, science, temples, pyramids and slaughter of his neighbors to extend the Farm and glory of the tribe, so that when this civilization is dug up in ten thousand years the people will cry out Yea! our ancients were the founders of this farm and we are inheritors of its glory, inheritors of the one true agronomic faith, and verily our agroscientists are Keepers of the Code. The Rainbow Serpent's farm was not like that.

From physical anthropology (Thorne and Wolpoff 1981, Thorne, Grün et al 1999) Mungo's family had been a mixed bag – some robust ('bearing of the mark of ancient Java'), some gracile (the mark of China). If this is accepted as indicating arrival of several and probably many pulses of people down through Indonesia – once they'd mastered bamboo rafts or got their glacial/interglacial timing right – it's reasonable to see these different peoples bringing many different Higgs-Jarman economic systems with them. They'd have found some things old (the rain and monsoon forests which once covered most of northern Australia, savannas, rich estuaries) and some things new (deserts, great inland swamps, cold mountains). These were the Garden of the Rainbow Serpent, best managed by reverence, fire and digging stick. For a long time it was thought they'd let it go at that, hunter-gatherers of the Dreamtime. Ignoring jibes at Greeks and regional patriots one could claim them as the world's first farmers (but won't because they weren't unique). To make this plausible four other conditions for agriculture would have to be added to the standard list – sustainability, risk minimization, leisure maximization and consistency with a governing purpose. When these attributes are considered, particularly in the Australian prehistoric system, they so dominate the standard technical conditions they lead us on from farming as technology to management of ecosystems; and these resource managers are better seen as systems analysts attempting to optimize a whole human material and spiritual system with its components inextricably linked, according to a cosmology.

Farm Management in the Dreamtime

Josephine Flood (1995), H.R. Allen (1974), R.A. Gould (1980), R. Jones (1975), N.B. Tindale (1974), B. Meehan (1977, 1989) and the 19th century explorers have described Aborigine plant and food management techniques which offer a checklist against which achievement of conditions for agriculture can be measured, and more important, reasons for their 'non achievement' can be considered. One keeps in mind that these are only *some* observed practices among variable human groups in some of many ecozones at some points over 60-100 000 years; ie any number of economic systems and practices were no doubt remembered, invented, applied, modified, abandoned, as climate and needs changed over this great period of time. Spriggs (1993) suggests consideration of the great yam (D. alata) in northern Australia, the gathering of which has been interpreted by

Yen as 'resistance to agricultural production' ie to farming. But as Spriggs points out it's equally reasonable to see this modern yam collecting *as a relic of a previous agricultural system*, transformed at some time to collecting as circumstances changed, as the Yuquí hunter-gatherers turned their backs on horticulture. Again as Stearman says, technology is not destiny.

Domestication Yen's conclusion (*The element of plant domestication is absent from the Australian hunter-gather system*) is still generally accepted, at least in relation to the grasses which could have but didn't become full granoculture, and with livestock. Flood suggests that while the large animals amenable to domestication were absent (cow, buffalo, goat...) the non farmers *could* have domesticated the duck, goose, pelican but didn't (to which one might add the iguana, as now in Costa Rica). Others are skeptical of this general failure, finding it difficult to believe that burning a grass plain or savanna regularly and in a controlled way for 60 000 years caused no change in the morphological structure of these plants, either directly or indirectly through the effect on other vegetation and faunal components of the ecosystem. Similar reservation extends to root crops/yams. It's well known that some of these were replanted from parts when harvested; and from Africa this simple operation is known to result in 'bigger yams', usually the object of formal domestication.

Doubt also extends to the fruit and nut trees. These are now being 'discovered' and domesticated and in the past 20 years a new Australian commercial bush foods industry has emerged (to complement the 'new' crocodile steaks and witchety grubs, with 1 grub = 1 pork chop – Flood), from on-going cultivar developments of these species. Again it's difficult to believe these now boutique trees and bushes were not already changed substantially by tens of thousands of years of fire, exposure and disturbance by human, presence (Rindos; Piperno in South America). Flood cites evidence from Meehan and Jones of the Anbara people of Arnhem Land who established orchards around their camps by spitting out fruit seeds into the debris of fish remains and shells in refuse heaps...*These midden soils with their compost and shell lime provided an ideal environment for tree growth, so in a few years the camp would be well supplied with fruit trees. Indeed there is a consistent association between old camp sites and trees with edible fruit* (which also proved a useful thing for archeologists who now know where to dig). If this sign of the origins was good enough for Edgar Anderson in Guatemala, why not in Arnhem Land?

Other critics follow Rindos. Food was generally so plentiful in Australia relative to a controlled population that the best way of obtaining it was not to 'bring it into the household' but integrate ManWoman into the ecosystem, and become part of it, part of the Rainbow Serpent's garden. It domesticated them. And since this was a mobile garden they had to be mobile too, more so than the Yuquí (though archeology doesn't think too highly of mobility).

From Flood's reconstruction of the economy of Willandra Lakes (where Mungo lived) the grasses – the granoculture that never was – became important only after 20 000 BP when the lakes and their aquatic resource dried up as they periodically did and the people had to consider this new possibility seriously. This is about the same time the first signs of wild grain harvesting (production?) appear in the Nile Valley, though there's little doubt it had been occurring in both places for tens of thousands of years previously. The Mungo people had a choice. Their agronomists wanted to settle down to domestication and invent proper farming. But fortunately their systems analysts talked them out of it: it would have wrecked the rest of their multi-enterprise system, and the Rainbow Serpent's garden too.

Clear-and-cultivate   These are rational production activities only if the clearing yields a greater product or equal product with less energy input than does uncleared uncultivated land, and does so sustainably. In the Rainbow Serpent's garden this was not the case. Another important difference was that any necessary disturbance of the ecosystem was provided strategically, temporarily and more efficiently by controlled fire. Especially in the tropical north the forests and swamps were already rich in food so why permanently clear or drain them? On the savannas such clearing would have been unnecessary had cultivation been intended; and since these already supported vast natural crops of Setaria and Panicum grass-grains, why dig them up to try to plant them better than nature had done? And again, whatever 'cultivation' was needed could best be done by timely and strategic fire – which also fertilized the fields, checked pests and brought fresh game, integrating livestock with the 'granoculture' part of the system. Later rejection of this systems management logic by the European clear-and-cultivate concepts which displaced it only proved how superior the first agroecology had been – this proof being soil erosion, salinization of soil and water, acid sulphate poisoning of the wetlands, destruction of fish breeding grounds on continental scale in just 200 years (Energy).

Plant-weed-pest control-irrigate   Apart from irrigation of the grassland grainfields by turning small streams onto them the Rainbow Serpent's gardeners had no need for this suite of practices. At worst – since farmers create their own weed problems by soil disturbance and monocropping and their own pests by invading and contracting their habitat – these operations would have been counter-productive and illogical (as modern field crop farming is now discovering through such practices as minimum tillage and integrated pest management).

Harvest-store Here at least the gardeners conceded orthodox agriculture. Flood cites Thomas Mitchell, exploring down the Darling River in 1835...*The grass had been pulled to a great extent and piled in hayricks which extended for miles...the grass was of one kind, a species of Panicum and was beautifully green beneath the heaps and was full of seed...* And further inland on Coopers Creek in 1886 the explorer A.C. Gregory found...*fields of 1 000 acres growing this cereal, cut down with stone knives half way down the stalk and the grain beaten out and winnowed in the wind.* In the tropical north Bancroft in 1884 found wild rice to be the...*most important grass food in (tropical) Australia, being little inferior to cultivated grain...* So why domesticate it?

Storage technology varied widely by ecozone and type of food (Flood). In some arid places the wild grain was stored in hollowed out wooden bins (1 000 kg was found in one cache); elsewhere in skin bags wrapped in straw and smeared with mud. The bunya-bunya pinetree nuts (a major seasonal food in SE Queensland, later with Europeans as well as Aborigines) were buried; the cycad nuts were wrapped in paper bark (tea tree, Melaleuca spp) and buried in large grass-lined trenches; in the northern tropics yams and water lily roots were pulled, stacked up and protected by their own toxicity. One livestock harvest-store technology was to drive game into one corner of the 'farm', the range, and hold them there until needed, conserved from foolish youths and over-harvesting by ritual. ('Only clan elders may take the honored emu'; or with the Sepik Sawiyonö, women could not partake of the cassawary unless they had excess of 'fight'.)

Sedentism, mobility and risk Unlike the technical branches of agricultural science, modern farm management systems analysis places much emphasis on uncertainty and risk minimization, skeptical of promised outcomes from research stations which are not stated on a reliability basis. The reason for this is obvious: farm economics finds little value in some highest yielding cultivar or crop if they're going to be wiped out by drought four years in ten, or in some New Crop which through generating New Pests wrecks other components of the system. There are five main strategies for handling droughts, the most common of these events. One is to sit them out on a cache of bunya nuts (or money). The second is bend to them, meaning stocking in good times, de-stocking in bad, requiring mobility. It's no accident that the most successful pioneer pastoralists in this harsh land were semi nomads, driving (now trucking) their herds along a chain of stations/ranches scattered over several thousand miles, following the seasons and the rain clouds. The third is to drought-proof the farm with irrigation or feed reserves, usually the most expensive in modern times in terms of money and in prehistoric times in terms of human energy. The fourth is through a multiplicity of enterprises (grain gathering, mussel and honey collecting, fishing, hunting...) and since the resources for this enterprise diversity were scattered around the 'farm' in many ecozones this again required mobility. The fifth would have been – as it was in Near East and upland Mexico – to fully develop granoculture; but under Australian conditions this, through the sedentism it requires would have meant sacrificing the other enterprise diversity and maximizing risk not minimizing it.

The theory of the leisure class   One important attribute of an agroeconomic system, which techno-agronomic approaches to the origins can't handle or do so awkwardly, is how much leisure it produces, with this offering the potential for pursuing either still more technology or developing a cosmology expressed through legend, art and ritual. Here the gardeners disagreed with Sauer: their existing system afforded so much leisure they didn't need further technical advance towards agriculture to give them more. (The same conclusion is arrived at if the system operating objective is taken as minimizing personal energy expenditure.) They devoted it instead to interpretation of their world and expressed this cosmology in art and oral literature, again with the great advantage that art, dance, stories of the Dreamtime – the essentials of civilization – are portable while grindstones, kangaroo-proof fences, stores of domesticated seed and similar paraphernalia which would have been needed by an advancing agro technology are not. (Aborigine art dates from at least 50 000 BP – Roberts et al 1990.)

Leisure was obtained by maximizing food supply from their economic system while minimizing demands on it. Supply, always partly subjective (Lapita), was maximized by defining a large number of possible uses of often poor resources, operating the widest possible suite of economic activities (hunting, collecting, fishing...) and managing these strategically which required mobility. Demand was minimized by a cosmology which found no personal merit in wealth accumulation and no need for a social surplus which would have and elsewhere did support an aristocracy and their Prince and the temples and pyramids and armies needed by this Prince to testify to the glory of the tribe. (It was also minimized by population control, including occasionally infanticide when things got really bad.) From first European contact almost to the present day there's ample evidence that Aborigines felt they lacked nothing (...*They may appear to be the most wretched people on earth but in reality they are far happier than we Europeans...They seemed to set no value upon anything we gave them* – Captain Cook, 1770). In the richest ecozones of the north, and in good seasons generally, they met their own and dependants' needs with only one or two hours work per day; in the worst of seasons in the arid zone it took 6-8 hours per day.

Global and local optimization Any human system must be operated consistently to a purpose. Without this it will drift opportunistically and soon collapse. The gardener's purpose was to optimize their place in a whole material and spiritual world, to find what economists would call a global optimum, with the separate components making sense only if they each contributed *most efficiently* to the whole. The separate material components of grass seed gathering, duck netting, fishing, hunting...are the equivalent of enterprises on a highly diversified farm. These material activities were in turn dominated by a religious force, for again the material part of the system made no sense unless it was sanctioned by the spirit ancestors. The Rainbow Serpent had made the land, coiling through it to form the mountains and direct the rivers and she gave birth to many of the people; and the people had to use the land and its creatures according to her plan (Tacy 1995).

Modern agrosciences don't understand this plan. They have no holistic plan of their own because they have no spirit reference. *White man got no Dreaming.* They see the world, like Inspector Gadget, as a multitude of separate problems to be reacted to and solved separately by defining for each a local optimum – how many hogs to run per farm? – the optimum amount of corn to feed these hogs? – the optimum machinery investment to grow this corn? – without asking whether hogs should be raised at all or if it has a purpose. Applied to Aborigine land management systems eg in the case of their 'failure' to domesticate grain and develop full granoculture, this preoccupation with such local or single enterprise optimization would almost certainly have prohibited the more important optimization of their wider survival and spiritual system. When prehistoric economic systems are viewed in this fragmented and technocentric way it's inevitable they will be judged non agricultural, something short of legitimate, because some of the standard bits (domesticate-clear-drain-cultivate-plant...) might be missing. Another reason why science—or more correctly agro technology—can't find agricultural legitimacy in these systems is because it can't find their boundaries; and to define or specify the boundaries of a system is the first necessary step in its evaluation or construction. But where do the Rainbow Serpent's boundaries begin and end? Truth is, they encompass such a universe of ecozones, seasons, resources, El Niño chance events, decision trees, risk probabilities, spiritual interpretations, that we're only dimly aware of what might have been the scope and structure of the Rainbow Serpent's garden.

But still, consensus is, it's not a proper garden and there was never much in it. Only time, ManWoman and the natural world. Only the Rainbow Serpent would claim it as a Garden of Complete Design.

References

Allen, H.R. (1974) The Bagundji of the Darling Basin: cereal gatherers in an uncertain environment. *World Archeology* 5(1974):309-322.

Allen, J. and J.P. White (1989) The Lapita homeland: some new data and an interpretation. *Journal of Polynesian Society* 98:129-146.

Ames, O. (1939) *Economic Annuals and Human Cultures.* Botanical Museum Harvard University, Cambridge, Mass.

Anderson, E. (1954) *Plants, Man and Life.* Little Brown, Boston.

Arenas, P. and G.C. Giberti (1987) The ethnobotany of Odontocarya asarifolia (Menispermaceae) and edible plant from the Chaco. *Economic Botany* 41(3):361-369.

Atchley, J. and P.A. Cox (1985) Breadfruit fermentation in Micronesia. *Economic Botany* 39(3):326-335.

Bahre, C.J. and D.E. Bradbury (1980) Manufacture of mescal in Sonora, Mexico. *Economic Botany* 34(4):391-400.

Barrau, J. (1959) The sago palms and other food plants of marsh dwellers in the South Pacific Islands. *Economic Botany* (1959) 13:151-162.

Barrau, J. (1965) L'Humide et le sec: An essay on ethnobotanical adaptation in contrastive environments in the Indo-Pacific area. *Journal of Polynesian Society* (1965) 74:329-346.

Bayliss-Smith, T. and J. Golson (1992) A Colocasian revolution in the New Guinea Highlands? Insights from phase 4 at Kuk. *Archeology Oceania* 27:1-21.

Bellwood, P. (1980) Plants, climate and people: the early horticultural prehistory of Austronesia. In J.J. Fox, R. Garnaut, P. McCawley, J.A. Mackie (eds) *Indonesia: Australian Perspectives.* Research School of Pacific Studies, Australian National University, Canberra.

Bender, B. (1975) *Farming in Prehistory: from Hunter-Gatherer to Food Producer.* John Baker, London.

Bohs, L. (1989) Ethnobotany of the genus Cyphomandra (Solanaceae). *Economic Botany* 43(2):143-163.

Bowdler, S. (1993) Sunda and Sahul: a 30 KYR BP cultural area? In M.A. Spriggs and B. Fankhauser (eds) *Sahul in Review: Pleistocene Archeology in Australia, New Guinea and Island Melanesia.* Papers in Prehistory 24; Research School of Pacific Studies, Australian National University, Canberra.

Bruman, H.J. (1940) *Aboriginal Drink Areas in New Spain.* Doctorate Dissertation, University of California, Berkeley.

Cox, P.A. (1980) Two Samoan technologies for breadfruit and banana preservation. *Economic Botany* 34(2):181-185.

Flood, J. (1995) *Archeology of the Dreamtime.* Angus and Robertson/Harper Collins, Sydney.

Glover, I.C. (1977) The Hoabinhian: hunter-gatherers or early agriculturists in South East Asia. In J.V.S. Megaw (ed) *Hunters, Gatherers and First Farmers Beyond Europe: an Archeological Survey.* Leicester University Press, UK.

Gorman, C. (1969) Hoabinhian: a pebble-tool complex with early plant associations in southeast Asia. *Science* 163:671-673.

Gorman, C. (1973) Excavations at Spirit Cave, north Thailand: some interim interpretations. *Asian Perspectives* 13:79-107.

Gosden, C. (1993) Understanding the settlement of Pacific islands in the Pleistocene. In M.A. Spriggs and B. Fankhauser (eds) *Sahul in Review: Pleistocene Archeology in Australia, New Guinea and Island Melanesia.* Papers in Prehistory 24; Research School of Pacific Studies, Australian National University, Canberra.

Gosden, C. (1995) Aboriculture and agriculture in coastal Papua New Guinea. *Antiquity* 69:807-817.

Gould, R.A. (1980) *Living Archeology.* Cambridge University Press, New York.

Gragson, T.L. (1997) Use of underground plant organs and use in relation to habitat selection among the Pumé Indians of Venezuela. *Economic Botany* 51(4):377-384.

Groube, L. (1988) The taming of the rain forests: a model for Late Pleistocene forest exploitation in New Guinea. In D.R. Harris and G.C. Hillman (eds) *Foraging and Farming: the Evolution of Plant Exploitation.* Unwin Hyman, London.

Guddemi, P. (1992) When horticulturists are like hunter-gatherers: the Sawiyanö of Papua New Guinea. *Ethnology* 31:303-314.

Haudricourt, A.C. and L. Hedin (1943) *L'Homme et les Plantes Cultivées.* Gallimard, Paris.

Higgs, E.S. and M.R. Jarman (1969) The origins of agriculture: a reconsideration. *Antiquity XLIII, Antiquity XLIII* (1969):31-41.

Higgs, E.S. and M.R. Jarman (1972) The origins of animal and plant husbandry. In E.S. Higgs (ed) *Papers in Prehistory.* Cambridge University Press.

Higham, C.F.W. (1977) Economic change in prehistoric Thailand. In C.A. Reed (ed) *Origins of Agriculture.* Mouton, The Hague, Netherlands.

Hope, G.S. and J. Golson (1995) Late Quaternary change in the mountains of New Guinea. *Antiquity* 69:818-830.

Hope, G.S., J. Golson, J. Allen (1983) Palaeoecology and prehistory in New Guinea. *Journal of Human Evolution* (1983) 12:37-60.

Irwin, G. (1989) Against, across and down the wind: a case for the systematic exploration of the remote Pacific islands. *Journal of the Polynesian Society* 98:167-206.

Jones, R. (1975) The Neolithic, Paleolithic and the hunting gardeners: man and land in the antipodes. *Royal Society of New Zealand Bulletin* 13:21-34.

Joyal, E. (1996) The use of Sabal uresana and other palms in Sonora, Mexico. *Economic Botany* 50(4):429-445.

Kirch, P.V. (1989) Second millennium BC aboriculture in Melanesia: archeological evidence from the Mussau Islands. *Economic Botany* 43(2):225-240.

Lancaster, P.A. and J.E. Brooks (1982) Cassava leaves as human food. *Economic Botany* 37(3):331-348.

Lathrap, D.W. (1977) Our father the cayman, our mother the gourd: Spinden revisited, or a unitary model for the emergence of agriculture in the New World. In C.A. Reed (ed) *Origins of Agriculture*. Mouton, The Hague, Netherlands.

Leal, F. and L. Amaya (1990) The curagua (Ananas lucides) crop in Venezuela. *Economic Botany* 45(2):216-224.

Loy, T.H., M. Spriggs, S. Wickler (1992) Direct evidence for human use of plants 28 000 years ago: starch residues on stone artifacts from Northern Solomon Islands. *Antiquity* 66:898-912.

Meehan, B. (1977) Hunters by the seashore. *Journal of Human Evolution* 6(4):363-370.

Meehan, B. (1989) Plant use in a contemporary Aboriginal community and prehistoric implications. In W. Beck, A. Clarke and L. Head (eds) *Plants in Australian Archeology*. Anthropology Museum, Univ. of Queensland, Brisbane.

Mont, J.J., N.R. Gallardo, D.V. Johnson (1994) The Pacaya palm (Chamaedorea tepejilote) and its food use in Guatemala. *Economic Botany* 48(1):68-75.

Morton, J.F. (1985) Indian almond (Terminalia catappa), salt-tolerant useful tropical tree with 'nut' worthy of improvement. *Economic Botany* 39(2):101-112.

Morton, J.F., E. Alvarez, C. Quiñonez (1990) Loroco, Fernaldia pandurata: a popular edible flower of Central America. *Economic Botany* 44(3):301-310.

Morwood, M.J. (1998) *Stone tools and fossil elephants: the archeology of eastern Indonesia and its implications for Australia*. Maurice Kelly Lecture, Museum of Antiquities. University of New England, Armidale, Australia.

Morwood, M.J., P. O'Sullivan, F. Aziz, A. Raza (1998) Fission track age of stone tools and fossils on the east Indonesian island of Flores. *Nature* 392:173-6.

Page, P.E. (1984) *Tropical Fruit Trees for Australia*. Queensland Dept Primary Industries, Brisbane.

Paijmans, K. (1976) *New Guinea Vegetation*. CSIRO, Canberra.

Pavlides, C. and C. Gosden (1994) 35 000-year-old sites in the rain forests of West New Britain, Papua New Guinea. *Antiquity* (1994) 68:604-610.

Peters, C. and E. Pardo-Tejeda (1982) Brosimum alicastrum: uses and potential in Mexico. *Economic Botany* 36(2):166-175.

Piperno, D.R. and D.M. Pearsall (1998) *The Origins of Agriculture in the Lowland Neotropics*. Academic Press, NY.

Pookajorn, S. (1984) *The Hoabinhian of Mainland Southeast Asia: New Data from the Recent Thai Excavation in Ban Kao Area*. Thammasat University Press, Bangkok.

Pyramarn, K. (1989) New evidence on plant exploitation and environment during the Hoabinhian (Late Stone Age) from Ban Kao Caves Thailand. In D.R. Harris and G.C. Hillman (eds) *Foraging and Farming: the Evolution of Plant Exploitation*. Unwin Hyman, London.

Reed, C.A. (1977) *Origins of Agriculture* (ed). Mouton, The Hague, Netherlands.

Rindos, D. (1984) *The Origins of Agriculture: an Evolutionary Perspective*. Academic Press, New York.

Roberts, R.G., R. Jones, M.A. Smith (1990) Thermoluminescence dating of a 50 000 year old human occupation site in northern Australia. *Nature* (1990) 345:153-156.

Robillard, A.B. (1992) *Social Change in the Pacific Islands*. Kegan Paul International, New York/London.

Russell, C.E. and P. Felker (1987) The prickly-pears (Opuntia spp): a source of human and animal food in semiarid regions. *Economic Botany* 41(3):433-445.

Sauer, C.O. (1952) *Agricultural Origins and Dispersals*. American Geographical Society, New York.

Sheldon, S. (1980) Ethnobotany of Agave lecheguilla and Yucca carnerosana in Mexico's Zona Ixtlera. *Economic Botany* 34(4):376-390.

Smith, C.E. (1986) Import of paleoethnobotanical facts. *Economic Botany* 40(3):267-278.

Solheim, W.G. (1972) An earlier agricultural revolution. *Scientific American* 226(4):34-41.

Solheim, G.G. (1973) Northern Thailand, southern Asia and world prehistory. *Asian Perspectives* 13:145-162.

Spriggs, M. (1993) Pleistocene agriculture in the Pacific: why not? In M. Spriggs and B. Fankhauser (eds) *Sahul in Review: Pleistocene Agriculture in Australia, New Guinea and Island Melanesia.* Research School of Pacific Studies, Australian National University, Canberra.

Spriggs, M. (1996) Early agriculture and what went before in Island Melanesia: continuity or intrusion? In D. Harris (ed) *The Origins and Spread of Agriculture and Pastoralism in Eurasia.* Univ. College London Press, UK.

Stearman, A.M. (1991) Making a living in the tropical forest: Yuquí forages in the Bolivian Amazon. *Human Ecology* 19(2):245-259.

Swadling, P. and G.S. Hope (1992) Environmental change in New Guinea since human settlement. In J.R. Dodson (ed) *The Naïve Lands: Prehistory and Environmental Change in the Southwest Pacific.* Longman Cheshire, Melbourne.

Tacy, D. (1995) *Edge of the Sacred: Transformation in Australia.* Harper Collins, Blackburn, Vic., Australia.

Thorne, A.G. (1980) The longest link: human evolution in Southeast Asia and the settlement of Australia. In J.J. Fox, R. Garnaut, P. McCawley, J.A. Mackie (Eds) *Indonesia: Australian Perspectives.* Research School of Pacific Studies, Australian National University, Canberra.

Thorne, A.G. and M.H. Wolpoff (1981) Regional continuity in Australasian Pleistocene hominid evolution. *American Journal of Physical Anthropology* (1981) 55:337-349.

Thorne, A.G. and M.H. Wolpoff (1992) The multiregional evolution of humans. *Scientific American* (Apr 1992) 266:28-33.

Thorne, A.G., R. Grün, G. Mortimer, N.A. Spooner, J.J. Simpson, M. McCulloch, L. Taylor, D. Curnoe (1999) Australia's oldest human remains: age of the Lake Mungo 3 skeleton. *Journal of Human Evolution* 36(5):591-612.

Tindale, N.B. (1974) *Aboriginal Tribes of Australia.* Australian National University Press, Canberra.

Toth, N., D. Clark, G. Ligabue (1992) The last stone axe makers. *Scientific American* July 1992.

Verhoeven, T. (1958) Pleistozäne funde in Flores. *Anthropos* 53:264-265.

Whistler, W.A. (1987) Ethnobotany of Tokelau: the plants, their Tokelau names and uses. *Economic Botany* 42:155-176.

Yen, D.E. (1974) Aboriculture in the subsistence of Santa Cruz, Solomon Islands. *Economic Botany* 28:247-284.

Yen, D.E. (1977) Hoabinhian horticulture? The evidence and questions from northwest Thailand. In J. Allen, J. Golson, R. Jones (eds) *Sunda and Sahul: Prehistoric Studies in Southeast Asia, Melanesia and Australia.* Academic Press, New York.

Yen, D.E. (1988) Domestication of the environment. In D.R. Harris and G.C. Hillman (eds) *Foraging and Farming: the Evolution of Plant Exploitation.* Unwin Hyman, London.

Yen, D.E. (1990) Environment, agriculture and colonization of the Pacific. In D.E. Yen and J.M.J. Mummery (eds) *Pacific Production Systems: Approaches to Economic Prehistory.* Papers in Prehistory 18, Research School of Pacific Studies, Australian National University, Canberra.

Yen, D.E. (1993) The origins of subsistence agriculture in Oceania and the potentials for future tropical food crops. *Economic Botany* 47(1):3-14.

Index

Printed and bound by CPI Group (UK) Ltd, Croydon, CR0 4YY

21/10/2024

01777082-0010